From Sticks and Stones

Personal Adventures in Mathematics

Paul B. Johnson
University of California at Los Angeles

SCIENCE RESEARCH ASSOCIATES, INC.
Chicago, Palo Alto, Toronto
Henley-on-Thames, Sydney, Paris
A Subsidiary of IBM

Library of Congress Cataloging in Publication Data
Johnson, Paul B date
From sticks and stones.

Includes index.
1. Mathematics--1961- I. Title.
QA39.2.J63 510 74-23322
ISBN 0-574-19115-1

Printed in the United States of America.

Preface

One reads a preface to see what the author thinks of the book after he has finished the manuscript and can't correct any more mistakes. Perhaps you want something about its background, and maybe hints on how to use it.

I hope you will find a lot of old ideas explained in new ways. Most of the ideas are discussed from several attitudes or approaches so that you may find new approaches that will give you greater power.

The book is intended for people who plan to teach in elementary and junior high schools, who are now teaching, or who wish to understand and use arithmetic but have been unhappy with it.

The ideas come from (1) books used in elementary schools and in colleges, (2) experience in teaching fundamentals of arithmetic at UCLA, (3) experience in planning curricula with and for schools in Africa, (4) teaching teachers in Africa, (5) talks by great mathematics teachers at meetings sponsored by NCTM, CMC, AMA, and other organizations, (6) trials with children. The African connection helped me observe mathematics learning by intelligent, cooperative people who happened to see things from a down-to-earth point of view. Their penetrating questions showed me many points that needed discussions different from what I usually gave.

This is not a sixth-grade book with a different cover. While most of the ideas and learning methods are used in the elementary grades, more is said about them than usual.

Certain themes run through the book. There are many valid approaches to arithmetic. Which one is best depends as much on who is learning and who is teaching as it does on the particular learning exercise. Even though a teacher has a preferred attitude that he uses most of the time for a particular topic, he benefits from knowing other attitudes for use when they are more powerful, and for the insights they give him about his preferred attitude.

Generalizing and extending are the great processes in mathematics. A single advanced idea may have started from observations in many places with no particular relation between them. It is useful to have a clear picture of both the early and the advanced stages of a concept such as number, addition, or fraction. This pattern of an idea growing and developing occurs over and over again.

Mathematics is a participating, not a spectator, sport. Most of the exercises are designed to make the ideas in the text clearer. There are not many questions suitable for flash cards; we encourage you to make them up for yourself where you think they will be helpful.

Manipulative materials are used throughout the book. Even more such math-lab type activities are recommended.

If some of the discussions seem wordy, just skim them to get any ideas that are new to you. Other readers will benefit from the greater detail.

"Who is buried in Grant's tomb?" was missed by several people on a TV show. While some of the exercises in this book are just as easy as that question, most of them don't hint the answer quite so much, and some are open-ended, with more than one correct answer. Sometimes an exercise will be interpreted differently from what was intended. In such a case, I have found it best to give the student full credit for an answer that is correct according to his interpretation.

Each chapter is relatively independent of the others. This has resulted in some repetition, but allows the reader to skip a chapter or two, or study them in a different order if he chooses. In particular the material after chapter 26 could be read earlier. The chapter on Problem Solving, included as an appendix, can be read early and referred to repeatedly.

I've received great help and insight from many people. It is a privilege to give credit to a few, even though some of them disagree with several of the points in the book: certainly the great modern curriculum-development groups such as UICSM and SMSG; Clarence Hardgrove, Grace Alele Williams, Shirley Hill, and Vincent Haag of the Entebbe Mathematics Program; Tim Nyae and teachers in Liberia, Ethiopia, Tanzania, and Ghana, and also in Thailand; Clark Lay, Roy Dubisch, Beryl Cochran, and the readers and teaching assistants at UCLA. Special thanks go to Carol Kipps who graciously read, used, and criticized the manuscript. But my greatest appreciation goes to the eager, hard-working people, often defeated by arithmetic, who gave me their trust, sharing their fears and anxieties in the hope that what I told them would be useful.

Contents

From Sticks and Stones

1

Introduction

1.1 HOW THIS BOOK CAME TO BE WRITTEN From personal experience and watching students the author has observed that learning by rote is an easy way to learn mathematical facts. Such learning is hard to use, however, and indeed vanishes after the first exam. It is more interesting and little, if any, more work to learn facts as arising from people's efforts to make their lives easier.

This book discusses the mathematics we all study, teach, and use from the human point of view. The essence of mathematics is abstraction from human experience. The number systems, calculations, diagrams, and measurements all arise from various events happening over and over again to everyone.

Mathematics is powerful, because it helps people see a common structure or thread in different situations. For example, the fingers on a hand, the time interval from Monday morning to Friday night, and the position of E in the alphabet are all described by the word *five*.

The title of the book suggests this theme. Mathematics arises from the adventures people have had with sticks (segments, lines), stones (sets, counting), and other objects. The form usually taken by mathematics comprises a set of symbols and rules of great simplicity and insight. In this book we see these symbols and rules as reflections of common attitudes and experience.

The truly amazing fact is that so many different human experiences lead to the same set of symbols and rules. Some people like to think of mathematics as an independent structure, which happens to fit many situations. They like to study this independent structure more or less by itself. We wish to recognize the structure in everyday life—in fact, we are forced to see it when we look closely.

The ideas of mathematics are discussed, not as something new, but to see how they arise in different ways. These different ways are called *attitudes*. Readers of the book are assumed to know the important facts; no one can complete elementary school without being able to count, add, multiply, and so on. Seeing these operations from several different attitudes makes them clearer, easier to use.

Most choices in life are made for the convenience of the decision maker. Being aware of multiple attitudes shows how mathematics was developed by efficient people who wanted a lot for the least possible effort. The motive shows even though intellectual power rather than money was the object.

The teacher who knows how an idea appears from several attitudes can recognize a pupil's difficulty sooner and give more effective help. Such knowledge is also a good foundation for advanced mathematics.

1.2 A SHORT HISTORY OF MATHEMATICS CURRICULA In the days of old, mathematics was concerned with "how many and how much." It still is, but the ideas and problems used to fit only the specific circumstances of their birth. The concept we call "two" was a different concept to the cattleman (yoke of oxen) than to the bird man (brace of pheasants). Two times three was a different problem when counting the number of children in two three-child families than when finding the area of a rectangle two steps wide by three steps long.

Perhaps, as time went on, people recognized 2×3 as being the same problem even though it arose in different circumstances. But they saw

less connection between 2×3 and 4×7 than modern man sees between 2×3 and $\sqrt{5 + 9}$.

It was a great step forward when people recognized that the same approaches would work on both $5 + 7$ and $8 + 23$. While teachers and leaders wanted young people to learn to use arithmetic in their lives, they saw it was more valuable to look for general patterns and to learn techniques that worked in many cases than merely to learn long lists of specific problems.

For example, many years ago a student learned "3 apples and 4 apples are 7 apples," "3 dogs and 4 dogs are 7 dogs," "3 fish and 4 fish are 7 fish" as unrelated ideas. Just as knowing apples grow on trees and dogs bark doesn't give you the slightest hint that fish swim, so knowing 3 apples and 4 apples are 7 apples gave no hint as to how many fish were 3 fish and 4 fish. So the teachers said, "If we teach $3 + 4 = 7$, then a student won't have to learn all these separate facts." But an unforeseen problem arose. Johnny knew that $3 + 4$ was 7, but he had difficulty applying this to the situation of 3 apples and 4 apples. This, of course, was overcome by Johnny seeing that $3 + 4$ is a model of 3 apples and 4 apples. We agree now that he needs to know about some apples and dogs before he understands $3 + 4$.

The process continued. In modern times we identify even more powerful techniques with wide application. These techniques or abstractions are emphasized these days in what is called Modern Mathematics. Sometimes the abstractions are emphasized before the student understands any of the special circumstances brought together by the abstraction. The abstraction is his special circumstance, learned independently of other things. This has led to grizzly gags such as "He knows that $6 + 7 = 7 + 6$, but he doesn't know whether the sum is 13 or 15."

Actually, this turns out to be a problem at all levels of mathematics. Even students of calculus and beyond learn manipulation by rote without being able to identify what they do with specific events in the physical world or with other basic concepts.

1.3 OUR GENERAL ATTITUDE Mathematics is a human activity engaged in by human beings to meet human needs. Mathematics is useful because it helps people make models of their problem situations. These models help people make better choices.

Perhaps the most powerful models are number systems. We shall look at number systems quite a bit—the situations they model, how their properties arise from these situations, and the basic operations.

"There's more than one way to skin a cat" is a marvelous idea, even if you have no desire to do such a thing. We consider several approaches to each of the main ideas. Significance and clarity come from multiple examples and attitudes. Look for counterexamples to test how good the ideas are.

Most choices in life are made for the convenience and benefit of the decision maker. We show that mathematics was probably developed by greedy, lazy people who wanted a lot for the least effort possible.

Mathematics is very down to earth. Just as the ideas arise out of simple situations, so they can be illustrated with simple situations. You are encouraged to look for such illustrations.

1.4 HOPES FOR OUR READERS It is popular these days to list behavioral objectives for a program. Here are some of the behaviors we accept as evidence of knowledge. You may wish to make up others.

Skills that show knowledge include—

1. working appropriate numerical problems (computation)

2. explaining algorithms, or "dances of the digits," in terms of the operation attitudes and properties and numeral systems

3. identifying and solving problems arising with stories or physical equipment

4. relating, comparing, and contrasting different ideas, notations, or techniques

5. making choices between alternatives on the basis of convenience, usefulness, and effectiveness, as well as validity

6. making up relevant and appropriate illustrations, questions, and problems

7. communicating with spoken and written words, diagrams, physical objects, and models

Every author and teacher wants the reader to like the subject, to do exercises because he wants to, not merely because he must. There has always been considerable compulsion in the teaching of arithmetic, as there is in all activities adults think good for the young. Learning arithmetic is considered only slightly less important than not playing in the middle of the highway, so it is not surprising that similar tactics are often used in their teaching. Here are some behaviors often accepted as evidence that a person likes arithmetic:

1. stating he likes arithmetic

2. doing arithmetic—thinking, observing, making conjectures, counting, estimating—when not required

3. asking questions and doing things beyond those required

4. doing arithmetic with vigor and enthusiasm

5. starting, joining, and participating in discussions on mathematics

6. including mathematics among the fun and games planned for social events

1.5 LEARNING How do you learn best? The author cannot know and perhaps you do not know. Here are some activities with good reputations:

1. Be aware. Be curious. What do you see? What is going on? Can you make a list? Is there a statement of a problem, a pattern, or a theory? Can you verify the statement? Does it make sense? Can you point out where it fits?

2. Explore ideas with concrete objects—sticks, beans, cutouts, wires, string, and so forth. Draw pictures and sketches. Count on your fingers if you wish.

3. Write your ideas in complete sentences as well as saying them. Solve problems with paper and pencil, and check your answers.

4. Do even more after you feel you understand and can work a problem.

5. Work with others. Mathematics is both an individual and a social activity. You learn greatly by phrasing and asking questions of others. You learn even more by answering their questions.

6. Look for the joyous "Ah ha!" feeling of successfully solving a problem or seeing the point to a discussion. This sense of achievement, this gaining a new power is what arithmetic is about. Discuss your successes with others.

1.6 LEARNING LEVELS Most arithmetic learning is on the memory or recall level. There are higher levels. One theorist has labeled six levels: knowledge, comprehension, application, analysis, synthesis, and evaluation. While everyone agrees there are different levels of cognitive activity, there is no complete agreement on the categorization. A particular problem may represent different levels for different learners. If a certain behavior is regarded as "high level," there is a tendency to drill on it and memorize it to get high-level credit for what is actually recall. Avoiding multiple levels, we identify only two: (1) recall of knowledge and (2) higher level. Students are encouraged to promote higher-level as well as recall learning.

1.7 RULES You may wonder at the lack of rules printed in contrasting colors for the operations of arithmetic. Many books have them and many people find them quite helpful. We are great believers in rules too. You will find many exercises, "Make a rule covering ______________." Show it to your classmate. Does he agree that your rule correctly states what is to be done? Is his rule better? Where does his rule fail? If needed, change your rule until you both agree your rule is correct. You may have different rules—both correct. Writing a rule is one way to clarify your understanding. Such a rule is a guide for your future action and a reminder of how you got it.

Rules are a great help. Where quick precise action is required, a rule is essential. Assembling a radio, baking a pie, being a member of a dance team or volleyball team are helped by clear rules. Why not have author-made rules of action for both teacher and pupil to follow? If you are in a spot where a rule might be helpful, we prefer you to simplify the problem and work it out from first principles. Then make your rule, with the help of others if needed. We are more interested in strengthening your long-run power than getting a specific answer quickly. If worst comes to worst, learn a rule from your neighbor. Following his rule, checking as you go, will get you a long way. Adapting his rule to your purposes is almost as good as making your own rule.

Here are some actual dangers in textbook rules:

1. No rule applies everywhere. You need to learn not only the rule, but the circumstances where it can and cannot be used. This appears to be more easily learned when you write the rule yourself. You know the circumstances that led to the rule and will be more inclined to check carefully before following the rule in places where it may not apply.

2. Rules are easily misread. The answers they lead to are sometimes right and sometimes wrong when they are read incorrectly. Such rules can be nerve-wracking torture. "I did what it said, so why didn't I get the right answer?"

3. Some people come to feel that knowing rules is what mathematics is about, rather than making idea models that simplify life.

4. Alas! Some widely published rules are almost pure myth, passed on from generation to generation like Tom Sawyer's belief that horsehairs turn into snakes.

Some people may want more emphasis on the abstract "definition, axiom, theorem, proof" chains of thinking that are so important in mathematics. You will notice such structures throughout the book. They are valuable in crystallizing and organizing ideas and things you see. We do not study such structures extensively as objects for fear that for many they will be like a magnificent bridge or freeway that is unconnected to the traffic of life. We urge you to make definitions, to write statements that are possibly true, and to see what it takes to convince yourself and others that they are true or false.

1.8 THE SECRET SOCIETY Some people believe that years ago a Secret Society was set up to keep people from learning mathematics. Its motto was, "A little knowledge is a dangerous thing—arithmetic is a ca-

tastrophe." This Society schemed to keep mathematics for a chosen few. Egyptian priests persuaded people that only they could predict the level of the Nile. People have died rather than eat foods such as wheat, tomatoes, and potato peel, because they had been tricked into believing them poisonous. The Secret Society felt it quite possible to do a similar job on mathematics. It is done by perpetuating four slogans.

1. *Mathematics is hard.* If you persuade someone he cannot do something, he will work very hard to keep from doing it. Being committed to failure, success is a threat to his ego.

2. *Mathematics will be good for you someday.* Usually this is said in the same tone as "This spanking is for your own good—it hurts me worse than it does you." This slogan keeps people from seeing that mathematics is good *now*.

3. *Mathematics is strange.* We use language and words not used elsewhere. We can manipulate symbols without thinking of any outside meaning for them. Traditionally, the stranger in town has been shunned, sacrificed to the gods, burned at the stake, or eaten. The Secret Society hopes to encourage negative attitudes toward mathematics.

4. *Girls cannot do mathematics.* This automatically keeps half the population out of the candy shop. Further, the Society knows that mothers will have a hard time encouraging their husbands and sons in activities they do not share.

There may be no such Society you can fight in courts or legislatures. As you read the rest of this book, work exercises, write up your ideas, and discuss ideas with your fellows, we hope you will find yourself improving in power and understanding. In return we ask you to share yourself with others so that their minds may be free from the Secret Society.

2

Common Uses of Numerals

2.1 INTRODUCTION People are fond of numerals and find many ways to use them. Everyone is so familiar with them that we just shift back and forth without thinking about the different uses. Following the theory that it is good to talk about things we do automatically, we consider some of these uses. We restrict ourselves to the counting numerals {. . .,4,5,6,. . .}. Numerals include symbols, words, and sounds, as well as 23 and XV.

The situations where numbers arise we classify into five sets labeled *sequences, sets, segments, symbols,* and *operators.* You are familiar with the ideas, but perhaps not by these names. We discuss them in order.

Our attitudes toward numbers and numerals are so dominated by the situations where they arise that we will use these same names to describe the corresponding attitudes. These attitudes are the basis for discussing the properties of numbers later in this book. The classifications *cardinal* and *ordinal* will be compared with these attitudes.

2.2 SEQUENCES Many children at an early age learn to count—that is, to repeat the standard sequence of names "one, two, three, ..." From this point of view, *five* is defined as the name or number that comes after four and before six. There is no particular relation to the fingers of the hand.

There are two standard sequences in use—the numbers and the alphabet. Often when things are lined up in a row, we name them with either letters or numbers as a convenient way of keeping track of them.

We see that a sequential use of number may be free of any "how many" meaning. In books, for example, we can be confident that

page 17 is between pages 16 and 18 (if there are these pages), but we do not know that there are 16 pages ahead of it. Most books have a series of introductory pages before page 1, usually marked with Roman numerals.

2.3 SETS Numbers such as 5 can be answers to the question "How many members are in a given set?" Our attitude toward numbers used in this way is called the *set attitude*. The set attitude is related to the sequential attitude, since we often count the members in a set using the counting names 1, 2, 3, 4, ... The counting process is not always necessary, since many people learn to tell at a glance how many dots are in patterns such as .·. and :·:.

2.4 SEGMENTS Five and seven are sometimes answers to "How long?" or "How much?" A unit is also given, such as feet or liters. The amount of a continuous quantity, such as length, weight, area, and volume, is often shown by a segment in a bar graph. Hence, we label our attitude toward such numbers as the *segment attitude*.

As we will see in chapter 17, segments are especially good as a basis for fractions, because segments can be separated into parts of any convenient size. The segment attitude is closely related to the set attitude, since a segment 5 inches long, although in one piece, is as long as 5 segments, each one inch long, placed end to end.

2.5 SYMBOLS The numeral 5 may be just a mark, a symbol to manipulate as our fancy leads us. Sometimes we use numbers as convenient names when we have many items to deal with and each is to have a separate name (telephone numbers, license numbers, athletes' jersey numbers). There is no natural order, or question of how many or how much, related to these objects and the numbers used.

In advanced arithmetic we think of numbers as symbols. While we may do anything we please with them, we usually restrict ourselves to actions suggested by the sequence, set, and segment attitudes.

2.6 OPERATORS Numbers used to report or to direct something to be done to some unknown number are used as operators. Some examples are:

"Triple the eggs in the omelet," "Add four to the sum," and "Stretch the segment to twice its size."

Some special curriculum development groups use the images of "stretchers" and "shrinkers" to help pupils visualize the operation. Some typical phrases representing an operator are: "Add four to ____," "Multiply ____ by seven," and "Half of ____."

2.7 RELATIONS AMONG THE NUMBER ATTITUDES The number uses are not necessarily independent. Consider the sentence "The graph is on page 47." One person regards this as a symbol use, because 47 names the page with the graph. Another says it is primarily a sequence use, because if the book is opened to page 53 we know the graph is earlier. A third says it is a set use, because there are 47 pages of interest at the front of the book. While all of this may be true, we can test these interpretations by the question "Of what use is the information?" To the reader who is looking for the graph, the order is of great value. So the sequence is an important use. We might also ask, "Could I give the information in another way?" If we named the page "Cedar," this would be a good name with no order properties. Would it be useful to say "The graph is on page Cedar"? As pagination is usually done, there may or may not be 47 pages of interest. Hence, in this case, the use is probably sequential more than any other.

Clearly, different people may have different attitudes toward a particular use of a number. For example, "three quarts of milk" may mean three separate and distinct cartons to some. Here the set attitude is being used. To another, it is the amount to put in a recipe and might come in one, two, or fifteen separate containers. Here the segment attitude is used.

The useful thing is to recognize that people legitimately have these different attitudes. When talking to someone it often helps if you recognize the attitude the other person has toward the numbers being used.

The words *number* and *numeral* are often used interchangeably. Where they are used differently, a numeral is a name for a number. One way of telling whether a numeral or a number is being considered is to see if changing the form makes any difference. Consider the statement, "There are three *e*'s in eleven." Now replace "three" with "3" and "eleven" with "11." Since the sense of the statement changes drastically when replacing "eleven" but not when replacing "three," we are using "eleven" as a numeral and "three" as a number. If it is a number use the meaning does not change when the name is changed.

2.8 CARDINAL, ORDINAL, AND NOMINAL Traditionally the set attitude is often called *cardinal*. The answer to "How many?" is called a cardinal number.

Adjectives like *first, second,* and *fifth* are often called *ordinal* numbers. In a phrase such as *the fifth boy in line,* the ordinal *fifth* serves three purposes. It names an individual, it locates him in a sequence, and it tells how many people are in the line up to and including him.

When using a numeral to name something ("Number 7 made a hit"; "48 intercepted the pass"), we speak of this as the *nominal*, or naming, use of a number.

Cardinal numbers are discussed at length in arithmetic. These are numbers that add and multiply nicely. Ordinal and nominal numbers—or uses of numbers—are not discussed much, because they do not readily support addition, multiplication, or other operations.

In this book we will not use *cardinal* and *ordinal,* since it appears that the important uses are better classified as sequences, sets, segments, symbols, and operators. We regret we know no synonym for operator that begins with s.

EXERCISES

In each of the following sentences a number appears. State whether the use is primarily sequence, set, segment, symbol, or operator. Different people may have different correct answers for some.

1. The teacher thinks 18 students is a good size for a class.
2. She lives on Thirteenth Street.
3. The track is 400 meters long.
4. His phone extension is 631.
5. The can contains 15 ounces of beans.
6. The store sold 18 cases of peanut butter.
7. She is staying in room 644 of your hotel.
8. Submarine U501 sailed under the Pole.
9. He ran a mile in eight minutes.
10. My number for service at the counter is 43.
11. Franklin Roosevelt was elected president in 1932.
12. The washing machine uses 22 gallons of water per load.
13. Look for car license 63245.
14. The library has more than 2763 books in it.
15. His home is at 439 Woodland Avenue.
16. The serial number of his typewriter is 468372581.
17. The house has 986 square feet of floor area.
18. High tide is at 4 P.M. today.
19. Highway 101 goes between Los Angeles and San Francisco.
20. Work the exercises on page 108.
21. Helen has 5 eggs in her right hand.
22. Robert is 177 centimeters tall.
23. One gross equals 144.
24. The lake level is 7 meters above normal.
25. The fullback is number 36.
26. The dog weighs 6341 grams.
27. Tuesday is day 3 of the week.

28. Maria was born on January 13.
29. The pie was separated into 4 parts.
30. The amount of yeast in the dough tripled while the dough was warming.

In 31–35 make up sentences showing the attitudes with the given numbers.

31. Use 15 in a sentence showing the set attitude.
32. Use 16 in a sentence showing the segment attitude.
33. Use 17 in a sentence showing the sequence attitude.
34. Use 18 in a sentence showing the operator attitude.
35. Use 19 in a sentence showing the symbol attitude.

Answers to selected questions: **1.** set; **3.** segment; **5.** segment; **7.** sequence; **9.** segment; **11.** sequence; **13.** symbol; **15.** sequence; **17.** segment; **19.** symbol; **21.** set; **23.** symbol; **25.** symbol; **27.** sequence; **29.** set.

3

Some Computational Situations

3.1 INTRODUCTION A little girl was asked to tell the class how she could tell whether to add, subtract, multiply, or divide in an arithmetic problem. "It's easy," she reported. "You just look at the numbers. If they are both long, you add them. If one is longer than the other, you subtract the short one from the long one. If they are both short, you multiply them. If one is short and the other long, you divide the short one into the long one."

While we won't come up with as satisfying a rule as this one, we will study typical human situations that lead to the basic operations of addition, subtraction, multiplication, and division. Each operation arises from several types of situations, or attitudes. The most common of these attitudes are named.

Each of these attitudes is widely used. Each is the best for some situations and for some people. While people may differ in their attitude toward a particular situation, it is good to know several possible ways of looking at the same problem.

We are interested in naming operations and seeing where they are appropriate. While questions are asked which appear to be calling for numerical answers, we are not interested in the number answer but in naming the operation to be used.

Merely naming the operation, such as division, is not enough. For example, we need to know which number is the divisor. Hence answers are complete sentences—often a sentence in symbols, such as an equation.

The three numbers arising in a computation show the operation. For example, we say that (2, 3, 5) shows addition, and (4, 5, 20) shows multiplication. The notation $(4, 7, n)$ with the third number, n, not yet named is used when you are asked to name the operation. In this case, one has to read the story carefully to determine the operation, since no clues, such as an operation symbol $\times$, $+$, $-$, $\div$ or the number $n = 11, 28$, or -3, give the operation away.

Most of these attitudes are old friends; you have used them, although you may not have given them a name.

3.2 ADDITION SITUATIONS Addition situations usually arise from continued counting or when things are put together. These things may be, for example, sets, segments, quantities, rates. We give illustrations of quite different situations, each of which requires addition and might be used as a typical experience leading to addition.

1. Mileposts along the highway are labeled 1, 2, 3, ... Starting with the first, Maurice counted 28. Then starting over, he counted 17. What was the label on the milepost he counted last? The numbers 28, 17, and the last label l form a triple (28, 17, l) which shows addition. We write $28 + 17 = l$.

2. After a party there were 28 glasses in the kitchen and 17 glasses in the dining room. How many glasses all together? We may imagine putting all the glasses together and counting them. We could form the union of a set of 28 with a set of 17. If we count the glasses in the combined set, we get a number, say n. The number triple (28, 17, n) shows addition and we write $28 + 17 = n$. The numbers that are added, 28 and 17, are called *addends*. At this time we are not interested in the name, but in writing the sentence showing addition.

3. Marie's table is 28 inches high and 40 inches long. Her chair seat is 17 inches above the floor. If she places the chair on the table, how far is the seat above the floor? Let us suppose that we measure this last distance and get h inches. We could think of this distance arising from putting two sticks or segments end to end, one being 28 inches long and the other 17 inches long. The desired distance is that from one end of the combined segment to the other. We say the number triple (28, 17, h) shows addition. We write $28 + 17 = h$. The length 40 inches has nothing to do with the distance wanted, so it is ignored. Such irrelevant information always exists in any real-life problem. One of the objectives of mathematics and science learning is to select the relevant and ignore the irrelevant numbers.

4. In one tank 30 inches long are 28 gallons of water. In another tank 26 inches long are 17 gallons. How much water is there all together?

While we aren't told how these measurements were made, we should have some feeling for how they might have been made. We might have had a dip stick or scale that tells how much water is in the tank from how deep it is. Or we might have had 28 one-gallon cans of water poured into the 30-inch tank. Or we might have filled one one-gallon can 28 times, emptying it into the 30-inch tank each time. Or we might have got the water from a tank truck with a meter that said we had removed 28 gallons. Or we might have had a cup that we filled and emptied an appropriate number of times. Or someone in whom we had confidence looked at the tank and told us it had 28 gallons. Or . . .

"How much water all together?" we can interpret as follows. Suppose all the water were put in one tank. There would then be g gallons in this tank. We suppose the water is measured by one of the reliable methods mentioned above. Four numbers were given in the problem. Only two are relevant here. The triple of numbers $(28, 17, g)$ is said to show addition, and we write $28 + 17 = g$.

5. A man who swims 28 inches per second in still water is swimming down a river 1000 inches wide that is flowing 17 inches per second. How fast is he moving downstream?

The method of measuring these speeds is not so clear. Perhaps the man has a speedometer that reads like the dial in a car, or perhaps a log is floating down the river that is seen from shore to move 170 inches in ten seconds. Perhaps a light fish line is fastened to the swimmer as he swims past a fisherman. The fisherman has a speedometer on his fish line or he measures v feet of line going out in one second. By some reliable method of measurement (you may think of better ones), we determine the swimmer is going v feet per second downstream. We say the number triple $(28, 17, v)$ shows addition. We write $28 + 17 = v$.

In our examples, two quantities have been joined to form a third. Each of the three quantities is counted or measured independently. The triple of numbers (a, b, c) we get is said to show addition and we write a statement of the form $a + b = c$.

We note that in each of our examples, $a = 28$ and $b = 17$, while $c = l$ (a label), n (a number of glasses), h (a number of inches), g (a number of gallons), and v (a speed in feet per second). Can we have much confidence that these five numbers—l, n, h, g, v—are equal? In general, numbers that arise in different ways will be different. Is there something the same in these five cases–in the structure or in the way the numbers arise–that guarantees that $l = n = h = g = v$? Observing and identifying such a common structure or invariance principle is an objective of arithmetic.

A commonly used invariance principle is that sets which can be matched exactly have the same number of members, and conversely, two sets with the same number of members can be matched exactly. Can we use this principle?

The phrase *match exactly* is perhaps clear, but we will talk about it a bit, since the ideas will be used again. By matching exactly we mean to set up a one-to-one correspondence (1:1) or pairing between the members. Each member of one set is paired with, or corresponds to, one and only one member of the other set. It may be necessary to force the set structure in order to use the principle. Let us examine each of the five examples.

1. *Continued counting.* The 28 labels or sounds 1, 2, 3, . . . , 28 used first form one set; the next 17 labels or sounds 29, 30, . . . , 45 form the second set; and l is the label for the number of members in the third set.

2. *Party glasses.* Here the set structure is obvious. The sets are sets of glasses: 28 in one set, 17 in the second, n in the third set. The 28 set and the 17 set are disjoint; that is, they do not overlap or have any members in common. The n set is the union of the other two sets.

3. *The chair on the table* presents no clear set structure. However, as was seen, this problem is equivalent to finding the length of a segment formed by joining a 28-inch segment with a 17-inch segment. We force a set structure by imagining the 17-inch segment as made up of 17 small segments, each one inch long. These one-inch segments can be matched one to one with the glasses in the 17 set. Similarly for the 28 one-inch segments and the 28 glasses. Putting the two segments end to end corresponds to forming the union of the two sets of segments. The two matchings before putting the segments together still hold, and together form a matching between the inch segments and the glasses in the united sets. The invariance principle then guarantees that $n = h$.

We draw a diagram of the matchings. Dashes represent inch segments; circles represent glasses. For convenience we don't leave the glasses scattered all over the house, and we imagine the segments separated from each other. We mentally form the union:

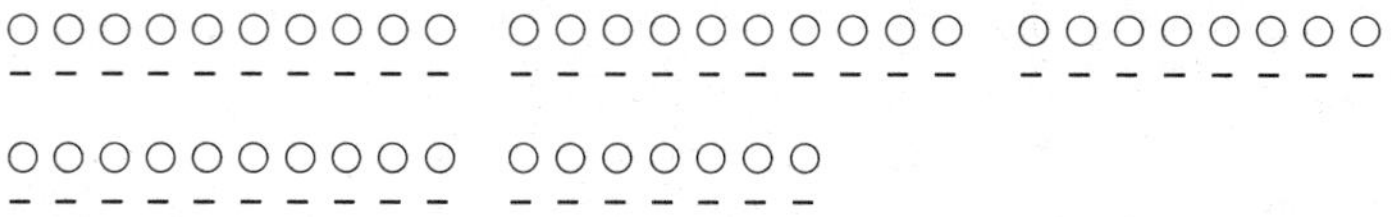

```
○○○○○○○○○○ ○○○○○○○○○○ ○○○○○○○○
---------- ---------- --------

○○○○○○○○○○ ○○○○○○○
---------- -------
```

4. *Water in the tanks.* Can you force a set structure on the water which matches the sets in the above examples? We leave this as an exercise for the reader.

5. *Speeds in the river.* We have called the speed of the swimmer moving downstream the sum of his speed in still water and the speed of the river. We shall match this situation with putting two segments end to end, and show that the union of sets attitude toward addition is not very helpful.

We recall that speed may be measured by the distance traveled in one second. For convenience we suppose that a rock in the river, a floating chip of wood, and the swimmer are at the same point at the start of observation. One second later the chip and the swimmer are both downstream, the swimmer further than the chip.

R
C • Rock, chip, and swimmer at start.
S

Rock, chip, and swimmer one second later.
R • C• S•

The segment $\overline{RC}$ is the segment traveled by the floating chip. Its length in inches gives the speed of the river in inches per second. The chip and the swimmer are both floating freely in the water, and, as far as they are concerned, might just as well be floating in a still lake. Hence, the segment $\overline{CS}$ is a segment traveled by the swimmer as though in still water and whose length gives the speed of the swimmer in still water. The segment $\overline{RS}$ is the segment traveled by the swimmer in the river and whose length gives his speed, v, downstream. The segment $\overline{RS}$ is the union of the segments $\overline{RC}$ and $\overline{CS}$. Hence, adding the speed 17 of the river to the speed 28 of the swimmer in still water matches with putting segments end to end.

It is more difficult to imagine a set of 17 speeds whose union is the speed of the river, as is needed if we wish to see a set structure in the speeds. Perhaps the following will do. Because of friction, the water at the bottom of a river runs slower than that at the top. We imagine the river separated into 17 layers, each moving one inch per second faster than the level below. To force the set structure, the speed of the top layer is called the "union" of these 17 unit relative speeds. The image is weak, because the layers are not of equal thickness and the water doesn't actually move in layers; the speed varies smoothly from zero at the bottom to a maximum a bit below the surface.

It is even harder to imagine 28 speeds whose union is the speed of the swimmer in still water. We shall not attempt, then, to match the adding of speeds directly to a union of sets.

Caveat: For very high speeds, velocities do not act this way. That is, if the swimmer were swimming at $\frac{1}{4}$ the speed of light (c) and the river were also traveling at $\frac{1}{4}c$, the speed of the swimmer with respect to the ground would not be $\frac{1}{2}c$, but somewhat less. Physicists discuss this under the theory of relativity.

To conclude our earlier discussion, we see that by using the same process on segments $\overline{RC}$ and $\overline{CS}$ that we used on the chair heights and table heights we can match one-inch segments in $\overline{RS}$ with glasses. Hence, the invariance principle (sets that can be matched one to one have the same number of members) tells us that $v = n = h = g = l$. Because these numbers are equal, we feel easier about calling the relation "addition" in each of these five examples.

Addition situations may involve more than two addends. In $3 + 5 + 7 = 15$, 3, 5, and 7 are called addends, and 15 is called the sum. "Peter had \$8 in his pocket. Then he was paid \$4 by Mary and \$5 by a 7-foot basketball player. How much money does he now have?" If d is the number of dollars he has now, then $8 + 4 + 5 = d$.

3.3 SUBTRACTION SITUATIONS Subtraction usually arises in situations like the following:

1. *Missing addend.* "Carol needs 8 eggs for a recipe. She has 5. How many more does she need?" If m is the number of eggs she needs, then we have an addition situation $5 + m = 8$; m and 5 are addends, and m is the unknown or missing addend. We say the same thing by $m = 8 - 5$. Carol can determine m by getting a supply of eggs and seeing how many more she has to take. She can also look in an addition table and see what number added to 5 yields 8 as a sum.

2. *Take away-difference.* "Rosada had 8 necklaces. She took 5 away. How many are left?" Let d be the number of necklaces left. We write $8 - 5 = d$. The answer d is also called the difference of 8 and 5.

3. *Comparison.* Subtraction arises when we ask how much bigger or smaller, more or less, one object is than another. "How much longer is an 8-foot cobra than a 5-foot cobra? We write $8 - 5 = c$. "How many fewer offspring has our pig Alice with 5 than Owaka's pig Louise, who has 8?" We write $8 - 5 = f$.

4. *Counting backward.* "Dana has playing cards lined up labeled 2, 3, 4, 5, 6, 7, 8, 9, 10. Starting with 8 he counts back 5, thinking "one" as he taps 7, "two" as he taps 6, and so on. What is p, the last playing card he taps?" We write $8 - 5 = p$.

These situations are all different. In each sentence relating the numbers, we have $8 - 5 =$ something. Can we be sure the numbers m, d, c, and f are all equal? We look for some common property in the structure of these three situations, some invariance principle. Since we have already discussed addition, we think it might be helpful if these situations were expressed in terms of addition. Can we phrase each situation in terms of "missing addend"?

Someone might ask, "Why these questions? Everyone knows that $8 - 5 = 3$. Why confuse us by calling it m, f, etc.?" We want to be sure that the processes of missing addend, take away, and comparison are sufficiently alike that we can use the same name (subtraction) and the same sentence ($8 - 5 = x$) in every case.

We may relate these examples to addition, to joining two disjoint sets, or to putting two segments together end to end. These sets and segments may have to be mental, since there may be no physical objects available. But human beings like to use their imagination.

1. *Eggs.* We imagine a set of 8 eggs needed for the recipe. We separate it into two sets, one with 5 eggs, the other with m eggs. The set of 8 is the union of the set of 5 and the set of m.

2. *Necklaces.* Rosada's 8 necklaces are separated into a set of 5 and a set of *d*. The set of 8 is the union of the set of 5 and the set of *d*.

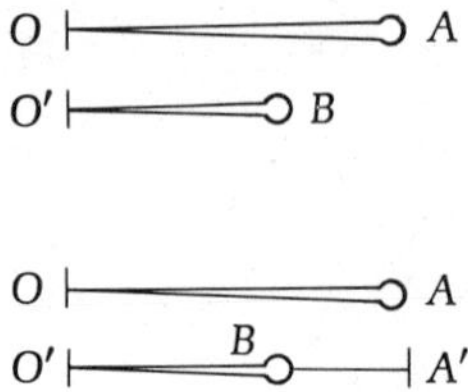

3. *Cobras.* We think of the cobras straightened out side by side, tails together. They can be represented by two segments as shown, $\overline{OA}$ and $\overline{O'B}$. We may think of O′ being on top of O and B in the interior of $\overline{OA}$. We form the segment $\overline{BA'}$ as shown (A′ on A). The length of $\overline{BA'}$ is *c*. This approach is a segment version of missing addend in the sense that $\overline{BA'}$ is the segment that must be joined to $\overline{O'B}$ to make $\overline{O'A'}$ congruent (same length) to $\overline{OA}$.

A take-away approach is to cut the long cobra into two parts, one just as long as the short cobra. Take this piece away. The length of the piece that is left is *c*. It is easy to see why people prefer to do this mentally.

We see that each subtraction can be regarded mentally as a missing addend situation. That is, given one addend and the sum, we wish to know the missing addend. Working with all sets shows that only one number is possible for 8 − 5. Hence the four numbers *m, d, c, f* are all equal.

A psycho-historical comment *Traditionally the take away–difference point of view was chosen as the invariance principle rather than the missing addend. People redescribed all subtraction problems in terms of take away. The name* subtract *comes from Latin words meaning "away take." Subtraction, from this attitude, appears quite different from addition. It is a new and separate operation to the learning child. Such a child might know that 7 + 8 = 15, but have no idea what 15 − 8 is. When it comes to thinking about manipulating digits in subtraction algorithms, the take-away attitude is more common than the missing addend. This is discussed later.*

3.4 MULTIPLICATION SITUATIONS Multiplication arises in a large number of situations that look surprisingly different. Each leads to an attitude or approach to multiplication. We will point out the advantages of the separate attitudes in different situations meeting human needs. No single attitude is best for all situations. The skilled mathematics educator will know when he is using a particular attitude, and, even more important, when he is shifting from one to another.

We will make a partial list of the more important attitudes, and describe the first three briefly. The discussion continues in chapter 10, where we will describe some of the problems associated with multiplication and discuss how some of the attitudes relate to the problems. To simplify the language, we will discuss 3×4 in a way which extends easily to any counting numbers. The attitudes we name are:

1. repeated addition
2. array
3. Cartesian product
4. tree
5. segment-segment-region
6. segment-segment-segment
7. operator-segment-segment
8. operator-operator-operator
9. formal manipulation

1. *Repeated addition.* By 3×4 we mean $4 + 4 + 4$. Note that we do not say $3 + 3 + 3 + 3$. In our definition the first factor gives the number of addends and the second factor gives the addend. It is a significant fact that $3 \times 4 = 4 \times 3$, and we do not wish to hide this fact in the definition. Repeated addition is the attitude most people give when asked to describe multiplication.

EXAMPLE Ruth did 4 problems on page 23, 4 on page 25, and 4 on page 30. How many did she do all together?

2. *Array.* By 3×4 we mean the number of members in an array of 3 rows and 4 columns.

$$\begin{matrix} a & b & c & d \\ e & f & g & h \\ i & j & k & l \end{matrix}$$

In the illustration the set $\{e,f,g,h\}$ forms the second row and the set $\{c,g,k\}$ forms the third column.

Eggs in cartons, seats in a room, boxes in a warehouse are among the situations where this attitude arises.

EXAMPLE Trees in an orchard are planted in 5 rows, 7 in each row. How many are there all together?

3. *Cartesian product.* By 3 × 4 we mean the number of members in the Cartesian product of a set with 3 members and a set with 4 members. The Cartesian product of the set $\{a,b,c\}$ and the set $\{w,x,y,z\}$ is the set $\{cw,cz,aw,bw,ay,bz,bx,by,ax,az,cy,cx\}$. The product set is the set of all ordered pairs of letters with the first letter from the 3 set and the second letter from the 4 set. No particular sequence in listing the pairs is required, so the list given here is random rather than, say, all the a's first.

A frequently used example of a Cartesian product is the set of outfits a girl can assemble if she has 3 skirts and 4 blouses. Many pattern-minded mathematicians like this attitude, because multiplication is described in terms of an operation on sets. This appeals when counting numbers have been defined in terms of sets and addition in terms of union of sets.

We give several cases where multiplication is used. We are not as interested in the numerical product as in the description of the product.

1. Eight boxes, each 13 inches high, are piled one on top of another. What does 8 × 13 measure or describe?

2. Peter bought 7 candy bars at 9 cents each. What does 7 × 9 represent?

3. A right whale swam for 8 hours at 5 miles per hour. What does 8 × 5 represent?

4. The bank pays 5 percent interest per year. Marie deposits $75 for one year. What does 75 × 5 measure?

5. Water ran into the pool for 24 minutes at a rate of 3 gallons per minute. What does 24 × 3 measure?

6. A right triangle has legs 3 and 5. What does $\frac{1}{2} \times 3 \times 5$ measure?

7. Peter put $1.50 in the church collection every Sunday for 6 Sundays. What does $1.50 × 6 represent?

8. John has 3 shirts and 5 ties. What does 3 × 5 represent?

9. Helen has 4 white blouses and 5 blue blouses. What does 4×5 represent?

10. Sally runs 100 yards per minute for 7 minutes. What does 7×100 represent?

(See page 26 for answers.)

3.5 DIVISION SITUATIONS Traditionally division arises from sharing (partitioning) or measurement (number of times a larger unit is contained) situations. Recently the missing-factor approach has been emphasized more, because of its greater generality—that is, it includes the other cases as special examples. Therefore, each approach to multiplication will give at least one approach to division, sometimes two. We look at some of these situations. Our examples will all come out even, because we are emphasizing approaches. Noneven division is discussed later.

1. *Sharing.* Four girls wish to share 20 cherries equally. How many does each girl get? We write $20/4 = c$, where c is the number of cherries each girl gets. Anyone who cannot recognize the common name for 20/4 (cannot divide) can still get the correct answer. He takes 20 cherries (or stones) and sorts them one at a time into 4 piles. The number in each pile is c. This is the sharing (partition) approach to division.

2. *Measurement.* Suppose a cafeteria has 36 prunes to serve. Four prunes go in each dessert. How many desserts are there? This is the measurement approach, so called because the prunes are being measured by 4s rather than individually. We write $36/4 = d$, where d is the number of desserts. Anyone who did not recognize 36/4 could get the correct answer by taking 36 prunes (or seeds), dishing them up 4 at a time, and counting the number of dishes.

It is not important to remember these two names; what is important is to know these two ways of interpreting, say, 48/12: (1) Sharing. Sort 48 stones one at a time evenly into 12 piles. 48/12 is the number of stones in one pile. (2) Measurement. Make piles of 12 stones each. 48/12 is the number of piles you get. This problem also arises as "How many feet wide is a piece of plywood 48 inches wide?" That is, we measure in feet rather than inches.

Each of these approaches may be regarded as a missing-factor problem in multiplication. We know that if 7 boys eat 3 apples each, they eat 21 apples. We write $7 \times 3 = 21$. If we ask how many apples each of 7 boys will get if they share 21 apples evenly, we see this is described by $7 \times a = 21$, where a is the number of apples each gets. This is also a division (sharing) problem. Another problem might be "How many boys can I treat with 3 apples each if I have 21 apples?" This is described by $b \times 3 = 21$ and is also called a division (measurement) problem. To find a or b in $7 \times a = 21$ or $b \times 3 = 21$ is called the missing-factor approach to division.

3. *Missing factor.* Each approach to multiplication yields approaches to division from the missing-factor attitude. Repeated addition leads to repeated subtraction. 48/12 is the number of times 12 can be subtracted from 48; this is very close to the measurement attitude.

Twenty rosebushes are to be planted 4 in a row. How many rows will there be? This is related to the array approach to multiplication. The segment-segment-region attitude leads to a division problem such as "How many feet of fence 6 feet high can be painted with one quart of paint if a quart will cover 150 square feet?" The Cartesian product approach to multiplication corresponds to a division problem such as "Robert wants a different outfit (shirt and tie combination) for every day in April. He has 5 shirts. How many ties does he need?" The operator-segment-segment division approach is illustrated by "Three times as much rice was grown in Phillipa in 1969 as in 1959. If 3600 bags were

---

Answers

1. The height in inches of the pile of boxes.
2. Total cost in cents.
3. Miles the whale swam.
4. Cents Marie received as interest.
5. Gallons of water that went into the pool.
6. Area of the triangle in appropriate units.
7. Total Peter gave in 6 weeks.
8. Number of different shirt and tie combinations John can wear.
9. This represents nothing here.
10. Yards Sally ran all together.

grown in 1969, how many were grown in 1959?" and "An 18-inch baby grew to be a 6-foot man. By what factor did his height increase?"

The student will find it helpful to determine how he would answer these questions if he did not remember his division tables. How would a person get the answers if he had not studied arithmetic? He could use chips, beans, sticks, rectangles, and so on, sorting them into arrays or piles, putting them end to end and side by side, and counting. It is good to think of one or two ways like this for solving each of these problems. It helps to ask what each bean or stick represents, and what approach to division you are using.

Remember that every problem can be worked in several ways. It makes things clearer to think through several approaches to a problem.

3.6 MIXED SITUATIONS The following situations may have several numbers. You are to determine what, if anything, is measured, shown, or described by different combinations of the numbers. Some combinations, most in fact, will be of no interest or relevance. You should so state. You are also to state what combinations to use to get certain requested measures. This is best shown by a sentence in equation form, such as the perimeter of an equilateral triangle of side 4 is $p = 3 \times 4$ or $p = 4 + 4 + 4$. Since we are emphasizing the processes of addition, subtraction, multiplication, and division, you should set up the sentences to show the process, not the numerical value of the final result.

1. One dish has 4 prunes and another dish has 12 prunes. What do the following represent? $4 + 12$, $12 - 4$, 12×4, $12/4$

2. Mary took a trip starting at 2 P.M. and ending at 5 P.M. What do the following represent? $5 + 2$, $5 - 2$, 5×2, $5/2$

3. A rectangular flower bed is 8 feet long and 2 feet wide. What questions do the following answer? $8 + 2$, $8 - 2$, 8×2, $8/2$

4. A steel rail 16 feet long weighs 480 pounds. What do the following represent? $16 + 480$, $480 - 16$, 480×16, $480/16$

5. Mary is 66 inches tall, Oscar is 70 inches tall. What do $66 + 70$, $70 - 66$, 70×66, $70/66$ represent?

6. The beam on a scale for weighing has an 8-ounce weight 4 inches from the balance point, or fulcrum. What do 4 + 8, 8 − 4, 8 × 4, 8/4 represent?

7. A grocery carton is a brick-shaped box 16 inches long, 12 inches wide, and 8 inches high. How would you find (1) the area of the largest face (side), (2) the number of faces, (3) the area of the smallest face, (4) the perimeter (distance around) of the largest face, (5) the perimeter of the smallest face, (6) how much longer the longest edge is than the middle-sized edge, (7) by what factor you would multiply the shortest edge to get the longest edge (this is the same as finding the ratio of the largest to the smallest edge), (8) the volume of the box?

8. A 12-pound box is lifted 6 feet in the air. What is measured by 12 + 6, 12 − 6, 12 × 6, 12/6? What is the work done?

3.7 SUMMARY Different attitudes and typical human situations leading to the basic computations are discussed. Knowing the different attitudes helps us recognize when addition, say, is helpful. While in practical affairs the numerical answer to a problem is required, in this chapter we emphasize recognizing what to do rather than finding the answer.

Knowing there are other attitudes toward multiplication than repeated addition gives us greater flexibility in solving problems. Similarly with the other operations. These attitudes have been widely used, although frequently they have not been named.

The attitude toward a problem depends on both the problem and the solver. Different people see the same problem differently. What looks like repeated addition to one person may look like an array to another. Both can be correct. We discourage claims that one or another attitude is "the" attitude or "the best" attitude in a given situation.

Formal manipulation is a common attitude toward each operation, although not always mentioned. Indeed, one of the main purposes of arithmetic is to learn formal steps in the dance of the digits. However, with calculators more common we may have people saying "Addition is what you get when you punch the + button." This reasonable attitude is not discussed.

Many of the topics are discussed more fully in the chapters on specific operations.

EXERCISES

In exercises 1–5 choose a letter for the number desired. Write a sentence relating the letter numeral to other numerals. You may not need to use all the numbers in the situation.

1. Antonio served four pizzas at 12 o'clock and seven at 6 o'clock. How many pizzas did he serve?

2. Sharon ate 7 of the 23 chocolate-chip cookies her mother made with 5 chips in each cookie. How many cookies remained?

3. How long a bicycle way was made by joining a 16-mile brick path to a 7-mile concrete road at one end and a 5-mile black-top road at the other end?

4. Seven pounds of almonds are mixed with five pounds of hazelnuts and put in packages weighing two pounds each. How many pounds of nuts are there all together?

5. Four gallons of water are in a tank with a capacity of 25 gallons. In ten minutes, eight gallons of water flow into the tank from a faucet and three gallons run out through a drain. How much more water is there in the tank?

In exercises 6–10, explain how the story shows the missing-addend attitude by describing the situation in terms of either union of sets or putting segments end to end.

6. A dozen pieces of fruit (apples and oranges) were packed for a picnic lunch. How many apples were there if four of the fruits were oranges?

7. Peter has 7 eggs. How many must he borrow from Aunt Agatha to make a 13-egg omelet?

8. Sandra cut 11 feet from a hose 42 feet long. How much is left for her father to use?

9. Starting at 10 o'clock Celia and John stood on one foot. At 10:15 Celia had to put her other foot down, and John lost his balance at 10:18. How much longer did John stand on one foot than Celia?

10. How many more girls than boys are there in a class having 17 girls and 12 boys?

In exercises 11–15, multiplication attitudes are considered.

11. Dorothy buys 6 pairs of shoelaces at 13¢ per pair.

(*a*) Can you see what an individual addend represents when adding six 13s?
(*b*) Similarly, when adding thirteen 6s?
(*c*) Which repeated addition process is easier to justify, adding thirteen 6s or six 13s?

12. An oriental prayer rug is 4 feet long and 3 feet wide. Draw a diagram to show how the segment-segment-region attitude leads to the area. Show what segment in your drawing represents 3 in $3 \times 4 = A$.

13. In exercise 12 the repeated addition attitude might say $A = 3 + 3 + 3 + 3$. What does an individual addend 3 represent? Point out in a picture what the underlined 3 represents in $A = 3 + \underline{3} + 3 + 3$.

14. In the town of Middleswamp there are 3 streets running east and west, and 6 running north and south. Show how an array attitude fits the problem of counting the intersections.

15. A cartoon book has 5 noseless faces *a, b, c, d, e,* on which can be put 3 noses x, *y, z* to make funny faces. Draw a tree and list the faces such as *ax, cz, dy,* . . . .

In exercises 16–20, division attitudes are presented.

16. Suppose you had several K & K's (candied nuts) to distribute equally among three children. Think how you would do this. Which process is this most like—sharing, subtraction, or making an array?

17. You have 63 cents to buy pencils at 7 cents each. In spending all your money, are you most likely to think of sharing, repeated subtraction, or building an array as a way of seeing how many pencils you get?

18. If there were 21 K & K's in problem 16, what can you conclude knowing that $3 \times 7 = 21$? Which attitude toward division would you be using—sharing, missing factor, or repeated subtraction?

19. Three treatments are necessary to cure green fever before it runs its natural course. If 90 people are ill, would sharing, making an array, or partition be the most likely attitude for determining how many people can be cured with 45 treatments?

20. A sidewalk 4 feet wide contains 32 square feet. Would missing factor, repeated subtraction, or sharing be the most likely attitude for finding the length of the walk?

Exercises 21–32 are studies in recognizing several ideas in the same story. The several ideas are seen by following six instructions for each story:

(*a*) Name the useful number operation: add, subtract, multiply, or divide.

(*b*) Write a number sentence (equation) relating some of the numbers given, including the answer to any question asked. Give a letter name *n*, *x*, or *y* to unnamed numbers. While questions may be asked in the story, you are not charged with answering them.

(*c*) Which attitude toward the operation you gave in *a* is most appropriate? Of those available, restrict your choice to the one of the following or "none of these."

For addition: joining of sets, joining of segments, continued counting.
For subtraction: take away, missing addend.
For multiplication: repeated addition, array, segment-segment-region.
For division: missing factor, repeated subtraction, sharing.

(*d*) Rephrase the story using the words in the attitude name.

(*e*) State another attitude toward the operation that someone else might think appropriate. If the other attitudes are not appropriate, so state.

(*f*) If another person might see a different operation as useful in *a*, state the other operation. Write a number sentence using this operation.

ILLUSTRATION: Ruben, who sat in chair number 15, did 4 problems more than Shelly. Shelly worked 18 problems. How many did Ruben work?

(*a*) Addition.
(*b*) $N = 18 + 4$.
(*c*) Joining of sets.
(*d*) Ruben worked a set of problems formed by joining a set of 4 problems with a set of 18 problems. How many problems did he work?
(*e*) Yes. Continued counting. Ruben counted 4 more than 18 to get the number of problems he worked.
(*f*) Subtraction. $N - 4 = 18$.

Here are the stories. For each story, follow instructions *a–f*.

21. Helen washed dishes for 13 minutes and vacuumed the 24 square yards of carpet for 18 minutes. How much time did she spend all together?

22. Peter spent his 75¢ allowance on his eighth birthday for a 55¢ kite and a 20¢ bag of popcorn.

23. John, with 17 marbles, has 4 more marbles than Peter. How many does Peter have?

24. Of the 23 men on the squad, 8 graduated, leaving only 15 for next season.

25. What is the cost of a dozen 7¢ candy bars if they are bought on special for 6¢ each?

26. The 60 students are seated in 6 rows of 10 each at tables arranged in 3s and 4s.

27. For 88¢, Jeanne bought 400 sheets of paper in 4 packages, 100 sheets in a package.

28. Six girls shared 480 milliliters of orange juice equally. How much did each girl receive?

29. Rocky has $2.80. How much more does he need for a $4 ticket to an 8:00 P.M. party with Harriett?

30. How many kinds of sandwiches can Sherry make if she has 3 kinds of bread and 5 kinds of fillings? (Only one kind of bread and one kind of filling to a sandwich.)

31. Starting from the 17th floor, Alice climbed 8 floors. What floor did she end up on if there are 20 steps between floors?

32. Cast adrift on a barely floating raft on February 13 with 32 ounces of water to drink, Robinson limited himself to 4 ounces a day. How many days did his water last?

Answers to selected questions: **1.** $n = 4 + 7$; **3.** $w = 16 + 7 + 5$; **5.** $g = 8 - 3$; **7.** 7 eggs joined with Aunt Agatha's eggs make a set of 13 eggs; **9.** Celia's time segment put end to end with the "how much longer?" time segment equals John's time segment; **11.** (*a*) yes—cost of one pair; (*b*) very obscure, one cent for each pair; (*c*) adding 6 13s; **13.** Each addend 3 is the area of a rectangle 3 × 1.

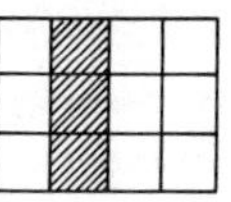

15.

a: ax, ay, az
b: bx, by, bz
c: cx, cy, cz
d: dx, dy, dz
e: ex, ey, ez

17. repeated subtraction; **19.** Partition shows 15 people can be cured; **21.** (*a*) addition; (*b*) $t = 13 + 18$; (*c*) joining segments; (*d*) Helen spent a time segment of 13 minutes followed by a segment of 18 minutes. How much time all together? (*e*) Yes. Union of sets. Helen spent a set of 13 minutes on dishes and a set of 18 on vacuuming. How much did she spend in their union? (*f*) subtraction, $18 = N - 13$; **25.** (*a*) multiplication; (*b*) $C = 12 \times 6$; (*c*) repeated addition; (*d*) How much are 12 bars at 6¢ each? (*e*) array, but not as good as repeated addition; (*f*) Multiplication is best; **27.** several stories; (*a*) multiplication; (*b*) $400 = 4 \times 100$; (*c*) repeated addition; (*d*) In 4 packages of 100 sheets there are 400 sheets; (*e*) repeated addition best; (*f*) division $\frac{400}{100} = 4$; (*a*) division; (*b*) $C = \frac{88}{4}$; (*c*) missing factor; (*d*) What is the cost per package if 4 packages cost 88¢? (*e*) sharing; (*f*) multiplication $4C = 88$.

4

Sets and Classification

4.1 INTRODUCTION Classifying is a most important scientific activity. Sets usually arise from classifying things according to some property such as sex, weight, height, cost, and so on. It is not surprising, then, that many mathematical ideas, such as counting numbers, are closely related to sets and can be discussed in terms of sets.

In recent years sets have been made the central theme of several "modern mathematics" curricula. While there is presently a swing away from an overemphasis on sets, people do find a good understanding of the ideas useful.

Good teachers have always used sets. Children played with counters—seeds, stones, chips, or other objects easy to handle. Number ideas such as counting and addition were taught by means of these physical objects. The ideas of sets were used unconsciously or were considered so obvious as not to be worth discussing.

Now we know we may have only partial understanding about an "obvious" idea we accepted early in life and never explored with mature judgment. Further, what is obvious and observed by one may not be obvious or observed by another. Anyone who watches a teenage girl and her younger sister enter a store with a popcorn stand next to the cosmetic counter will have a vivid demonstration of this.

In this chapter we will point out some of these obvious ideas with the current standard ways of treating them. We will have a common language and pay particular attention to some ideas of great future value. In particular we will consider ways of describing sets and classification in words and Venn diagrams and certain relations such as union, intersection, and complement.

The intuitive definition of a set is a collection of objects with a common property. A set of books and a set of dishes are excellent examples. We shall give further illustrations and then show how it is more useful not to require the objects to have a common property. Pragmatic pressures have even forced people to extend the concept to include singletons and the empty set. This process is discussed in detail, not only to explain the concept of sets, but also to show how ideas grow and generalize.

4.2 CLASSIFICATION BY ONE PROPERTY Sorting objects into two sets as they do or do not have a specified property is the most basic classification process. Some examples are shown by these contrasting terms: men, women; living, dead; old, young. We ask for a general description of this process that will apply in all cases.

What is involved in a classification is (1) a set of distinguishable, individual objects—a collection or population; (2) a property that each object either has or lacks; (3) a mechanism, process, or method of doing the sorting.

For example, we might sort the set {4,5,6,7,8,9} into odd and even integers by listing or by underlining: {5,7,9}, {4,6,8} or $\{4,\underline{5},6,\underline{7},8,\underline{9}\}$.

We might classify English letters as vowels or nonvowels (consonants) by listing as before. Before we finish we meet a difficulty. The set of objects is clear (alphabet), the process of classifying (listing) is clear, but it is not completely clear which letters are vowels. Consider *y* in the words *my* and *year,* and *w* in *won* and *new.* The classification cannot be completed until we decide about *y* and *w,* thus satisfying the second requirement.

The properties may have two dissimilar names, as old–young. Or one name may merely be the negative of the other. Perhaps in the word *woman wo-* is a prefix meaning "not" in some patriarchal language.

If we have only a few objects, listing is a reasonable way of showing the classification. For many sets, listing may be inconvenient or impossible. Here we can show the set by accurately describing the common property. The set {5,7,9} is characterized as "the odd counting numbers between 3 and 10." While listing is easy in this case, the set "the odd counting numbers less than 10 million" would make a long list. Indeed, we could not list "the set of odd counting numbers."

We define a set, then, either by (1) listing or (2) describing. It is conventional to enclose the list in braces. We write "the set {4,6,8}" rather than "the set (4, 6, 8)." Many people place the description inside braces in so-called set-builder notation. We write "{x|x is an even number between 3 and 9}" or "{even numbers between 3 and 9}."

In the expression {x|x is ...}, x is a variable standing for an element of the set. To the right of the vertical bar is given the condition(s) characterizing x. This is a flexible notation allowing the conditions to include x. This set-builder notation is read "the set of all members x, such that x is ____." The objects forming a set are called *members* or *elements* of the set.

Classification is shown vividly by a Venn diagram:

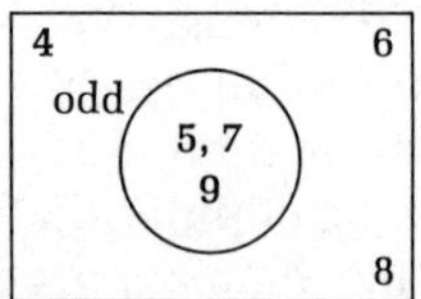

The universal set—all the elements under discussion—is included in the rectangle. Those members classified as having the property are placed inside the circle. Those lacking the property are outside the circle. A word or letter is placed on the circle to show the property the members inside the circle have.

The figures do not have to be a circle and rectangle. Here are two other satisfactory diagrams:

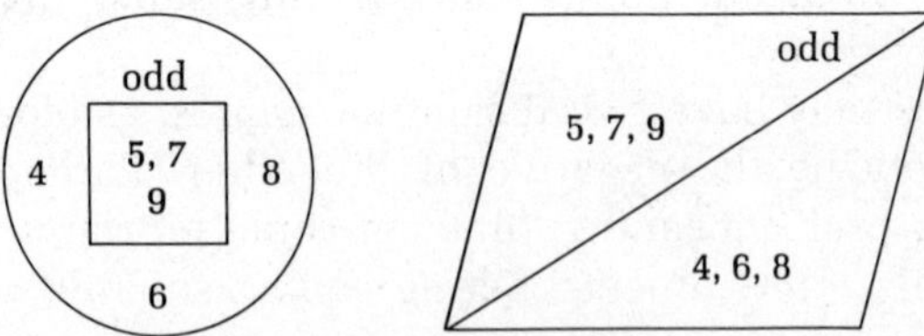

4.3 CLASSIFYING ON THREE PROPERTIES A used-car dealer classifies his cars according to whether or not they are made by American Motors, whether they are pre-1971 models, and whether they are sedans. A Venn diagram illustrates this classification. We may imagine his used-car lot in the form of a rectangle, with 3 large painted circles.

He parks the cars so all the AM and only AM cars are inside the A circle. Similarly pre-1971 models are in the B circle and sedans are in the C circle. We list some of his cars.

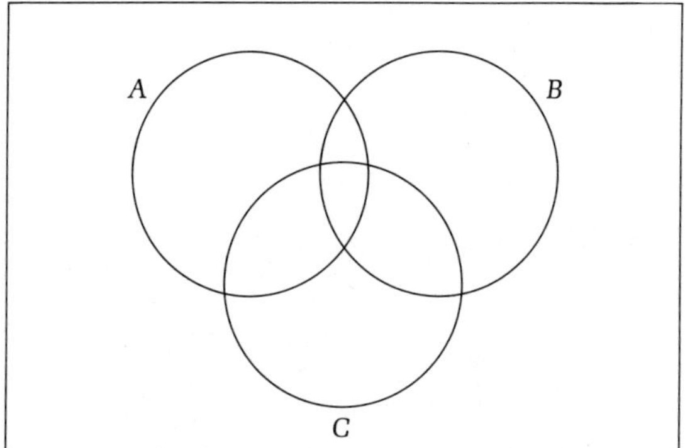

a.	1971	white	Rambler	station wagon
b.	1969	beige	Chevrolet	pickup
c.	1973	ivory	Ford	4-door sedan
d.	1972	black	Plymouth	convertible
e.	1967	cream	Ambassador	station wagon
f.	1971	beige	Rambler	2-door sedan
g.	1973	blue	American	convertible
h.	1968	white	Chevrolet	sedan
i.	1970	green	Ford	station wagon
j.	1972	black	Dodge	4-door sedan
k.	1963	ivory	Rambler	2-door sedan
l.	1964	blue	American	convertible
m.	1973	black	Rambler	sedan
n.	1965	maroon	Plymouth	4-door sedan
o.	1970	black	Ambassador	4-door sedan
p.	1971	ivory	Ford	pickup
q.	1968	white	Rambler	station wagon
r.	1974	green	Mercury	sedan

We draw the diagram (parking lot) with some of the cars located, leaving it as an exercise to locate the others.

Consider car a, a 1971 white Rambler station wagon. 1971 is not before 1971, hence a is outside circle B. A Rambler is an American Motors car, so a is inside circle A. A station wagon is not a sedan, so a is outside circle C.

Car l, a 1964 blue American convertible, is an American Motors car, pre-1971, and is not a sedan. Hence l goes inside A, inside B, but outside C.

Car d, a 1972 black Plymouth convertible, goes outside all three circles. Car k, a 1963 ivory Rambler sedan, goes inside all three circles.

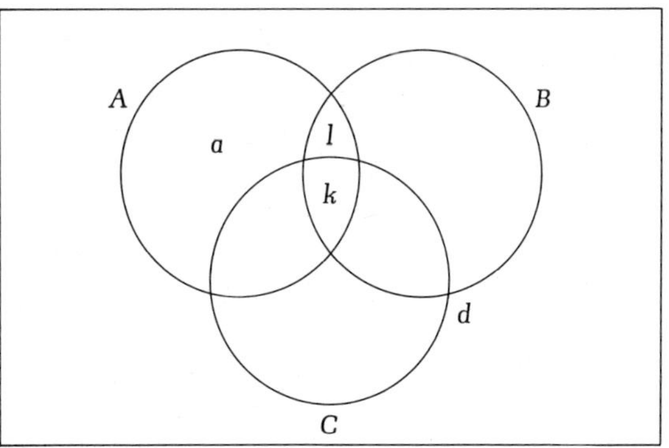

It may be good to practice classifying on two properties first. An appropriate Venn diagram is shown in the margin.

You may want to try classifying on four properties. For example, the cars are dark-colored or not. A useful diagram for this classification is also shown.

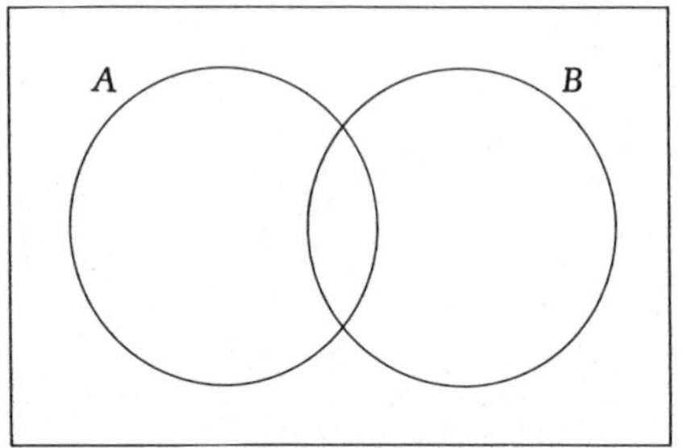

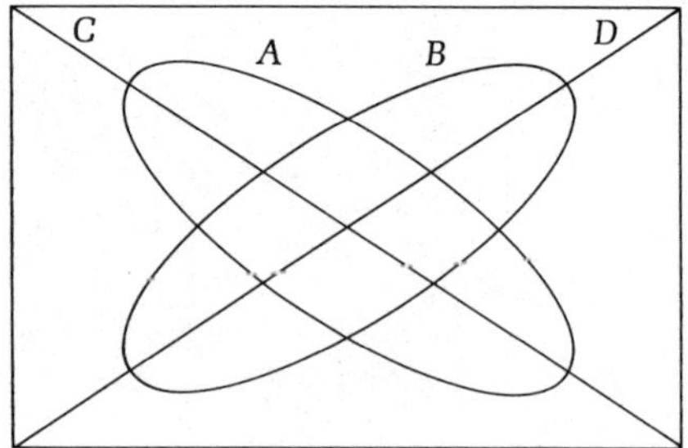

4.4 CLASSIFICATION TABLE When we wish to classify on more than three qualities, the Venn diagrams get complicated. It is useful to have a classification table. Suppose we wish to classify a certain set of numbers as to whether they are divisible by 2, 3, 5, or 7, and to see these results in a vivid way. We show the resulting table for the set {2,3,15,17,30,35, 42,630}.

	2	3	5	7
2	1	0	0	0
3	0	1	0	0
15	0	1	1	0
17	0	0	0	0
30	1	1	1	0
35	0	0	1	1
42	1	1	0	1
630	1	1	1	1

We see that the divisors head the columns. A 1 or 0 is placed in the column as the number at the left does or does not have the quality described at the top.

Alternatively, checks could be put in the columns. It is good, however, to put something in each column for each number. Seeing a blank, one does not know whether the number does not have the divisor or if the blank was merely overlooked.

The table is a convenient way of listing all the possible categories based of these qualities. Suppose a girl wishes to classify her boy friends as to whether they are rich (R), handsome (H), and socially concerned (S). What are the possible categories? We make a table, with one of her friends in each category as an example.

	R	H	S
Al	1	1	1
Bob	1	1	0
Carl	1	0	1
Doug	0	1	1
Eugene	1	0	0
Frank	0	1	0
George	0	0	1
Harry	0	0	0

Can you describe the sequence of categories? Can you make a similar list on two qualities? On four qualities?

This listing of possible categories by table is closely related to tree diagrams. We illustrate:

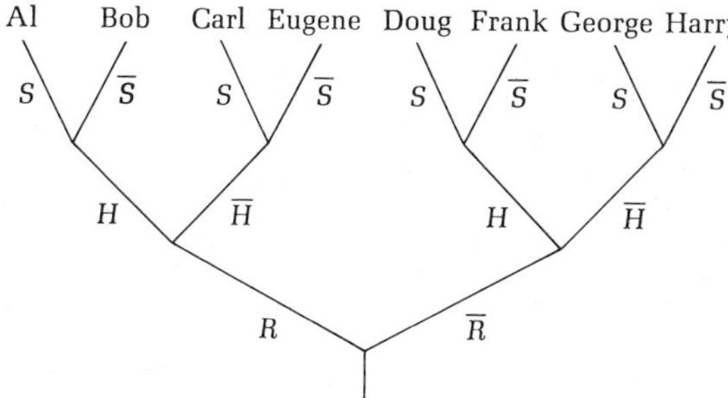

The lowest branches from the trunk of the tree separate by rich and not rich. Each of these branches separate by handsome and not handsome. Each of these separate by socially concerned and not socially concerned. We note that the order of listing of the boys is slightly different at the top of the tree—some people prefer the order produced by the tree diagram.

4.5 SET RELATIONS AND OPERATIONS Suppose the Venn diagram for the two sets T and F is

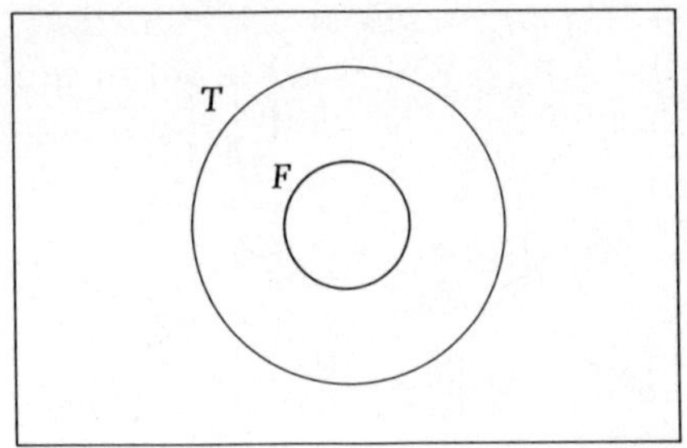

Some phrases describing the relation between T and F follow. The set F is contained in set T. F is a part of T. Every member of F is a member of T. An element is a member of F only if it is a member of T. If an element is a member of F, it is a member of T. F is a subset of T. T is a superset of F.

An example of two sets with this relation is $T = \{\text{set of multiples of } 2\}$, and $F = \{\text{set of multiples of } 4\}$. If the universal set is the counting numbers less than 10, we get

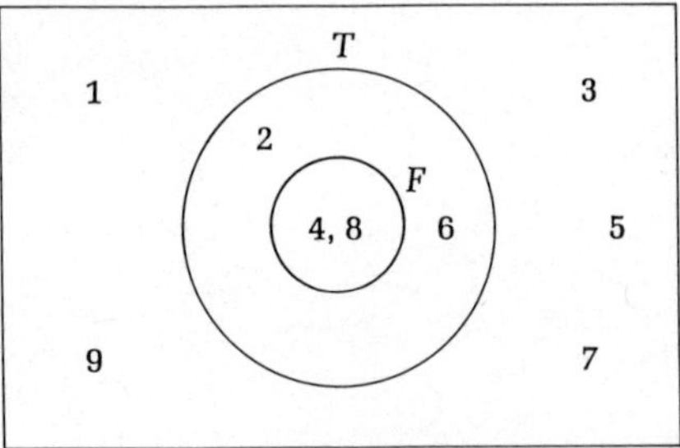

Intersection of sets Suppose sets A and B have the Venn diagram

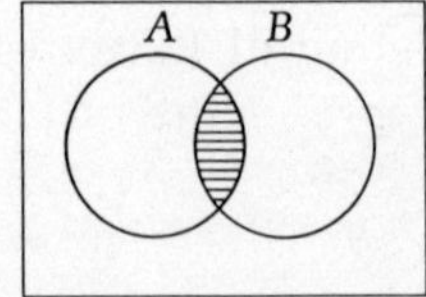

and we are interested in the hatched, or shaded, set. This set is called the *intersection* of A and B. It consists of those members in both A and B.

We also write "$A \cap B$" and say "A cap B" for the intersection. Suppose A is the set of multiples of 2, and B is the set of multiples of 3, and the universe is the counting numbers less than 20; then we would have

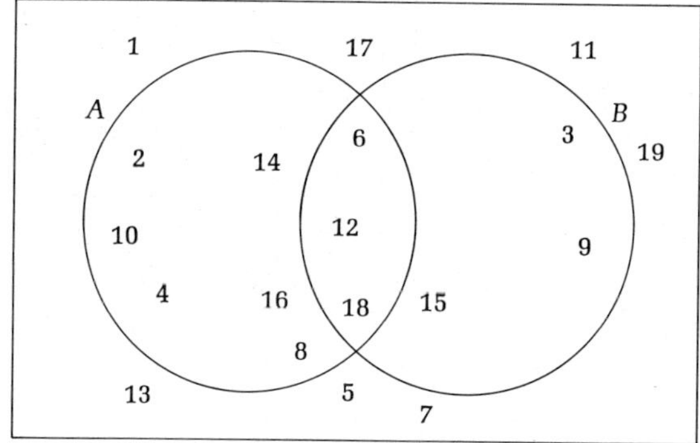

If we call the intersection C, we write $C = A \cap B = \{6,12,18\}$, where $A = \{2,4,6,8,10,12,14,16,18\}$ and $B = \{3,6,9,12,15,18\}$.

Union If two sets F and S have Venn diagrams and we are interested in the set D shown hatched, we say "D is the union of F and S" and write $D = F \cup S$ and also say "D equals F cup S." If F is the set of multiples of 4 and S is the set of multiples of 6, and the universal set is the counting numbers less than 25, we get

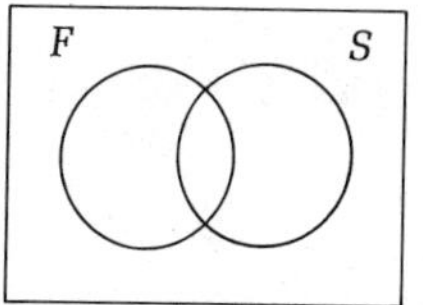

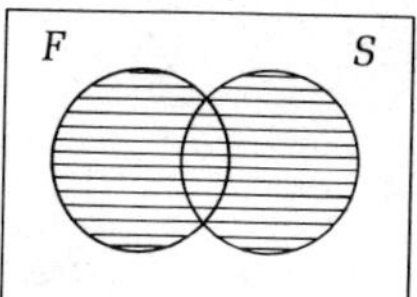

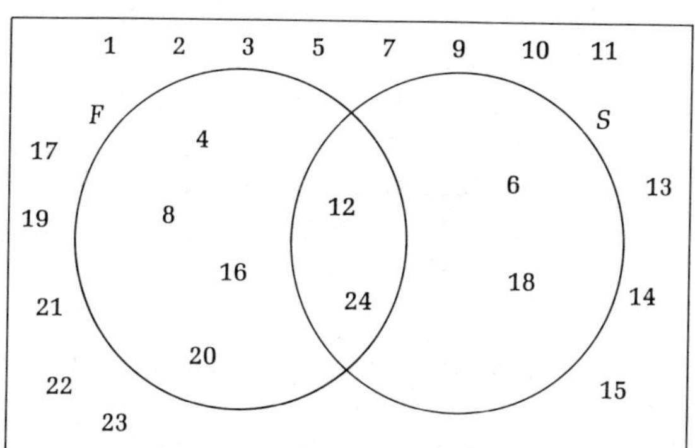

The numbers are written in sequence in the diagram so you can find them easily. We note that $D = F \cup S = \{4,6,8,12,16,18,20,24\}$, where $F = \{4,8,12,16,20,24\}$ and $S = \{6,12,18,24\}$.

We note certain subset relations: $(A \cap B) \subset A$ and $(A \cap B) \subset B$. The symbol $\subset$ means "is a subset of." The intersection of two sets is a subset of each.

Similarly, $A \subset (A \cup B)$ and $B \subset (A \cup B)$. Any set is a subset of its union with another set.

If two sets have no elements in common, they are called *disjoint.* The set of boys is disjoint from the set of girls. Sets *overlap,* or are not disjoint, if they have common elements.

4.6 EXTENDING THE CONCEPT OF SET Extending or generalizing an idea is one of the most powerful ways of gaining greater understanding, as well as being a favorite pastime. We are fortunate that an opportunity to show this process comes early in set theory. We use the process to introduce singleton and empty sets.

The concept of set extends in four stages.

1. In the first stage, a set is a "collection of objects with a common property that can be readily understood." The words *set, group, bunch, herd, flock, collection* show that this concept is widespread; it goes far back in history and appears in many forms with many names. If you make up some sentences using these words in common, ordinary ways, you will note that they "unify a multitude." That is, several objects are classified together as one. The set concept is a sort of mental bag into which several things are placed so that they can be treated as one thing, yielding a convenience like putting a dozen oranges in a bag and carrying them with one grasp. E pluribus unum.

You may note that the members of the sets are easily identifiable, distinct physical objects. A person can see the whole set at a glance. Some typical phrases are *a set of books, a set of dishes, a flock of sheep.* The language is formed to fit reality.

2. In the second stage, the concept is extended to allow (*a*) the members to be dispersed or widely separated, (*b*) the members to lack any common property, and (*c*) the members to be nonphysical.

It is a distinct advance in awareness from Mr. Brown's herd of cows and Mr. Lynch's herd to the set of all cows in the state. We classify into one mental set objects which are not grouped together so that they can be seen at a glance.

The concept of "common property" is widely used and intuitively obvious. The task of explicitly defining this concept is a tangle of turns. It seems that whether an observer says two objects have a common property or not depends as much on the observer as it does on the objects. Further, this concept is quite remote from "unifying a multitude." The value of a bag doesn't depend on its containing only oranges, and not containing an orange, a pencil, a book, and an apple. Since the unifying power is highly useful, the common property requirement for a set is discarded at this stage.

Clearly, sets with no common property are given by listing, and not by describing the common property of members.

Man is doing a lot of thinking at this stage. So it is not surprising that he collects ideas, as well as physical objects, into sets. Indeed, some hairsplitters will argue that the set of cows in California is a set of ideas, since at any given instant the observer can only be aware of the ideas. More useful is a set such as {sweet,sour,salty}, which are properties rather than objects. The words are used more frequently as adjectives than as nouns.

Even with these extensions, the concept of set still unifies a multitude, and language is formed to fit reality.

At this stage, there is no upper bound to the number of members in a set. It is easy to imagine putting one more idea into a set of ideas. At stage 1, an observer may reasonably believe there is a largest number. Just as he runs out of cows in counting a herd, he might run out of physical objects in counting any collection he can see at a glance.

3. In stage 3, the concept of set is extended to include singletons, or sets with one member. Clearly, the observer is not "unifying a multitude," the major purpose of sets in earlier stages, since there is only one object. A bag is a nuisance if you are only going to carry one orange.

Yet the concept of a singleton set must be useful and convenient, or people would not use it. What human problems are made easier by this concept?

Consider a club of stamp collectors. Naturally, each member has a stamp collection, a set of stamps. Shari would like to join the club, but he has only one stamp—a "one-penny black," the first stamp ever printed. His one stamp is worth more than all the stamps of the other members combined. At the moment Shari cannot afford other stamps he would like to own. Does Shari have a stamp collection? Should he be a member

of the club? The most convenient thing to do is to extend the concept of set to include singletons, and admit Shari.

A census taker who is counting the herds of cows in a county may find it more convenient to include Mrs. Reva's one cow as a herd than to set up a new category. It is more convenient to report "There are fifteen herds in the county," than to say "There are fourteen herds in the county and Mrs. Reva has one cow."

In general, when one is classifying objects (stamps, cows) into many sets (who owns them), it is highly convenient to allow a set with only one member.

As we extend the concept of set from stage 2 to stage 3, the convenience gained is not that of unifying a multitude but of making our descriptions simpler. Language fits reality but singletons are accepted to simplify our words.

4. In stage 4, the concept of set is extended to include the empty set. Does it serve a useful purpose to say "There is an empty set of elephants in this room"? Silly, isn't it?

However, there are situations where life would be greatly simplified if there were an empty set. Consider the following.

(*a*) Yoris is forming the intersection of sets. He has

$$\{a,b,c\} \cap \{b,c,d\} = \{b,c\}$$
$$\{p,q\} \cap \{a,q,t,r\} = \{q\}$$
$$\{a,b,c\} \cap \{p,q\} = Z$$

When asked what Z is, he says "Z isn't anything. In the other cases where the sets overlap, I have $A \cap B = C$, where A, B, C are sets. If M and N are disjoint, $M \cap N$ is meaningless. I write $M \cap N = Z$ to remind me of this, rather than make a separate column for the cases where the two sets are disjoint. I agree it is misleading to have a symbol that looks like it might represent a set, when really all it does is remind me that the expression does not really describe a set."

(*b*) Sallie is making a list of sets by describing the common property held by all members. She lists four descriptions:

the girls enrolled in algebra
the boys enrolled in English
the boys enrolled in sewing
the girls enrolled in history

Marie says she has listed only three sets. "There are no boys enrolled in sewing," says Marie; "therefore there is no set."

Sallie says, "Maybe there is no set, but there might have been." They argue. Finally Sallie says, "We are wasting time. Let's just mark the entry with a Z. It sounds like a set, and we'll just remember it is not."

(*c*) Mr. Cantor is thinking about sets. He says "I wish I had a set Z with the property that $A \cup Z = A$ for any set A. It would be easier for me to keep track of what I'm doing if there were one. There isn't—but I'll just write Z wherever it is helpful, with the understanding that Z by itself is not a set, but $A \cup Z = A$."

Yoris and Sallie and Mr. Cantor were bewailing their troubles. Suddenly Mr. Cantor said, "Look, we are acting as though Z stands for a set. Let us just assert that it does. We'll say that Z is a new kind of set. We'll call it a 'null set.' Agreed?" Yoris and Sallie agreed, and so does everyone else who has advanced to stage 4. A special symbol, $\emptyset$, a letter in the Norwegian alphabet, is used for a null set.

The convenience brought by a null set allows us to assert that—

1. the intersection of two sets is always a set

2. a phrase that appears to describe a set by giving the common property actually does so even if there are no objects with the property

3. there is a set $\emptyset$ where $A \cup \emptyset = A$ for all sets A. $\emptyset$ is thus the "identity set for union," just as in numbers zero is the identity number for addition and 1 is the identity number for multiplication

The convenience of having a null set is very real. It is a convenience of language, which saves us from having to make qualifying phrases such as "providing there are any members" or spend energy figuring out if there are members. It confers reality on a symbol that is useful in keeping ideas straight. But, even more than singleton sets, the null set does not "unify a multitude." Further, instead of choosing language to fit reality, we change our idea of reality to fit our language. We assert the existence of the null set in order to make our language convenient.

Having introduced $\emptyset$ as a new type of set to simplify language, we must state how it is to relate to other sets. In this section, we discuss number of members, union, intersection, and order. Uniqueness (how many sets $\emptyset$ are there) and subset-superset relations are discussed in section 4.8.

There are no members in $\emptyset$, for if $\emptyset$ had members it could not do what we want it to do. We define 0 as the number of members of $\emptyset$, using the symbol used earlier as a spacer, or placeholder, in numerals such as 205.

The union of two sets consists of all members in either set. The only members in $\emptyset \cup A$ are those in A. This supports our definition that $\emptyset \cup A = A$.

Since $A \cap B$ is the set of members of both A and B, we define $\emptyset \cap B = \emptyset$ consistent with this property.

In putting sets in natural order, we match the members 1:1, and then say that A has more members than B if there are elements of A left unmatched, with no unmatched elements of B. It seems natural to say that every nonempty set has more members than $\emptyset$.

Can we continue the extension? Is there a set with a negative number of members? What would this mean? Would it be useful to extend the set concept this way? Hardly any description or language problems are helped by negative sets, so the concept of set is not extended beyond $\emptyset$ in elementary practice.

Some people may feel there were more than four stages in the development of the modern concept of sets. This may be so, but in any case it seems clear that, in the stages of development, people progressed from language closely following existence to existence following language. Empty sets did not exist while the language was being formed for the early concepts of sets. The empty set was created later, to ease some of the difficulties of that language.

Some teachers believe no mention of this concept-development process should be made to children. Such authors say, "A set is a collection of objects. The empty set is a set with no objects"—and children are taught to repeat these obvious contradictions. Such authors say children are not able to understand concept development, and need to be taught concepts as being complete and finished, arising mature at birth. They want children to accept the empty set without reservations, and feel this will be done only if the empty set is presented in the same way as other sets are presented.

There are differences of pedagogic opinion. Other teachers feel that forcing children to assert obvious contradictions as being true either weakens their ability to think logically or causes them to believe that arithmetic is fantasy and not to be taken seriously outside the classroom. To insist that concepts are complete makes learning very difficult. People love their concepts. It is usually easier and better to regard a concept as growing and improving rather than as being rejected and replaced.

4.7 SOME COUNTING PROBLEMS A society reporter stated that 10 women attended a party; 7 wore long dresses, 8 had long hair, and 4 long-haired women had long dresses. She also said the host was making love to the high school mathematics teacher. The host's wife sued for divorce on the evidence of the reporter. The host defended himself by saying the reporter was obviously a poor observer. Do you see any reason in the above figures to support the host?

Diagrams help analyze the counting. Suppose that all the women were placed in a square, with the long-haired women in the upper half, and the long-dressed women in the left half. Let H represent long-haired women, $\overline{H}$ short-haired women, D women with long dresses, and $\overline{D}$ women with short dresses.

	D	$\overline{D}$
H		
$\overline{H}$		

On the unmarked sides we write the numbers reported for H and D; at the lower right is placed the total number of women reported. In the empty cells we place any figures we have been given.

	D	$\bar{D}$	
H	4		8
$\bar{H}$			
	7		10

Next we write numbers in the empty cells and along the sides so that the numbers in a row add to the sum on the right, and the numbers in a column add to the sum at the bottom.

	D	$\bar{D}$	
H	4	4	8
$\bar{H}$	3		2
	7	3	10

But now—catastrophe! We have more than 10 women all together! $4 + 4 + 3 > 10$! And we cannot fill the empty cell. Some mistake has been made. The reporter was inaccurate. Further investigation showed that it was the science teacher who was making love to the mathematics teacher, and he was her husband.

This suggests that while the reporter may have had the right numbers, she had them in the wrong places. Can you find a way to set things right? Here is one way:

4 long-haired women in long dresses
7 long-haired women in short dresses
8 short-haired women in short dresses
10 short-haired women

	D	$\bar{D}$	
H	4	7	11
$\bar{H}$	2	8	10
	6	15	21

Can you find another way?

A Venn diagram with circles can also be drawn for this last case. This diagram shows the number in the four subsets, but there is no convenient way to record the totals.

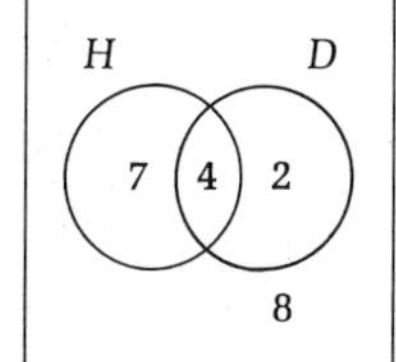

A mattress manufacturer makes several kinds of mattresses. In particular, he makes king size and standard size, with foam or inner-spring interiors, and with either plastic or cloth covers. He makes them in several colors. One day a clerk brings in the following inventory:

50 mattresses in stock
14 king size
21 foam interiors
22 plastic-covered
6 king size with plastic covers
7 king size with foam interiors
8 foam with plastic covers
2 king size with foam interiors with plastic covers

After the clerk leaves, the manufacturer finds he wants to know how many cloth-covered, standard-size, innerspring mattresses he has in stock. Can he find out from the data he has, or must he send the clerk out to count?

We draw a Venn diagram, and put the number of mattresses in each region. We first draw several small diagrams, shading the region where we have the corresponding count:

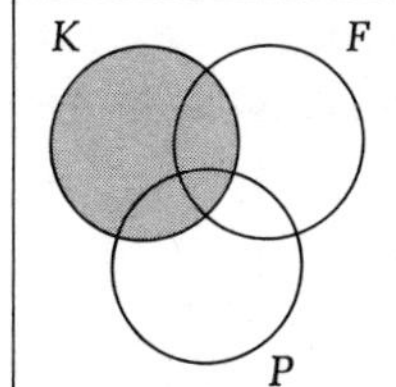

14 king size

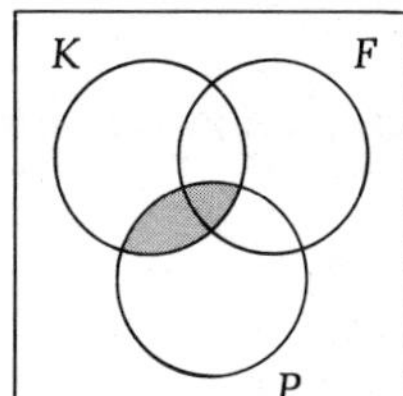

6 king size, plastic-covered

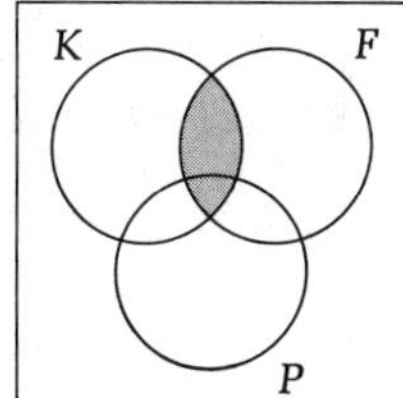

7 king size, foam interior

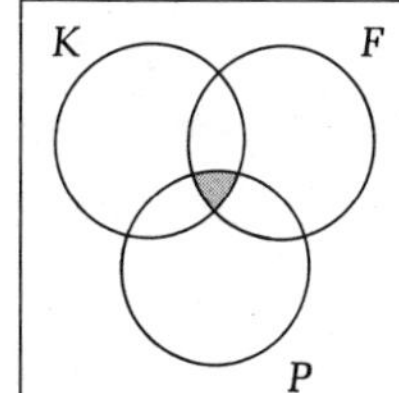

2 king size, foam interior, plastic-covered

It would be helpful to draw the other four diagrams corresponding to the other four counts given by the clerk. We want the number outside the three circles, which will represent the number of cloth-covered, standard-size, innerspring mattresses.

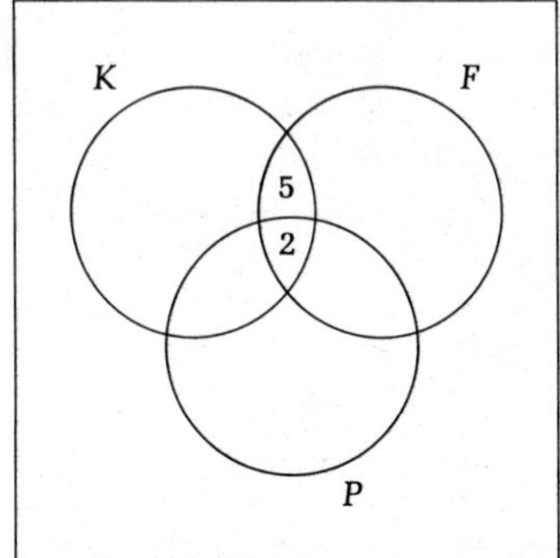

From the set with 2 and the set with 7, we deduce that there are 5 mattresses in the region shown. Similarly there are 4 king-size, innerspring, plastic covered mattresses, and 6 standard-size, foam interior, plastic-covered mattresses. We have

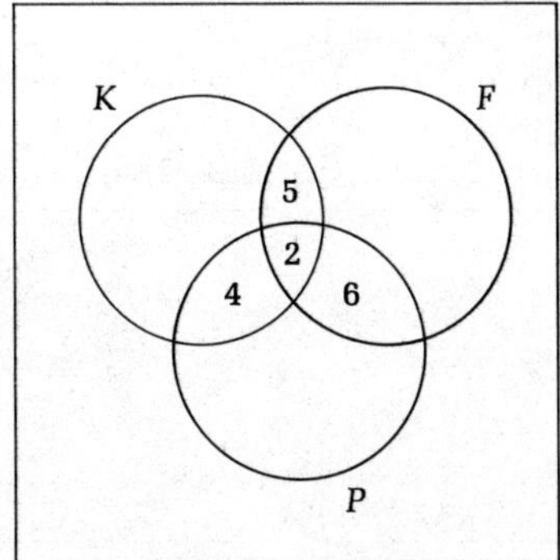

Since there are 14 king size, there are 14 – (5 + 2 + 4) = 3 king-size, innerspring, cloth-covered mattresses. Similarly, we get the other numbers shown:

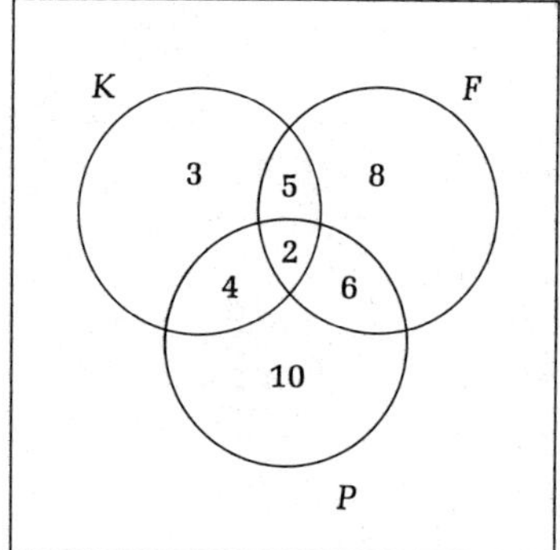

Now the standard-size, innerspring, cloth-covered mattresses are outside all the circles drawn. Since there are 50 all together, the number desired is $50 - (3 + 5 + 8 + 4 + 2 + 6 + 10) = 50 - 38 = 12$. The manufacturer doesn't need to send the clerk back again.

4.8 SAME, EQUAL These words describe concepts that give rise to a question. When do two descriptions or lists describe the same set? We have seen that the same set can be given in different ways. Here are several descriptions of the same set:

A = {Washington,Adams,Jefferson}

B = {The first three presidents of the United States}

C = {Adams,Jefferson,Washington}

D = {American presidents from 1788 to 1808}

We say, "Two sets are the same, or two statements or lists give the same set, if the members are the same in both." Wishing to avoid repeating the word *same,* we also say, "Two sets are the same if the members of each are also members of the other."

We note that a single set can have many different descriptions, or orders of listing. Sometimes the order in which members are listed is important, as in the set of people lined up in a queue at a ticket office. This is called an *ordered* set.

Changing the names of members in the list does not change the set. In talking about numbers, the set {2,3,4} = {four,two,three}.

A perhaps unforeseen conclusion is that we accept only one empty set, ∅. The set of lions at the North Pole, the set of rocks floating down the river, and the set of living people who knew Lincoln are all the same set. These sets have no members. We say these sets satisfy "vacuously" the condition that they have the same members. There is no object that is a member of one and not a member of another. Since man invented the empty set to simplify his language and thinking, he is free to assert that there is only one of them.

In all human communication there is room for misunderstanding. Sometimes we need to take special pains to point out the universal set, and the individual members. For example, we form the set of the letters used in the words *rectangle, circle,* and *line*. This will be the union of the sets {a,c,e,g,l,n,r,t}, {c,e,i,l,r} and {e,i,l,n}. If these symbols are put together, we get {a,c,e,g,l,n,r,t,c,e,i,l,r,e,i,l,n}.

We note that some symbols appear more than once. Are these symbols for different objects or for the same object? We see they are symbols for the same letter, so more than one is unnecessary, and redundant. Eliminating the redundant or duplicated symbols, we list the set by {a,c,e,g,i,l, n,r,t}.

On the other hand, we may find that the girls in a club are {Carol,Ruth, Carol,Mary,Zelda}. The name Carol appears twice. It may be that there are two girls with the same name. Or it may be that one girl's name is listed twice. We cannot say here which is the case. In fact, there may be any number from one to five girls in the club. Can you see how this could be? Suppose the club roster were drawn up by listing the names of the chairmen of the executive, social, scholarship, financial, and membership committees. Does this help show what might happen?

To avoid these difficulties, in naming objects and forming lists we try to have a single name for each object, and then list each name only once in a listing of a set.

The author on several occasions has held up a familiar open book and asked a group to estimate the number of letters on a page. Everyone agrees that the question is quite clear, but the estimates vary widely. Many estimates are less than fifty, and many are in the hundreds or thousands. It is often a surprise for people to realize that there is a correct answer less than fifty and a correct answer of several hundred.

Consider the set {six,seven,eight}. Are the members of the set words or numbers? It is not clear. Further information is necessary.

4.9 SUBSET AND SUPERSET When two sets are related as A and B in the diagram, we say B is a subset of A, and A is a superset of B. "Chairs" is a subset of "furniture." "People" is a superset of "women." In each case there are members in the subset, and the subset is not the superset.

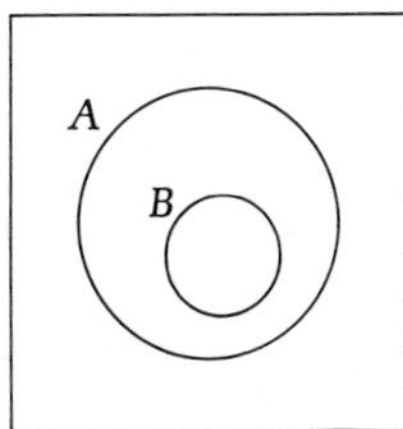

This suggests that the empty set is not a subset, and a set is not a subset of itself. Do we want it this way? The following situations show these restrictions are quite inconvenient. Hence, we shall say the empty set is a subset of every set, and every set is a subset and superset of itself.

Women form a subset of the set of people. But do the women in a specific room form a subset of the people in that room? If there are some men and some women, the women clearly form a subset. If there are no men in the room, the women are the people. If all the people are men, the set of women is the empty set. Now clearly it is a great inconvenience to have to run over to the room to see who is there before we can truthfully say that the women form a subset of the set of people in the room. For simplicity's sake we would like the empty set to be a subset of any set, and every set to be a subset of itself.

Further, consider the intersection of two sets, $C = A \cap B$. Suppose A and B overlap as shown in the Venn diagram.

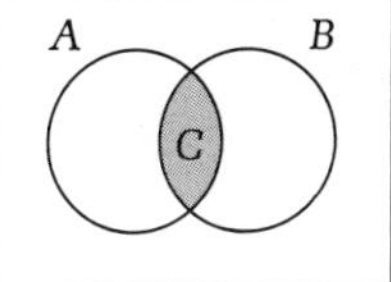

Then C is a subset of both A and B. If $A \subset B$, then $C = A$. Similarly, if A and B are disjoint, then C is the empty set. For convenience people would still like to say that C is a subset of A. They want to say that $A \cap B$ is a subset of A without having to look first to see the extent of the overlap, if any.

As a third preference for counting $\emptyset$ and the superset as subsets, let us list the subsets of several supersets M.

M	**Subsets**
{a,b}	{a} {b} {a,b} $\emptyset$
{a,b,c}	{a} {b} {a,b} $\emptyset$ {a,c} {b,c} {a,b,c} {c}

We make a table. Let N be the number of members of the superset. Let S be the number of subsets, counting both the superset and the empty set.

Then we get the table

N	2	3	4	5 . . .	m
S	4	8	16	32 . . .	2^m

We can derive this formula by observing that the subsets of $\{a,b,c\}$ are just the subsets of $\{a,b\}$ together with those sets formed by joining c with the subsets of $\{a,b\}$. When we join one more element to the superset, we double the number of subsets. The formula 2^m is slightly more elegant than $2^m - 1$ or $2^m - 2$, as it would be if the empty set and the superset were rejected as subsets.

For these reasons we accept the superset itself and the empty set as subsets of the superset.

Can we make a concise definition of subset? Three statements suggest themselves:

1. A subset is a part of the superset.
2. A subset is a set of members of the superset.
3. A subset is a set, each of whose members is a member of the superset.

Statement 1 is the clearest, most intuitive, and shortest. Alas! We must reject it, because neither the empty set nor the superset would then be a subset. Statement 2 includes the superset but not the empty set as a subset. Statement 3 labels both the superset and the empty set as subsets. The empty set satisfies statement 3 vacuously: there is no member of $\emptyset$ that is not a member of the superset. Hence we take statement 3 as our definition of the subset-superset relation.

Still, many people feel that "sub" clearly implies "part of" or "less than all." Recognizing this feeling, we say that the superset is an *improper* subset of itself. All other subsets are called *proper subsets* of the superset.

Some people wish to call $\emptyset$ an improper subset, since it satisfies definition 3 vacuously. However, we call $\emptyset$ a proper subset whenever $\emptyset$ is not the superset.

4.10 COMPLEMENT OF A SUBSET AND n(A) The complement of a subset A with respect to a superset B is the set of elements in B not in

A. For example, the complement of $A = \{y,z\}$ with respect to the superset $B = \{v,w,x,y,z\}$ is $\overline{A} = \{v,w,x\}$, where $\overline{A}$ is one of the symbols used for the complement of A. Other frequently used notations for the complement of A are A' and $\sim A$.

The complement of A is also defined by $\overline{A} \cap A = \emptyset$ and $\overline{A} \cup A = B$. When B is the universal set, we speak of the complement of the set A rather than subset A. The complement idea is illustrated by boys and girls as subsets of children. Odd and even are subsets of the counting numbers. We see that $\overline{\overline{A}} = A$. The complement of the complement of a set is the set itself.

You already expect the idea of "the number of members in a set A" to play an important role. We represent this by $n(A)$. For example, $n \{v,w,x,y,z\} = 5$. These ideas will be used in later chapters. Can you guess how?

4.11 TEACHING AIDS Sets of things to classify form excellent aids. Blocks of different colors, sizes, shapes, and thickness are good. Rings of colored string, or plastic hoops—like hula hoops—make good Venn diagrams.

EXERCISES

1. Let U, the universal set, be the numbers 26–36 inclusive. Let T be the numbers in U divisible by 3. List T.

2. Draw a Venn diagram of U and T in exercise 1.

3. Let D be the even numbers in U. Draw a Venn diagram showing U, T, and D.

4. Let F be the numbers divisible by 5. Draw a Venn diagram showing U, T, D, F.

5. *Let* U be the numbers 1–500 inclusive, and D, T, F be the numbers divisible by 2, 3, 5, respectively. Draw a Venn diagram. Instead of writing in all 500 numbers, write in each region the smallest two of the numbers in that region.

6. Make a classification table classifying numbers as in exercise 5. List and classify at least the numbers 1–20, and those you placed in the Venn diagram in exercise 5.

7. Let U be the set of foods, and C the set of foods sometimes eaten cooked. Describe $\overline{C}$, the complement of C.

8. Let U be the set of foods. Let F be fruits, B the foods often eaten at breakfast, C the foods often eaten cooked. Make a Venn diagram showing F, B, C. Enter at least one food in each of the eight regions if possible.

9. Let $U = \{a, b, c, d, e, f\}$, $A = \{a, b, d, f\}$, $B = \{b, c, d, f\}$, $C = \{a, c, d, e\}$. List or characterize in some other way:

$A \cup B =$	$A \cap (B \cup C) =$	$A \cup \emptyset =$
$A \cap B =$	$(A \cap B) \cup C =$	$B \cap \emptyset =$
$\overline{A} =$	$\overline{A \cup \overline{B}} =$	$\overline{B \cup C} =$
$A \cap \overline{B} =$	$\overline{\overline{C}} =$	

10. The telephone book classifies names into sets by various criteria. Name three qualities that categorize sets in the book. Draw a Venn diagram, and list one name in each of the elementary regions, or mark it as empty.

11. Suppose there is no empty set. Each of the following phrases purports to describe a set. Does it? Write "yes," "no," or "maybe."

(a) The set of cows in a meadow.
(b) The set of water in a glass.
(c) The set of birds in a tree.
(d) The set of lions in the animal show.
(e) The set of butter on a piece of bread.
(f) The set of living people.
(g) The set of purple in a picture.

12. Let U be the alphabet. Let $A = \{a, b, c, d\}$.

(a) Give two disjoint, nonempty subsets of A.
(b) Give two supersets of A, neither of which is a subset of the other.

13. List the subsets of {girl, dog, bone}.

14. How many subsets are there of {apple, pear, peach, ice cream}?

15. How many members are in the set $\{2, 3, 4, 3, 2\}$?

16. If Barbara, Mary, and Helen are the girls in club A, and Mary, Helen, and Sallie are the girls in club B, how many girls are in $A \cup B$?

17. Could there be more than one correct answer to exercise 16? If so, list the possible correct answers.

18. How many girls are members of both clubs in exercise 16?

19. List 5 properties of people according to which they have been sorted, or classified, at one time or another.

20. How wide is your acquaintance? Make a Venn diagram based on 3 of these 5 properties. Name a personal friend for each of the 8 regions in the Venn diagram.
(If the properties separate people into approximately equal sets, relatively few people can name a personal friend in all 32 of the elementary sets determined by the 5 properties.)

21. Sandra had a set of A's on her report card, yet she failed. How could this happen?

22. A set of lambs was placed in a cage of hungry lions, yet no lamb was eaten. How could this happen?

23. In a set of 20 people, $n(A) = 15$, $n(B) = 14$, $n(A \cap B) = 13$. What is $n(A \cup B)$? $n(A)$ is the number of members in A.

24. In a set of 30, $n(P) = 18$, $n(Q) = 12$, $n(P \cup Q) = 21$. Find $n(P \cap Q)$.

25. In a set of 25, $n(A) = 23$, $n(B) = 24$. List the possible values for $n(A \cup B)$.

26. List the possible values for $n(A \cap B)$ in exercise 25.

27. In a group of 12 people, 8 can bake a cake and 7 can scramble eggs. What can you say about the number who can do both?

28. In exercise 27, what can you say about the number who can neither bake a cake nor scramble eggs?

29. Without using set-builder notation, describe $\{y \mid y \text{ is brown}\}$. What is $\{y \mid y \text{ is brown}\} \cap \{z \mid z \text{ is a dog}\}$?

Answers to selected questions: **1.** 27, 30, 33, 36.
3.

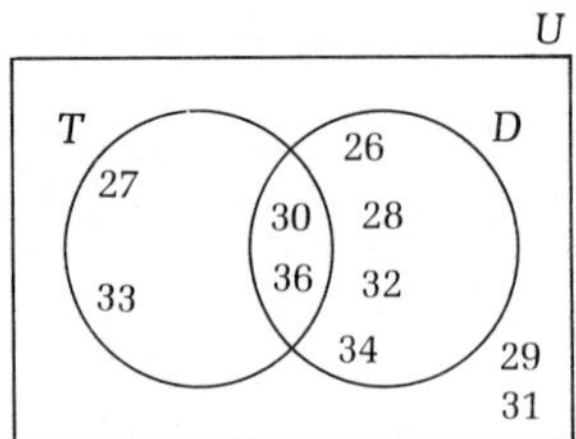

5.

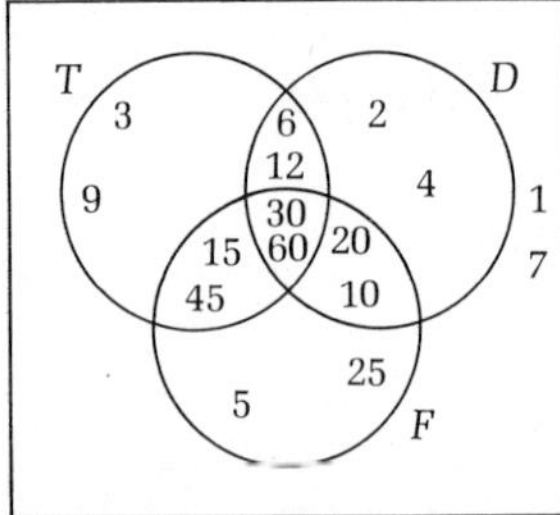

7. foods always eaten raw; **11.** (*a*) maybe; (*b*) no, or maybe; (*c*) maybe; (*d*) maybe; (*e*) no, or maybe; (*f*) yes; (*g*) no, or maybe; **14.** 16; **16.** 4, assuming same name goes with same girl and different names with different girls; **17.** yes, 1, 2, 3, 4, 5, 6; **19.** age, sex, residence, year in school, major—more specifically, < 18, M, lives in Los Angeles, underclassman, studies history; **21.** Her set of A's was $\emptyset$; **23.** 16; **27.** from 3 to 7 inclusive; **29.** set of brown things; set of brown dogs; **30.** No. Sometimes they want a set to be a subset of itself, so the definition is made to allow this; **33.** (1) Phrases like "the set of books on the table" actually describe a set even if there are no books; (2) the intersection of 2 sets is always a set; (3) $A \cup X = A$ for all A if $X = \emptyset$; (4) used to define 0; **35.** yes.

30. Joseph says that a subset is only part of a superset. Is he right? If not, why have people decided to disagree?

31. If you flew in a time machine back to 1700, you would find the empty set useful, but no one around you would know about it. Why would these people be doing things the hard way?

32. Describe a situation in 1700 where the empty set would be useful, and tell how the local people would handle the problem.

33. Give three reasons why life is more convenient because we have an empty set.

34. Give two reasons why life is slightly simpler because the empty set is a subset of the set of chairs in a classroom.

35. Is the set of elephants in a teacup the same set as the set of people 970 years old?

36. Make up and work a problem showing classifying. Use a classification table and a Venn diagram.

37. Make up and answer a problem concerning subsets.

5

The Numbers . . . , 4, 5, 6, . . .

5.1 INTRODUCTION You are familiar with the counting numbers, 1, 2, 3, ... We have already used them in earlier chapters. Just as some great teachers claim that education is learning what you already know, in this chapter we look at the counting numbers more closely to see the physical situations that bring them forward and where they apply. The ellipsis (three dots) at the start and at the end of the list is an invitation to explore how the sequence is extended to the right and to the left before our notions of number are extended as well.

Counting as a process is discussed, with the number line introduced and used as an aid to skip counting. We close showing that counting may be done by weighing or other forms of scale reading.

5.2 PHYSICAL ORIGINS AND ATTITUDES The counting numbers arise in naming properties of sets, segments, and sequences. They tell us how many members are in a set, compare a segment with a unit segment, tell how much, and locate members in a sequence. The properties we associate with numbers are properties arising from these physical situations.

Of course, numbers are ideas, not physical objects. Yet the meaning of the idea of number is born in our experience with physical objects. We shall say we are following a set attitude toward numbers when we think of them as associated with sets. Similarly, we follow a segment or sequence attitude when we consider these ideas as arising from segment or sequence situations. Since the same numbers are abstracted from all three attitudes, we look closely at the three situations to see what they have in common and to be clear about how they differ.

As discussed in chapter 4, a set is a collection of objects, each element clearly defined and distinctly different from the other elements. A segment is part of a line, such as this segment $\overline{AB}$:

A segment can be separated into as many parts as we please, each of which is also a segment. This is not true of sets. This separation may be done mentally or physically. (Strictly speaking, all mathematical concepts are ideas and cannot be treated physically. We mean that a picture of a segment can be separated mentally or physically.)

Many things in nature can be separated just like a segment. Sticks, quantities of water and other fluids, chunks of butter and clay, and time intervals are examples. The segment will be used as a model for each of these when we consider measuring them.

Counting numbers are applied to segments as follows. The segment is separated into a set of well-known parts of equal size, which are counted. This is called measuring and results in phrases such as "6 inches" and "4 pounds." Each part is called a *unit.*

A sequence is a set arranged in some order. Children lined up according to height form a sequence. Runners at the start of a race, days, pages in a book, and cars in a train are other examples. The problem is how to describe this order and to locate a particular member.

The usual way of describing a sequence is to match the members to

some standard sequence. A standard sequence is any thoroughly familiar sequence. There are two commonly used, the alphabet {a,b,c,. . .,z} and the counting numbers {. . . 4,5,6 . . .}. Some memory experts say that the secret of a well-developed memory is to tie ideas to some standard sequence.

We note that in using letters or numbers for a sequence, one can begin anywhere. It is customary when talking about sequences of years, for example, to use numbers such as 1972, 1973, 1974, 1975, ... There is a certain "one" somewhere in the past that is of no relevance now—only of historical interest. Scholars even doubt that Jesus was born in the year 1. We shall talk more about order later.

It is clear that one can have order ideas such as "before," "after," "first," "last" with no concept of "how many."

Many people feel keenly that one or the other of these attitudes is the only correct approach to the study of numbers. Actually one can start with any of the attitudes—set, segment, or sequence—and develop the properties of the others. Which approach to take is a matter of taste or convenience. Euclid developed numbers from segments. Recent number theorists such as Peano base their theory on sequential ideas. Many mathematics educators wish to introduce numbers as based on sets.

Among the reasons for preferring sets are (1) the individual numbers are more obvious; (2) counting is more obvious; (3) the operations of addition and multiplication are easier to describe; (4) the algorithms, or steps in computation, are easier to explain.

There are some disadvantages. Extending counting numbers to fractions and real numbers is much easier from segment ideas. No one even suggests discussing 2/3 or $\sqrt{2}$ without using the number line and right triangles as physical models. Still, on balance, we prefer to introduce numbers by means of sets, bringing in segments and sequences later as they are useful and needed. You may wish to imagine another way.

5.3 SETS AS THE BASIS FOR COUNTING NUMBERS When we say that the counting numbers {. . .,4,5,6,. . .} can be introduced as based on sets, we mean that a counting number can be introduced as a set property. The notation {. . .,4,5,6,. . .} means that we are sure that 4, 5, 6 are counting numbers, and the dots say there are others. We will need to determine how to put in numbers where the dots are.

While the origins of mathematics are hidden in the misty shrouds of antiquity, most people believe these numbers arose to help people communicate the idea "how many?" To think about and feel this situation, we imagine that we have no ideas of numbers or numerals and wish to carry the "how many" idea in the next example.

EXAMPLE Suppose each of a set of people is to be given a book. The books are in a storeroom some distance away. How do we get the books to the people?

1. All the people go to the storeroom. The storekeeper hands each person a book. The group returns. This method clearly works. However, it requires a lot of time by the people in traveling and perhaps waiting at the storeroom.

2. One person goes to the storeroom to get the books for all. But how does he know how many books to get?

3. One person counts the people in the set. He writes the number on a piece of paper so he won't forget it, carries the paper to the storeroom, and so on. But what if he has no knowledge of numbers or numerals? The process clearly fails.

4. One person prepares a portable set that matches with the people in the class. This can be done by putting marks on a sheet of paper, one for each person, or by placing pebbles or other small objects in a bag, one for each person. The messenger carries this prepared set to the storeroom. The storekeeper gives out the books, one for each mark or pebble.

We observe that by matching in one-to-one correspondence there are as many people in the room as there are members in the "carrying set," and as many books as there are members in the "carrying set," so there are as many books as there are people. The idea of "how many" has been carried without the use of number!

The notion of "as many as" through matching extends readily to "more than" and "less than." By quick observation a lecturer in a room can tell if there are "as many chairs as listeners," "more chairs than listeners" or "less chairs than listeners." Can you describe what the lecturer must observe to make this decision?

In general one can compare sets in this way: As far as possible, match the members of one set with those of the other so for each member of the first set there is one and only one member of the second set. Such a matching is also called a 1:1 correspondence between the sets. If there are no members left over in either set—that is, no members left unmatched in one set while all those of the other are paired up—then there are as many members in one set as in the other. The two sets are called *equivalent*. If every time they are matched, there are some members left over in the first set, that set is said to have more members than the other, or second, set. The second set is then said to have less, or fewer, members than the first.

We note that we can in theory compare two sets in this fashion without counting, or using numbers at all. Anyone who would compare the boys with the girls on a grade school playground during recess will recognize certain practical problems.

Whether one should match or count in comparing sets depends on the circumstances. One does whichever is easier. We discuss the problem without numbers—

1. to show that comparisons do not depend on numbers

2. to lay a foundation for discussing numbers

Numbers arise when our messenger is asked, "How many books do you want?" He will show his portable set and say, "This many," or, "As many as there are in this set." Basically, this is all he can say, even when he says something like "Thirty-two." To speed up communication it becomes valuable to have certain standard sets which one uses, rather than a new set for each occasion like marks on a paper or pebbles in a bag. For example, if asked how many stars in the set {∗∗∗∗∗}, we would say, "As many as the fingers on a hand." Whenever there was an equivalent set to describe we would agree to always say the same thing.

We can imagine that as time went on the answers would be shortened. People shortened "as many as the fingers on a hand" to maybe just "hand."

Something like this actually happened. The word for "five" in some languages is the same or very close to the word for "hand."

A certain confusion might arise. When the word *hand* is used, is the speaker referring to a part of the body or giving an answer to "how many?" As a result there is a tendency to change the language. Perhaps the word is pronounced differently when used for its different meanings. Or perhaps a word from the language of a nearby trading tribe would be used. At any rate, the modern result is that words for answers to "how many?" have no other meaning in the current language at all.

The words for the small sets like *one, two, five, three* are probably contractions in some language of phrases like *as many as. . . .* The words for larger sets like *fifteen* and *forty-eight* are artificially made up from the smaller names. We consider the naming problem in the next chapter.

We note that at this stage of our language development we answer "How many in this set?" by recognition. We name a few:

{* * *} three {* *} two {* * * * *} five

Today these names are associated with their sets in the same way we associate Alice or Bob or Carol with certain individuals. There is no counting or muttering to ourselves "one, two, three, four, FIVE." We recognize that this name goes with its particular set, using whatever clues such as shape or pattern that we use to recognize old friends.

From one point of view, counting is the way we determine how many members are in a set. This may be merely by recognition. Counting is a well-known process, at least for small sets. To count the toes on a baby's foot is a task enjoyed by many. One indicates the toes one at a time and repeats the sounds "one, two, three, four, five." The last word uttered as the last toe is pinched is then said to be the answer to "How many toes does the baby have?"

This process is a way of determining and stating that "there are as many toes as there are members in the set {one,two,three,four,five}."

We see that the standard sets used in counting are {one}, {one,two}, {one,two,three}, {one,two,three,four}, {one,two,three,four,five}, and so on. The counting process sets up a matching between the objects in the set counted and some counting set; which counting set is automatically determined by the process. The name of the set is the name of the last element in the set.

5.4 ORDER AND SEQUENCE AS THE BASIS FOR COUNTING NUMBER Traditionally, small children are taught to count by repeating the sounds {one,two,three,...} in order. They count to five, ten, . . . hundred, depending on the interest of parent, teacher, or child. While there may be sets involved, frequently this is only a standard sequence being learned.

Five, the child is taught, is the number after four and before six. Nineteen is the number before twenty. There is no set with nineteen members for which the question "how many?" is being asked. The sequence is learned in much the same way the alphabet sequence {a,b,c, . . .} is learned. Rhyme, rhythm, and song are often used. The child is told he has erred if he says "One, two, five, four, three" instead of "One, two, three, four, five."

In many parts of life the sequential, or order, properties of the counting numbers are the ones used, as discussed in chapter 2. Numbers are used to impose an artificial order for people forming queues for service at ticket counters, for listing names of players in sports, and so on.

The sequential use is changed to cardinal use by forming the standard counting sets. While people can learn to identify small numbered sets (up to five or seven), hardly anyone has the ability to recognize the number in a set of twenty, say, without some use of the sequential properties.

There is a certain chicken-and-egg (which came first?) argument about how people learn order in the counting numbers. Some people say the order in the sounds is first learned as a standard sequence; then the use of these sounds to tell "how many" in a set is learned. Others say that the number names are learned first as associated with sets; then the order in the numbers is established based on set-order from 1:1 matching. This is not a mathematics controversy. The teacher needs to know that both routes are possible and to use both as teaching tools.

It is useful to explore these routes by making up number names using the alphabet sequence {a,b,c, . . .} rather than the digit sequence {1,2,3, . . .}.

Abstractly, there is no reason (except convenience) why the accepted order for naming numbers should be the same order as the natural order of sets. Those who wish to explore this might try naming numbers with letters at random, say:

1–C	5–J	9–I	13–H
2–F	6–M	10–N	14–D
3–E	7–K	11–L	
4–A	8–B	12–G	

Now try to count the number in a set of chips or beans. The numerical or set order is different from the alphabetical order. Then try naming the numbers with ordered letters:

1–A	4–D	7–G	10–J	13–M
2–B	5–E	8–H	11–K	14–N
3–C	6–F	9–I	12–L	

Here the numerical or set order is the same as the alphabetical order. Note the greater convenience.

Set order comes from matching. Set A is larger than set B if, whenever they are matched, there is always at least one member of A left over. If we use "square," "thumb," "eyes," "hand," "triangle" as names of numbers, and order them in set order, we get "thumb," "eyes," "triangle," "square," "hand."

When numbers are arranged in the order of sets, they are said to be in *natural order*. This is also the order we teach as the standard order of the sequence.

5.5 SEGMENTS AS THE BASIS FOR NUMBER. NUMBER LINE Phrases such as "three pounds," "4 centimeters," and "eight minutes" come up frequently. We compare them briefly with "three apples," "4 eggs," and "8 chairs."

"Four centimeters" might come up as follows. A segment $\overline{AB}$

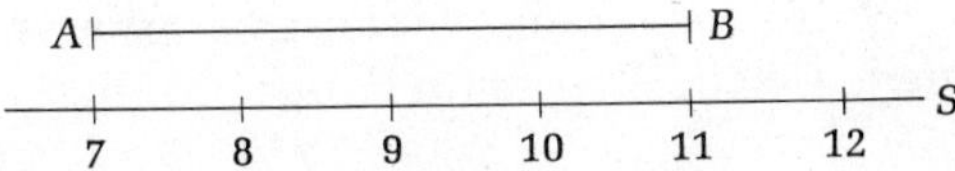

is placed alongside a scale S as shown, where the numbers on the scale are one centimeter apart. Point A coincides with 7, point B coincides with 11. We say that $\overline{AB}$ is 4 centimeters long.

Or $\overline{AB}$ might be broken up into shorter segments, each one centimeter long. This separation is done mentally.

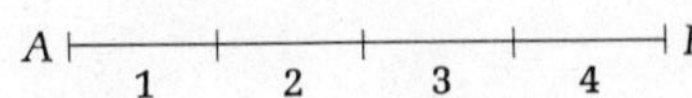

We count the set of segments, getting 4 as shown. A number is placed under each segment to help count.

Now the answer to "How long is $\overline{AB}$?" did not have to be "4 centimeters." It could have been "1.6 inches." But "4 eggs" cannot be called 1.6 in any reasonable way. There are natural units in sets, like eggs. There is no natural unit segment. In answering "how long?" we always choose a unit, and the answer is in terms of this unit.

This is also true in measuring weight, time, area, and many other things. We note that weight and time are often measured by means of scales using a pointer on a dial.

5.6 EXTENSION OF THE NUMBERS A natural question is "Is there any beginning or end to the numbers {. . . ,4,5,6, . . .}?" We look at the right end, or increasing side, first. We consider the set meaning. Since one more object can always be joined with a given set, we see that there is an infinity of such numbers. We can always get another number. Keeping the numbers in natural order, we see that the set continues indefinitely to the right as far as we please.

Going to the left is different. Taking one member away from 4 leaves 3, then 2, then 1. Do these numbers end here? If we take this last member away, do we have a set for which we wish to say "This set has as many members as _____"? Do we wish to ask the question "How many?" Do we have another name that we can use in the process of counting?

The answers are ambiguous. We have already accepted the empty set. Since we answer the question "How many?" about all other sets, we would like an answer here. On the other hand, in the counting process we always start with 1.

It is typical in mathematics that whenever in doubt, we do both. We say that "Zero is the number of members of the empty set. Zero precedes one in the natural order." Many people find zero easier to accept as the cardinal number of the empty set than as a "placeholder" or some other meaning sometimes given it. The numbers used to express the cardinality of sets, then, are 0, 1, 2, 3, ... This set of numbers is called the *whole numbers.*

On the other hand, the empty set is not as natural a set as the others. Further, we do not use zero in the counting process where we say to ourselves "1, 2, 3, ..." Hence the set without zero, {1,2,3,...}, is called the *natural numbers* or the *counting numbers.*

One does not make a big thing about these names. The main idea is that sometimes one wishes to consider the set {0,1,2,...} and sometimes the set {1,2,3,...}. These sets differ solely in that 0 is a member of the first and is not a member of the second. We wish different names for these sets. Why the names were chosen is of no great importance. Indeed, some authors like to use *counting numbers* as a second name for the whole numbers.

Can the numbers be extended to the left even more? Is there a set we can get by taking one away from the empty set? This seems preposterous. But then we recall we developed the empty set as a notion for a set that did not seem to exist. Why not the notion of a set with fewer elements than the empty set?

We ask if this notion will serve a useful purpose. It brings in the notion of signed sets, similar to signed numbers. The notion can be applied to debts, positive and negative electrons, and other situations. Just as the empty set is a useful introduction to zero, so the negative sets could be a useful introduction to negative numbers.

However, it is not customary to extend the set idea to negative sets. We say that no set has fewer members than the empty set.

When we think of the ordinal use of the numbers {...,4,5,6,...}, their barrier to the leftward extension is somewhat different. Either the set begins with some number or it extends indefinitely to the left. If you have ticket 5 in a queue, this could mean that four other people will be served, and then you. Or it might mean only that you will be served after the man with ticket 4. You will actually be the next person served if the man with 4 has already been served.

If the numbers used in an ordered sense also tell us how many are ahead, then the set begins with 1. If we use the numbers to show position in a sequence, the numbers can go back indefinitely. This is discussed in more detail in the next section on the number line.

5.7 THE NUMBER LINE It is helpful to think of the whole numbers as labeling points equally distant apart on a line. This arrangement is called a *number line.* While a number line is an idea, a picture of this idea is so vivid we often call a picture a number line as well.

The number line vividly shows set, sequence, and segment ideas of the counting numbers. First we discuss sequence, or order, ideas. The type of order shown by counting numbers is called linear order.

We start with three numbers:

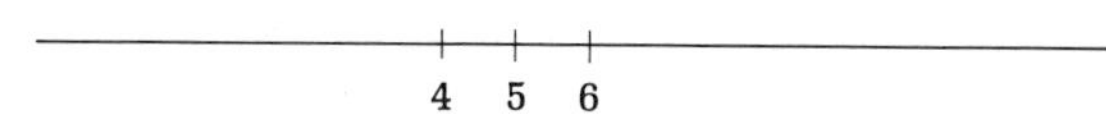

Clearly, we can extend this picture:

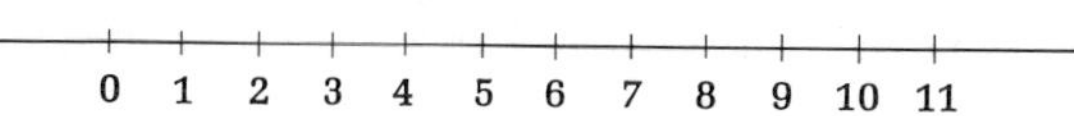

The distance between adjacent points is called a unit distance. It can be any convenient length for a particular picture.

It is not necessary for the line to be straight, or for the distances between adjacent points to be equal. Nor need the line be horizontal. The essential order ideas can still be shown.

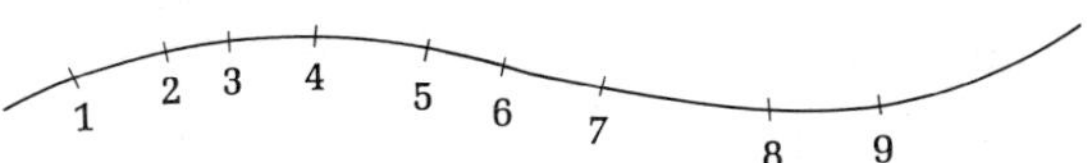

Mileposts and numbered stations along a highway are applications of the number line. On some highways, numbered mileposts are put in by auto clubs or highway departments exactly one mile apart. They are useful for checking speedometer readings.

When the line is straight and the distances are equal, the number line forms a scale. Such scales appear on rulers, yardsticks, and weighing machines.

The picture suggests strongly that the numbers can be extended indefinitely in both directions. The appropriate names to apply as one moves to the right with increasing numbers will be discussed in the next chapter on numeration. In extending to the left of zero, we assign the same names as to the right with a dash or negative sign. The symbol "−2" is read "negative two," "minus two," or "opposite two."

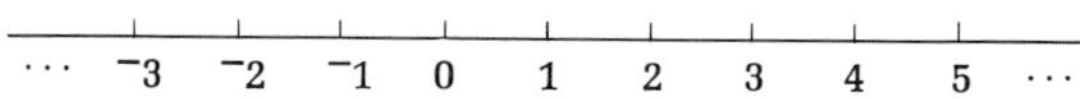

The number line is used extensively to teach and display number relations, such as order and size, skip counting, and the number operations of addition, subtraction, multiplication, and division. Fractions and real numbers are introduced by means of the number line, and their properties discussed.

Any teaching tool or model as rich as this one may also lead to students thinking the teacher means one thing when he actually means another. When we consider that *five* can refer to at least four separate geometric objects we can see this. Let us list these four meanings.

1. Five is the name or label of a point after four and before six. This relates strongly to the sequence idea. Any other name, such as *Alice*, with no "how many?" meaning, could also be used for this purpose. However, *five* is superior. Note that we need not know how far the number line goes to either right or left.

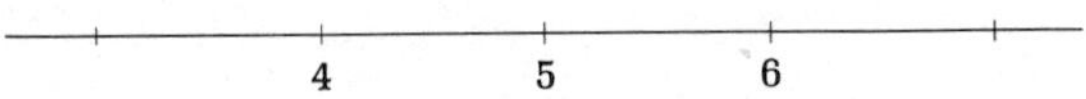

2. Five is the segment (0, 5). (0, 5) means the segment from the point labeled 0 to the point labeled 5. Five is also used for the segments (1, 6), (10, 15), and so on.

3. Five is the length of any of the segments in (2).

4. Five is the name of the set, as well as the number of members in the set of segments (0, 1), (1, 2), (2, 3), (3, 4), (4, 5). We think of the segment in (2) as separated into these unit segments, thus forming a set. Any set of five such segments can be named *five*, but this is the one of greatest interest.

Some people like to think of the number line as generated by segments, like toothpicks—each segment one unit tall and standing one unit apart.

1 2 3 4 5 6 7

| | | | | | | ...

The segments all fall to the right, just fitting together.

1 2 3 4 5 6 7 ...

Children like to skip-count and to play games. Counting by 2s and by 5s are examples of skip counting. One game is to imagine a rabbit that hops two units at a time. If he starts at 7, where does he end if he hops 4 jumps forward? 3 jumps back? and so on.

For these games, a row of stepping stones may replace the number line. These emphasize that we are using the sequence and not the segment attitude toward numbers.

. . . ② ③ ④ ⑤ ⑥ ⑦ . . .

5.8 COUNTING AND SCALE READING Counting is used in several ways. "Count to twenty by ones" means to repeat the number names in natural order, beginning with "one" and ending with "twenty." There is no set with twenty members involved. This is an exercise in number names and order. The exercise is varied by changing the starting number, stopping number, interval of counting, and direction of counting. "Count backward by threes from 34 to 7." Such exercises are good. They can be illustrated nicely on the number line.

"Count the people in the room" may also mean "Find the number of people by any means." The method above is a fine way. We mention some others. Skip counting by 2s, 3s, 4s, or 5s is a very good technique if the set is arranged in such a pattern, or in counting a large set rapidly, such as people walking down a corridor. Scale counting is often done, by lining the members up against a scale and reading the scale at the end of the line. Coin counters work on this principle. If the objects being counted all weigh the same, they may be counted by weighing. There are 150 members in 300 pounds of two-pound sausages. Sometimes small errors creep in, but errors also arise in one-by-one counting. Further, the extra cost of a more accurate count may not be worth the extra expense from a practical viewpoint. Weighing is often a scale-reading process, either by reading a dial or by reading the position of a sliding weight on a rod.

5.9 THE NUMBER PLANE Two number lines, usually intersecting at right angles at their common zero, form axes for a coordinate system, or grid, in a plane. Every ordered pair of numbers represents a point, and every point is labeled with an ordered pair of numbers.

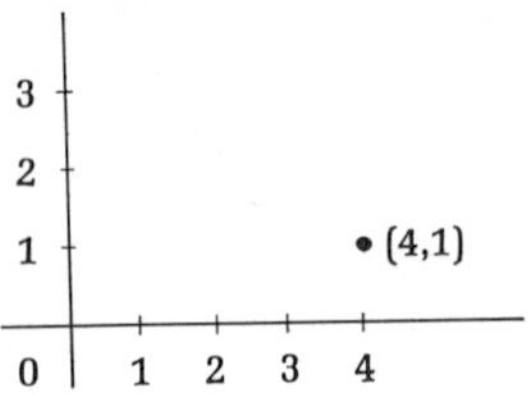

Just as the set of single numbers forms a number line, we say that the set of ordered pairs of numbers forms a number plane. The first number of the pair matches the number on the horizontal axis; the second matches the number on the vertical axis, as shown in the figure. Sometimes the letters of the alphabet are used on one scale, as in maps.

Many ideas in geometry can be shown with a number plane. Graphs are usually drawn on a number plane. When whole numbers are used, the number plane is pictured by a grid. Many useful and intriguing games such as tictactoe, Five in a Row, Go, checkers, and chess are played on a grid or its equivalent.

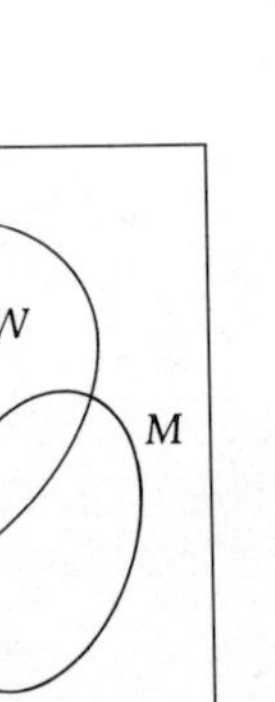

5.10 SUMMARY The counting numbers 1, 2, 3, . . . we have seen arise from three types of physical situations, which we label sets, sequences, and segments. In each case the idea of a set of numbers was probably not worth thinking about until people were able to think of several at once, say 4, 5, 6. We showed that this sequence can be continued forward indefinitely; but backward, there is a sensible beginning at one or zero. The set of numbers {1,2,3,...} is called the counting numbers, or *natural numbers.* Beginning with zero, the set {0,1,2,...} is called the whole numbers.

The number line and number plane are introduced.

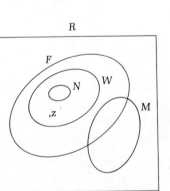

In later chapters, the idea of number is further extended to recognize negatives, rational numbers, and real numbers. In the margin is a Venn diagram relating these sets, where R is the real numbers, F the rationals, W the whole numbers, N the natural numbers, M the minus numbers, and z is zero, the only number in W not in N.

Is there an even greater superset than R? There is: the complex numbers, discussed very briefly in chapter 29.

EXERCISES

In exercises 1–6 you are asked about some uses of numbers in your life. According to your recollection, without long searching—

1. What is the largest number to which you have counted 1, 2, 3, . . . ? When and why did you do this?

2. What is the largest number you have used in a measurement? What was the measurement?

3. What is the largest count or measurement you remember seeing?

4. Describe an occasion when you used the number zero (other than as a digit in a larger number). Could you have done what you wanted to do without using zero?

5. Suppose a law were passed tomorrow making all numbers greater than 1000 illegal. Describe a situation where you would be inconvenienced. How would you meet the difficulty?

6. What number, other than routine numbers such as your telephone number, do you remember best? How was it used?

7. Suppose you had made a list of all the four-digit numbers that could be made with 3, 5, 7, 8, using each digit only once. Which number is next larger than 5738? Next smaller?

8. Give illustrations showing 2 from the set, segment, and sequence attitudes.

9. Give illustrations (if possible) showing zero from the set, segment, and sequence attitudes.

Exercises 10–21 show games on the number line, a great game board.

10. We imagine a menagerie of animals whose home is the counting numbers 2, . . . , 99 on the number line. The rabbit jumps 3 spaces with each hop, the frog 4 spaces, the mouse 2, and the kangaroo 5. For example, if all animals are on 5 and make one jump forward, the rabbit lands on 8, the frog on 9, the mouse on 7, and the kangaroo on 10. Give two other places the frog will land.

11. The kangaroo is on 88. Will he ever get to 2? To 99? What is the smallest number he can reach? The largest?

12. The rabbit is on 29. He makes 4 hops forward and 7 backward. Where does he end?

13. The mouse starts on 10, the rabbit on 20, the frog on 30, and the kangaroo on 40. Show that the mouse can go everywhere the frog goes.

14. Starting as in exercise 13, what is the smallest number on which both the mouse and rabbit can land?

15. Starting as in exercise 13, what is the largest number on which both the frog and the kangaroo can land?

16. Starting as in exercise 13, is there a number they can all jump to? If so, give the smallest such number.

17. Starting as in exercise 13, is there a number none of them can reach? If so, find the largest.

18. Suppose the mouse, rabbit, frog, and kangaroo start on 11, 22, 34, and 47, respectively. Show that the mouse and the frog can never be on the same number.

19. Starting as in exercise 18, find the smallest number that will not be reached by any of them.

20. If Hurry Harry starts at x, he moves to $3x$. For example, starting at 16, he goes to 48. Starting at 2, where does he end after 3 steps?

21. Make up and solve a question about jumps on number lines.

22. Draw a picture of a number plane with x and y both between 0 and 5 inclusive.

(*a*) Draw a small circle about the point (2, 3) where $x = 2$ and $y = 3$.
(*b*) Draw crosses on the points where $x = y$.
(*c*) Draw a small triangles about the points for which y is 2 more than x.

23. When the band is lined up on the edge of the football field, the players stand 3 feet apart. The band extends for 40 yards. How many people are in the band?

24. If Slideback Sam starts at point x on the number line, he moves to point $2x - 8$. Draw a number line from 1 to 20. Draw arrows showing how Sam moves starting at 5, 6, 10, and 12.

25. Do you think there is any point where Sam just stands still? If so, what point?

26. Estimate how much a lima bean weighs. Suppose you are right. How can you use scale counting to get about 7000 lima beans to put in a jar for a "guess the number of beans" contest?

Answers to selected questions: **7.** 5783, 5387; **11.** no, no, 3, 98; **15.** 90; **16.** yes, 50; **18.** mouse and frog on odd and even numbers respectively; **20.** 54; **23.** 41; **25.** yes, 8.

6

Numeration or Naming Numbers

6.1 INTRODUCTION A bright young student looked up at the astronomy professor and said, "Sir, I know all about how you figure what the stars are made of, and how far away they are, but I could never understand how you discovered their names."

Even more astonishing was devising names for the numbers. This was one of the truly great acts of human genius. Consider the problem of inventing a large number of different names, so that (1) any number that came up would have its own name; (2) the names would be easy to learn and everyone would remember them; (3) the naming system would show not only the cardinal (how many) properties of numbers but also the order properties.

We don't apply these criteria to naming people. Many different people have the same name. But imagine the confusion if different numbers had the same name.[1] The value of an orderly naming system can be seen if you imagine using a disorderly system. Suppose you write down two hundred words—any words—as fast as you can, and let them be the names of the numbers 1 to 200. Try counting, measuring, adding, multiplying with these names.

Historically, many peoples were unable to develop a numeral system that met these three criteria. They had only a few number names, unorderly and hard to remember.

We first consider numeral names, such as 347, then literal names, such as three hundred forty-seven. The literal name is a reading of the numeral name. The real thinking of the numeral system goes into making up the numeral name.

There are many numeral systems. Some go back to the dawn of history; others have been invented in the last few years for special

1. In the Blackwood convention in contract bridge one says "Five clubs," meaning no aces or four aces. Here is a modern example of two numbers, 0 and 4, with the same name.

reasons, such as usefulness with electronic computers. Every name is spelled with digits, just as words are spelled with letters. The number 359 is spelled with the three digits 3, 5, and 9. The test of whether or not we really know the number being named is whether we can describe it in some other way. One important way is to produce a set with that number of members. Another way is to give the name of the number in another numeral system.

The three most important principles in many numeral systems are the additive, place-value, and base principles. The additive principle says that each digit stands for a number, and that the whole numeral stands for the sum of the numbers represented by the digits. The place-value principle says that the number represented by a digit depends on the position of the digit in the numeral—at the right, at the left, or in the middle. The base principle tells how we assign value to place.

We will discuss Roman numerals (without the subtractive convention) as an example of an additive, no-place-value system, and then Hindu-Arabic numerals (our common system) as one using both systems. We will also examine using the dish abacus to illustrate this concept. The polynomial form of a base numeral will be given, along with methods of translating from one numeral system to another. We close with an account of the literal English names of numbers and a numeral system annoyingly different from Hindu-Arabic.

6.2 ADDITIVE NUMERAL SYSTEMS—ROMAN NUMERALS Roman numerals (without the subtractive convention) are a delightful example of an additive numeral system. A symbol stands for the same value wherever it appears in the numeral—at the left, in the middle, or at the right. The value of a numeral with several digits, such as XVII, is the sum of the values of the individual digits. XVII equals X + V + I + I = 10 + 5 + 1 + 1 = 17.

It is customary to write the digits in a Roman numeral with the value

decreasing from left to right. For example, we write XVII, not VXII or IXIV.

We use Roman numerals as a non-place-value, additive numeral system. Later we discuss how to perform various operations with Roman numerals as a way of understanding the operations better. It is not our purpose to promote great skill with Roman numerals, but rather, by considering them, to get a greater insight into numerals in general and their influence on the operations of arithmetic.

The Roman numeral digits with their values are I(1), V(5), X(10), L(50), C(100), D(500), and M(1000). A bar above the digit multiplies its value by 1000. Hence we have $\overline{\text{V}}$(5,000), $\overline{\text{X}}$(10,000), $\overline{\text{L}}$(50,000), $\overline{\text{C}}$(100,000), $\overline{\text{D}}$(500,000), and $\overline{\text{M}}$(1,000,000).

Clearly, the same numbers can often be written in several ways; 30 = VVVVVV = XXVV = XXX. It is considered good form to replace VV with X, XXXXX by L, and so on.

Numbers can be easily translated into Roman numerals. Consider 768 = 7 hundreds, 6 tens, 8 ones. In Roman we write CCCCCCC XXXXXX IIIIIIII. Repeating the symbols gives us the number of hundreds, tens, and units needed. We write 7 C's, 6 X's and 8 I's. We would now convert to fewer digits using CCCCC = D, XXXXX = L, IIIII = V to get DCCLXVIII.

You may recall a subtractive convention as "smaller digits to the left of larger digits are subtracted," so that "IV = 4" and "IX = 9." The convention appears to have been used only in certain special cases and becomes very awkward if widely used. (What does IVX mean?) It is difficult to describe accurately how the convention was defined and used. Since this convention does not advance our purposes, we do not use it, even though you will see IV and IX on clocks and monuments.

Roman numerals are closely related to sets. If we ask how many children there are in a playground, we may make a set of strokes, one for each child, | | | | | | | | | | | | | | | | |. This set of strokes is not only a carrying set, to carry the idea of "how many," but also the Roman numeral for the number. In an intermediate step in a calculation we may have some digits not in the standard descending order, with the value of the numeral still the sum of the values of the digits. We may reduce the number of digits by substituting other digits for subsets, obtaining VVVII. They can be condensed further to XVII.

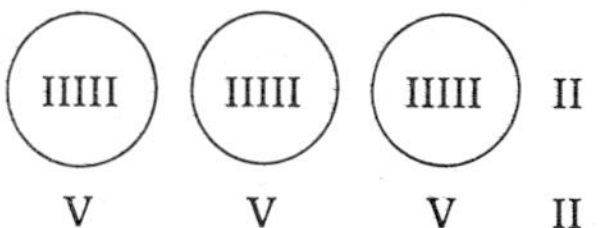

Hence, we see that Roman numerals are excellent for going from sets to number ideas.

Some students may wish to investigate ancient Egyptian numerals. These formed an additive system with no subtractive practices. Sumerian, Mayan, Chinese, modern Arabic, and various African numerals are also interesting.

6.3 THE DISH ABACUS "How many bolts do you want?" asked the storekeeper. "That many," said the buyer, and he put a pile of nuts in a dish. Was this an adequate answer? It was, because the storekeeper could put out one bolt for each nut. He did not need a name for the number of bolts.

Let us consider our storekeeper a bit more. (This will be an introduction to bases and place-value numerals, but that won't come out until later.) He has four dishes in front of him, with all the nuts in the dish at the right.

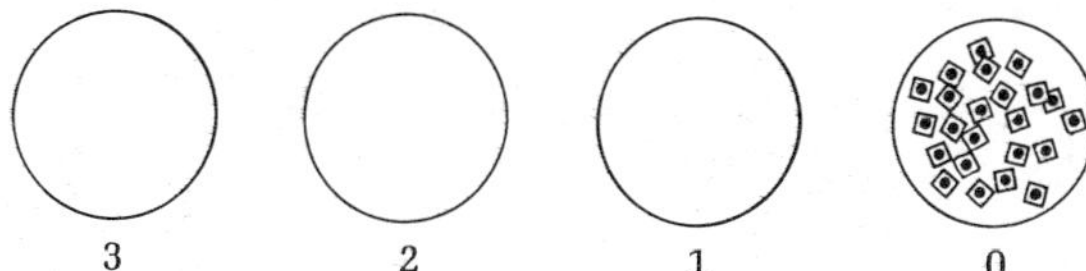

He takes the nuts from the dish 3 at a time. Each time he takes 3 nuts from dish 0, he puts one nut in the dish on its left, dish 1. Suppose after doing this four times he has 4 nuts in dish 1, and 11 nuts in dish 0, as shown:

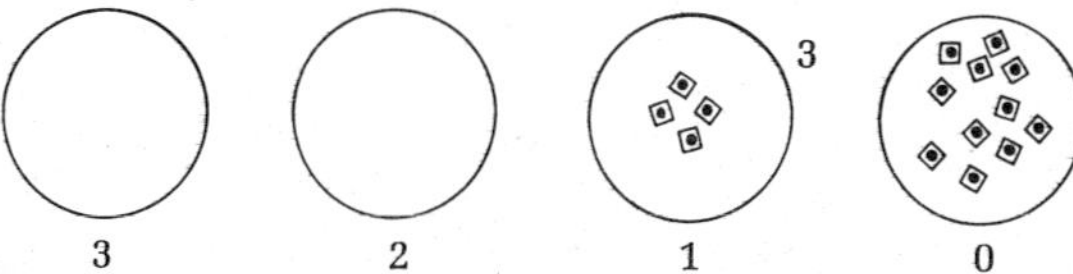

The small 3 is placed between dishes 0 and 1 to recall that for every 3 nuts removed from dish 0, 1 nut is placed in dish 1.

Could the storekeeper, starting from this $(0, 0, 4, 11)_3$ position, determine the needed supply of bolts? Can you describe how he would do it, without naming the number? The subscript 3 on the right gives the number of nuts in any dish equal to one nut in the dish on its left.

The notation $(0, 0, 4, 11)_3$ gives the numbers of nuts in the four dishes. We assume this information for readers of the book. The storekeeper and buyer, we assume, have not made this observation.

The storekeeper continues by removing 3 nuts from dish 0 and putting one nut in dish 1, and so on. Can you determine what the last position will be when no more threes can be taken from dish 0?

The storekeeper now continues by taking 3 nuts from dish 1 and putting one nut in dish 2. He does this twice, and finds the dishes are as shown:

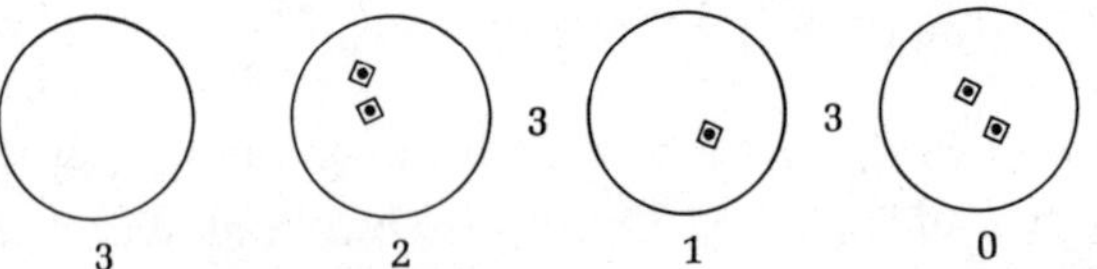

The notation for this situation is $(0, 2, 1, 2)_3$.

Can you show how the storekeeper can get the correct supply of bolts from the notation $(0, 2, 1, 2)_3$? Exploring the notation further, we see that $(0, 0, 1, 0)_3$ and $(0, 0, 0, 3)_3$ represent the same number, since 3 in dish 0 are removed and one put in dish 1. Further, $(0, 1, 0, 0)_3 = (0, 0, 3, 0)_3 = (0, 0, 0, 9)_3$ and $(1, 0, 0, 0)_3 = (0, 3, 0, 0)_3 = (0, 1, 6, 0)_3 = (0, 0, 9, 0)_3 = (0, 0, 0, 27)_3$. These equalities can be checked by using dishes and beans, or stones, or chips, or nuts. We see that some equivalent representations for the storekeeper are

$$
\begin{aligned}
(0, 2, 1, 2)_3 &= (0, 2, 0, 5)_3 = (0, 1, 1, 11)_3 = (0, 0, 4, 11)_3 \\
&= (0, 0, 1, 20)_3 = (0, 0, 0, 23)_3
\end{aligned}
$$

By counting, we see that these representations use 5, 7, 13, 15, 21, and 23 nuts, respectively. You can find other representations as well. You might like to convince yourself that every representation uses an odd number of nuts, that 5 is the smallest number and 23 the largest. Only one representation (0, 2, 1, 2) uses 5 and only one (0, 0, 0, 23) uses 23. There are several that use 15 nuts.

There is a certain elegance about "smallest," "most," and "only one." Hence, we call (0, 2, 1, 2) and (0, 0, 0, 23) standard forms. We note that in the fewest nuts form every dish has fewer than 3 nuts, and in the most nuts form, only dish 0 has any nuts in it.

Suppose now the storekeeper used a ratio of 2 rather than 3 in his dishes. Two in one dish equals one in the dish on the left. That is, if he started with

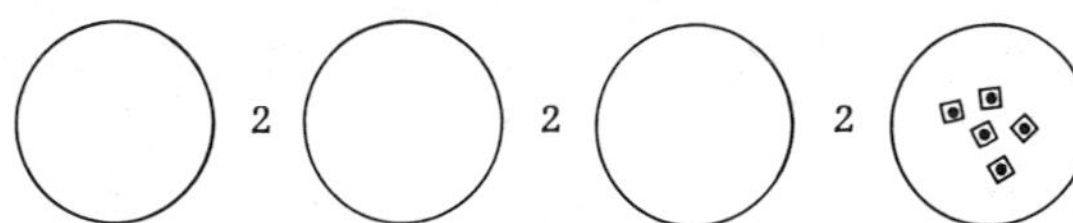

he would get

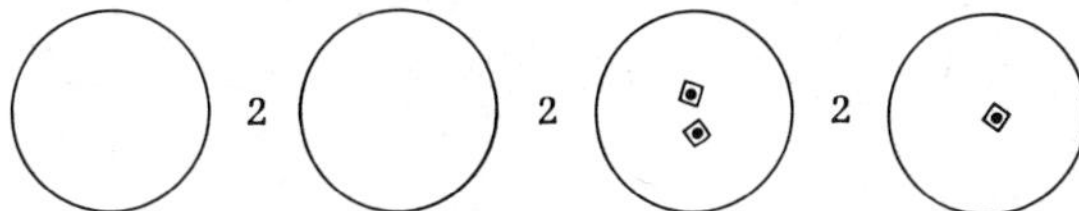

and

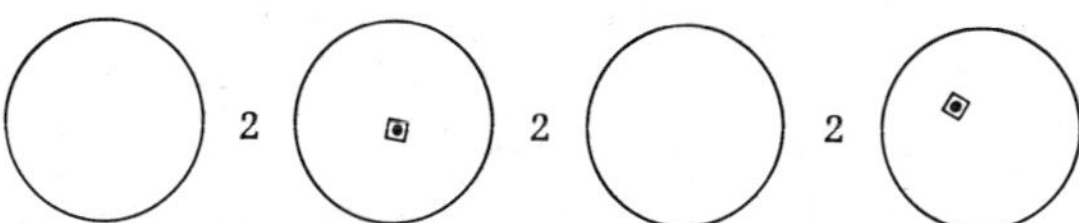

These dishes are represented by

$$(0, 0, 0, 5)_2 = (0, 0, 2, 1)_2 = (0, 1, 0, 1)_2$$

If he started with the original number of nuts, he might have

$$\begin{aligned}(0, 0, 0, 23)_2 &= (0, 0, 1, 21)_2 = (0, 0, 3, 17)_2 = (0, 0, 9, 5)_2 \\ &= (0, 0, 11, 1)_2 = (0, 2, 7, 1)_2 = (0, 5, 1, 1)_2 = (2, 1, 1, 1)_2\end{aligned}$$

We need another dish or two:

$$(0, 0, 2, 1, 1, 1)_2 = (0, 1, 0, 1, 1, 1)_2$$

Using a two-ratio, we see that

$$\begin{aligned}
(0, 0, 0, 0, 1, 0)_2 &= (0, 0, 0, 0, 0, 2)_2 \\
(0, 0, 0, 1, 0, 0)_2 &= (0, 0, 0, 0, 2, 0)_2 = (0, 0, 0, 0, 0, 4)_2 \\
(0, 0, 1, 0, 0, 0)_2 &= (0, 0, 0, 2, 0, 0)_2 = (0, 0, 0, 0, 4, 0)_2 \\
&= (0, 0, 0, 0, 0, 8)_2 \\
(1, 0, 0, 0, 0, 0)_2 &= (0, 0, 4, 0, 0, 0)_2 = (0, 0, 0, 0, 16, 0)_2 \\
&= (0, 0, 0, 0, 0, 32)_2
\end{aligned}$$

It is easy to convince yourself that

$$\begin{aligned}
(0, 1, 0, 1, 1, 1)_2 &= (0, 1, 0, 0, 0, 0) + (0, 0, 0, 1, 0, 0) + (0, 0, 0, 0, 1, 0) \\
&\quad + (0, 0, 0, 0, 1) \\
&= (0, 0, 0, 0, 0, 16) + (0, 0, 0, 0, 0, 4) + (0, 0, 0, 0, 0, 2) \\
&\quad + (0, 0, 0, 0, 0, 1) \\
&= (0, 0, 0, 0, 0, 23)
\end{aligned}$$

It is customary to say that $(0, 2, 1, 2)_3$ and $(0, 1, 0, 1, 1, 1)_2$ are numerals, or names, of this number in base three and base two, respectively. Frequently, the parentheses, commas, and zeros on the left are omitted and we write 212_{three} and 10111_{two}. The base values, three and two, are given their literal rather than their numeral names to avoid confusion. The base is the value ratio between dishes. One bean in any dish equals the base number of beans in the dish to the right.

Suppose the storekeeper used a base of ten rather than two or three. It is easy to see that

( ) ten ( ) ten (23) = ( ) ten (1) ten (13) = ( ) ten (2) ten (3)

or $(0, 0, 23)_{\text{ten}} = (0, 1, 13)_{\text{ten}} = (0, 2, 3)_{\text{ten}}$.

With a ten ratio we also have

$(0, 1, 0)_{ten} = (0, 0, 10)_{ten}$
$(1, 0, 0)_{ten} = (0, 10, 0)_{ten} = (0, 5, 50)_{ten} = (0, 0, 100)_{ten}$
$(1, 0, 0, 0)_{ten} = (0, 10, 0, 0)_{ten} = (0, 0, 5, 950)_{ten} = (0, 0, 0, 1000)_{ten}$

These facts show what is meant when we say that one in dish 1 stands for 10, one in dish 2 stands for 100, and one in dish 3 stands for 1000. Hence, we call the dishes the tens, hundreds, and thousands dishes respectively. Dish 0 is called the units dish.

Similarly, in 23 we say we have 2 tens and 3 units, because we have 2 nuts in the tens dish and 3 nuts in the units dish.

Similarly, a larger number of nuts, say 463, would give us dish representations of

$$(0, 0, 0, 463)_{ten} = (0, 0, 46, 3)_{ten} = (0, 4, 6, 3)_{ten}$$

So we say we have "4 hundreds," "6 tens," and "3 units."

These ideas appear in other base numerals as well: $212_{three} = (0, 2, 1, 2)_{three}$ represents 2 nines, 1 three, and 2 units. In general, the numeral $MARY_{three}$ has Y units, R threes, A nines, and M twenty-sevens. If this is the standard form, then each of the digits, M, A, R, Y is less than 3—that is, each digit is one of {0,1,2}.

This last example shows a difference between literal and numeral naming. $MARY_{three}$ is the name of a number. Each digit—M, A, R, Y—means something by itself. When MARY is the name of a girl, the letters by themselves have no meaning.

A set of dishes used in this way is called a dish abacus. It compares with the oriental abacus, which has counters on wires or rods. A dish and a rod are similar in that their single counters each have a value equal to that of a base number of counters in the dish or rod to the right. The dish abacus is more flexible, because any number of counters can be put in any dish. The dish abacus is also cheaper and more readily available, since one can be made in a moment from objects close at hand. If extensive calculations are to be made, an oriental abacus is faster. Both can be used to explain numerals and number operations. The dish abacus is preferable here, because there is no built-in base. Any base can be used. Since in this book we are interested in numbers and how they act rather than in developing speed on the abacus, we emphasize the dish abacus.

6.4 HINDU-ARABIC NUMERALS Number names like those developed above with the dish abacus are part of the Hindu-Arabic numeral system. The digits commonly used are 1, 2, 3, 4, 5, 6, 7, 8, 9, 0 (*t* and *e* are added for ten and eleven when single-digit names are desired for these numbers in base-twelve numerals).

The standard name of a number in a given base is spelled by the digits representing the number of counters in the corresponding dishes of the dish abacus having that base, or ratio of values. The common numerals use base ten. Most people believe that this is related to the ten fingers.

The Hindu-Arabic system is both an additive and a place-value system. It is additive because each digit represents a value, and the number named is the sum of the values of all the digits. In 3092, for example, the 3 stands for 3 thousands (as can be seen if we use a base-ten abacus), the 0 for no hundreds, the 9 for 9 tens, and the 2 for 2 units. The numeral stands for $3000 + 0 + 90 + 2$. It is a place-value system, because the value of the digit depends on the place it occupies in the numeral. Thus $3092 \neq$ (is not equal to) 3902; the 9 represents different values—90 and 900—in the two numbers.

When a number is recorded from the dish abacus, one needs to show the dish and the number of counters in the dish. The number of counters is shown by a single digit, such as 3, 9, or 2. The dish is shown by the place of the digit in the numeral. It could also be done by naming the dish (9 in dish 1, etc.), but our practice is more convenient. Zero is recorded for an empty dish. We say it is the "number of counters in an empty dish," just as we say it is the number of members in the empty set.

The dish abacus can have many dishes. There is no confusion if we do not record empty dishes on the left. We would have no need to record any empty dish if we identified each dish directly. But by using place value to identify the dish, we need to record all empty dishes not on the left.

Note that in representing 4023 in Roman numerals, MMMMXXIII, there is no recording of C's, or hundreds.

6.5 NUMBER ORDER—LINEAR, CIRCULAR A line continues indefinitely in both directions, without end, and the two paths never meet. A circle continues indefinitely in both directions, without end, but the two paths meet. A set of objects that are appropriately lined up on a line are said to be in linear order. Those appropriately lined up on a circle are said to be in circular order.

The counting numbers can be lined up on a line. They show linear order. The days of the week show circular order. For example, Tuesday comes both before and after Friday. Do the holidays show circular or linear order? Which comes first, Christmas or New Year's?

Suppose some eager person were to print a large number of numbers in increasing order. You look at the units digits. What kind of order do they show? Here is a number with units digit 4. Is it before or after a number with units digit 7?

From these considerations we see that the digits with the same place value form a circular order as the numbers increase, even though the numbers are in linear order.

The distance traveled in an automobile is recorded by the odometer. It is usually mounted with the speedometer, which tells how fast the car is going. The digits are printed on wheels that turn, showing the circular order of the digits.

Many calculating machines work similarly. The digits are on wheels geared so that the wheel on the left moves 1/10 of a revolution while the wheel next to it on the right concludes one revolution.

Given the names of two numbers, say 3475 and 3745, how do we determine which is larger? We note that we do *not* use the basic process defining larger. The basic process is one of the following: (a) Form two sets with 3475 and 3745 members. Match the sets and observe which set has some members unmatched when the other members are all matched. (b) Form two segments 3475 and 3745 units long. Put them side by side with the same endpoint and see which segment sticks out. (c) Start counting, 1, 2, 3, . . . and see which number, 3475 or 3745, you come to last.

What we do is (1) order the digits 5, 3, 0, 9, 4, 1, 2, 8, 6, 7 by one of these methods, obtaining 0, 1, 2, . . . , 9, then (2) count the number of digits in each numeral. The numeral with the larger number of digits is the larger. If they have the same number of digits, we go to step 3. (3) We look at the leftmost digit. The number for which this digit is larger is the larger number, regardless of the other digits. (4) If the leftmost digits are equal, we look at the second digit from the left, and so on.

Such a method of determining which of two numbers is larger is called lexicographic. *Lexicographic* is a word referring to dictionaries. To determine which of two words comes later in a dictionary, one looks first to see which pages they are on. If they are on the same page, one looks at their position on the page. If they are on different pages, one doesn't care about the position on the page.

For our two numbers 3475 and 3745, we see both numerals have 4 digits. We cannot tell which is larger by this criterion. The left-hand digit is 3 for both numerals; this tells us nothing. The next digits are 4 for one numeral, 7 for the other. Since any numeral 37xx is larger than any numeral 34xx, we say that 3745 is larger than 3475. We don't care about the 5s.

6.6 POLYNOMIAL NAMES In many cases we wish a number name closely related to the standard name, such as 342, but not involving place value. We need this when we wish to use the digits singly without losing their proper value. The Roman numeral name CCCXXXXII does this nicely, since it has no place value, but it looks too different from the Hindu-Arabic numeral name: we have lost the digits 3, 4, and 2.

We introduce the polynomial name $342 = 3t^2 + 4t + 2$ (t stands for ten). We recall the exponential notation. $t^3 = t \times t \times t$. $t^5 = t \times t \times t \times t \times t$. $t^1 = t$, $t^0 = 1$. $3t^2$ means $3 \times t \times t$. And so on.

We can define the power t^n as 1 multiplied by n factors t. n is called the exponent of the power. t is called the base. This definition of power, exponent, and base holds for all whole numbers.

An expression like $3t^2 + 4t + 2$ is called a polynomial in t. Polynomial names for numbers in base ten are of the form $at^n + bt^{n-1} + \ldots + kt + r$. The numbers $a, b, \ldots, k, r$, called coefficients, are all digits from the set $\{0,1,2,\ldots,9\}$.

In algebra the idea of polynomial is extended so that the coefficients can be any numbers and some terms can be subtracted rather than added. We do not make that extension here.

Let us consider a few examples of polynomial names:

$$53 = 5t + 3$$

$$6075 = 6t^3 + 0t^2 + 7t + 5 = 6t^3 + 7t + 5$$

$$4020 = 4t^3 + 2t = 4t^3 + 0t^2 + 2t + 0$$

$$03456 = 0t^4 + 3t^3 + 4t^2 + 5t + 6 = 3t^3 + 4t^2 + 5t + 6$$

We note that 0s, which are so important in 4020, are not important in the polynomial name $4t^3 + 2t$.

Using $t^1 = t$ and $t^0 = 1$, we can have a t with an exponent in each term in a polynomial name if we wish: $53 = 5t^1 + 3t^0$, $6075 = 6t^3 + 7t^1 + 5t^0$.

The polynomial name is called *expanded notation* by many authors.

Polynomial names are available for numbers in other bases as well. For example:

$$212_{\text{three}} = 2b^2 + b + 2, \text{ where } b \text{ stands for three}$$
$$3201_{\text{four}} = 3f^3 + 2f^2 + 1, \text{ where } f \text{ stands for four}$$
$$= 3f^3 + 2f^2 + 1f^0.$$

The sharp reader will note that the exponent of the base is the name given to the corresponding dish of the dish abacus. No *invisible* assumptions such as place value, base, or addition are needed in the polynomial name.

6.7 TRANSLATING Frequently, people wish to translate a number name from one system to another. For example, what is the base-ten name for MDCCXVI? What is the base-four name for 365_{ten}? And so on. We will consider a number of special cases that will give several methods. We will assume we are working with additive systems.

If the values of the digits in the old system are easily known in the new, write these numbers and add:

$$\text{MDCCXVI} = 1000 + 500 + 100 + 100 + 10 + 5 + 1 = 1716$$

Similarly,

$$768 = \text{CCCCCCCXXXXXXIIIIIIII}$$
$$= \text{DCCLXVIII}$$

where certain simplifications are made after the direct conversion from Hindu-Arabic base ten to Roman.

In working with numerals to bases other than ten, it may be convenient to write the number in expanded notation, then work it out. For example, translate 3201_{four} to a base-ten numeral:

$$\begin{aligned} 3201_{\text{four}} &= 3 \cdot f^3 + 2 \cdot f^2 + 1 \\ &= 3 \cdot 64_{\text{ten}} + 2 \cdot 16_{\text{ten}} + 1 \\ &= 192_{\text{ten}} + 32_{\text{ten}} + 1 \\ &= 225_{\text{ten}} \end{aligned}$$

NOTE: $f^2 = 16_{\text{ten}}$
$f^3 = 64_{\text{ten}}$

In writing numerals to different bases, it is customary to show the base by a *word* subscript. We note that "10" is not a clear notation, since 10 = base in every system. That is, 10_{ten} = ten, 10_{five} = five, 10_{three} = three. Hence 225_{10} is ambiguous, while 225_{ten} is clear.

For those brought up with base-ten numerals, this way of converting to base-ten numerals is quite satisfactory. Converting from base ten to another base is not so easy. For example, convert 36_{ten} to a base-three numeral. Now, $36_{\text{ten}} = 3 \cdot t + 6$, but we aren't quite clear about these values in base-three numerals. It is best to go back to the dish abacus in our thinking. Base three means that ratio is three. We start with

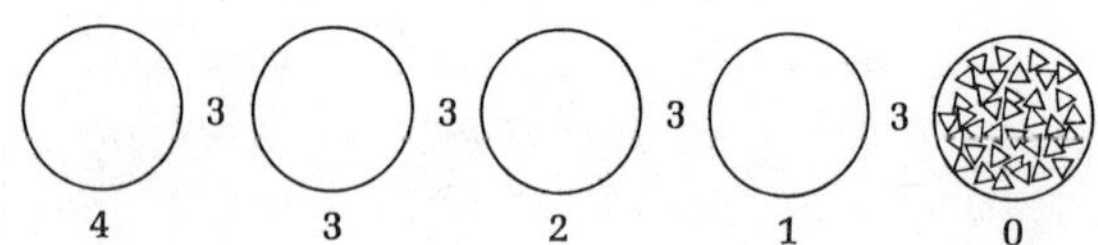

$$(0, 0, 0, 0, 36)_{\text{three}}$$

The numerals within the parenthesis are base-ten numerals with which we are familiar. This notation indicates thirty-six counters in the units dish. We remove 3 from the units dish and put 1 in the threes dish (dish 1). We get

$$(0, 0, 0, 0, 36)_{\text{three}} = (0, 0, 0, 1, 33)_{\text{three}}$$

As we continue moving counters from dish to dish, we get

$$(0, 0, 0, 12, 0)_{\text{three}} = (0, 0, 4, 0, 0,)_{\text{three}} = (0, 1, 1, 0, 0)_{\text{three}}$$

Of course, along the way we may have had situations like $(0, 0, 0, 4, 24)_{\text{three}}$ and $(0, 0, 3, 3, 0)_{\text{three}}$. We could also have had $(0, 1, 0, 1, 6)_{\text{three}}$.

A look at this process shows it to be equivalent to the following: We put the number in the units dish (dish 0). We divide by the base and put the quotient in dish 1, the remainder in dish 0. We then divide the number in dish 1 by the base, putting the quotient in dish 2, and the remainder in dish 1; and so on. We illustrate:

$$\begin{aligned} 36/3 &= 12 \quad 0 \text{ remainder} \\ 12/3 &= 4 \quad 0 \text{ remainder} \\ 4/3 &= 1 \quad 1 \text{ remainder} \\ 1/3 &= 0 \quad 1 \text{ remainder} \end{aligned}$$

Hence $36_{\text{ten}} = 1100_{\text{three}}$.

EXAMPLE Convert 228_{ten} to base four.

$$\begin{aligned} 228/4 &= 57 \quad \text{with } 0 \text{ remainder} \\ 57/4 &= 14 \quad 1 \\ 14/4 &= 3 \quad 2 \\ 3/4 &= 0 \quad 3 \end{aligned}$$

Hence $228_{\text{ten}} = 3210_{\text{four}}$.

You may wish to check this by converting 3210_{four} to a base-ten numeral.

It is suggested that the reader work out the details on paper.

To convert 2120_{three} to a base-two numeral, we would find either of these methods unusual. If $b = 3$, $2120_{\text{three}} = 2b^3 + b^2 + 2b$. If we know b^3, $2b^3$, b^2, $2b$ in base two, we substitute and add.

Or if we know how to divide base-three numerals by two, we can compute $2120_{\text{three}}/2 =$ ______ remainder ______.

If we can't do either of these easily, it is probably best to convert to a base-ten numeral, then to base two.

$$\begin{aligned} 2120_{\text{three}} &= (2 \times 27) + 9 + (2 \times 3) \\ &= 54 + 9 + 6 = 69_{\text{ten}} \end{aligned}$$

$$\begin{aligned} 69_{\text{ten}}/2 &= 34 + 1 \text{ remainder} \\ 34/2 &= 17 + 0 \\ 17/2 &= 8 + 1 \\ 8/2 &= 4 + 0 \\ 4/2 &= 2 + 0 \\ 2/2 &= 1 + 0 \\ 1/2 &= 0 + 1 \end{aligned}$$

Hence, $2120_{three} = 69_{ten} = 1000101_{two}$. It helps to do this with a dish abacus in front of you.

6.8 NUMBER NAMES IN ENGLISH The English, or verbal, names for numbers, just as the French and German names, developed independently from the digital names. The fact that different countries use the same digital names but different verbal names shows this. We make a partial list of the names.

1 one	**13** thirteen	**30** thirty	**60** sixty
2 two	**14** fourteen	**31** thirty-one	**70** seventy
3 three	**15** fifteen	**39** thirty-nine	**80** eighty
4 four	**16** sixteen	**40** forty	**90** ninety
5 five	**17** seventeen	**41** forty-one	**100** one hundred
6 six	**18** eighteen	**49** forty-nine	**143** one hundred forty-three
7 seven	**19** nineteen	**50** fifty	**315** three hundred fifteen
8 eight	**20** twenty	**51** fifty-one	**5004** five thousand four
9 nine	**21** twenty-one	**59** fifty-nine	
10 ten	**22** twenty-two		
11 eleven	**23** twenty-three		
12 twelve	**24** twenty-four		
	29 twenty-nine		

As we look at the number names listed, we see certain patterns appearing.

The first few numbers, from one to twelve, have individual names.

All the names from thirteen to nineteen have *-teen* in them.

Any folklore interpreter makes guesses about how words developed. Most people guess that the name *sixteen* is a contraction of *six* and *ten*. Does this seem reasonable? How about *seventeen? eighteen? nineteen? fourteen?* We do not have *fiveteen* and *threeteen,* but if you say these words very fast they sound like *fifteen* and *thirteen.* From these name patterns we can make various conjectures—which may not be true. Our ancestors thought of the smaller numbers as being completely different, so gave them different names. They doubtless knew that $2 + 3 = 5$, but they did not use names like *tunthree* for *five.*

When our ancestors considered larger numbers like sixteen, they were not so sure they were independent numbers. They thought of sixteen as "six and ten," a sum combination of two earlier numbers. Or they may have tired of thinking up and teaching children new and different sounds. Far easier to have the names of larger numbers be combinations of the names of smaller numbers.

Our forebears placed heavy emphasis on ten. We say "six and ten" rather than "eightneight" or "doubleight," "seven and ten" rather than "fiftwelve," and so on. Each name is some number and ten, and not some number and nine or twelve. Despite this, eleven and twelve must have been very important numbers to have independent names rather than oneteen and twoteen.

Numbers smaller than 20 have one-word names, starting with the units digit. Numbers greater than 20 generally have two-word names (or longer), which begin with the largest value digit. Anthropologists and folklorists may want to guess why. Were scholars who felt easy with multiword names the only ones who used numbers larger than 20? Were scholars appreciative that the largest portion was most significant and hence read largest digits first? Did ordinary people in counting a set see at a glance that it was more or less than ten, and hence, for accuracy, give the units digit first on the numbers they used most? Conjectures such as these on attitudes in the misty past are hard to prove.

Many number names end in *-ty:* twenty, thirty, forty, fifty, sixty, seventy, eighty, ninety. We recall that sixty is the number for a set that can be grouped into six subsets of ten each. Sixty stands for six tens. It is easy to suppose that the name arose with *-ty* representing ten. We just remember that sixteen stands for six and ten, while sixty stands for six tens. It is easy to believe that the other names were formed the same way, with slight spelling changes. Twenty stands for two tens, for example.

Consider *thirty-four.* This name has two words, like *Peter Smith.* The first word stands for three tens, the second for four. When we put these words together, side by side, the name stands for their sum. On comparing with the numeral 34, we see that the number is read from left to right—a contrast with the *teen* convention.

Other numbers to one hundred are named the same way. Each word represents a number. When a name contains two words, the number is the sum of the two numbers.

It is quite clear that the verbal number names in this range are not new, independent names for each new, independent number, but are made by combining the names of numbers whose sum is the number of interest.

By this process we can make up as many different names as we need. Ten plays a peculiar role in the naming. There are many sets of numbers whose sum is 34. For example, {17,17}, {15,17,2}, {25,9}, {9,9,9,7}, {10, 10,10,4}. Only the last set is chosen for the naming. This special role of ten relates to the base ten.

Do you see a different pattern for names over one hundred? Consider the examples five hundred forty-three, two hundred thirteen, three hundred nine. The numerals are 543, 213, 309.

There is one difference. Where "five tens" is contracted to "fifty," "five hundreds" is not. Otherwise, the additive principle holds. Two names put side by side stand for the sum of the two numbers. It is not good form to read *and* in the numeral. We do not say five hundred and forty-three.

Larger numerals are punctuated with commas, separating the digits into sets of three. We write 37,483,962 rather than 37483962. We give the number a name that reflects this punctuation. Its name is thirty-seven million, four hundred eighty-three thousand, nine hundred sixty-two.

Each set of three digits is called a period. Just as the places have values, such as units, tens, and hundreds, the periods have names. These names, reading from right to left, are units, thousands, millions, billions, trillions. The need for numbers of this size has arisen only recently, so names were assigned only recently, and in this highly patterned manner. The names are compounded from Latin syllables with number meanings not used otherwise in English. *Mil* means one thousand in Latin: a million is a thousand thousands. The prefixes *bi-, tri-, quadri-, quinti-, sexti-, septi-, octi-, noni-, decti-,* are Latin forms meaning two, three, four, five, six, seven, eight, nine, ten, respectively. Followed by *-illion* these prefixes give the names of the periods. Each period is named in the hundreds. Zeros in a numeral are not named. For example, the verbal name of 863,000,309,050,003,170,028 is "eight hundred sixty-three quintillion, three hundred nine trillion, fifty billion, three million, one hundred seventy thousand, twenty-eight."

The point to this discussion is to show the pattern by which more and more names are generated for numbers, not to learn the common names used. Actually numbers this size are so rare, and so few people are familiar with the names, that it is most common to spell the numeral. Many would read this number "eight six three comma zero zero zero comma three zero nine comma zero five zero comma zero zero three comma one seven zero comma zero two eight." Some people read "oh" rather than "zero."

6.9 ORDINAL NUMERALS Words such as *fifth* and *seventh* are called ordinal numerals. They are adjectives that locate the position in a sequence that begins with *first*. The beginning numerals are *first, second, third, fourth*. Later numerals are generally formed by joining "th" to the end of the cardinal name for the corresponding number, with some smoothings. The following list of selected names will illustrate:

fifth (not fiveth)
sixth
eighth (not eightth)
ninth (not nineth)
tenth
eleventh

twelfth (not twelveth)
thirteenth
twenty-first
twenty-second
twenty-third
forty-fourth

Sometimes we say "ordinal numbers" rather than "ordinal numerals." These words are never used in arithmetic operations, however. We do not add the fifth boy to the seventh boy and get the twelfth boy. For this reason, many people do not call them numbers.

6.10 A NO-CARRY NUMERAL SYSTEM The following numeral system has the property that no carrying is needed in addition or multiplication. It is a place-value system without a base. The system given here is fine for numbers under 210. Different numbers will have the same name if we go above this limit. A fuller discussion of this system requires a knowledge of primes and the solution of congruences, and other topics usually considered in a theory of numbers course. However, we can describe the system, how it works, and some of its properties.

Every numeral has four digits. These digits are the remainders when the number is divided by (reading from left to right) 7, 5, 3, and 2. We make a table for the names in the Hindo-Arabic system, and in this new, prime-remainder system. We note that 2, 3, 5, and 7 are prime numbers.

To see how to get these prime-remainder numerals, look at 17. 17/7 = 2 with 3 remainder. Hence, 3 is the left digit. 17/5 leaves a 2 remainder. 17/3 leaves a 2 remainder, and 17/2 leaves a 1 remainder. Hence 3221_{pr} is the numeral for 17.

H-A	P-R			
	7	5	3	2
0	0	0	0	0
1	1	1	1	1
2	2	2	2	0
3	3	3	0	1
4	4	4	1	0
5	5	0	2	1
6	6	1	0	0
7	0	2	1	1
8	1	3	2	0
9	2	4	0	1
10	3	0	1	0
11	4	1	2	1
12	5	2	0	0
13	6	3	1	1
14	0	4	2	0
15	1	0	0	1
16	2	1	1	0
17	3	2	2	1
18	4	3	0	0
19	5	4	1	1
20	6	0	2	0
21	0	1	0	1
22	1	2	1	0
23	2	3	2	1
24	3	4	0	0

Look at 22. 22/7 has a 1 remainder. 22/5 has a 2 remainder. 22/3 has a 1 remainder. 22/2 has 0 remainder. Hence the numeral for 22 is 1210_{pr}. The subscript *pr* means "prime remainder."

You may wish to extend this table. If you do so to 209 you will see that each number has a different *pr* name. With 210, however, the names start repeating: $210_{ten} = 0000_{pr}$, $211_{ten} = 1111_{pr}$, $212_{ten} = 2220_{pr}$. These are duplicates of the names for 0, 1, 2.

Let us try the addition and multiplication for small numbers.

1. $1 + 2 = 1111 + 2220$. If we just add the corresponding digits we would get 3, 3, 3, 1. These numbers are replaced by their remainders on division by the corresponding prime. The remainders on dividing are $R(3/7) = 3$, $R(3/5) = 3$, $R(3/3) = 0$, $R(1/2) = 1$, where $R(a/b)$ means the "remainder on dividing a by b". Hence, $1 + 2_{ten} = 3301_{pr}$. We check the table and see that this is 3.

2. $(3 + 4)_{ten} = 3301_{pr} + 4410_{pr} = 7, 7, 1, 1 = 0211_{pr}$, since $R(7/7) = 0$, $R(7/5) = 2$, $R(1/3) = 1$, $R(1/2) = 1$. We see that 0211_{pr} is the name of 7.

3. $(11 + 13) = 4121_{pr} + 6311_{pr} = 10, 4, 3, 2 = 3400_{pr}$, since $R(10/7) = 3$, $R(4/5) = 4$, $R(3/3) = 0$, $R(2.2) = 0$.

4. $2 \times 3 = (2220)_{pr} \times (3301)_{pr}$. If we multiply corresponding digits we get 6, 6, 0, 0, which reduces to 6100_{pr}.

We note that the order of the digits in each place is circular order. No carrying is necessary in either addition or multiplication. The rule for multiplication is to multiply the corresponding digits, and put down the remainder when divided by the number going with that column. That is, $abcd_{pr} \times wxyz_{pr} = R(aw/7)\, R(bx/5)\, R(cy/3)\, R(dz/2)_{pr}$. $11 \times 13 = (4121)(6311)$ = remainders in 24/7 3/5 2/3 1/2 = 3321_{pr}. Now, $11 \times 13 = 143$, whose remainders are $R(143/7) = 3$, $R(143/5) = 3$, $R(143/3) = 2$, $R(143/2) = 1$. Hence, $143_{ten} = 3321_{pr}$.

Here is a way of getting the base-ten numeral from the prime-remainder numeral, illustrated with 3321_{pr}. Since N leaves remainder one when divided by two, N is one of the numbers 1, 3, 5, 7, 9, ... We ask which of these leave remainder two when divided by three. The remainders on dividing by three are 1, 0, 2, 1, 0, 2, 1, ... The remainders go through cycles of three. So the candidates with remainder two will increase by

$2 \cdot 3 = 6$ and are 5, 11, 17, 23, 29, 35, 41, ... The remainders of these on dividing by 5 are 0, 1, 2, 3, 4, 0, 1, . . . , going in cycles of five. The difference between two numbers with remainders one on division by 2, two on division by 3, and three on division by 5 is $2 \cdot 3 \cdot 5 = 30$. Hence, the candidates for N are 23, 53, 83, 113, 143, 173, 203, ... The remainders on division by 7 are 2, 4, 6, 1, 3, 5, 0, ...

Hence we see that 143 has the required remainders. Further, we note the remainders on division by 7 go in cycles of 7. Hence numbers that have the required remainders will differ by $30 \cdot 7 = 210$. Other candidates for N besides 143 are then 353, 563, 773, ...

Each of these numbers could be the number 3321_{pr}. This shows that while each number has only one prime-remainder name, several numbers will have the same name. However, if we are in a restricted situation, such as knowing all numbers under discussion are less than 210, then there is only one number of interest to us: 143. Over a larger range the system can be expanded to give uniqueness by taking more primes and/or larger primes.

EXERCISES

1. Write the Roman numeral names (without using the subtractive principle) for (*a*) 3, (*b*) 9, (*c*) 30, (*d*) 305, and (*e*) 2222.

2. Write Hindu-Arabic names for (*a*) VIIII, (*b*) XXVII, (*c*) CCLXXVI, (*d*) MDCCLXXVI, and (*e*) LXXXXVIIII.

The dish abacus. We recommend that you work several problems actually manipulating beans or other counters in actual dishes or substitutes. After working several problems physically, you will see the pattern and be able to work others mentally or with pencil and paper. Many people who see patterns before moving the beans find nevertheless that working with the beans helps them see patterns they did not notice before.

3. Make a dish abacus ratio 3. List the different arrangements of beans that show the number 7.

4. Give three other arrangements of beans in a ratio 3 abacus that show the same number as $(0, 4, 0)_{\text{three}}$.

5. Repeat exercise 4 for $(0, 4, 2, 7)_{\text{three}}$, with one arrangement using the most beans and another using the fewest.

6. Give the two standard arrangements (most beans, fewest beans) for each of the following, adding any needed dishes:

(*a*)	$(0, 6, 3, 0)_5$	(*f*)	$(1, 0; 0, 0)_3$
(*b*)	$(6, 3, 0)_2$	(*g*)	$(1, 0, 1)_5$
(*c*)	$(6, 3, 0)_{\text{ten}}$	(*h*)	$(1, 0, 0, 0)_2$
(*d*)	$(6, 3, 0)_{\text{twelve}}$	(*i*)	$(0, 0, 23, 0)_{12}$
(*e*)	$(14, 7, 7)_5$	(*j*)	$(1, 1, 1)_4$

7. Find the base-five numerals for (a) 76, (b) 194, and (c) CLV.

8. Which number is the largest: 22_3, 22_4, 22_5, 22_6?

9. Compute $33_6 - 33_5$. (HINT: If needed, convert to base ten.)

10. Find base-ten numerals for (*a*) 1101_{two}, (*b*) 342_{five}, and (*c*) $\text{TE7}_{\text{twelve}}$.

11. Match the steps taken in converting 42_{five} to base ten with the steps in moving the beans to find the most beans form of $(0, 4, 2)_{\text{five}}$.

12. Match the steps in converting 213_{four} to a base-ten numeral with the steps in moving beans to find the most beans form of $(2, 1, 3)_{\text{four}}$.

13. Match the steps in converting 67_{ten} to base three with the steps in moving beans to find the fewest beans form of $(0, 0, 0, 67)_3$.

14. Match steps in converting 88_{ten} to base four with the steps in moving beans in a dish abacus.

15. With $t = 10$, give the polynomial forms of (*a*) 493 and (*b*) 6005.

16. With $d = 2$, give the base-ten form for $d^3 + d^2 + 1 =$. Also for $d^5 + d^3 + d =$.

17. Give two numerals for 46_{ten} that do not involve place value.

18. Does the right side of $4062 = 4 \times 10^3 + 6 \times 10 + 2$ involve place value? How?

19. Write the English word names for (*a*) 373, (*b*) 47,809, and (*c*) 169,208,000,200,057.

20. Write Hindu-Arabic numerals for (*a*) two hundred five thousand sixty, and (*b*) eighty-seven quadrillion one hundred twenty-three million seventy thousand sixteen.

21. We say "*J* is the tenth letter in the alphabet." Why don't we say "*U* is the twenty-oneth letter?"

22. What adjective do we use to show position for the number 2 man in line? For the number 80 man?

23. What is the prime-remainder numeral for 63?

24. What is the base-ten numeral for 2401_{pr}?

25. We know that $2401 > 2301$ in base-ten numerals. Is $2401_{pr} > 2301_{pr}$?

26. Make up a numeral system of your own with four number names. Let the names of the four things you usually see first when you wake up in the morning be the number names, with their first letters being the digits. Write the names for 1 to 10 in your system. What is your name for 24? 16? Give the name both in words and in digits.

27. Is there a number whose digit name in your system spells a familiar English word? What is it?

Answers to selected questions: **1.** (*a*) III; (*b*) VIIII; (*c*) XXX; (*d*) CCCV; (*e*) MMCCXXII; **3.** $(0, 0, 7)_{three}$ $(0, 1, 4)_{three}$ $(0, 2, 1)_{three}$; **5.** $(1, 1, 2, 7)_3$ $(1, 2, 1, 1)_3$ $(0, 0, 0, 49)_{three}$; **6.** (*b*) $(0, 0, 30)_2$ $(1, 1, 1, 1, 0)_2$; (*e*) $(0, 0, 392)_5$ $(3, 0, 3, 2)_5$; **10.** (*a*) 13; (*b*) 97; (*c*) 1, 579; **12.** match mentally $(2 \times 4 + 1) \times 4 + 3 = 39$; **14.** match mentally $88 = 22 \times 4 = [5 \times 4 + 2] \times 4 = [(1 \times 4 + 1) \times 4 + 2] \times 4$; **17.** $4t + 6$, $t =$ ten; $5n + 1$, $n = 9$; **19.** (*a*) three hundred seventy three; (*c*) one hundred sixty-nine trillion, two hundred eight billion, two hundred thousand fifty-seven; **23.** 0301; **25.** no.

7

Operations and Relations

7.1 INTRODUCTION Squaring, multiplying, finding the volume of a box, and choosing the largest are examples of operations with numbers. Concrete mixing, judging a beauty contest, getting married, and translating the Bible are operations in other sets.

Operations are also called operators, functions, mappings, and transformations. The idea is so widespread in life that different names may be used in different circumstances without people recognizing that the central ideas are the same, or at least very similar. We cheerfully expect to gain greater power by considering how operations in general act before we consider the specific operations in later chapters.

It is helpful to classify operators according to the number of inputs. We may call the operators unary, binary, ternary, . . . for 1, 2, 3, . . . inputs. We will see that unary operations may have inverses. Binary (2-input) operators may be used to construct unary operators by fixing one input. Two binary operators may be hooked up to form a ternary operator. Since the work is simpler if the binary operator is commutative and associative, we examine these properties, which are also basic for algebra.

When comparing and describing numbers, we use words such as *less than, equals, one more than,* and *has the same units digit.* These words apply to relations between numbers. Some relations (*has same units digit*) are similar to the relation *equals.* We look at this similarity. Some relations are "order" relations such as *less than.* Finally, we show how relations may be represented as subsets of Cartesian products of sets. In some ways these comparisons are like *is a brother of* or *is a mother of*, whence comes their name. Some people like to think of relations as operations having possibly more than one outcome.

7.2 UNARY OPERATIONS One number goes in. The operation is defined by a rule that tells what to do to the number going in to get the number coming out. We also show the operation by a table of ordered pairs of numbers, where the right (second) number in the pair is what comes out when the left (first) number is operated on or placed in the operation machine. We picture the machine as a funnel, with the input at the top and output at the bottom. Often in a table the column of inputs is headed "x" and outputs "y." Care is needed not to confuse x, y for one operation with x, y from another when thinking of both at once.

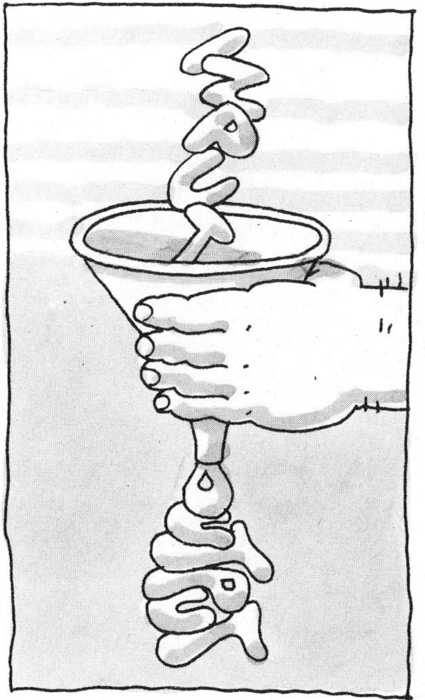

EXAMPLE 1 Squaring

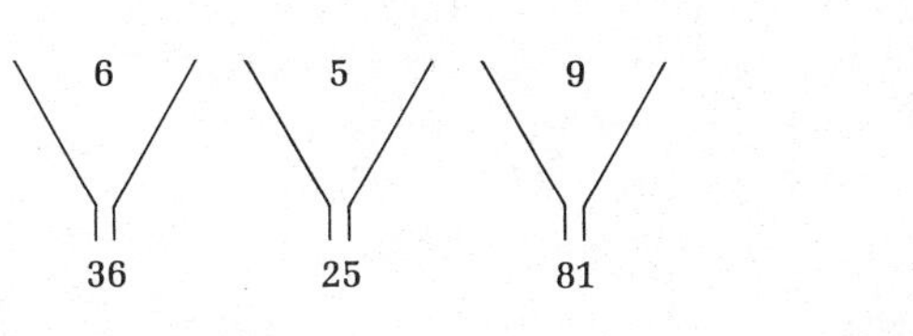

x	y
1	1
2	4
3	9
8	64
9	81

The set of pictures and the table may be made as large as needed.

EXAMPLE 2 Add 5. The number that comes out is 5 more than the number that goes in.

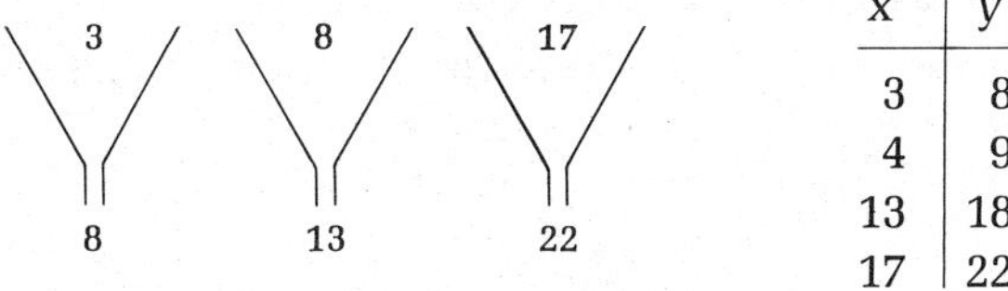

x	y
3	8
4	9
13	18
17	22

The rule is given by the formula, $y = x + 5$. We might also think of x as the opening or top of the funnel where an input number is placed, and y as the bottom of the funnel where the outcome appears.

EXAMPLE 3 Multiply by 3 and add 2.

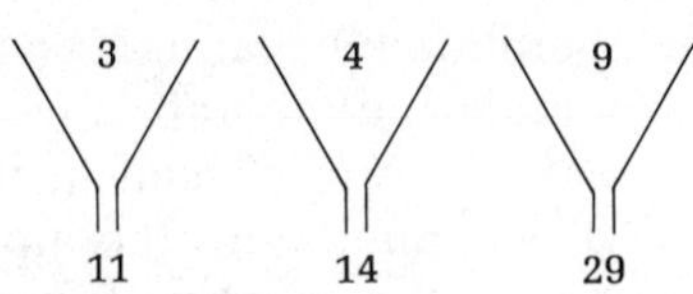

x	y
3	11
4	14
8	26
11	35

The formula is, $y = 3x + 2$.

In these illustrations we have put only counting numbers into an operation. We note that while we might put any counting number in and that every number coming out will be a counting number, not all counting numbers will come out. For instance, in example 3, 15 will never come out. There is no counting number put in for x that will produce 15.

If we actually had real machines, we would find that children would soon be hooking them up or stacking them. Suppose, for example, that the machine in example 1 were placed above the machine in example 2. If 3 is placed in the top machine, can you see that 14 comes out the bottom? We make a table to show the effect of these hooked-up machines. To simplify ideas, we make a three-column table with x heading the column of inputs of the first (or top) machine, y the output of the top machine (which is also the input of the bottom machine), and z the output of the bottom machine. The combined machine has the two-column table with x and z.

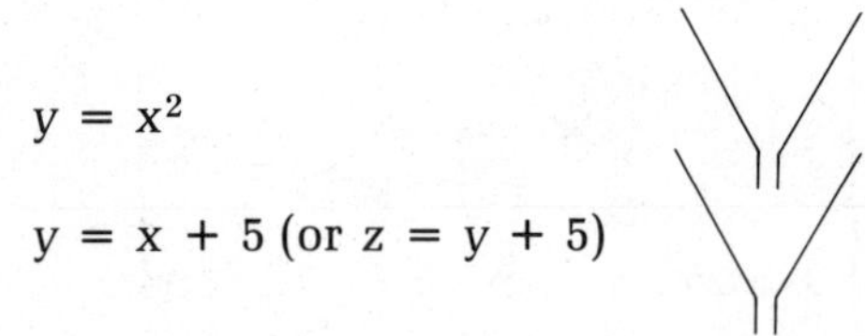

x	y	z
2	4	9
3	9	14
5	25	30
6	36	41
8	64	69

x	z
2	9
3	14
5	30
6	41
8	69

7.3 WHAT'S MY RULE? An interesting game is to look at a table and see if you can guess the rule of the operation. Can you guess more entries for these tables? Their rules and formulas?

EXAMPLE 4

x	y
2	8
3	12
5	20
6	24
8	32

EXAMPLE 5

x	y
2	1
4	2
8	4
16	8
20	10

EXAMPLE 6

x	y
4	5
5	7
8	13
10	17
11	19

EXAMPLE 7

x	y
3	97
5	95
20	80
22	78
35	65

As a class game, the leaders would reveal the table one pair at a time. The players who guess the rule first, win. It is quite possible, after seeing just a few pairs, to guess different rules that would give the pairs seen so far. Each guesser is inclined to insist that his rule is the correct one. Seeing more pairs will probably show who is right.

Did you guess $y = 4x$, $y = x \div 2$, $y = 2x - 3$, and $y = 100 - x$ as the rules for these examples? We may call them "multiply by 4," "divide by 2," "multiply by 2, then subtract 3," and "subtract from 100."

7.4 INVERSE OF UNARY OPERATORS The inverse of an operator is an operator that "undoes what the first operator did."

EXAMPLE 8 Suppose the operator of example 1 is followed by (hooked up ahead of) an operator *R*, with the resulting triple-column table:

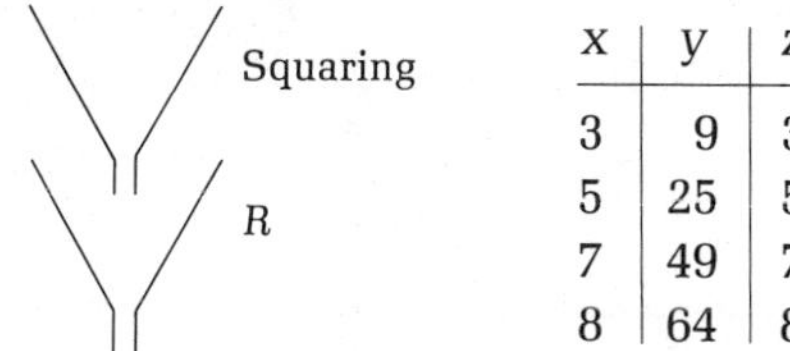

x	y	z
3	9	3
5	25	5
7	49	7
8	64	8

We think of x going into the squaring operator, y coming out and immediately going into the *R* operator, and z coming out of the *R* (and the combined) operator. The values for the z column are the same as the values for the x column. The *R* operator is called the inverse of the squaring operator. It has the table shown, where labels x and y are used in the standard way for inputs and outputs of this operator, instead of y and z.

x	y
9	3
25	5
49	7
64	8
16	4
0	0

It is easily seen that *R* is the square-root operator. If we restrict our attention to whole numbers, then only numbers from a select set will be accepted by R. Indeed, one of the great advances in mathematics came when people thought of allowing any whole number to go into R (that is, taking the square root of any whole number).

x	y	z
2	7	2
3	8	3
4	9	4
13	18	13
17	22	17

EXAMPLE 9 To find the operator inverse of Add 5, the operator of example 2, we make a triple table in which the first two columns are those of Add 5 and the third column repeats the first. A partial table for the operator inverse of Add 5 is shown below. Can you guess more pairs? Can you guess the rule?

x	y
7	2
8	3
9	4
18	13
22	17

It appears that the inverse of Add 5 is Subtract 5.

x	y	z
2	8	2
3	12	3
5	20	5
6	24	6
8	32	8

EXAMPLE 10 From the triple-column table generated from example 4, we see that D4, the inverse of the operator of example 4, has a partial table given below. Can you see that the rule for D4 is Divide by 4 and the formula is $y = x \div 4$?

D4

x	y
8	2
12	3
20	5
24	6
32	8

x	y	z
3	97	3
5	95	5
20	80	20
22	78	22
35	65	35

EXAMPLE 11 Some people are startled by the inverse of example 7, the operation Subtract from 100. We make a triple table with the third column repeating the first column.

The table for E7, the inverse of example 7, shows that E7 has the rule Subtract from 100 and the formula $y = 100 - x$.

E7

x	y
97	3
95	5
80	20
78	22
65	35

Mirabile dictu! Example 7 and its inverse are the same operation: Subtract from 100. In the other examples, the inverse operator was always different from the operator itself. Example 11 shows this is not always the case.

EXAMPLE 12 Not all unary operations have inverses. Consider the table in the margin. Inputs 2 and 4 both yield outputs of 3. This is quite all right for a unary operation. But if there were an inverse operation, an input of 3 would have to yield an output of both 2 and 4. This is not allowed. For a given input, an operation can have only one output. Hence the operation in example 12 has no inverse operation.

x	y
1	6
2	3
3	2
4	3
5	6

Do you see that each output of an operation must come from only one input if the operation is to have an inverse?

The formula for example 12 is $y = (x - 3)^2 + 2$. Can you show how to hook up an A2 (add 2) with an S3 (subtract 3) and an Ex1 (squaring) operator to make the operator Ex12?

7.5 BINARY OPERATIONS Two numbers (or things) go into the operator and one comes out. Some examples are addition, subtraction, multiplication, division, union, intersection, raising to a power, finding the largest number, diluting acid, getting married, breeding cattle, translating the Bible, and blending gasoline. Certainly the objects considered are quite different and range over all of human experience. Yet they all involve binary operators.

We represent binary operators by diagrams, formulas, and/or triples. The diagram shows a machine with two inputs and one outlet at the bottom. *A* goes in the left input, *B* goes in the right input, and *C* comes out the bottom.

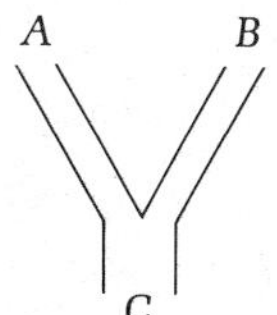

We may also say $A * B = C$ or (A, B, C). For example, if the binary machine "divides," we might have $A = 42$, $B = 6$, and $C = 7$. Then, $42 * 6 = 7$ and (42, 6, 7). Since we have a special sign for division, we might say $42 \div 6 = 7$, not using $*$, the general symbol for operator.

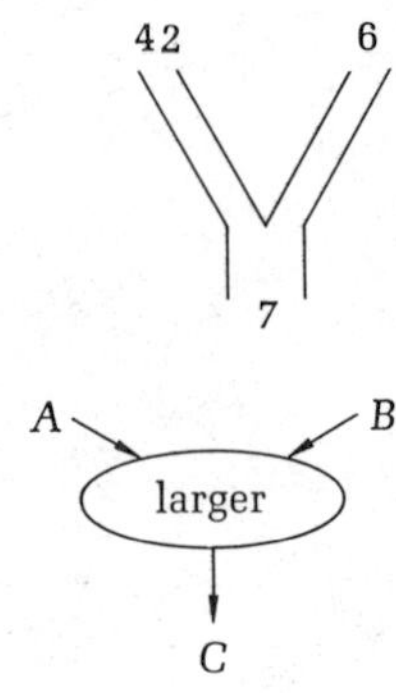

Sometimes the diagram has distinctly different-looking inputs to emphasize that for most machines it is important to get the correct thing in the correct opening. We comment that, in general, A, B, C may all be different, or two or all three may be the same.

Consider choosing the larger number of A, B. The numbers are compared, and the larger is called C. To get E, the largest of three numbers A, B, D, follow this pattern:

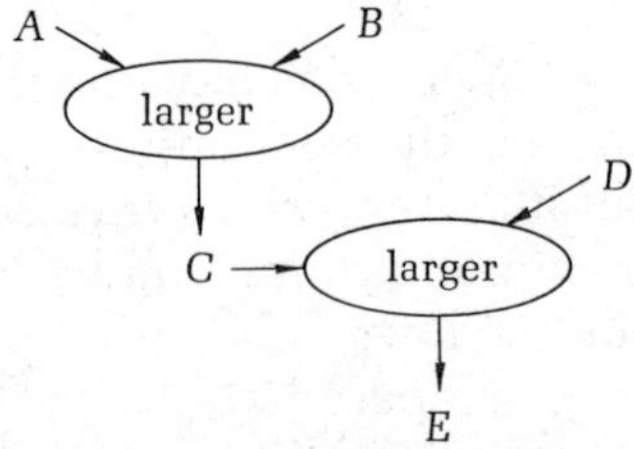

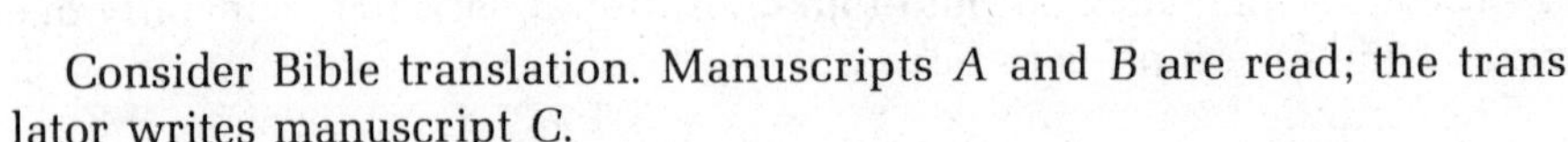

Consider Bible translation. Manuscripts A and B are read; the translator writes manuscript C.

Consider cattle breeding. A bull of strain A is mated with a cow of strain B. The offspring will be of strain C.

Consider making gravy. Meat juice A and flour B are stirred in a pan to form a mix C. Mix C and liquid (water or milk) D are stirred into the pan to form gravy E.

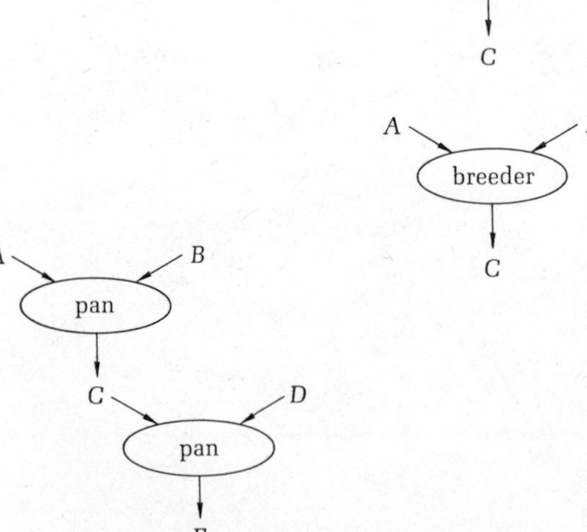

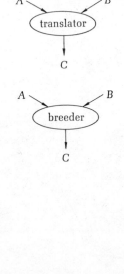

Consider addition. Addends (numbers) A and B are put in the adder. The result is the sum C.

Binary operations are also called functions of two variables.

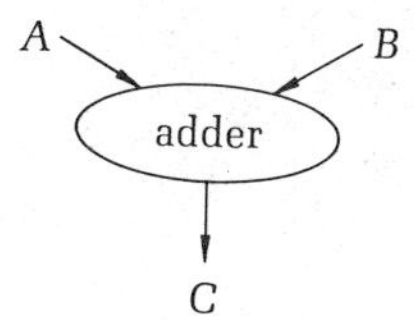

7.6 HOOKING UP BINARY OPERATORS Just as unary operators can be hooked up so that the output from one is the input for another, so binary operators can be hooked up so that the output of one is one of the inputs for another. Here are diagrams showing how two operators might be hooked up. The operator diagrams have been drawn lopsided (asymmetric) to show more vividly that we need to state specifically whether a given input goes into the left opening or the right. The formula for each hookup is written beside the diagram. Other hookups were illustrated in the preceding section.

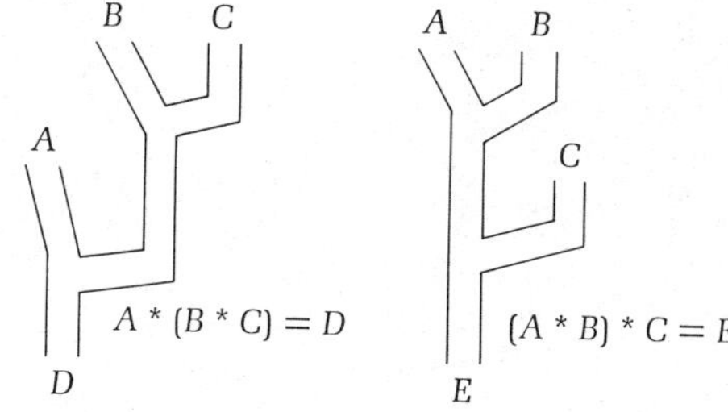

Suppose these operators are subtraction. That is, $*$ is replaced by $-$. If we set $A = 33$, $B = 20$, $C = 5$, the left hookup yields $D = 18$ and the right hookup yields $E = 8$.

Two hooked-up binary operators have three inputs and one output. Together they make a ternary operator.

Suppose the operators in the diagram were addition rather than subtraction. Then these inputs into the left hookup would yield $D = 58$ and into the right hookup would also yield $E = 58$.

Three binary operators hooked up make a "fourholer." Here are two possible patterns. You may think of others.

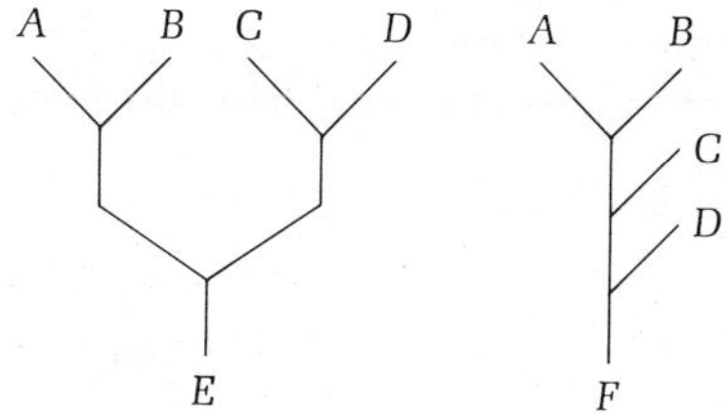

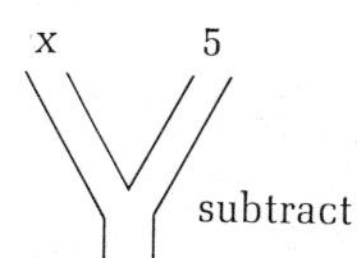

A unary operator may be made from a binary operator by specifying one of the inputs. If the operator is subtraction and 5 is always placed in the right input, only the left input is open. The resulting unary operator is Subtract 5. The corresponding formula is $x - 5 = y$. If 100 is always placed in the left input, leaving only the right input open, the resulting unary operator is Subtract from 100. The corresponding formula is $100 - x = y$.

Different binary operators may be hooked up. Suppose a subtract operator is placed ahead of a multiply operator.

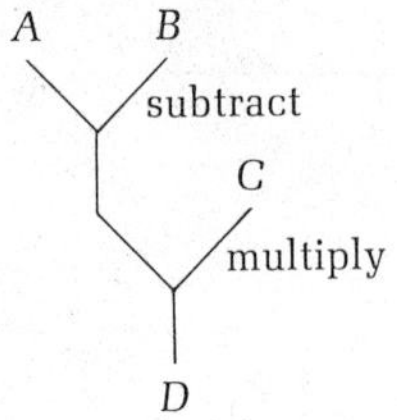

A B
subtract
C
multiply
D

If $A = 10$, $B = 7$, $C = 5$, then $D = 15$.
If $A = 8$, $B = 2$, $C = 3$, then $D = 18$.

Can you determine D, the output, if $A = 7$, $B = 3$, $C = 6$? The formula is

$$D = [(A - B) \times C]$$

7.7 PROPERTIES OF BINARY OPERATORS Does it make a difference if the inputs are interchanged? Is $(A * B) = (B * A)$? Is the output from 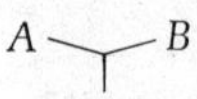 A B the same as the output from B A?

If the operator is addition of numbers or union of sets, it makes no difference. If the operator is subtraction or putting on clothes (with inputs shoes and stockings), it does make a difference. Indeed, the subtraction operator chokes up and refuses to work if only counting numbers are allowed.

Those binary operators for which it makes no difference, for which $(a * b) = (b * a)$ for all possible inputs, are called *commutative*. The others are called noncommutative. Here is a short table classifying some of the operations listed.

Commutative	**Noncommutative**
Union of sets	Subtraction of numbers
Intersection of sets	Raising to powers
Multiplication	Division of numbers
Blending of gasoline	Breeding of cattle
Choosing the smaller child	Dilution of acid

To illustrate the idea, for each operator pick a pair of inputs and imagine putting them into the operator in different order. Are the results the same for the commutative operators and different for the noncommutative? It may be that for a particular pair of inputs the results will be the same even for noncommutative operators. Then one looks for another pair of inputs.

The commutative property is especially useful when one way is easier than another. Some people find 182 + 7 easier to do than 7 + 182, or 3 × 15 easier than 15 × 3. One famous mathematician says he hasn't multiplied 7 × 8 for thirty years. He always replaces it with 8 × 7, which he happens to find much easier.

We saw earlier that two identical operators can be hooked up to form a ternary operator in two ways, as illustrated. When the inputs *A*, *B*, *C* are put into the holes in the order shown, does the difference in hookup make any difference in output?

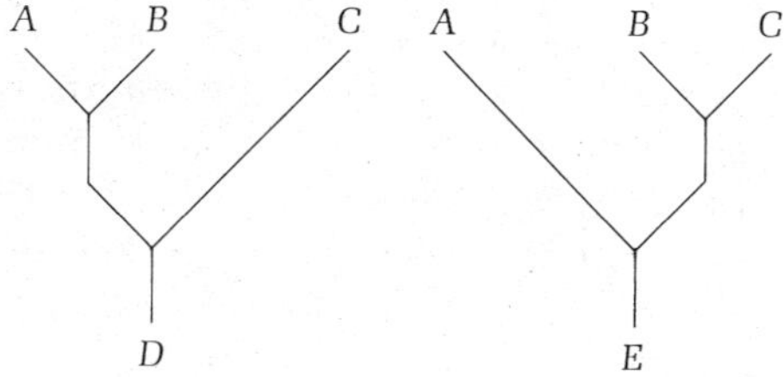

If the operators are addition of numbers or intersection of sets, there is no difference in the outcome. If the operators are subtraction or making gravy or mixing flavors, the outcomes may be different. For example, let $A = 33$, $B = 20$, $C = 5$. If the operator is addition, the output is 58 for both hookups. If the operator is subtraction, the outputs are different—8 for one and 18 for the other.

Let *A* = meat flavorings, *B* = thickening (flour), and *C* = water. Let the operator be "mixing in a hot pan to make gravy." Some gourmets may think the results taste different.

Those operators for which different hookups make no difference in the outcome are called *associative*—that is, how the operators are associated makes no difference in the outcome. Expressed in formulas, the operator $*$ is associative if $[(A * B) * C] = [A * (B * C)]$ for all possible inputs *A*, *B*, *C*.

Here is a short table classifying some operators. You will find it valuable to make examples for each of these cases and test the classification. Also, you may want to classify other operators.

Associative	**Nonassociative**
Union of sets	Division of numbers
Intersection of sets	Making gravy
Addition of numbers	Raising to a power
Blending gasoline	Cattle breeding
Judging beauty	Getting married

The associative property is helpful when one hookup is easier than the other. For example, $(17 + 3) + 9$ is easier than $17 + (3 + 9)$.

Some expressions are easier to write if the operation $*$ is associative. We may write $A * B * C$ rather than $(A * B) * C$ or $A * (B * C)$ as is needed for nonassociative operators.

We see that the associative property is a commutative property of the order of performing the operations. However, as mentioned above, the title "commutative property" is used when the order of the objects in one performance of the operation makes no difference—that is, $A * B = B * A$.

When it does not make sense to perform the operation twice, we say the associative property does not hold. Getting married falls into this category.

It is easy to mistakenly think the associative property is a ternary operator, or at least a property of ternary operators. This is because the binary operator $(a \times b) = q$ is tested for associativity by the question "Is $a \times (b \times c) = (a \times b) \times c$ for all a, b, c?" Even though ternary operators appear in the test question, it is the binary operator that is being tested. It is the binary operator that is, or is not, associative. This error is like that of the little boy who thought the color yellow had two sleeves because all the examples of yellow things he had been shown had been shirts.

7.8 DISTRIBUTIVE PROPERTY When two different binary operations are hooked up, it almost always makes a big difference which operator comes first. We illustrate the combination of an addition and a multiplication operator.

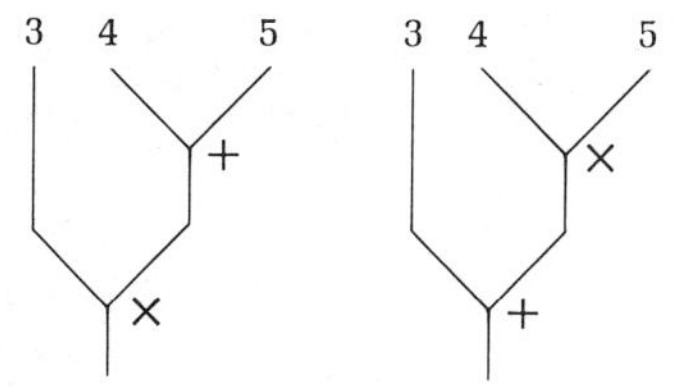

Following the pattern of the discussion of commutative and associative properties, we are tempted to ask about operators where this interchange makes no difference. However, it turns out this is not a useful question.

Instead we look for some type of hookup with the order of operations changed that yields the same result. Can you see that the combinations in the margin are equivalent? On the left we have an addition operator followed by multiplication. On the right we have two multiplications followed by addition. The formulas for these two hookups are

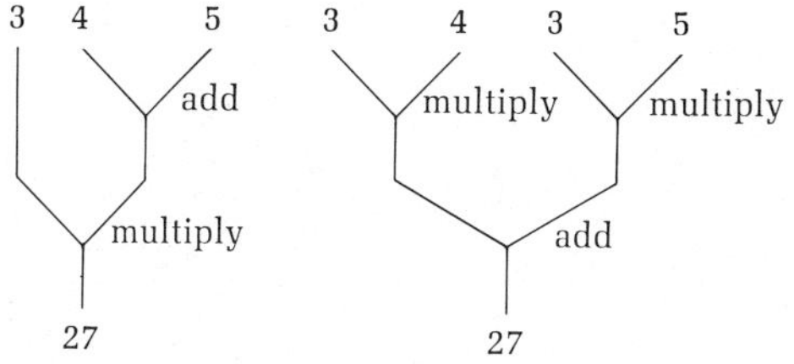

$$[A \times (B + C)] = D \qquad \text{and} \qquad [(A \times B) + (A \times C)] = E$$

It is shown in chapter 10 that D and E must be the same number for all number inputs A, B, C. Can you substitute other numbers for A, B, C and get the same outputs from both hookups?

Because we get the same result, we say that multiplication is *distributive* over addition. In general, whenever there are any two operations that we indicate by a and m, such that the hookups yield the same output, we say that m is distributive over a. The formula for the law is $[Am\ (BaC)] = [(AmB)\ a\ (AmC)]$.

We use this property when we say $3 \times 46 = 3 \times 40 + 3 \times 6$. We thus reduce the task to single-digit multiplications. A problem involving several additions and multiplications can be rephrased using the distributive property so that all the multiplications may be done before the additions. This is often useful in algebra. That intersection of sets is distributive over union is easily seen. In $A * (B \cdot C)$ and $(A * B) \cdot (A * C)$, we interpret $*$ as intersection and $\cdot$ as union, and draw Venn diagrams.

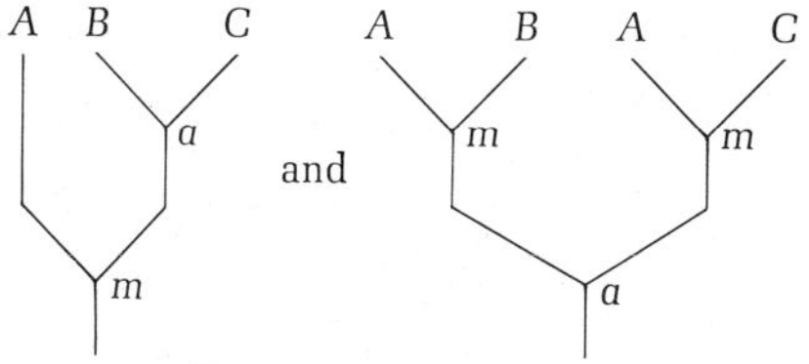

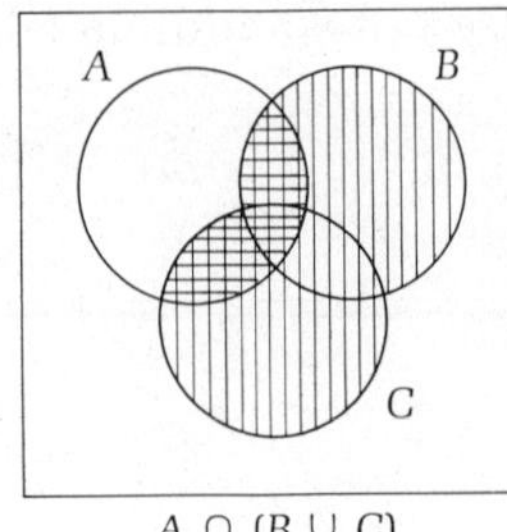

$A \cap (B \cup C)$

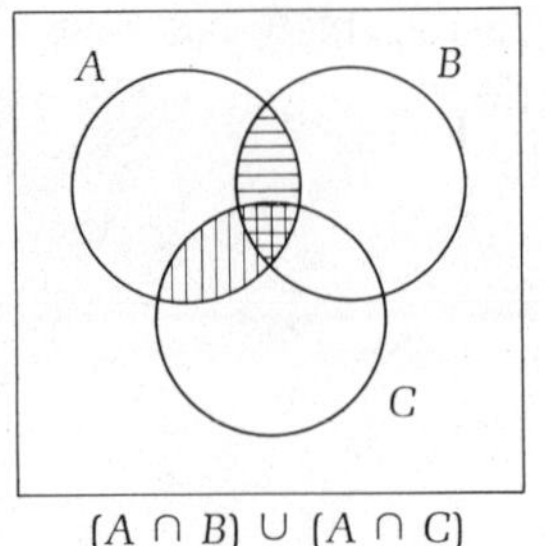

$(A \cap B) \cup (A \cap C)$

Since $A \cap (B \cup C) = (A \cap B) \cup (A \cap C)$, intersection is distributive over union.

How about the other way? Is union distributive over intersection? Is $A \cup (B \cap C) = (A \cup B) \cap (A \cup C)$? We draw Venn diagrams.

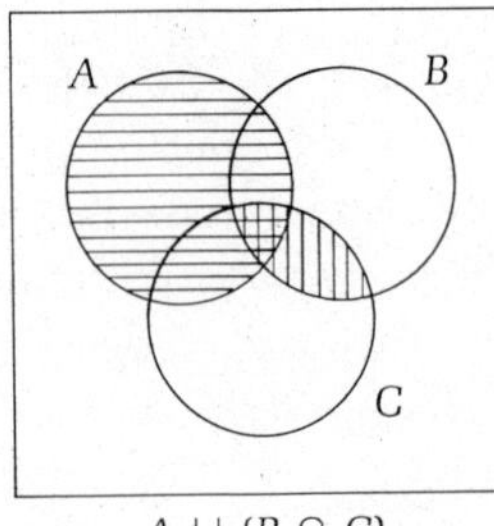

$A \cup (B \cap C)$

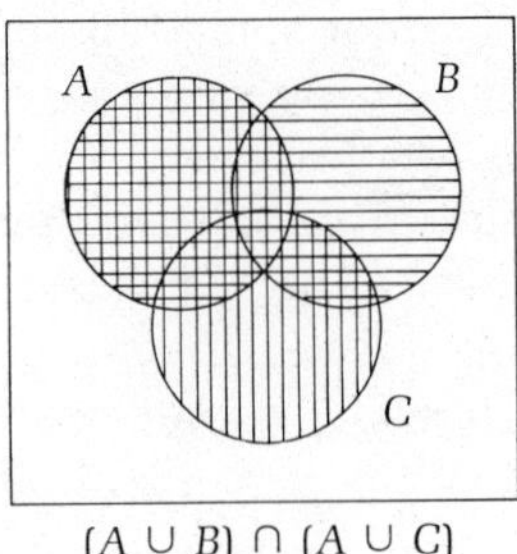

$(A \cup B) \cap (A \cup C)$

Anything shaded at all in one figure is compared with the cross-hatched region in the other. Do you conclude that union is distributive over intersection?

These three properties—commutative, associative, and distributive—are called the CAD properties. We will see them in more detail when we talk about algorithms and algebra.

7.9 RELATIONS In life and mathematics there are ties that connect people and things. Phrases like *is the uncle of, is congruent to, contains, is less than, is to the left of, has eyes the same color as, is in the same set as, sells for more than, is the brother of* show some of these connections. They are called *relations*, as a generalization of blood ties.

We can represent relations by listing the members of the sets and drawing curves connecting the related members, by graphs, and by description. Suppose we have four children, Alan, Betty, Carl, and Dot, and three men, Robert, Sam, and Tom. Robert is the uncle of Alan, Betty, and Carl. Sam is the uncle of Betty, Carl, and Dot. Tom is the uncle of Alan and Dot. We could represent these facts by the following diagram:

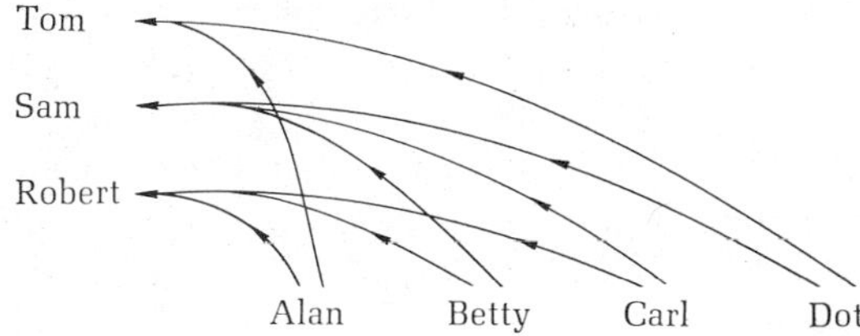

The curves are directed—that is, have an arrow on them—to show that Tom is the uncle of Alan and not Alan the uncle of Tom. We represent the relation and directed curve by an arrow, such as Alan $\rightarrow$ Tom.

The diagrams can get rather crowded, so we often use dots in a rectangular grid, or on graph paper.

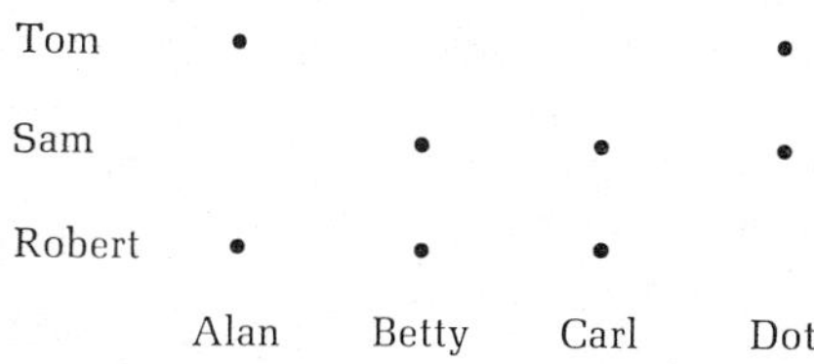

The members of the set on the left have the relation with, in this case are the uncles of, the members of the set on the bottom where there is a marked point on the graph paper.

Because all the possible points of the grid would represent the Cartesian product of the two sets, children and men, and because the marked points are a subset of these, we see that the relation is represented by a subset of the Cartesian product. Some say the relation is a set of ordered pairs (x, y).

We give some examples of the most important types of relations between a set X = {1,2,3,4,5,6} and a set Y = {3,4,5,6,7,8}.

1. *Equality, or sameness.* Identical. This concept is useful when we have different names for the same thing. The 5 in set Y names the same number as the 5 in set X. We have drawn a graph in the margin.

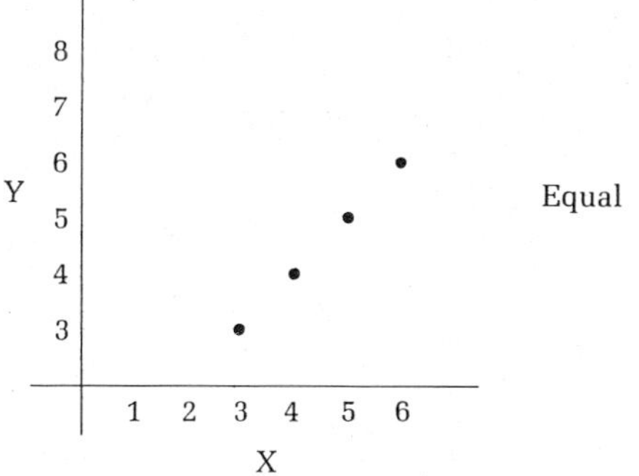

People use the word *equal* in many ways. Here we say $A = B$ only if A and B are names of the same thing. For example, "the number of letters in *shave* equals the number of letters in *grows*." "The candy bar has the same price as the ice-cream cone."

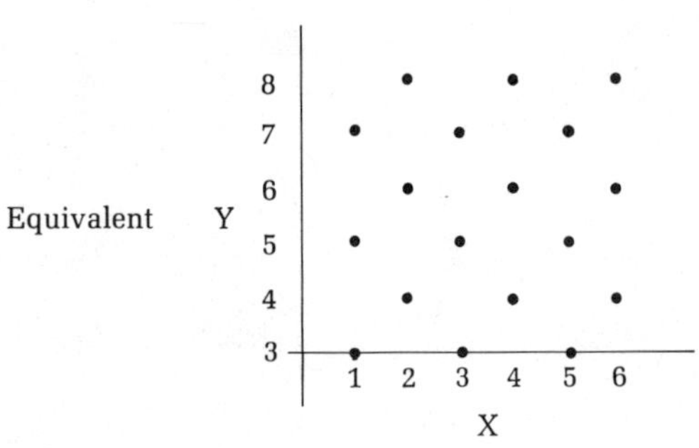

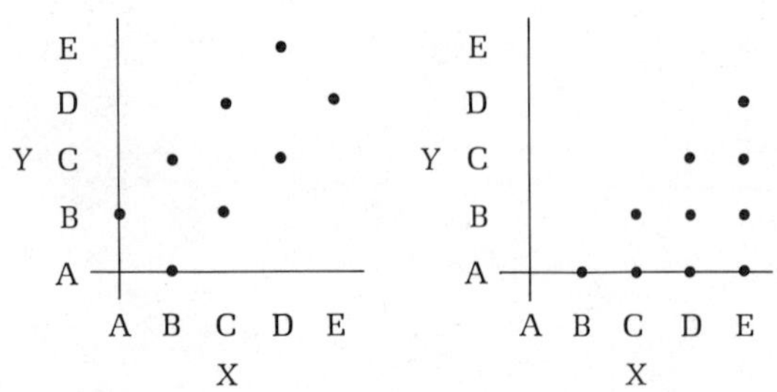

Y is next to X. Y is smaller than X.

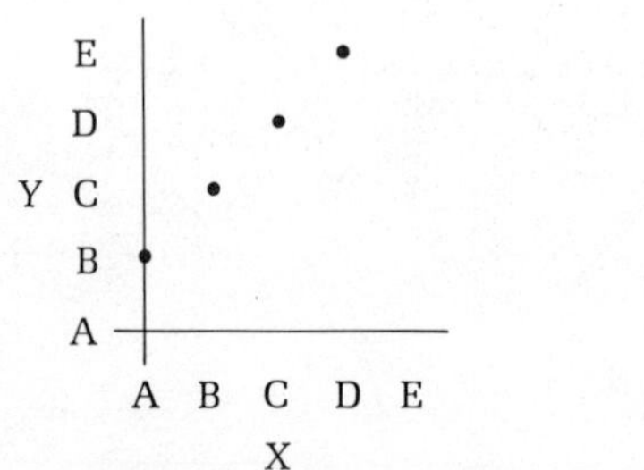

Y is the first person to the right of X.

Often, as in "all men are created equal," people call different things equal that are classed together in the same set. Such a relation we call *equivalence*.

2. *Equivalence*. Some examples of sets of equivalent objects are a set of dimes, candy bars from the same box, paper clips from a box, members of the Senate, even numbers, equilateral triangles, two squares with one-inch sides, and the names 5 + 2, seven, and VII. Equivalent things may be different, but are in the same set because they have the same property.

Two equal objects are equivalent for all times, places, and purposes. Two equivalent objects may be equivalent for those purposes which put them in the same set, but not for others. All even numbers are equivalent, for example. The graph in the margin shows when two numbers are equivalent if they are both even or both odd.

These two graphs show a difference between "equal" and "equivalent." Suppose we have a set *S* with several members, with some members being equivalent—that is, having a common property, or being members of a subset. If we graph this equivalence relation with *S* being both the X, or horizontal set, and Y, the vertical set, we find plotted points all over. A member of the X set may be related to several members in the Y set. There may be several plotted points for a given X member.

In the equality or sameness relation, however, an X member is connected to only one Y member: itself. The plotted points fall on a line.

3. *Less-than, or linear, order*. A property, such as height, enables us to arrange people along a line in linear order. When this is done, there are several interesting relations among the people. Among these are (1) is next to, (2) is smaller (shorter) than, (3) is the first person to the right of. We graph these relations, supposing Alfred, Bob, Caron, Dolores, and Evelyn are lined up by increasing height. (See graph in margin.) If the graphs were made with the members in some other order, we might get the layout shown on the opposite page. While these graphs are by no means as pretty as the earlier ones, we could probably line the letters up correctly using the graphs alone.

"Is smaller than" is the most important of these relations, because we can get this graph merely by comparing the people in pairs, without lining them up. If we had a group of people, we could not get "is next to" without lining them up (or measuring them and lining the numbers up).

4. *Functions.* A function was defined in section 7.1 as an operation. It is also a special kind of relation. If each member of the X set is related to only one member of the Y set, then we say that Y is a function of X. In terms of the ordered pairs (x, y), the relation is a function if no two pairs have the same first member with a different second member. Important functions arise from indirect measurement. The area of a square is related to the side. For each length of side there is only one area. So we say that area is a function of length of side. The height of a ball tossed in the air is a function of time. For each time there is one height.

We note that going the other way, the inverse relation need not be a function. Each height of the ball is related to two times, one going up and one going down. In the square function, the inverse relation is also a function. For each area there is only one side of a square.

The inverse relation is the relation we get when the ties or threads are regarded as connected in the opposite direction.

It is useful to think of a function as shifting our attention from one object to another; for example, from side to area or from time to height. If the inverse of a function is also a function, then the inverse shifts our attention back to where it started.

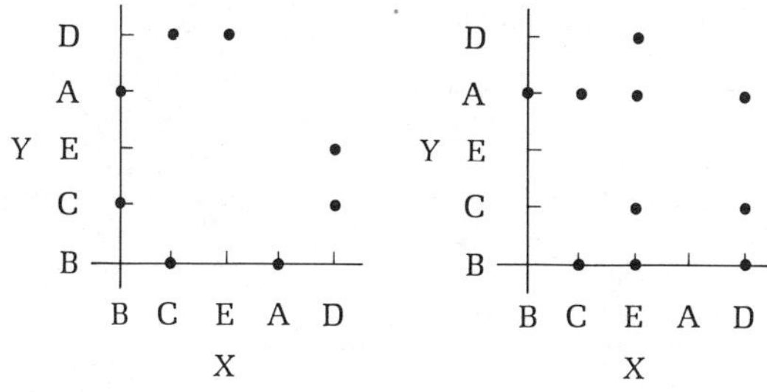

Y is next to X. Y is smaller than X.

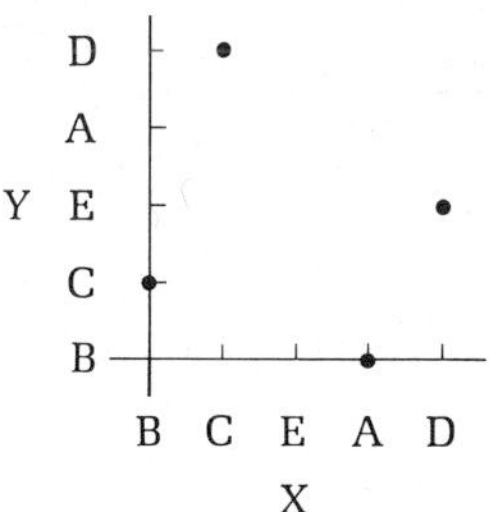

Y is the first person to the right of X.

7.10 RECOGNIZING EQUIVALENCE AND ORDER RELATIONS Objects are equivalent if they have a given property—that is, are members of a set. Objects have linear order if there is a relation similar to "less than" that we use to line them up on a line. Now we may have a set with a relation and not know whether or not the relation is an equivalence or order relation. We may just not see the common property, and no one has corralled the objects into subsets, or lined them up. Are there any observations on the threads of the relation and their connections that will tell us whether the relation is an equal, equivalence, or order relation? We give some. The X set and the Y set in a graph are both the universal set under discussion.

1. If each member is tied to itself and to no other, the relation is equal.

2. The relation is an equivalence if it is reflexive, symmetric, and transitive.

(*a*) A relation is *reflexive* if each member is related (tied) to itself.
(*b*) A relation is *symmetric* if, whenever A is related to B, then B is related to A.
(*c*) A relation is *transitive* if, whenever A is related to B and B is related to C, then A is related to C.

You may like to test these qualities in relations such as "have the same property" and "are members of the same set"; in a universal set of triangles "has a right angle" and "has an angle the same size as"; in the counting numbers "leaves the same remainder when divided by 3."

3. A relation $\rightarrow$ is an order relation if–

(*a*) The trichotomy, or 3 part, property holds–that is, for any two members A, B of the set, either $A \rightarrow B$, $B \rightarrow A$, or $A = B$. (This compares with the reflexive and symmetric properties.)
(*b*) The transitive property holds–that is, if $A \rightarrow B$ and $B \rightarrow C$, then $A \rightarrow C$.

You may like to see if any of the following relations are equivalence or order relations: "is the brother of," "is older than," "is defeated in tennis by," "is congruent to." You need to be clear about what your universal set is. These are physical-world illustrations of the above ideas.

EXERCISES

1. Classify these operations as unary, binary, ternary, or quaternary:
(*a*) finding the height of a building
(*b*) finding the volume of a brick from h, l, w
(*c*) finding the difference of two numbers
(*d*) making pie with prepared crust, berries, and sugar
(*e*) finding the eldest son of two people
(*f*) finding the mother of a lost child
(*g*) finding the area of a rectangle, one of whose sides is 8 inches

2. If 8 is put into an "add 5" unary operator, what comes out?

3. If 7 is put into a "times 4" operator, what comes out?

4. An "add 2" operator is hooked up ahead of a "times 3" operator to form operator F. Make a table of outputs of F when 2, 3, 5, 7 are inputs.

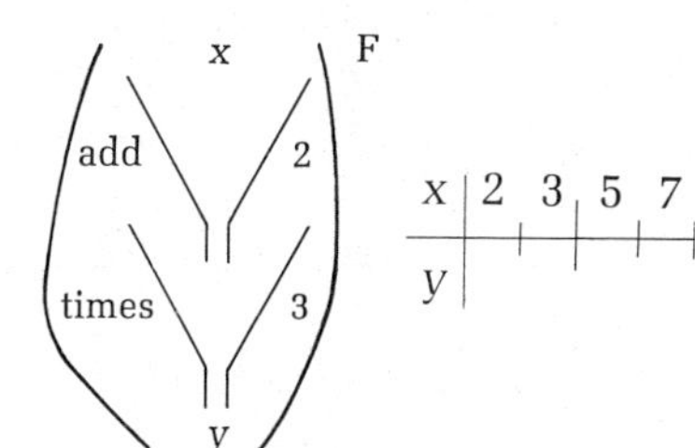

x	2	3	5	7
y				

5. Complete a table for an operator with input a and output m if $m = 2a - 8$.

input a	6	7	10			
output m				4	8	12

6. Guess a rule for an operator A with table in margin.

input x	2	7	13	14	16
output y	7	12	18	19	21

7. Guess a rule for operator B with table in margin.

input x	4	5	7	8	9	10
output y	6	15	39	54	71	90

8. Guess a rule for operator C with table in margin.

input x	7	9	12	18	20	21
output y	2	4	7	13	15	16

9. Operator E is formed by hooking up operators A and C in exercises 6 and 8. Make a table for E.

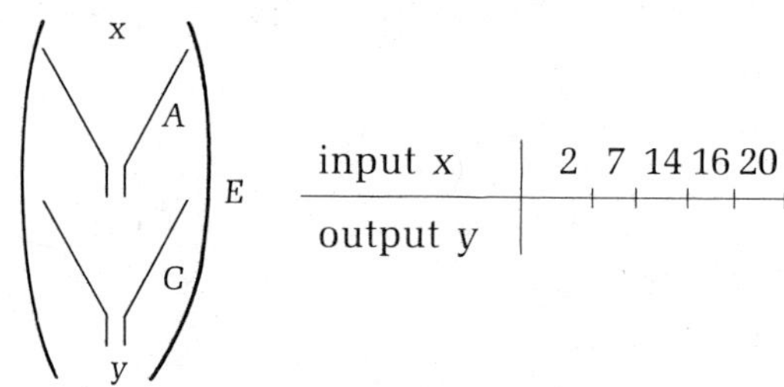

input x	2	7	14	16	20
output y					

10. Make a table for F, the inverse of B in exercise 7. Call the inputs w and the outputs z.

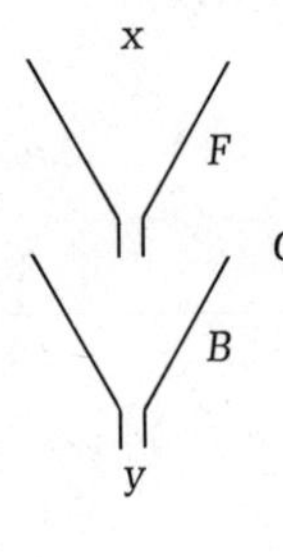

11. Make a table for G, formed by hooking B in exercise 7 after F in exercise 10. Guess a rule for G.

12. Make a table for P, the operator $y = 3x + 5$ for $x = 2, 3, 5, 8$. Make a table for Q, the inverse of P, for four values of inputs. Guess a formula for Q. Substitute 11, 14, 20, 29 as inputs in the formula for Q. Do you get 2, 3, 5, 8 as outputs?

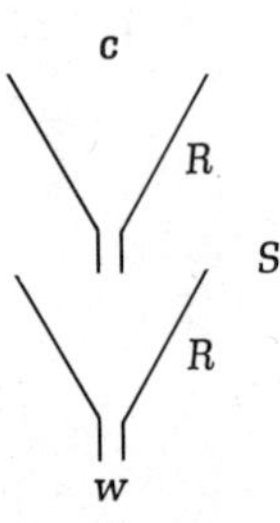

13. Make a table for R, the operator $y = 70 - x$, for $x = 5, 10, 30, 50$. Hook up two R operators to form operator S. Make a table for S with inputs $c = 5, 10, 30, 50$. Guess a formula for S.

14. The following are formulas for binary operators with x, y as inputs and z as output. Complete the table for each operator if possible.

(a) $z = 5xy$
(b) $z = 5x + y$
(c) $z = 2(x + y)$
(d) $z = (x + y^2/x)x$

(a)

x	y	z
2	3	
3	2	
4	1	
1	4	

(b)

x	y	z
2	3	
3	2	
4	1	
1	4	

(c)

x	y	z
2	3	
3	2	
4	1	
1	4	

(d)

x	y	z
2	3	
3	2	
4	1	
1	4	

15. Which of the binary operators in exercise 14 are commutative?

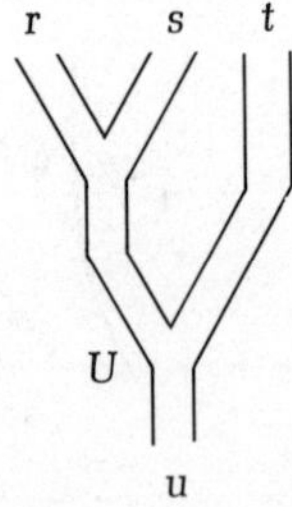

16. Four identical binary operators follow the formula $z = 5xy$. They are hooked up in pairs to form D and U as shown. Complete the tables.

	a	b	c	d
	2	1	3	150
D	3	4	2	600
	1	3	4	
	4	2	4	

	r	s	t	u
	2	1	3	150
U	3	4	2	
	1	3	4	
	4	2	4	

17. If you put the same ordered triple of numbers for *abc* in *D* and *rst* in *U*, would you get the same output values *d* and *u* regardless of the input numbers *abc*?

18. Four identical binary operators with rule $z = 2(x + y)$ are hooked up in pairs to form two ternary operators *D* and *U* as in exercise 16. Complete the tables.

	a	*b*	*c*	*d*
	4	2	4	
D	2	1	3	
	1	3	4	
	3	4	2	

	r	*s*	*t*	*u*
	4	2	4	
U	2	1	3	
	1	3	4	
	3	4	2	

19. Is the binary operator $z = 2(x + y)$ associative?

20. Form two hookups as in exercise 16 from 4 identical binary operators $z = x + y$. Is this operator associative?

21. Test enough values in exercise 16 to determine whether the operator $z = x - y$ is associative.

22. Try several triples of values in exercise 16 to test whether the operator $z = xy$ is associative.

23. Is the operator $z = x/y$ associative? Make hookups as in exercise 16 and test.

24. A graph of the relation "Y is larger than X" is shown below for Alvin, Bob, Carter, David, and Eugene. Who is the largest? List the boys according to size with the largest first.

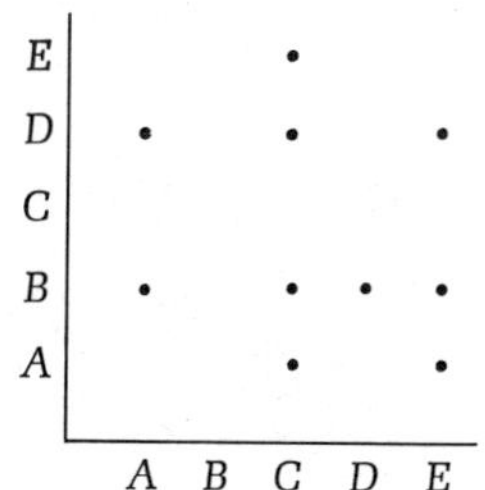

25. Draw a graph of the data in exercise 24 with the boys listed along the axes by size rather than alphabetically.

26. As part of a study of learning in a course on the structure of arithmetic, 8 students reported those whom they had helped prepare for an exam. They did not report who, if anyone, had helped them. The graph of the relation "X helped Y" is shown below.

(*a*) Whom did *E* help?

(*b*) Is this an equivalence relation?

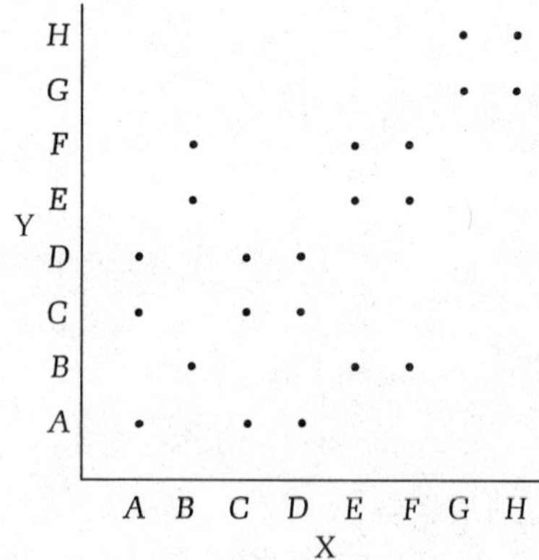

27. Make up and solve a question involving hooking up two operators.

Answers to selected questions: **1.** (*a*) unary; (b) ternary; (c) binary; (g) unary; **3.** 28.

5. input *a* 6 7 10 6 8 10
output *m* 4 6 12 4 8 12

7. $y = x^2 - 10$; **9.** 2 7 14 16 20.

11. *G*

x	6 15 39 54 71 90
y	6 15 39 54 71 90

y = x identity

13. *R*

x	5 10 30 50
y	65 60 40 20

S

x	5 10 30 50
y	5 10 30 20

y = x

16.

	a b c d		r s t u
	2 1 3 150		2 1 3 150
D	3 4 2 600	U	3 4 2 600
	1 3 4 300		1 3 4 300
	4 2 4 800		4 2 4 800

17. yes; **19.** no; **23.** no; **24.** BDAEC; **26.** (*a*) B, E, F; (b) yes.

8

Addition—The Operation and the Algorithms

8.1 INTRODUCTION We are all familiar with statements such as $2 + 3 = 5$ and $23 + 18 = 41$. Addition was mentioned in earlier chapters as a computational situation and as one of the main operations. In this chapter we review the human situations leading to addition and show how the properties of addition reflect the properties of these situations. Our attitudes toward addition, or interpretations of the meaning of the operation, are named from the physical situations that suggest it. We think of addition as being associated with putting sets together, continued counting, and joining segments. These then are our names.

The commutative and associative properties are explained as coming from these attitudes. Some people prefer to say that we postulate that addition is commutative and associative rather than that these properties come from our attitudes. This is a perfectly sound approach. The important question is "Why do people assert this postulate?" rather than "Why is addition commutative and associative?" The properties are postulated so that addition will reflect the attitudes. The student's problem of understanding addition is very similar in both cases.

A process by which, given the names of two addends, one gets the name of the sum we call an addition algorithm. We call an algorithm that comes directly from the attitude a natural algorithm. Most algorithms reflect the numeral system used and are thus more efficient than natural algorithms.

We give the usual algorithms, then show that they actually do produce what they say they do. This verification always involves some natural algorithm or addition attitude with the numeral system being used. Algorithms are also called "rules of reckoning" and, in this book, "addition dance of the digits." People wish to learn these algorithms by rote, but they also want to know why they work.

8.2 ADDITION FROM SET, SEGMENT, AND SEQUENCE IDEAS We define addition from the putting-sets-together attitude. By 2 + 3 we mean the number of members in the union of a set having 2 members with a disjoint set having 3 members. The sum of any other two numbers is similarly defined. For example, since

$$\{a,b\} \cup \{c,d,e\} = \{a,b,c,d,e\}$$

we define

$$n\{a,b\} + n\{c,d,e\} = n\{a,b,c,d,e\}$$

or

$$2 + 3 \qquad = 5$$

The notation n {. . .} means "the number of members in the set {. . .}." "Disjoint" means that the two sets have no common elements; their intersection is the empty set. While we often find the union of overlapping sets, we define addition in terms of disjoint sets only. The members of the sets may be anything—marbles, beans, ideas, ...

From the continued-counting-in-sequence attitude toward addition, we define 2 + 3 as meaning the number reached when, starting at 2, we count 3 more. Similarly, from the joining-of-segments attitude, we define 2 + 3 as meaning the length of a segment formed by joining a segment 2 long with a segment 3 long. Again, similar definitions (by example) hold for adding pairs other than 2 and 3.

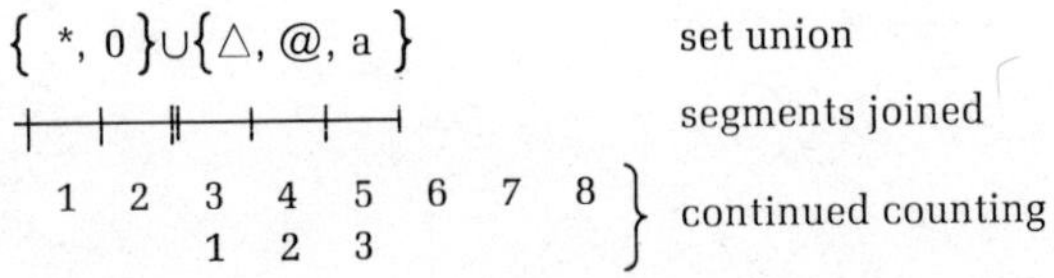

Do these three definitions always give the same result? Will the number $a + b$ from the union-of-sets attitude always be the same as $a + b$ from the continued-counting and the joining-segments attitudes? The matching shown by the diagram in the margin illustrates this for 2 + 3.

{ *, o }∪{△, @, a } set union

segments joined

1 2 3 4 5 6 7 8
    1 2 3 } continued counting

We see, for example, that the sign @ in the set of 3 is matched with the middle of 3 unit segments in the segment 3 long, and that both are matched with 2 in the continued counting on the second line, "1, 2, 3." Similar matching shows that $a + b$ is the same for any attitude and for any numbers.

Ordinarily in mathematics we define addition from one attitude and then show that the other attitudes give the same results. Here we speak of defining from all three attitudes to indicate that each attitude is a valid starting point.

Note that we did not use the name of the sum (5) at any point. We have shown that the three processes will yield the same result without knowing the standard name of the sum.

Some people wonder if it would not be better to emphasize one attitude, say sets or continued counting, and always start from there in discussing properties of addition. This has particular appeal to those who desire "one right way" of doing things, and hence is valuable. However, certain properties are easy to show from one attitude and difficult from another. Each attitude is the easiest starting point for some property or problem. For example, union of sets is best for discussing the standard addition algorithm. Joining of segments is best for showing the commutative and associative properties (only slightly better than union of sets) and for addition of fractions. Continued counting is best for problems such as "When is 17 days after March 8?"

Having several approaches is not only easier, but each attitude supports and throws light on the others. For many people, addition is rote manipulation of symbols according to a pattern or form. We call their attitude *formal manipulation.*

The fact that zero adds the way it does follows easily from the concept of the empty set $\emptyset$. For any number a, $a + 0 = a$, because the union of the empty set $\emptyset$ with any set A is just the set A. This is often easier to see than trying to discuss $0 + 5$ in terms of successive counting, since counting usually begins with 1.

8.3 THE COMMUTATIVE PROPERTY OF ADDITION Does $3 + 5 = 5 + 3$? Why? For such small numbers it is easy to see that the sum is 8 in either case. For larger numbers the sums are not so obvious. Therefore, we would like to know if the sums are equal without actually finding the sums.

We may associate $3 + 5$ with putting a set of 3 members in a dish, and then putting a set of 5 members in the dish to form the union of the sets. We associate $5 + 3$ with putting the 5 set in the dish and then putting in the 3 set. Since in either case the union is the same set of members, the numbers must be the same. Hence, we can assert that $3 + 5 = 5 + 3$ without knowing the name 8. The process is general. If a and b are any two whole numbers, then $a + b = b + a$.

Some people feel that it is obvious that the union of two sets is commutative. That is, the union is the same regardless of which set is placed in the dish first. It is not so obvious that counting to a and then counting b more will lead to the same number as counting to b and then counting a more. This is one of the reasons why defining addition by sets is sometimes preferred to defining addition by successive counting. You may wish to look again at chapter 7.

8.4 THE ASSOCIATIVE PROPERTY OF ADDITION So far we have defined only the sum of two numbers. What is $4 + 5 + 7$, the sum of three numbers? Which addition do we do first? Does it make a difference? Is $(4 + 5) + 7$ different from $4 + (5 + 7)$? We use parentheses to show the order of adding. Parentheses are a symbol of grouping. We work from inside out, doing what is indicated inside the parentheses first. In this instance, we have $(4 + 5) + 7 = 9 + 7 = 16$ and $4 + (5 + 7) = 4 + 12 = 16$, with the same result. Would we get the same result with any three numbers? Here again the fact that addition is defined in terms of union of sets makes us confident. In putting three sets together it makes no difference in the final outcome which two we put together first. Thus $(a + b) + c$ and $a + (b + c)$ describe the same set for any three whole numbers. Hence $(a + b) + c = a + (b + c)$. We say that addition is associative, just as union of sets is associative.

The commutative and associative properties are perhaps even more obvious from the joining-of-segments attitude. The joined segments looked at from P show $a + b$ (reading from left to right). Looked at from Q, they show $b + a$. Since they form the same long segment, $a + b = b + a$.

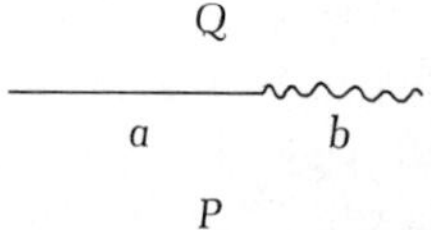

We may think of the three segments below as being joined first at A and then at B, or first at B and then at A.

x y z

A B

In the first case, the long segment shows $(x + y) + z$; in the second, $x + (y + z)$. Since there is only one long segment with only one length, $(x + y) + z = x + (y + z)$.

It is quite difficult to show the commutative and associative (CA) properties from continued counting. While specific examples check out, there appears to be no obvious structure about the process that will convince us that changing the order of the addends or the order of the associations leaves the outputs unchanged.

The CA properties help shortcut computations in arithmetic and are a basis for algebra. Here are some valuable uses:

1. Checking. Change the order in adding.

2. Reorder for personal ease. You may find 7 + 69 hard to do mentally, but 69 + 7 easy.

3. Which is easier, 197 + 53 or 97 + 3 + 100 + 50?

The origin of the word *commutative* is the same as that for *commute,* to go back and forth. This suggests the structure $a + b = b + a$, where the addition is done in two directions, so to speak. Similarly, in the associative property $a + (b + c) = (a + b) + c$, it makes no difference whether b associates with a first, or with c. These ideas help some people keep the names and the properties straight.

Not all operations are commutative or associative. You may recall that division is not commutative; 5/3 is not the same as 3/5. In putting on shoes and stockings, the order makes a great difference. Similarly, subtraction is not associative, $9 - (6 - 2) \neq (9 - 6) - 2$.

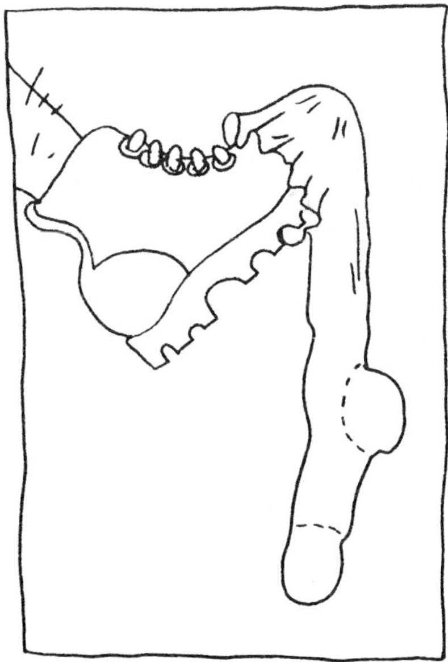

8.5 ADDITION ALGORITHMS: THE DANCE OF THE DIGITS An algorithm for addition is a process whereby, given the names of two numbers, we get the name of their sum. The attitudes themselves indicate algorithms that we call "natural algorithms." Here are details for finding the standard name for 38 + 54.

1. We suppose we have a supply of disjoint sets labeled with the number of members. We find a set that matches the union of sets with 38 and 54 members. The number on the label is the sum.

2. We suppose we have a supply of segments labeled with their lengths. We find a segment that matches a segment formed by joining segments labeled 38 and 54. The number on the label is the sum.

3. We have a sequence of counting numbers: 1, 2, 3, . . . , 101, 102, . . . Starting from 38, we count 54 more: 1 (39), 2 (40), ...

Strictly speaking, counting and measuring as in (**3**) are not directly involved in (**1**) and (**2**), although these are most often the ways we determine the number of members in a set or the length of a segment. Note too that these three algorithms do not depend on the numeral system. Any scheme for naming the numbers could be used in all three.

4. In actual practice with paper and pencil, people seldom use any of these methods. What most people actually say and do is something like the following:

Write 38. Write 54 under it, with the 5 under the 3 and the 4 under the 8. Draw a line. 4 and 8 are 12. Write a 2 below the line under the 4 and a 1 over the 3. 1 plus 3 plus 5 is 9. Write a 9 below the line and under the 5 and next to the 2. The sum is 92.

It is clear that process (**4**) is just a list of steps done with digits. A dance is a list of steps. Hence, this process is sometimes called the addition dance of the digits. More generally it is called the addition algorithm.

Natural questions about (**4**) are (*a*) just what are the steps in the dance that apply to any two addends? and (*b*) can the dance be relied on to give the same answer as processes (**1**), (**2**), (**3**)? and (*c*) can the dance be extended to three or more addends? We recall that addends are the numbers being added. In $3 + 4 + 5 = 12$, 3, 4, 5 are addends.

We shall answer these questions by relating the dance to two ideas: (1) forming the union of sets, the process used to define addition, and (2) the numeration system used. We consider several cases typical of different situations.

1. $3 + 5 = 8$. A set of 3 members is joined with a set of 5 members. The resulting union is counted. The name is 8. This is something to be memorized and is called an *addition fact.* Even though 3 looks like the right half of 8, there is no relation between the names.

2. $5 + 7 = 12$. 5, 7, 12 have verbal names that are independent. This addition fact must be memorized as in case 1. However the numeral 12 consists of two digits (1, 2) and stands for one ten, two ones. We think of the union of 5 and 7 as being more than 10. Ten of the members are separated out, leaving two. Graphically,

$$\{\text{xxxxx}\} \cup \{\text{xxxxxxx}\} = \left\{\begin{matrix}\text{xxxxx} \\ \text{xxxxxxx}\end{matrix}\right\} = \left\{\boxed{\begin{matrix}\text{xxxxx} \\ \text{xxxxx}\end{matrix}}\,\text{xx}\right\}$$

A circle is drawn around the ten to show the relation.

We may think of this as done in a base-ten dish abacus. Five beans and then seven beans are put in dish 0. Next, 10 beans are removed and one bean put in dish 1:

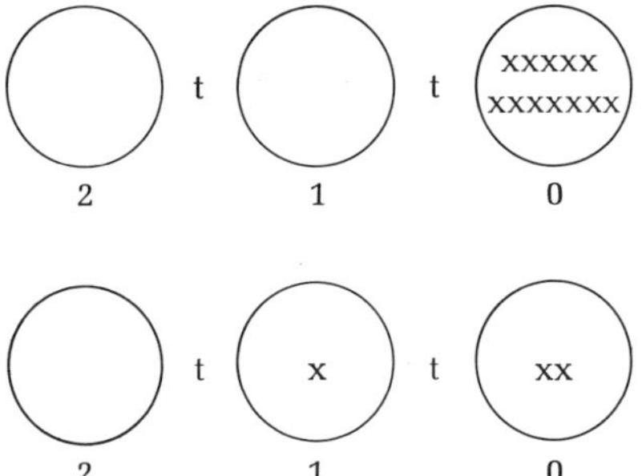

The so-called "addition facts" are the 100 sums where each of the two addends are single digits. These sums are learned individually.

3. 12 + 5 = 17.

$$\{\text{xxxxxxxxxxxx}\} \cup \{\text{xxxxx}\} = \left\{\boxed{\begin{matrix}\text{xxxxx} \\ \text{xxxxx}\end{matrix}}\,\text{xxxxxxx}\right\}$$

A circle has been drawn around 10 members in the union, leaving 7. This shows that there are 10 and 7 members in the union. Another interesting diagram is

$$\{\boxed{\text{xxxxxxxxxx}}\,\text{xx}\} \cup \{\text{xxxxx}\} = \left\{\begin{matrix}\boxed{\text{xxxxxxxxxx}}\,\text{xx} \\ \text{xxxxx}\end{matrix}\right\}$$

This diagram suggests that the 10 in the union is the same subset of 10 as in the 10 and 2, and that the 7 is the union of the 2 with the 5. It doesn't have to be done this way, but why not? This leads to the layout

$$\begin{array}{r} 12 \\ \underline{5} \\ 17 \end{array}$$

where the 5 and 2 are added, and the 1 brought down. This is the common way of thinking about the algorithm.

We think of two base-ten dish abacuses, the first with 1 in dish 1 and 2 in dish 0. The second has 5 in dish 0. The beans are combined in a third abacus, which then has 1 in dish 1 and 7 in dish 0. Earlier we said that the associative law was important in the algorithm. We now show how this addition can be based on expanded notation, the associative property, and the addition facts. That is, without going *all* the way back to sets, $12 + 5 = (t + 2) + 5 = t + (2 + 5) = t + 7 = 17$. ($t$ stands for ten.)

We have used expanded notation, the associative property, a basic fact, and expanded notation backward. Can you see how one takes these four steps in using the layout?

$$\begin{array}{r} 12 \\ \underline{5} \\ 17 \end{array}$$

4. $16 + 8 = 24$.

{XXXXXXXXXXXXXXXX} ∪ {XXXXXXXX} = {[XXXXX XXXXX] [XXXXXXX XXX] XXXX}

This time we have two sets of 10, with 4 left over. This diagram suggests that a set of 10 in the union may be any 10 members. This is true. But this diagram does not suggest the algorithm. Think of two dish abacuses:

( )	t	(1)	t	(6)	and	( )	t	( )	t	(8)
2		1		0		2		1		0

Putting the beans into one abacus, we get

( )	t	(1)	t	(14)
2		1		0

Taking 10 beans from dish 0 and putting 1 bean in dish 1 yields

( )	t	(2)	t	(4)
2		1		0

We show how this dance may be explained by the use of expanded notation and the associative property

$$16 + 8 = [t + 6] + 8 = t + [6 + 8] = t + [14]$$
$$= t + [t + 4] = [t + t] + 4 = 2t + 4 = 24$$

Here is the justification for each step:

(1) Statement of problem
(2) Expanded notation
(3) Associative
(4) Basic fact
(5) Expanded notation
(6) Associative
(7) Basic fact
(8) Expanded notation

5. 31 + 42 = 73.

$$\{- - - x\} \cup \{- - - - xx\} = \{- - - x \; - - - - xx\}$$

In this diagram we have drawn a dash to indicate a subset of ten members. In the union there are 7 dashes (or subsets of ten) and 3 ones. The algorithm reflects this process exactly. We say "2 plus 1 is 3, and 3 plus 4 is 7." In the algorithm we leave out words.

$$\begin{array}{r} 31 \\ \underline{42} \\ 73 \end{array}$$

We don't say "3 tens plus 4 tens is 7 tens." We allow the place to show the value of tens.

It is interesting to compare this with Roman numerals:

XXXI + XXXXII = XXXXXXXIII = LXXIII

Here we add merely by putting the numerals together according to size. The Romans then regroup 5 of the X's and rename them L. We note that in Roman numerals the dance of the digits is the same as forming a set union, with some regrouping and renaming.

Using abacuses, we have

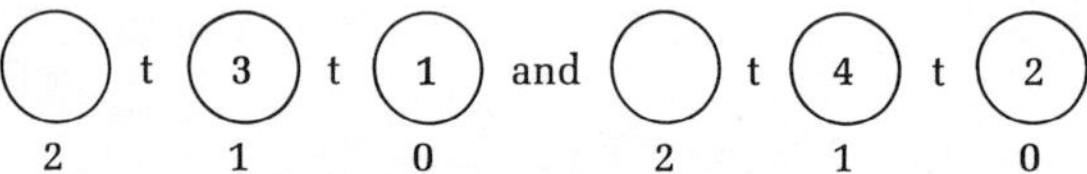

Putting the 3 beans and 4 beans from dishes 1 into a new dish 1, and similarly for dishes 0, we get

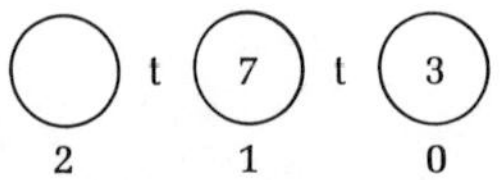

Again we look at the dance based on the algebraic properties of associativity and commutativity. We call these properties algebraic because they also play a powerful role in algebra.

Step	**Reason**
$31 + 42$	Statement of problem
$(3t + 1) + 42 =$	Expanded notation
$(3t + 1) + [4t + 2] =$	Expanded notation
$3t + (1 + [4t + 2]) =$	Associative property of addition
$3t + ([1 + 4t] + 2) =$	Associative property of addition
$3t + ([4t + 1] + 2) =$	Commutative property of addition
$3t + (4t + [1 + 2]) =$	Associative property of addition
$3t + (4t + 3)$	Basic addition fact

Using the commutative, associative, and distributive (CAD) properties makes a fairly long argument. The process involves changing from a place-value to a non-place-value notation, and the use of basic facts as well as the CAD properties. An equality such as $3 + 4 = 7$ or $42 = 4t + 2$ can be read from left to right or from right to left. We use the same name either way. That is, whether we replace 42 by $4t + 2$ or $4t + 2$ by 42, we give expanded notation as the reason. Some people may prefer to call $7t + 3$ replaced by 73 "contracted notation." Continuing:

$(3t + 4t) + 3$	Associative property of addition
$(3 + 4)t + 3$	Distributive property
$7t + 3$	Basic addition fact
73	Expanded notation

Here we used the associative property and the distributive property. The distributive property says $(3t + 4t) = (3 + 4)t$, interchanging the order of multiplying and adding. It is too bad we need to use multiplication before we discuss it at length. Some people like to justify $3t + 4t = 7t$ as "30 plus 40 is 70." This is true. However, in the dance of the digits one usually says "3 plus 4 is 7," so it seems good to show how this exact step appears.

6. $37 + 45 = 82$. Using sets,

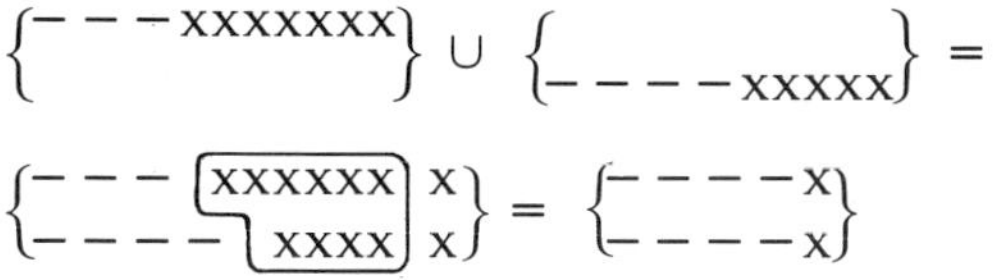

$$\{- - -\text{xxxxxxx}\} \cup \{- - - -\text{xxxxx}\} =$$

$$\left\{\begin{matrix} - - - \ \boxed{\text{xxxxxx}}\ \text{x} \\ - - - - \ \boxed{\text{xxxx}}\ \text{x} \end{matrix}\right\} = \left\{\begin{matrix} - - - -\text{x} \\ - - - -\text{x} \end{matrix}\right\}$$

Again each dash represents ten units. Any ten of the units could have been circled, and the dash placed anywhere.

Using dish abacuses we get $(0, 3, 7)_{\text{ten}} + (0, 4, 5)_{\text{ten}} = (0, 7, 12)_{\text{ten}}$. Ten of the beans in dish 0 are removed and one is placed in dish 1. This yields $(0, 8, 2)_{\text{ten}}$.

This leads to the algorithm in the margin. We say, "7 and 5 is 12. 3 and 4 is 7. 1 and 7 is 8," putting down the symbols in the layout as shown. Sometimes this is shortened to the layout with a crutch number. We say "7 and 5 is 12. Put down 2, put a crutch number, a small 1 above the 3. 4 and 3 is 7 and 1 is 8. Put down 8." We might also say that the 2 goes in the units place under the 5 and 7.

```
 37
 45
 --
 12
 7
 --
 82
```

```
 1
  37
  45
  --
  82
```

Often the 1 is not written but merely remembered. We might add "1 and 3 is 4, 4 and 4 is 8" rather than "3 and 4 is 7, 1 and 7 is 8."

7. 39 + 56 + 78 = 173. We recall the algorithm.

$$\begin{array}{r} 39 \\ 56 \\ \underline{78} \\ 173 \end{array}$$

(*a*) Write the numbers in a column, with the units digits lined up, and the tens also, forming two columns. Draw a bar under them.
(*b*) Add the units digits, getting 23. Put down 3 in the units column. Remember the 2.
(*c*) 2 and 3 and 5 and 7 is 17. Put 17 next to the 3. A slightly lengthier layout is

$$\begin{array}{r} 39 \\ 56 \\ \underline{78} \\ 23 \\ \underline{150} \\ 173 \end{array}$$

Here the sum of the units digits is written, then the sum of the tens. The algorithms are basically the same. We discuss this problem in four ways —using sets, using the abacus, using the algebraic CAD properties, and using Roman numerals.

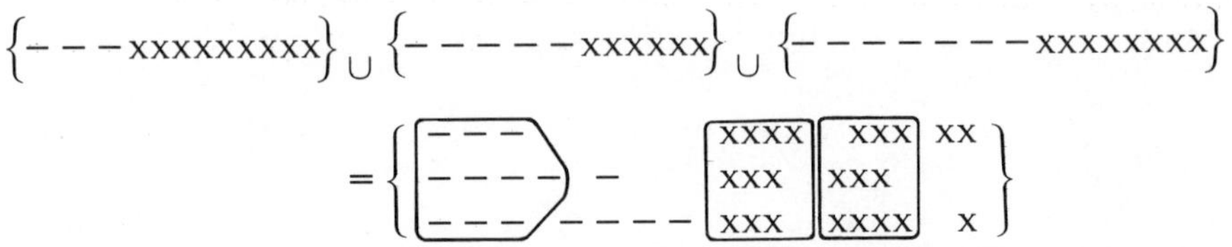

{– – –xxxxxxxxx} ∪ {– – – – –xxxxxx} ∪ {– – – – – – –xxxxxxxx}

= { [– – – / – – – – / – – –] – – – – – [xxxx / xxx / xxx] [xxx / xxx / xxxx] xx x }

We see that two subsets of 10 each can be formed from the units, with 3 left over. A subset of 10 tens, or 100, can be formed from the tens, leaving 5 tens and the 2 tens formed from the units. Hence, the set is partitioned (separated) into one set of a hundred, 7 sets of ten, and 3 units. Hence, the sum is 173.

Three dish abacuses are used: $(0, 3, 9)_{ten}$ $(0, 5, 6)_{ten}$ and $(0, 7, 8)_{ten}$. Putting the beans all in one abacus gives $(0, 15, 23)_{ten}$. Taking ten beans from dish 0 and putting one bean in dish 1 gives $(0, 16, 13)_{ten}$. Repeating this gives $(0, 17, 3)_{ten}$. Taking ten beans from dish 1 and putting one bean in dish 2 gives $(1, 7, 3)_{ten}$.

Adding these same numbers in Roman numerals, we get

39 + 56 + 78 =	Statement
XXXVIIII + LVI + LXXVIII =	Translate to Roman
LLXXXXXVVVIIIIIII =	Form the union of the symbols, and list according to size
LLXXXXXVVVVIII =	Regroup and rename
LLXXXXXXXIII =	Regroup and rename
LLLXXIII =	Regroup and rename
CLXXIII	Regroup and rename

Recall that we do not use the subtractive principle.

We see that the process in Roman numerals combines the set operations and the digital renaming. "List according to size" is just a way of using the associative and commutative properties. Regrouping and renaming are basic facts.

Do we say "5 and 5 is 10" or "5 and 5 are 10?" If we are referring to numbers we use *is*, since there is only one number ten. If we are referring to counters, members of sets, we use *are*. *Is* is the more advanced phrase, but people use both words.

Addition with more addends and more digits is a straightforward generalization of these steps. We conclude with one longer example.

8.

$$\begin{array}{r} 3692 \\ 4738 \\ \underline{620} \\ 9050 \end{array} \quad \text{or} \quad \begin{array}{r} 3692 \\ 4738 \\ \underline{620} \\ 10 \\ 140 \\ 1900 \\ \underline{7000} \\ 1050 \\ \underline{8000} \\ 9050 \end{array}$$

We shall not write down explicitly what a person would say to himself as he writes one or the other of these layouts (or perhaps a different one).

Using three dish abacuses, we have $(0, 3, 6, 9, 2)_{ten}$, $(0, 4, 7, 3, 8)_{ten}$ and $(0, 0, 6, 2, 0)_{ten}$. Putting the beans in one abacus gives $(0, 7, 19, 14, 10)_{ten}$. Converting 10 in one dish to 1 in the next gives $(0, 7, 19, 15, 0)_{ten}$, $(0, 7, 20, 5, 0)_{ten}$, $(0, 9, 0, 5, 0)_{ten}$.

Instead of analyzing by means of sets, we use Roman numerals, since the operations are so close.

MMMDCLXXXXII + MMMMDCCXXXVIII + DCXX =
MMMMMMMDDDCCCCLXXXXXXXXXVIIIII =
MMMMMMMMDCCCCLLXXXXX =
MMMMMMMMML

8.6 ADDITION IN DIFFERENT BASE SYSTEMS

No one expects to become expert in computing in other bases and with Roman numerals. They are included here to help us understand the process of addition and, by contrast, the algorithms for base-ten addition. It is amazing how much we learn about something we know thoroughly if we look at it in an entirely new way.

The dance of the digits for addition in base five and other bases is similar to that in base ten. We illustrate and explain using the CAD properties, sets, and abacuses (f stands for five).

$$
\begin{aligned}
&34_{five} + 42_{five} = 131_{five}\\
&34 + 42 =\\
&(3f + 4) + (4f + 2) =\\
&(3f + 4f) + (4 + 2) =\\
&(7f) + 6 =\\
&(5f + 2f) + (f + 1) =\\
&(f^2 + 2f) + (f + 1) =\\
&f^2 + 3f + 1 =\\
&131
\end{aligned}
$$

Using sets, the union of the two sets is

{(xxxxx)(xxxxx) (xxxxx) x xxx
(xxxxx)(xxxxx)(xxxxx) (xxxxx) xx}

Can you draw the individual sets?

We circle a set of 5 fives, and a set of 5 ones. There is one set of 5 fives, three sets of 5 ones, and one single. The numeral is 131_{five}.

We use base-five abacuses. The beans in $(0, 3, 4)_{five}$ and $(0, 4, 2)_{five}$ are put in one abacus, giving $(0, 7, 6)_{five}$. Converting by taking 5 beans from one dish and putting one bean in the dish on the left gives $(0, 8, 1)_{five}$ and $(1, 3, 1)_{five}$.

It is not necessary to go back to sets to add in base five, just as it is not necessary for adding in base ten. We develop a table of "addition facts." Addition facts are the set of additions of one-digit numbers. Then place value, and adding digits with the same place value, allow addition of numbers of more than one digit.

Here is a table of base-five addition facts with numbers in base five:

+	0	1	2	3	4
0	0	1	2	3	4
1	1	2	3	4	10
2	2	3	4	10	11
3	3	4	10	11	12
4	4	10	11	12	13

A layout for the example is

$$\begin{array}{r} 34 \\ +\ 42 \\ \hline \end{array}$$

Adding the ones, 4 + 2 = 11, as can be seen from the table or directly. After writing this down, the layout is

$$\begin{array}{r} 34 \\ \underline{42} \\ 11 \end{array}$$

Adding the fives in the second places, 3 + 4 = 12. Writing either with or without 0 in the ones place, the layout is

$$\begin{array}{r} 34 \\ \underline{42} \\ 11 \\ 12(0) \end{array}$$

The addition problem has been changed to adding 11 and 120. This addition is easier. To simplify, we write the sum of the units and the sum of the fives in different rows. The layout is now

$$\begin{array}{r} 34 \\ 42 \\ \hline 11 \\ 120 \\ \hline 131 \end{array}$$

with the sum 131_{five} appearing at the bottom.

This layout can be shortened. After adding the ones, 1 is placed in the ones place and the five is represented by a 1 placed in the fives column above the 3, as shown in this layout:

$$\begin{array}{r} {}^{1}34 \\ \underline{42} \\ 1 \end{array}$$

Now the five from the sum of the ones is added with the fives in the given addends (1 + 3 + 4) to produce the layout

$$\begin{array}{r} {}^{1}34 \\ \underline{42} \\ 131 \end{array}$$

1 + 3 + 4 = 13 by using the table twice, by using sets, or some other convenient way.

Experienced computers remember the one five from the ones sum rather than writing it in the layout. Because writing them is a help when learning, such numbers are called *crutch* numbers. Three possible layouts for the next addition example in base five are given in the margin. The last layout is perhaps the most elegant, but each is satisfactory.

$$\begin{array}{r} 34 \\ 20 \\ 41 \\ \underline{32} \\ 12 \\ \underline{220} \\ 232 \end{array} \qquad \begin{array}{r} {}^{1}34 \\ 20 \\ 41 \\ \underline{32} \\ 232 \end{array} \qquad \begin{array}{r} 34 \\ 20 \\ 41 \\ \underline{32} \\ 232 \end{array}$$

Numbers with more than two digits can be added. For example:

$$\begin{array}{r} 3243 \\ \underline{4323} \\ 11 \\ 110 \\ 1000 \\ \underline{12000} \\ 13121 \end{array} \qquad \begin{array}{r} 111 \\ 3243 \\ \underline{4323} \\ 13121 \end{array} \qquad \begin{array}{r} 3243 \\ \underline{4323} \\ 13121 \end{array}$$

Any number of numbers of any size can be added. Various layouts similar to these may be used.

Dish abacuses are useful in explaining the algorithms:

$$(0, 3, 2, 4, 3)_{\text{five}} + (0, 4, 3, 2, 3)_{\text{five}} = (0, 7, 5, 6, 6)_{\text{five}}$$

Reducing the number in each dish,

$$(0, 7, 5, 6, 6)_{\text{five}} = (0, 7, 5, 7, 1)_{\text{five}} = (0, 7, 6, 2, 1)_{\text{five}} = (0, 8, 1, 2, 1)_{\text{five}}$$
$$= (1, 3, 1, 2, 1)$$

8.7 CHECKING ADDITION Nobody checks a computation if he doesn't care whether the answer is correct. On the other hand, if the correctness of the result is important, people check quite carefully. Quite commonly people will spend as much time checking work as they do in performing the task originally. Sometimes they spend much more. Hence any discussion of arithmetic operations should include discussions of checking.

Every checking operation is basically doing the problem over again. Sometimes it is done more roughly, sometimes more fully in the check. It may be done more than once.

1. The most obvious check is to look at the answer and see if it seems right. An answer of 15 feet for the height of a girl is wrong. You know that no matter how you do the problem, you will not get that answer. Still, "seems right" is a process that comes mainly with experience. Clearly, an answer may seem right and still be wrong.

2. Do the problem over again, just as it was done the first time. This is all one can do, if only one way is known. If two different answers are obtained, the mathematician looks for an error. He does not flip a coin. The major difficulty with this check is that a person often repeats the same error when he does a problem the second time. There is some kind of subconscious screw temporarily loose that causes the mistake to be repeated. Hence, the mathematician varies the process as much as he conveniently can, and does something else between the original performance and the check.

Here are some common ways to vary the process of addition:

$$\begin{array}{r} 37 \\ 42 \\ +\ 85 \\ \hline 164 \end{array}$$

3. Add *up* the first time, but *down* for the check. Suppose one says originally, "5 and 2 is 7 and 7 is 14. 1 and 8 is 9 and 4 is 13 and 3 is 16." Then, in checking, one says, "7 and 2 is 9 and 5 is 14. 1 and 3 is 4 and 4 is 8 and 8 is 16."

4. Associate or arrange the numbers differently. These layouts show this process:

$$\begin{array}{r} 37 \\ \underline{42} \\ 79 \\ \underline{85} \\ 164 \end{array} \qquad \begin{array}{r} 42 \\ \underline{85} \\ 127 \\ \underline{37} \\ 164 \end{array} \qquad \begin{array}{r} 85 \\ 37 \\ \underline{42} \\ 164 \end{array}$$

5. Subtract one addend from the sum and compare with the sum of the remaining addends. For the example above:

$$\begin{array}{r} 164 \\ \underline{-\ \ 85} \\ 79 \end{array} \qquad \begin{array}{r} 37 \\ \underline{+\ 42} \\ 79 \end{array}$$

6. Change the numerals to a different base, add in this base, and convert back to the original base.

Often in schools only one or two of these possible checks are mentioned; (1) and (3) are perhaps most common. However, in cases where errors are very expensive, all these and others too may be done.

8.8 FRONT-END ADDITION The above algorithms all begin with the digits place at the right end of the numeral. The most important part of the numeral is at the left. Consider the following example:

$$\begin{array}{r} 368{,}435 \\ \underline{+\ 471{,}692} \end{array}$$

It may be that the computer is interested only in the first two or three digits in the sum. The algorithms so far discussed require him to start at the right, performing additions he is not interested in.

Consider two base-ten abacuses. If we combine the beans in $(0, 3, 6, 8, 4, 3, 5)_{ten}$ and $(0, 4, 7, 1, 6, 9, 2)_{ten}$ in the appropriate dishes, we get $(0, 7, 13, 9, 10, 12, 7)_{ten}$. Now we start converting. But there is no reason we must begin at the right. Actually, we can start at any dish. In particular, we can start at the left, or front, end. We get

$$\begin{array}{ll} (0, 8, 3, 9, 10, 12, 7)_{ten} & \text{then} \\ (0, 8, 3, 10, 0, 12, 7)_{ten} & \text{then} \\ (0, 8, 4, 0, 0, 12, 7)_{ten} & \text{then} \\ (0, 8, 4, 0, 1, 2, 7)_{ten} & \end{array}$$

One of the advantages of the dish abacus is that it clearly separates the uniting aspect of addition from the converting aspect. In the algorithm these two are mixed together. We imagine that we are combining front-end dishes first, converting as we go. Suppose we are interested only in the largest values. We might stop after two or three steps.

$$\begin{array}{l} 368435 \\ \underline{471692} \\ 7 \\ 13 \\ \underline{\ \ 9\ \ \ \ } \\ 839xxx \end{array}$$

Adding the partial sums so far, we get 839 thousand. Adding the terms in the smaller-valued places cannot change the thousands digit by more than 1. Hence, the sum will be 839 or 840 thousand something. The worker has an excellent idea about the sum with only half the work needed for an exact answer.

It is not necessary to write the partial sums on separate rows for a final adding. Using scratches one can add as one goes along. The layout might be

$$\begin{array}{l} 368435 \\ \underline{471692} \\ \not{7}39 \\ 8 \end{array}$$

The complete addition can be done from the left, as shown:

368435
471692
739027
8401

Because of the scratching that is necessary, front-end addition is not recommended if the complete sum, 840127, is desired. However, it does make an excellent check. We illustrate with a longer addition, using crutch numbers and then scratches.

121	
4832	4832
6789	6789
5096	5096
16717	15507
	671

EXERCISES

In exploring addition from joining sets, joining segments, and continued counting attitudes, a collection of beans or other counters such as stones, chips, or paper clips is needed. A collection of colored rods or sticks is also needed. Without them diagrams may be drawn, but this is neither as good nor as efficient as using the actual counters.

1. Using beans, show 2 + 3. Did you count or recognize a pattern to see the standard name for 2 + 3?

2. Use beans to show 8 + 6 as in exercise 1.

3. Arrange the beans in exercise 2 in a pattern that will show at a glance without moving any beans that 8 + 6 = 10 + 4.

4. Arrange beans to show at a glance that 4 + 9 = 10 + 3.

5. Choose rods of two different colors, say red and yellow. If possible, find a single rod so that you can say "Red + yellow = ______." Did you do any counting? If numbers are matched with the rods, write the corresponding number sentences.

6. Is it necessary to count to be able to add? If not, show how to find that $4 + 6 = 10$ without counting.

7. Find rods associated with 8, 6, 10, and 4. Arrange the rods to show that $8 + 6 = 10 + 4$. Did you use counting or sets to see that $8 + 6 = 10 + 4$?

8. Using continued counting along a number sequence, find $8 + 6$. Did you use counting, pattern recognition, or label reading to find the standard name for $8 + 6$?

9. Arrange sets to show at a glance, without counting, that $5 + 12 = 12 + 5$.

10. Arrange segments to show that $8 + 7 = 7 + 8$.

11. Tell how to arrange two segments to show that $763 + 891 = 891 + 763$ without moving the segments or finding the standard name for $763 + 891$.

12. Is it obvious, using continued counting and without finding the standard name, that

$$763 + 891 = 891 + 763$$

13. Arrange beans in a pattern to show at a glance that

$$(5 + 3) + 7 = 5 + (3 + 7)$$

14. Compute $(25 + 17) + 83$ and $25 + (17 + 83)$. Which is easier?

15. Arrange rods to show at a glance that

$$(5 + 3) + 7 = 5 + (7 + 3)$$

16. Without finding the sum, is it obvious that $13 + 8 = 6 + 15$?

17. Find three numbers such that $13 + 8$ and $6 + 15$ are the intermediate steps in testing the associative property of addition.

18. Find three numbers such that $9 + 30$ and $22 + 17$ are intermediate steps in testing the associative property of addition.

19. Describe the natural algorithms for finding 407 + 362 from (*a*) the set, (*b*) segment, and (*c*) continued-counting attitudes.

20. Roberta makes the layout at left for computing 78 + 94. A motion picture is made as she shows this process with three dish abacuses. Five successive frames show

$$\begin{array}{r}78\\+\ 94\\\hline 12\\16\\\hline 172\end{array}$$

(0, 7, 8)	(0, 7, 0)	(0, 0, 0)	(0, 0, 0)	(0, 0, 0)
(0, 9, 4)	(0, 9, 0)	(0, 0, 0)	(0, 0, 0)	(0, 0, 0)
(0, 0, 0)	(0, 0, 12)	(0, 16, 12)	(0, 17, 2)	(1, 7, 2)

Using a layout similar to Roberta's, compute

$$\begin{array}{r}47\\+\ 89\\\hline\end{array}\qquad\begin{array}{r}66\\+\ 78\\\hline\end{array}\qquad\begin{array}{r}487\\+\ 695\\\hline\end{array}$$

21. Set up abacuses and move the beans to illustrate the three addition exercises in exercise 20.

22. Successive frames of a motion picture of three dish abacuses Arnold uses in adding $21_{\text{three}} + 22_{\text{three}}$ show

$(0, 2, 1)_3$	$(0, 2, 0)_3$	$(0, 0, 0)_3$	$(0, 0, 0)_3$	$(0, 0, 0)_3$
$(0, 2, 2)_3$	$(0, 2, 0)_3$	$(0, 0, 0)_3$	$(0, 0, 0)_3$	$(0, 0, 0)_3$
$(0, 0, 0)_3$	$(0, 0, 3)_3$	$(0, 4, 3)_3$	$(0, 5, 0)_3$	$(1, 2, 0)_3$

Set up three abacuses and manipulate the beans as Arnold did.

23. Set up an addition layout with scratches which reflects the method Arnold used in addition.

24. Set up abacuses as in exercise 22 and manipulate beans as Betty did in the frames shown below. Make an addition layout in base-3 numerals for Betty.

$(0, 2, 1)_3$	$(0, 2, 0)_3$	$(0, 3, 0)_3$	$(0, 0, 0)_3$	$(0, 0, 0)_3$
$(0, 2, 2)_3$	$(0, 2, 0)_3$	$(0, 2, 0)_3$	$(0, 0, 0)_3$	$(0, 0, 0)_3$
$(0, 0, 0)_3$	$(0, 0, 3)_3$	$(0, 0, 0)_3$	$(0, 5, 0)_3$	$(1, 2, 0)_3$

25. Charlotte followed even another series of steps with the abacuses to work the problem in exercise 22. Find another series of steps, which might have been Charlotte's.

26. Dorothy set up five abacuses as in the margin. As she moved the beans in the steps while adding, the fifth, or sum, abacus successively showed the following:

(*a*) $(0, 0, 0, 0, 0)_3$
(*b*) $(0, 2, 0, 0, 0)_3$
(*c*) $(0, 2, 5, 0, 0)_3$
(*d*) $(0, 3, 2, 0, 0)_3$
(*e*) $(1, 0, 2, 0, 0)_3$
(*f*) $(1, 0, 2, 5, 0)_3$
(*g*) $(1, 0, 3, 2, 0)_3$
(*h*) $(1, 1, 0, 2, 0)_3$
(*i*) $(1, 1, 0, 2, 7)_3$
(*j*) $(1, 1, 0, 4, 1)_3$
(*k*) $(1, 1, 1, 1, 1)_3$

$(0, 0, 2, 2, 1)_3$
$(0, 0, 1, 2, 2)_3$
$(0, 2, 0, 0, 2)_3$
$(0, 0, 2, 1, 2)_3$
$(0, 0, 0, 0, 0)_3$

Set up five abacuses and move the beans through the eleven steps she took.

27. Set up an addition layout with scratches showing the addition steps Dorothy took with the abacuses.

28. A layout with scratches for front-end addition in base 5 is shown in the margin. Set up three dish abacuses and move the beans corresponding to this layout.

$$\begin{array}{r} 342_{\text{five}} \\ \underline{443_{\text{five}}} \\ 1\not{2}\not{3}0 \\ \underline{31} \\ 1310_{\text{five}} \end{array}$$

29. Using front-end addition in base 10, make layouts with scratches for the sums

$$\begin{array}{r} 36 \\ \underline{+\ 87} \end{array} \qquad \begin{array}{r} 694 \\ \underline{+\ 848} \end{array} \qquad \begin{array}{r} 69 \\ 78 \\ \underline{+\ 86} \end{array}$$

30. Use front-end addition to find the first three digits in the following sum. Place x's for the other digits in the sum.

$$\begin{array}{r} 7{,}658{,}965{,}438{,}924 \\ 6{,}983{,}708{,}290{,}763 \\ \underline{+\ 9{,}787{,}643{,}098{,}760} \end{array}$$

31. (*a*) Without converting to base ten, add CCCLXXVI + CLXVIII.
(*b*) Convert and add these numbers in base-ten numerals.
(*c*) In each layout, identify (circle) $6 + 7 = 13$.
(*d*) For each layout, what is done with the ten in the thirteen arising in step *c*?

Answers to selected questions:

1. o o
o o o

3. oooooooo
oo oooo

6. No. Use rods of known length; **10.** Look at $\underline{.\ 8\ .\ 7\ .}$ from left and right; **12.** no **14.** 125, 125, the second; **17.** 6 + 7 + 8; **19.** (*a*) Count the union of two sets of 407 and 362 members; (*b*) Find the name of the segment as long as a 407 segment placed end to end with a 362 segment; (*c*) Read the label in a number sequence after counting 362 starting after 407.

20.

47	66	487
89	78	695
16	14	12
12	13	17
136	144	10
		1182

23.

21	13	
22	021	
3	022	
40	3	
50	50	
120	120	Other versions also satisfactory.

27.

221		221	
122		122	
2002		2002	
212		212	
2		12557	
5		3221	
32	or	034	or . . .
102		101	
5		1	
32		11111	
10			
47			
111			
11111			Fewer scratches all right.

29.

36	694	69
+ 87	+ 848	78
113	1432	+ 86
2	54	213
		3

32. In polynomial notation, add $(3t^2 + 7t + 6) + (t^2 + 6t + 8)$. Do c and d as in exercise 31.

678
794
+ 579

33. Set up four dish abacuses and move the beans matching the steps you take to add the three numbers in the margin using back-end addition.

34. Make up and work a problem showing back-end addition with dish abacuses.

35. Make up and work a problem showing the steps in front-end addition with dish abacuses.

36. Professor B. Gordon has invented the following game for two players. Three pieces are placed on the number line on three counting numbers. At each turn a player chooses any one of the pieces he wishes and moves it to the left at least one number, but not past another piece or past zero. Play continues until the three pieces are on 1, 2, 3 and the next player cannot move. The player making the last move wins. Make a number line, and play the game with your neighbor, starting with pieces on 5, 7, 11.

37. A winning position in the game in exercise 35 is a set of three numbers such that if you leave the pieces on those three numbers at the end of your move you know you will be able to win regardless of the moves your opponent makes. A trivial winning position is (1, 2, 3). Is (1, 3, 4)?

38. Find five winning positions.

39. Do you see any pattern in the winning positions?

---

31. (*a*) and (*c*)

CCC	LXX	VI
C	LX	VIII
CCCC	CXXX	XIIII

(*b*) and (*c*)

1	1	
3	7	6
1	6	8
5	4	4

(*d*) The ten (tens) becomes one hundred.

38. (2, 3, 5), (1, 4, 5), (1, 7, 8), (3, 6, 9), (4, 7, 11).

9

Subtraction

9.1 INTRODUCTION Several physical situations with corresponding human attitudes lead to subtraction. Each of these suggests a natural algorithm. Several algorithms using the properties of Hindu-Arabic and Roman numerals are developed.

The two widely used algorithms, called here "take away–convert" (or standard) and "missing addend–equal additions" (or Austrian) are discussed and compared.

We shall not take sides as to whether take-away or missing addend is the "better" attitude. Perhaps you will have an opinion after reading the chapter. Did you know you could subtract from the left? You'll see how in front-end subtraction.

9.2 SUBTRACTION ATTITUDES While subtraction arises in many places, we list five that are somewhat different. Different people may have different attitudes toward the same situation, so one should not argue too strongly for one attitude as the "correct" one. We name "take away," "missing addend, or need," "comparison," "inverse addition," and "counting backward" attitudes. Here are some examples.

1. *Take away.* John has 9 pencils. He takes away 3. How many are left?

The same idea appears in other forms. Helen goes to the store with $2.00. She pays 50 cents for apples. How much money remains?

A freeway eliminates 50 trees from a farmer's orchard of 300 trees. How many remain to raise fruit?

Three crows are sitting on a fence. A hunter knocks one off with an arrow. How many are left? "Two" on this trick question brings

the rejoinder "You may know arithmetic, but you sure don't know crows."

Can we describe the general pattern of a take-away situation?

The take-away pattern in terms of sets arises when a superset whose number is known is separated into two sets. The number of one of these subsets is known. How many members are in the other? The fact of the separation of the superset is made vivid by taking away the subset that is counted. To see that this general pattern fits a particular case, one identifies the superset, the known subset (which is taken away), and the remaining subset.

In Helen's case, the superset is the 200 pennies she had at the start, and the known subset is the 50 pennies she separated and took away for the apples. The remaining set is the pennies she had left. Since she probably had the $2 in larger coins or currency, the set in this case may be a mental image of equivalent value to the actual coins in her purse.

The crow example shows that it is important to see that the general pattern or model really does fit the situation. Can you make the general pattern fit the other examples?

Very closely related to these situations is the following. Peter opens a five-pound bag of sugar and weighs out two pounds for some candy. How much sugar remains in the bag? Here the material being discussed comes in such tiny pieces we do not think of it as being a set of so many members. Instead we think of the sugar as being continuous. We compare with a two segment taken from a five segment.

2. *Missing addend* (*need*). Mrs. Roberts is giving a party for 8 people. She has place cards for 5. How many more does she need?

Peter has saved $4 toward a book costing $9. How much more does he need?

Bill has a pole 7 feet long. How much must he add to it to touch Jane, who can just barely be touched with a 10-foot pole?

Jasper has served 17 days of a 30-day jail sentence. How much longer will he be in jail?

Mrs. Washington needs 7 books of colored stamps to get a set of dishes that sells for $10 in a local store. She has 5 books. How many more does she need?

Can you see a general pattern in these situations? Can you describe the pattern?

We describe the pattern in terms of sets. A superset is the goal. A subset is at hand. What subset must be joined with the subset to match the superset? It is perhaps easier to describe the pattern in terms of numbers and addition. What number must be added to a given number to make the goal?

Because we are looking for a number to add, this situation is called a missing-addend situation.

In the last example more data was given than was actually used in the description. This usually happens in real-life problems. One must sort relevant facts from irrelevant ones.

3. *Comparison.* John and Dan worked 23 and 27 problems, respectively, from a given set. How many more did Dan work than John?

Alfred is 43 inches tall while Marie is 48 inches tall. How much taller is Marie than Alfred?

Can you see the pattern? Can you describe it?

We note there is no take-away operation. There is no obvious separation of a set into two subsets. Nor is someone trying to add to what he already has, to meet a set goal. In the comparison cases, we have two sets that we wish to compare by saying how many more or how many fewer members there are in one set than in another.

4. *Inverse addition.* The clerk in a store overcharged Paula by $5. What must be done to correct the bill?

The balloon rose 50 feet to an elevation of 1320 feet. What was the original elevation?

Fully dressed in 6 pounds of clothes, Agnes weighed 120 pounds. How much did she weigh without clothes?

If 4 is added to a number by mistake, what is necessary to correct this error?

Can you describe the pattern here?

The pattern of subtraction as inverse addition arises where one wishes to undo the effect of a previous addition. We say that subtracting 5 is the inverse of adding 5, because these two operations cancel each other out. Two operations that undo each other are called inverse operations.

Warning The name *inverse addition* or *inverse of addition* may lead some people to think that addition, a binary operator, has an inverse. As we saw in chapter 7, only unary operators have inverses. These examples use unary operators made by fixing one of the addends in an addition operator.

5. *Counting backward.* Mary counts 15 backward, starting from 38: 1 (37), 2 (36), . . . Where does she end?

At what milepost does Nathan end if he starts at milepost 137 and travels 13 miles towards milepost 100?

The temperature dropped 8 degrees, starting from 54 degrees. What was the final temperature?

These examples are quite similar to the earlier ones. The difference between this attitude toward subtraction and the earlier ones is more in the process by which one arrives at a final number than in the type of situation.

As we consider the different situations where subtraction arises, the human search for consistency and understanding urges us to decide which situation we wish to regard as the fundamental, or defining, situation. We then describe the other situations in terms of this one.

We choose a defining situation as one that is intuitively easy to explain and understand, one that leads to the algorithms, or dance of digits, and one that relates or illumines other properties of numbers and other parts of mathematics.

Traditionally, the take-away attitude was considered most satisfactory. Subtraction is squarely based on sets, and the sets considered are all clearly in view of the student. The concept leads easily to the subtraction dance.

In recent years it has been popular to base subtraction on the missing-addend and inverse-addition ideas. This relates subtraction closely with addition. Children learn the addition and subtraction facts more quickly. This approach leads more quickly to algebra and other advanced ideas. It can lead to the algorithms, but perhaps not as vividly as the take-away or separation-of-sets approach.

Counting in reverse, while a useful teaching device, does not lead to the algorithms as easily as the above. If addition is defined as continued counting, then reverse counting is the natural basic definition for subtraction. By and large, this approach is used more as an exercise and as an alternative approach to subtraction than as a primary basis.

Comparison is considered an application of subtraction. The ideas do not lead to the algorithms. Comparison leads to the ideas of order, greater than, and less than, which do extend significantly to other parts of mathematics.

9.3 SOME TERMINOLOGY In a subtraction problem such as 17 – 8 = 9, the number in the position of 17 is called the *minuend,* or the number "to be diminished." The number in the position of 8 is called the *subtrahend,* meaning "to be subtracted." The 9 is called the *difference.* These names themselves come from Latin and were very good when introduced. Most people who did subtraction knew Latin. If a person forgot their meaning he could make a good guess. Because these names are often hard to remember some people say "sum," "given addend" and "missing addend" for 17, 8, and 9. This emphasizes the relation to addition. Some people say "first number," "second number," and "third number" or "answer." Some might call 17 the "top number," 8 the "bottom number," and 9 the "answer," reflecting the layout to the right.

$$\begin{array}{r} 17 \\ -\ 8 \\ \hline 9 \end{array}$$

9.4 NATURAL ALGORITHMS Each of the attitudes implies a natural algorithm or technique whereby, given the names of two numbers, we can find the name of their difference. A natural algorithm comes from any attitude and may be used with any numeral system. We illustrate 47 – 23 with three natural algorithms.

1. *Take away.* Form a set with 47 members. Take 23 members away. The difference is the number of members left.

2. *Missing addend.* Can you see what to add to 23 to make 47? No? Try guessing. How about 30? (23 + 30 = 53, too much) How about 22? (23 + 22 = 45, too small) How about 24? (23 + 24 = 47, just right) Going back to sets, one would form a set A with 23 members. Then put counters, say beans, with A until the enlarged set has 47 members. The number of beans put with A is the difference 47 – 23.

3. *Counting backward.* Count "46, 45, 44, . . ." until you have named 23 numbers. The last number named is 47 – 23.

You may notice that every subtraction attitude matches an addition attitude. Similarly, every addition attitude has a matching subtraction attitude. Some say two attitudes, since we think of 23 + □ = 47 or □ + 23 = 47—that is, as the missing addend is the second or the first number.

We emphasize that all attitudes yield the same result, as can be seen by setting up appropriate sets or by appropriate counting.

```
  901
- 658
  243
```

9.5 STANDARD SUBTRACTION ALGORITHM For the layout in the margin, we ask, given 658 to be subtracted from 901, what rules or routines did we follow to persuade us to write 243? How are these rules related to a subtraction attitude and the numeral system? We probably did not follow any of the natural algorithms. We may have muttered something like "8 from 1 won't go, borrow 1 from 0. Can't do it. Borrow 1 from 9, leaving 8 and making 10. Borrow 1 from 10, leaving 9 and making 11. 8 from 11 is 3. 5 from 9 is 4, and 6 from 8 is 2."

We see that to get 243, we worked with the individual digits without considering the numbers as wholes. We paid attention to the place of the digit. We paid little attention to the meaning of subtraction, except single-digit subtraction. Does this dance of the digits work? Can working with the individual digits in this way actually give the results from counting backward, or separating sets? We shall see how the spelling of the number combined with place value and the meaning of subtraction as separating sets and missing addend will explain this dance (algorithm). While algorithm is the proper term, we call it a dance of the digits, because the subtraction process is a series of instructions about what to do with specific digits—just as a dance is a series of instructions about what to do with hands, feet, knees, hips, shoulders.

Corresponding to the 100 addition facts are 100 subtraction facts. The addition facts are the sums $a + b = c$, where a and b form all possible pairs of single digit numbers, such as $8 + 5 = 13$, $3 + 3 = 6$. There are 100 such sums. Now suppose in each sum we regarded a and c as known, and we wished to find b, as in $8 + b = 13$, $3 + \square = 6$. We write $b = 13 - 8$, $\square = 6 - 3$, and $b = c - a$. These 100 statements are called *subtraction facts*.

Subtraction is often clearer in Roman numerals, which are formed by joining symbols. Set unions and separations are reflected in symbol unions and separations.

27 – 3 may be written	
XXVII – III.	It is convenient to rename
XXIIIIIII – III.	Now separate
XXIIII + III – III	which simplifies to
XXIIII.	

The same writing may be interpreted as manipulating sets (with X standing for a subset of ten) or as subtraction in Roman numerals. The student may compare this with the earlier Hindu-Arabic tableau, or layout, and identify the different parts (2 corresponds to XX, and so on). You may also use the dish abacus.

To explain 42 – 5 is different since we cannot "take away," or separate, a set of 5 ones from the 2 ones in 4 tens 2 ones shown by 42. The trick is to regroup the 4 tens 2 ones into 3 tens 12 ones. Now we have 3 tens 12 ones – 5 ones. 12 ones – 5 ones is a subtraction fact, so we get 3 tens 7 ones as the difference. Sometimes people use this layout, with scratches and a little 3 and a little 1 to show that the 4 tens 2 ones had been renamed 3 tens 12 ones.

$$\begin{array}{r} \overset{3}{\cancel{4}}\,{}^{1}2 \\ -\quad 5 \\ \hline 3\ 7 \end{array}$$

Using the dish abacus shows the process clearly. We wish to take $(0, 0, 5)_{ten}$ from $(0, 4, 2)_{ten}$. If there were 5 beans in dish 0, this could be done nicely. There are not. So we convert $(0, 4, 2)_{ten}$ to $(0, 3, 12)_{ten}$. Now we have the necessary 5 beans in dish 0.

Roman numerals show the process. 42 – 5 becomes

XXXXII – V =

XXXIIIIIIIIIIII – IIIII =

XXXIIIIIII + IIIII – IIIII =

XXXIIIIIII

Do you see that X is converted to IIIIIIIIII matching the conversion of 40 to 30 plus 10 in Hindu-Arabic numerals? Can you identify "take away 5"?

There are, of course, alternative steps, such as

XXXXII – V = XXXVVII – V = XXXVII

Roman numerals allow X to be renamed as either 10 ones or 2 fives. This allows more alternative steps than the Hindu-Arabic system. Example: 901 − 658. The standard layout is

$$\begin{array}{r} 901 \\ \underline{658} \\ 243 \end{array}$$

Some think that using the dish abacus, converting, and take away is the easiest way to explain this. We start with $(0, 9, 0, 1)_{\text{ten}}$ and wish to take $(0, 6, 5, 8)_{\text{ten}}$ away. That is, we wish to take 8 from dish 0, 5 from dish 1, and 6 from dish 2. This cannot be done, so we convert some of the beans in the higher dishes into more beans in the lower. We convert $(0, 9, 0, 1)_{\text{ten}}$ to $(0, 8, 10, 1)_{\text{ten}}$ to $(0, 8, 9, 11)_{\text{ten}}$. Now we can take $(0, 6, 5, 8)_{\text{ten}}$ away, leaving $(0, 2, 4, 3)_{\text{ten}}$.

Scratches and crutch numbers can be used with the standard layout to reflect these actions. We show how this can be done in steps.

$\begin{array}{r} 9\ 0\ 1 \\ \underline{-6\ 5\ 8} \end{array}$	Initial layout. We cannot subtract 8 ones from 1 one. We convert and rename.
$\begin{array}{r} 8\phantom{{}^{1}0\ 1} \\ \not{9}\,{}^{1}0\ 1 \\ \underline{-6\ 5\ 8} \end{array}$	There are 10 tens, but still not enough ones. Continue converting and renaming.
$\begin{array}{r} 8\ \ 9 \\ \not{9}\,{}^{\not{1}}\not{0}\,{}^{1}1 \\ \underline{-6\ 5\ 8} \end{array}$	Now there are enough ones.
$\begin{array}{r} 8\ \ 9 \\ \not{9}\,{}^{\not{1}}\not{0}\,{}^{1}1 \\ \underline{-6\ 5\ 8} \\ 3 \end{array}$	11 − 8 = 3, a subtraction fact. We might also think "11 ones, take away 8 ones, leaves 3 ones."
$\begin{array}{r} 8\ \ 9 \\ \not{9}\,{}^{\not{1}}\not{0}\,{}^{1}1 \\ \underline{-6\ 5\ 8} \\ 4\ 3 \end{array}$	9 tens minus 5 tens is 4 tens, or 9 − 5 = 4, a subtraction fact, with the tens value understood as shown by the place.

```
 8 9
  1̸ 1
 9̸ 0̸ 1
−6 5 8
 2 4 3
```

Take away 6 hundreds from 8 hundreds leaves 2 hundreds, or 8 − 6 = 2, a subtraction fact, with the hundreds value understood by the place. The difference is now read off.

The same steps are easily seen and identified when the subtraction is in Roman numerals. We match steps. The circles indicate digits that are matched and taken away.

```
     9 0 1       D CCCC          I
   − 6 5 8      −D C       L V III

    8 1
     9̸ 0 1       D CCC LXXXXX   I
   − 6 5 8      −D C    L        V III

    8 9 1
     9̸ 0̸ 1       D CCC LXXXX(V)III(III)
   − 6 5 8      −D C    L    (V)    (III)
         3                       III

    8 9 1
     9̸ 0̸ 1       D CCC(L)XXXX III
   − 6 5 8      −D     C(L)
       4 3                 XXXX III

    8 9 1
     9̸ 0̸ 1      (D)C C(C)XXXX III
   − 6 5 8     −(D)    (C)
     2 4 3          CC    XXXX III
```

This subtraction algorithm is sometimes called the take away-convert algorithm, because these two steps play such an important and repeated role in the dance of the digits. There are two types of actions repeated in the algorithm: (1) subtraction fact steps, such as 8 − 5 = 3, and (2) converting steps in the upper number, such as 47 being replaced by $^{3}\not{4}{}^{1}7$, or 3 tens 17 ones. We think of the subtraction fact step as a take-away operation—that is, 5 from 8 leaves 3. The algorithm used to be called take away-borrow, but *borrow* is being replaced by *convert* in the subtraction language. Some authors use *regroup* instead of *convert* or *rename* as an even more general description. We also call this algorithm *standard* because it is most widely taught.

9.6 SUBTRACTION IN OTHER BASES The take away–convert algorithm can be applied to subtraction of numerals in bases other than ten, with appropriate changes in the facts and the converting. We illustrate with examples in base five.

EXAMPLE

$$\begin{array}{r} 43_{\text{five}} \\ -\ 21_{\text{five}} \\ \hline 22_{\text{five}} \end{array}$$

Considering the ones, there is "3, take away 1, leaves 2," which is a subtraction fact in both base five and ten. Similarly, "2 from 4 leaves 2" describes what is happening in the fives column. The difference is 2 fives 2.

To continue with more examples a table of the subtraction facts may be helpful. Here is such a table.

0	1	2	3	4	2	3	4	10	11
0	0	0	0	0	2	2	2	2	2
0	1	2	3	4	0	1	2	3	4
1	2	3	4	10	3	4	10	11	12
1	1	1	1	1	3	3	3	3	3
0	1	2	3	4	0	1	2	3	4
					4	10	11	12	13
					4	4	4	4	4
					0	1	2	3	4

You should be sure that you understand this table and can explain each example in terms of separating sets and the rules for spelling the numbers in digits. (For example, 12 means 1 five and 2 ones. It's the base-five numeral for 7.)

EXAMPLE

$$\begin{array}{r} 32 \\ -\ 14 \\ \hline \end{array} \qquad \begin{array}{r} {}^{2}\not{3}\,{}^{1}2 \\ 1\ 4 \\ \hline 1\ 3 \end{array}$$

4 can't be taken from 2. There aren't enough ones, so the upper number is renamed. Convert the 3 fives to 2 fives and five ones. Now the upper

number is grouped as 2 fives and 12 ones. 4 from 12 is 3, 1 from 2 is 1. The difference is 13.

This can be done with Roman numerals to recall the set structure.

```
  VVV II
–     V IIII
-------------
  VV  IIIII II
   V  IIII          Regrouping and, in the third line, writing the difference.
-------------
   V    III
```

EXAMPLES

$$\begin{array}{r} 413_{five} \\ -\ 224_{five} \\ \hline 134_{five} \end{array} \qquad \begin{array}{r} 341_{five} \\ -\ 23_{five} \\ \hline 313_{five} \end{array}$$

The operation is even clearer using base-five abacuses.

In base two, the same algorithm is successful. Here are some subtraction facts and simple examples:

$$\begin{array}{r} 1_{two} \\ -\ 1_{two} \\ \hline 0_{two} \end{array} \qquad \begin{array}{r} 10_{two} \\ -\ 1_{two} \\ \hline 1_{two} \end{array} \qquad \begin{array}{r} 11_{two} \\ -\ 1_{two} \\ \hline 10_{two} \end{array} \qquad \begin{array}{r} 11_{two} \\ -\ 10_{two} \\ \hline 1_{two} \end{array} \qquad \begin{array}{r} 100_{two} \\ -\ 1_{two} \\ \hline 11_{two} \end{array}$$

EXAMPLE

$$\begin{array}{r} 1100101_{two} \\ 110110_{two} \\ \hline 101111_{two} \end{array}$$

The student should work this example out, then compare his work with the following discussion.

We work the problem in steps, using the dish abacus and the standard layout with scratches and crutch numbers side by side. Recall that we express the number of beans in a dish as a base-ten numeral. In the layout, base-two numerals are used.

$$\begin{array}{r} (0,\ 1,\ 1,\ 0,\ 0,\ 1,\ 0,\ 1)_{two} \\ -\ (0,\ 0,\ 1,\ 1,\ 0,\ 1,\ 1,\ 0)_{two} \\ \hline \end{array} \qquad \begin{array}{r} 1100101_{two} \\ -\ 110110_{two} \\ \hline \end{array}$$

Step 1 In dish 0 (units place), we take no beans from 1 bean.

$$\begin{array}{r} (0,\ 1,\ 1,\ 0,\ 0,\ 1,\ 0,\ 1)_{\text{two}} \\ -\ (0,\ 0,\ 1,\ 1,\ 0,\ 1,\ 1,\ 0)_{\text{two}} \\ \hline (\qquad\qquad\qquad\qquad 1)_{\text{two}} \end{array} \qquad \begin{array}{r} 1100101 \\ \underline{110110} \\ 1 \end{array}$$

Step 2 We convert the bean in dish 2 to two beans in dish 1, and take 1 away.

$$\begin{array}{r} (0,\ 1,\ 1,\ 0,\ 0,\ 0,\ 2,\ 1)_{\text{two}} \\ -\ (0,\ 0,\ 1,\ 1,\ 0,\ 1,\ 1,\ 0)_{\text{two}} \\ \hline (\qquad\qquad\qquad 1,\ 1)_{\text{two}} \end{array} \qquad \begin{array}{r} \overset{0}{} \\ 1100\not{1}{}^{1}01 \\ \underline{1101\ 10} \\ 11 \end{array}$$

Step 3 To get any beans in dish 2, we make a cascade, or chain of conversions. $(0, 1, 1, 0, 0, 0, 2, 1)_{\text{two}} \to (0, 1, 0, 2, 0, 0, 2, 1)_{\text{two}} \to$ $(0, 1, 0, 1, 2, 0, 2, 1)_{\text{two}} \to (0, 1, 0, 1, 1, 2, 2, 1)_{\text{two}}$. We get

$$\begin{array}{r} (0,\ 1,\ 0,\ 1,\ 1,\ 2,\ 2,\ 1)_{\text{two}} \\ -\ (0,\ 0,\ 1,\ 1,\ 0,\ 1,\ 1,\ 0)_{\text{two}} \\ \hline (\qquad\qquad 1,\ 1,\ 1)_{\text{two}} \end{array} \qquad \begin{array}{r} 0\ \ {}^{1}\ {}^{11}\ \ 0 \\ 1\ \not{1}{}^{1}\not{0}{}^{1}\not{0}\ \not{1}{}^{1}0\ 1 \\ \underline{1\ 1\ 0\ 1\ 1\ 0} \\ 1\ 1\ 1 \end{array}$$

Step 4 We take none from 1 bean in dish 3.

$$\begin{array}{r} (0,\ 1,\ 0,\ 1,\ 1,\ 2,\ 2,\ 1)_{\text{two}} \\ -\ (0,\ 0,\ 1,\ 1,\ 0,\ 1,\ 1,\ 0)_{\text{two}} \\ \hline (\qquad 1,\ 1,\ 1,\ 1)_{\text{two}} \end{array} \qquad \begin{array}{r} 0\ \ {}^{1}\ {}^{11}\ \ 0 \\ 1\ \not{1}{}^{1}\not{0}{}^{1}\not{0}\ \not{1}{}^{1}0\ 1 \\ \underline{1\ 1\ 0\ 1\ 1\ 0} \\ 1\ 1\ 1\ 1 \end{array}$$

Step 5 We take 1 bean from dish 4.

$$\begin{array}{r} (0,\ 1,\ 0,\ 1,\ 1,\ 2,\ 2,\ 1)_{\text{two}} \\ (0,\ 0,\ 1,\ 1,\ 0,\ 1,\ 1,\ 0)_{\text{two}} \\ \hline (\qquad 0,\ 1,\ 1,\ 1,\ 1)_{\text{two}} \end{array} \qquad \begin{array}{r} 0\ \ {}^{1}\ {}^{11}\ \ 0 \\ 1\ \not{1}{}^{1}\not{0}{}^{1}\not{0}\ \not{1}{}^{1}0\ 1 \\ \underline{1\ 1\ 0\ 1\ 1\ 0} \\ 0\ 1\ 1\ 1\ 1 \end{array}$$

Step 6 We cannot take 1 bean from dish 5, so we convert one bean from dish 6 and then take 1 bean from dish 5.

$$\begin{array}{l} (0,\ 0,\ 2,\ 1,\ 1,\ 2,\ 2,\ 1)_{\text{two}} \\ \underline{(0,\ 0,\ 1,\ 1,\ 0,\ 1,\ 1,\ 0)_{\text{two}}} \\ (\qquad\ \ 1,\ 0,\ 1,\ 1,\ 1,\ 1)_{\text{two}} \end{array} \qquad \begin{array}{ccccccc} 0 & {}^{1}0 & {}^{1}\cancel{1} & {}^{11}\cancel{1} & 0 & & \\ \cancel{1} & \cancel{1} & \cancel{0} & \cancel{0} & \cancel{1} & {}^{1}0 & 1 \\ & 1 & 1 & 0 & 1 & 1 & 0 \\ \hline & 1 & 0 & 1 & 1 & 1 & 1 \end{array}$$

Step 7 and 8 We take no beans from dishes 6 and 7.

$$\begin{array}{l} (0,\ 0,\ 2,\ 1,\ 1,\ 2,\ 2,\ 1)_{\text{two}} \\ \underline{(0,\ 0,\ 1,\ 1,\ 0,\ 1,\ 1,\ 0)_{\text{two}}} \\ (0,\ 0,\ 1,\ 0,\ 1,\ 1,\ 1,\ 1)_{\text{two}} \end{array}$$

The curious may want to convert this example to base-ten numerals as a check.

9.7 CHECKING SUBTRACTION Some people work crossword puzzles in ink. Others never look to see if their hair is combed or if there is gas in the gastank. However, most of us find it worthwhile to check what we've done to make sure it is right.

Every check process is, essentially, doing the problem over again, sometimes in the same way, but usually in a different way. The check may or may not be faster or more accurate than the first solution.

The usual procedure for checking the take away–regroup algorithm for subtraction is to add the lower number and the difference. The sum should be the upper number. This is because of the way subtraction can be related to separating sets and addition to joining sets. Or subtraction is finding the missing addend. Having found it, the addition can be done with the sum known.

We recall that $32 - 17 = 15$ and $17 + 15 = 32$ are two ways of expressing the same relation between the three numbers 32, 17 and 15. The same set diagram

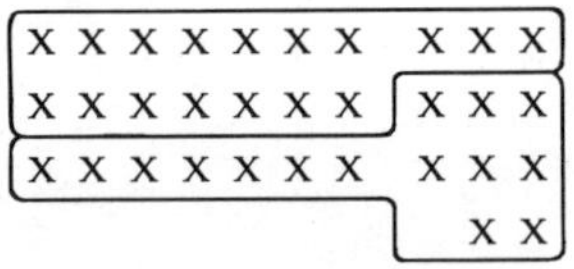

```
x x x x x x x   x x x
x x x x x x x   x x x
x x x x x x x   x x x
                  x x
```

shows both these relations, and also a third, $32 - 15 = 17$.

EXAMPLE

```
32
17
--
15
--
32 √
```

In this layout a line is drawn below the difference, 15. The difference and the lower number are added, with sum 32. The sum and the upper number are compared, and a check, √, made to show that their sameness is recognized.

EXAMPLE

```
  725
- 486
  ---
  239
  ---
  725 √
```

EXAMPLE

```
   DCCXXV
 - CCCCLXXXVI
 ------------
   CC  XXX  V  IIII
 ------------------------------------
   CCCCC  C  L  XXXXX  X  VV  IIIII
     D    C  L     L   X  X     V
     D    C     C      X  X     V √
```

EXAMPLE

$$\begin{array}{r} 32_{\text{five}} \\ \underline{14_{\text{five}}} \\ \underline{13_{\text{five}}} \\ 32_{\text{five}} \end{array} \quad \checkmark$$

EXAMPLE

$$\begin{array}{r} 1100101_{\text{two}} \\ \underline{110110_{\text{two}}} \\ \underline{101111_{\text{two}}} \\ 1100101_{\text{two}} \end{array} \quad \checkmark$$

An entirely different algorithm may be used on the subtraction as a check. We now explore some possible alternative algorithms for subtraction.

9.8 THE AUSTRIAN ALGORITHM The take away–convert algorithm is efficient and easy to understand in terms of sets, the dish abacus, and the place-value numeral system used to give names to numbers. The Austrian, or missing addend–equal addition, algorithm is a bit more sophisticated, and some people think it smoother and easier. It appears to develop intense partisans.

The two main types of steps in a subtraction algorithm are (1) using the basic facts, and (2) "bridging the tens" or setting up a basic fact. The two algorithms have different attitudes toward these steps.

For example, $7 - 3 = 4$ is a basic fact. People who have the take-away attitude are likely to think or mutter "3 from 7 is 4" or "7, take away 3, leaves 4" or "7 minus 3 equals 4." In working they might say "three from seven is four" and "one from six is five."

$$\begin{array}{r} 67 \\ \underline{-\ 13} \\ 54 \end{array}$$

A person steeped in the missing-addend attitude might think, "What added to 3 gives 7? 4." With slight guidance, he identifies $7 - 3 = 4$ as just another way of writing his old friend $3 + 4 = 7$. He knows his addition facts so well he recognizes them from two numbers. When he sees 7 − 3 = 4, he says "3 and 4 is 7," and he writes 4 for the difference. In working

$$\begin{array}{r} 67 \\ -\ 13 \\ \hline 54 \end{array}$$

he might say "3 and 4 is 7" and "1 and 5 is 6," writing 4 and 5 as he does so.

In "71 − 28," one cannot say "8 from 1" or "8 and ____ is 1." Somehow

$$\begin{array}{r} 71 \\ -\ 28 \\ \hline \end{array}$$

we must get another 10 with the 1 so that we can set up the basic fact, 11 − 8 = 3.

In the take away–convert algorithm, this extra 10 is found by converting one of the 7 tens to 10 ones, replacing (70 + 1) − 28 with (60 + 11) − 28. In the Austrian algorithm, we add the extra 10 and compensate by adding an equal 10 in the minuend or second number in the tens place. Thus 71 − 28 is replaced by (70 + 11) − (30 + 8) or,

$$\begin{array}{r} 7\,{}^{1}1 \\ {}^{3}\not{2}\;8 \\ \hline 4\;\;3 \end{array}$$

The crutch numbers indicate that 1 has been increased to 11 and 28 has been increased to 38. This step we call *equal additions.* The difference is unchanged if the same number, 10, is added to both subtrahend and minuend.

$$\begin{array}{r} 9\,{}^{1}0\,{}^{1}1 \\ -\ {}^{7}\not{6}\,{}^{6}\not{5}\;8 \\ \hline \end{array}$$

$$\begin{array}{rrr} 9 & 10 & 11 \\ \underline{7} & \underline{6} & \underline{8} \\ 2 & 4 & 3 \end{array}$$

In computing 901 − 658, the person using the missing addend–equal addition algorithm might first set up basic facts subtractions by adding 10 to 1, making 11, then compensating by adding 1 to 5 making 6. He adds 10 to 0 making 10 and 1 to 6 making 7, as shown in the layout in the margin. He then would say "8 plus 3 is 11, 6 plus 4 is 10, 7 plus 2 is 9," writing 3, 4, 2 as he did so. Of course, he might do the basic fact 11 − 8 = 3 and write 3 before setting up the basic fact 10 − 6 = 4 in the tens place, and so on.

A slightly different interpretation that moves very smoothly is the following, which emphasizes finding the missing addend. We imagine replacing the subtraction layout by an addition layout with a big blank as the second addend.

$$\begin{array}{r} 6\;5\;8 \\ + \\ \hline 9\;0\;1 \end{array}$$

Then, knowing our addition facts, we mutter, "8 plus 3 is 11. Write 3, carry 1 to the 5 making 6."

$$\begin{array}{r} 6\;{}^{6}\!\not{5}\;8 \\ +\;3 \\ \hline 9\;0\;1 \end{array}$$

For the tens digit we mutter, "6 plus 4 is 10. Write 4, carry 1 to 6 making 7."

$$\begin{array}{r} {}^{7}\!\not{6}\;{}^{6}\!\not{5}\;8 \\ +\;4\;3 \\ \hline 9\;0\;1 \end{array}$$

For the hundreds digit, we say "7 plus 2 is 9" and write 2. This places our answer, 243, in the middle of the layout, somewhat hiding it.

$$\begin{array}{r} 6\;5\;8 \\ 2\;4\;3 \\ \hline 9\;0\;1 \end{array}$$

So, writing the layout in the standard form, we mutter something like "8 plus 3 equals 11. Write 3, carry 1 to 5, making 6; 6 and 4 is 10. Write 4, change 6 to 7; 7 and 2 is 9. Write 2."

$$\begin{array}{r} 9\;0\;1 \\ -\;\not{6}\;\not{5}\;8 \\ \hline 2\;4\;3 \end{array}$$

Some people may wish to look at this last interpretation as somewhat different from the preceding one, while others think of them as essentially the same. Is a cup half full or half empty? Our purpose is more to show how people may look at subtraction rather than to list only clearly different algorithms.

A number of years ago there was quite a controversy over whether the take away–convert or the Austrian algorithm was better. Studies attempted to answer (1) which algorithm is easier to teach and understand, (2) which algorithm leads to greater speed, and (3) which algorithm leads to greater accuracy.

The studies seemed to show that take away–convert was easier to understand. Perhaps this is because it is more closely related to set ideas. On the other hand, the Austrian method seemed to be faster and to cause fewer errors.

Most school authorities have felt that most children should be taught just one method in the primary and elementary grades. They feel that the extra time to teach a second method is better spent otherwise. Some feel a child is confused with too many alternatives too young. As a result most American schools teach the take away–convert algorithm. Understanding is most important, while errors can be corrected by checking, and speed is not terribly vital. People with many subtractions to make will usually use a machine.

The author met the Austrian method only recently. He likes its advantages, especially in balancing a checkbook where one wishes to subtract the sum of several checks from a previous balance. One adds the checks, then the missing addend to get the previous balance. The new balance or missing addend appears to come much easier and more accurately than with his previous methods.

Some people may like to test for themselves which method they find best.

9.9 OTHER ALGORITHMS. FRONT-END SUBTRACTION The two practical attitudes toward basic facts, take away and missing addend, may be paired in other ways with the two useful ways of setting up basic facts, converting and equal addition. Besides take away–convert (standard) and missing addend–equal addition (Austrian) pairings, we can easily get a take away–equal additions and a missing addend–convert algorithm. We leave it as an exercise for you to determine appropriate mutterings, or steps in the dance of the digits for these algorithms.

Front-end subtraction, like front-end addition, is useful for giving an approximate answer in a few steps. It is also good as another algorithm to help us understand subtraction. Scratches make it easy. We illustrate with multiple layouts showing the steps.

EXAMPLE

```
 45      45
- 3     - 3
---     ---
 4       42
```

EXAMPLE

```
34      3¹4     3¹4
 8        8       8
--      ---     ---
3       3̸       3̸ 6
        2       2
```

EXAMPLE

```
83      8¹3     8¹3
24      2 4     2 4
--      ---     ---
6       6̸       6̸ 9
        5       5
```

EXAMPLE

```
825     8¹2 5     8¹2¹5
379     3 7 9     3 7 9
---     -----     -----
5       5̸ 5       5̸ 5̸ 6
        4         4 4
```

EXAMPLE

```
8 9 6 3 4 5 2
6 3 8 3 6 4 7
-------------
2 6
```

Just two steps show the difference is about 2,600,000.

```
  8 9 6 3 4 5 2      8 9 6 3 4 5 2
- 6 3 8 3 6 4 7    - 6 3 8 3 6 4 7
---------------    ---------------
  2 6̸ 8              2 6̸ 8̸ 0̸ 8
    5                  5 7 9
```

Front-end subtraction can be used with any of the four earlier algorithms. Hence we have eight different possible subtraction algorithms discussed so far. Any of these algorithms can be used with numerals in different bases or with Roman numerals.

Checking these algorithms follows the same principles as for any algorithm: do the problem over in a different way. Some people believe that adding the lower number and the difference is not as good a check for the Austrian algorithm as it is for take away–convert. They feel this is doing the problem over in practically the same way.

EXERCISES

Exercises calling for beans, rods, and so on are best done with these aids. Diagrams may be used but are not as good. Where a problem clearly requires too much time, describe how you would solve it. Such exercises are included to emphasize structure rather than answers.

xxxxx xxx

1. The diagram in the margin illustrates 8 − 3 from a take-away attitude. Point out what shows 8, what shows 3, what shows "take away," and what shows the difference.

2. Using beans, illustrate 7 − 5 from the take-away attitude.

xxxxxxxxx
xxxyyyyyy

3. The diagram shows 9 − 3 from a missing-addend attitude. Point out what shows "9," "3," "addition," "missing addend," "difference."

4. Use beans to illustrate 8 − 5 from the missing-addend attitude.

5. From 17 + 45 = 62, we deduce 62 − 45 = 17. What other subtraction statement do we deduce from 17 + 45 = 62? Are we using take away, missing addend, or counting backward?

6. Use counting backward to find 73 − 4.

7. If the purple rod is longer than the red rod, find purple minus red. Use other colored rods if necessary. What subtraction attitude did you use?

8. If the rods are associated with numbers, what number relation is shown in exercise 7?

9. Use rods to show 7 − 5 and 9 − 3.

10. Did you count, read labels, or recognize something familiar in determining the results in exercises 1, 2, and 9?

As you do exercises 11–16, mutter to yourself words that go with the take-away attitude. In 12 − 5 = , mutter something like "12 take away 5 is 7," or "5 taken from 12 is 7."

11. $\begin{array}{r} 9 \\ -\ 3 \\ \hline \end{array}$ **12.** $\begin{array}{r} 12 \\ -\ 4 \\ \hline \end{array}$ **13.** $\begin{array}{r} 7 \\ -\ 2 \\ \hline \end{array}$

14. 8 − 3 = **15.** 10 − 4 = **16.** 17 − 9 =

As you do exercises 17–22, mutter words that go with the missing-addend attitude. Thinking may be better than muttering to avoid disturbing your neighbor. In 12 − 5 = , mutter something like "five added to what, seven, equals twelve" or "five plus seven equals twelve," writing 7 as you mutter "seven."

17. $\begin{array}{r} 5 \\ -\ 2 \\ \hline \end{array}$ **18.** $\begin{array}{r} 8 \\ -\ 5 \\ \hline \end{array}$ **19.** $\begin{array}{r} 13 \\ -\ 4 \\ \hline \end{array}$

20. 12 − 7 = **21.** 16 − 8 = **22.** 15 − 6 =

23. Of the 100 basic subtraction facts in base ten—
(*a*) Give the fact using the largest number.
(*b*) Give the fact for which the sum of the three numbers is least.
(*c*) Give the set of facts for which the sum of all 3 numbers is 14.
(*d*) Give the set of facts for which the sum of all 3 numbers is 19.

24. In the exercise in the margin, you change 4 to 14 to make a basic fact 14 − 5 = . Give two adjustments you might make to compensate for changing 4 to 14.

$\begin{array}{r} 64 \\ -\ 25 \\ \hline \end{array}$

In exercises 25–27 put in crutch numbers to show the bridging-the-tens attitude, and mutter words that reflect the basic-facts attitude used.

25. Subtract, using convert and take-away attitudes:

$\begin{array}{r} 43 \\ -\ 17 \\ \hline \end{array}$ $\begin{array}{r} 82 \\ -\ 39 \\ \hline \end{array}$ $\begin{array}{r} 644 \\ -\ 157 \\ \hline \end{array}$

Answers to selected questions:

1. separation shows takeaway

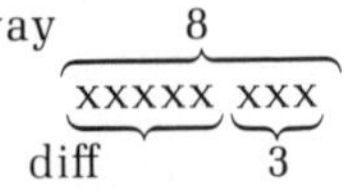

8
XXXXX XXX
diff 3

3. forming union 9 ⟶ XXXXXXXXX

shows addition 3 ⟶ XXXYYYYYY

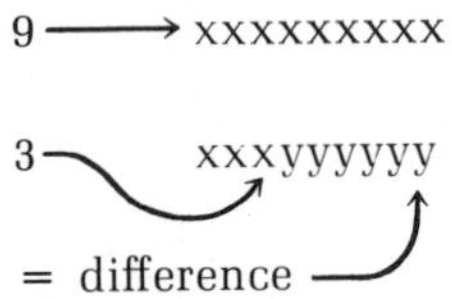

missing addend = difference

5. 62 − 17 = 45, missing addend

9. 7: 5 | 7 − 5 9: 3 | 9 − 3

13. $\begin{array}{r} 7 \\ -\ 2 \\ \hline 5 \end{array}$ **19.** $\begin{array}{r} 13 \\ -\ 4 \\ \hline 9 \end{array}$

24. Change 6 to 5 or 2 to 3.

26. (*b*) $\begin{array}{r} 6 \\ 75 \\ -\ 28 \\ \hline 47 \end{array}$ **27.** (*c*) $\begin{array}{r} 407 \\ 24 \\ -\ 138 \\ \hline 269 \end{array}$

30. XX X X V V II I
X V II
XX V I

D CL L XXX X V V II II
CCCC C
C L V II
CCC C L XX X V II

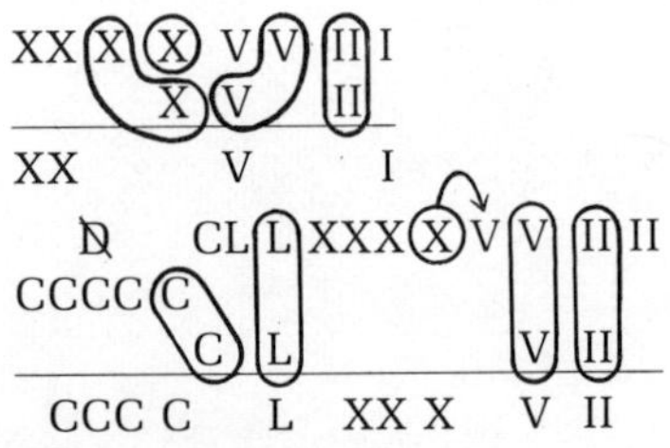

33. (*c*) $\begin{array}{r} 723 \\ -\ 346 \\ \hline 4\not{X}7 \\ 37 \end{array}$

26. Subtract, using convert and missing-addend attitudes:

$\begin{array}{r} 63 \\ -\ 47 \\ \hline \end{array}$ $\begin{array}{r} 75 \\ -\ 28 \\ \hline \end{array}$ $\begin{array}{r} 736 \\ -\ 288 \\ \hline \end{array}$

27. Subtract, using equal-additions and missing-addend attitudes (Austrian method):

$\begin{array}{r} 36 \\ -\ 19 \\ \hline \end{array}$ $\begin{array}{r} 43 \\ -\ 18 \\ \hline \end{array}$ $\begin{array}{r} 407 \\ -\ 138 \\ \hline \end{array}$

28. Subtract in the bases indicated, using take away–convert.

$\begin{array}{r} 32_{five} \\ -\ 13_{five} \\ \hline \end{array}$ $\begin{array}{r} 421_{five} \\ -\ 123_{five} \\ \hline \end{array}$ $\begin{array}{r} 10110_{two} \\ -\ 1101_{two} \\ \hline \end{array}$ $\begin{array}{r} 101000_{two} \\ -\ 10101_{two} \\ \hline \end{array}$

29. Subtract in the bases indicated using the Austrian method.

$\begin{array}{r} 42_{five} \\ -\ 13_{five} \\ \hline \end{array}$ $\begin{array}{r} 431_{five} \\ -\ 134_{five} \\ \hline \end{array}$ $\begin{array}{r} 10110_{two} \\ -\ 1101_{two} \\ \hline \end{array}$ $\begin{array}{r} 100100_{two} \\ -\ 10101_{two} \\ \hline \end{array}$

30. While you will never make change in Roman numerals at the bookstore, you may find the subtraction process clearer than in Hindu-Arabic numerals. Subtract in Roman numerals, using take away–convert:

XXXXIII − XVII =
DCXXXXIIII − CLVII =

31. Subtract in Roman numerals, using missing addend–equal additions:

XXXVI − XVIIII =
CCCCVII − CXXXVIII =

32. Use front-end subtraction:

XXXXIII − XVIII =
MDCLI − DCCLV =

33. Use front-end subtraction:

$\begin{array}{r} 65 \\ -\ 24 \\ \hline \end{array}$ $\begin{array}{r} 83 \\ -\ 46 \\ \hline \end{array}$ $\begin{array}{r} 723 \\ -\ 346 \\ \hline \end{array}$

34. Show take away–convert attitudes on dish abacuses by computing

$$\begin{array}{r} 413_{\text{five}} \\ -\ 143_{\text{five}} \\ \hline \end{array}$$

35. Show missing addend–equal additions on dish abacuses by computing

$$\begin{array}{r} 322_{\text{five}} \\ -\ 234_{\text{five}} \\ \hline \end{array}$$

36. What fourth (or missing) addend would you add to 34, 26, and 42 so that the sum is 528? Can you find this missing addend without finding the sum of 34, 26, and 42?

37. The following entries are listed in a checkbook:
Balance 87.63

groceries	17.42
car repair	14.35
shoes	12.72

What is the new balance? What numbers would you add to total 87.63? Do you need the sum of 17.42, 14.35, 12.72?

38. Check answers to exercises 36 and 37 on a calculator, if available.

39. Make up and work an exercise comparing subtraction with dish abacuses, in Roman numerals, and in base-ten numerals.

40. Which attitude, missing addend or take away, do you prefer toward subtraction? Why?

41. What number base is used in the subtraction problem at right. Check your answer by (1) adding and (2) converting to base-ten numbers.

$$\begin{array}{r} 5324 \\ -\ 3144 \\ \hline 2150 \end{array}$$

34. (4, 1, 3) (4 1 0) (3 6 0) (3 2 0) (2 2 0)
(1, 4, 3) (1 4 0) (1 4 0) (1 0 0) (0 0 0)

37. 43.14
17.42 + 14.35 + 12.72 + 43.14
no

41. base seven

10

Multiplication of Whole Numbers

10.1 INTRODUCTION Examples such as $3 \times 5 = 15$, $4 \times 13 = 52$, and $0 \times 8 = 0$ show that multiplication is a binary operation of the form $a \times b = c$. Two numbers go in and one number comes out. The major problems for the arithmetician are (1) why should students study multiplication? (2) how can one explain the steps in the multiplication dance of the digits, or algorithm? and (3) why do the CAD properties hold, and how are they useful?

Multiplication arises in many situations, resulting in attitudes and natural algorithms. Each has advantages and disadvantages. We shall discuss several algorithms and why they work. The CAD properties will be proved and shown to be useful in simplifying some computations and in helping understand algorithms. The nature of multiplication is clarified by considering lattice and Russian peasant algorithms, as well as some uncommon algorithms that use numbers in different bases and Roman numerals.

10.2 MULTIPLICATION ATTITUDES Traditionally, multiplication was defined as repeated addition. This attitude is still popular, but it is being sharply challenged by arrays and Cartesian products for the title of best attitude for introducing multiplication. Trees, and attitudes reflecting numbers as segments and operators, are useful for later discussions. Since some people think multiplication is merely a formal manipulation of digits, this too is recognized as an attitude; it can be based on attitudes presented earlier.

We shall define these attitudes by giving simple examples that obviously generalize to any counting numbers. Each attitude leads to a natural algorithm that will be valid in any numeral system, but is usually most effective for a particular numeral system.

In $a \times b = c$ or $5 \times 7 = 35$, a or 5 is called the multiplier, b or 7 is called the multiplicand, and c or 35 is called the product. Both a and b are called factors (as are 5 and 7).

10.3 REPEATED ADDITION By 3×4 from the repeated addition attitude, we mean the sum of 3 fours, or $4 + 4 + 4$. To find 95×37 by the corresponding natural algorithm, one writes a column of 95 thirty-sevens and adds them. Rather dull, but possible for anyone who can add but doesn't know how to multiply.

We note that the factors 3 and 4 play quite different roles. The addend is 4, while 3 is an operator that tells how many 4s to add together. Because of this difference it is not obvious that $3 \times 4 = 4 \times 3 = 3 + 3 + 3 + 3$.

Traditionally popular because it is easy to understand and because it appears sophisticated, repeated addition is under attack because it is not a good starting point either for showing the CAD properties or for proving the steps in the algorithm.

10.4 ARRAY By 3×4 from the array attitude, we mean the number of members in an array with 3 rows and 4 columns.

```
x x x x
x x x x
x x x x
```

To find 95×37 by the natural algorithm, form a large array with 95 rows and 37 columns. Then count the members. An orchard, cornfield, or pegboard may help in making the array.

We see that $a \times b$ is the same number whether defined by arrays or by repeated addition. The repeated-addition sum arises by counting the array by rows.

```
x x x x    4
x x x x    4
x x x x    4
```

Strictly speaking, any arrangement of things is an array. But we use array here as meaning a rectangular array where objects are arranged in rows and columns perpendicular to each other. Here are some possible arrays: seats in a classroom, autos in a parking lot, feet on a line of children, tiles on the floor. Imaginary arrays can be invented, corresponding to situations that don't appear as arrays at first glance. For example, the pieces of candy in 5 boxes, 16 pieces in each box, can form an array 5 by 16. The pennies needed to pay for 6 pencils at 4 cents each can easily be arranged in a 6-by-4 array. Whenever we use multiplication defined by arrays, we need to imagine an array somewhere.

Arrays are very good for showing the steps in multiplying two-digit numbers. For example, to find the product of 69 and 45, we would form an array of beans, say, that had 69 rows and 45 columns. Then we would count the beans. This is a perfectly straightforward, valid process, and it works for any numbers in any numeral system. If we followed the repeated-addition definition, we would write down 45 sixty-nine times and add them.

10.5 THE MULTIPLICATION DANCE The multiplication dance of the digits, or algorithm, is the process by which, given the names of two factors, one gets the name of the product. For example, how does one get the name 3105 for 69 × 45?

What we do is work with the digits in the names in a highly stylized ritual:

1. Write 45 with 69 under it, with 9 under 5 and 6 under 4. Draw a line under the 69.

$$\begin{array}{r} 45 \\ \underline{69} \end{array}$$

2. Say "9 × 5 = 45." Write 5 below the line under the 9. Write crutch number 4 above the 4 in some convenient place.

$$\begin{array}{r} 4 \\ 45 \\ \underline{69} \\ 5 \end{array}$$

3. Say "9 × 4 = 36, and 4 is 40." Write 0 below the line under the 6. Write 4 to its left.

$$\begin{array}{r} 4 \\ 45 \\ \underline{69} \\ 405 \end{array}$$

4. Say "6 × 5 = 30." Write 0 under the 0 under the 6 below the line. Write 3 in some convenient place near the crutch number 4.

```
 34
  45
  69
 ---
 405
   0
```

5. Say "6 × 4 = 24, and 3 is 27." Write 7 under the 4 below the line. Write 2 to its left. Draw a horizontal line under everything.

```
 34
  45
  69
 ---
 405
270
----
```

6. Say "Bring down the 5." Write 5 below the last line under the 5. Say "0 + 0 = 0." Write 0 below the last line under the 0s and to the left of the 5.

```
 34
  45
  69
 ---
 405
270
----
  05
```

7. Say "4 + 7 = 11." Write 1 below the last line under the 7. Write crutch number 1 above the 2. Say "2 + 1 = 3." Write 3 below the last line under the 2.

```
  34
   45
   69
  ---
  405
 1
 270
 ----
 3105
```

8. Say "The product is 3105."

Different people may vary somewhat in their approach from the above. The crutch numbers may not be written, for example. They may use phrases such as *tens place* and *hundreds place*. In any case, one ignores the factors as a whole and considers only the digits. One multiplies single-digit numbers, and writes the digits in the product in certain specified positions. Such a sequence of intimate steps is like the sequence of steps in a waltz or a ballet. This similarity has led to the phrase *dance of the digits* to describe the algorithm.

The steps in the dance may be clear, but a person who can follow them perfectly might still wonder if the product from the algorithm would always be the same as the product from counting beans in the array. Or from adding sixty-nine 45s. We will show that the products are the same by counting beans in a way that reflects the steps in the algorithm. We form a 69-by-45 array in a framework, with 69 dots down the left side and 45 dots across the top marking the 69 rows and 45 columns.

For clarity, we have drawn segments at the sides, one for every 10 dots. We may count the dots in any order we please. It is convenient to group them by tens and hundreds as shown in the next figure. Each dash is 10 dots in the body. Each square is 100 dots.

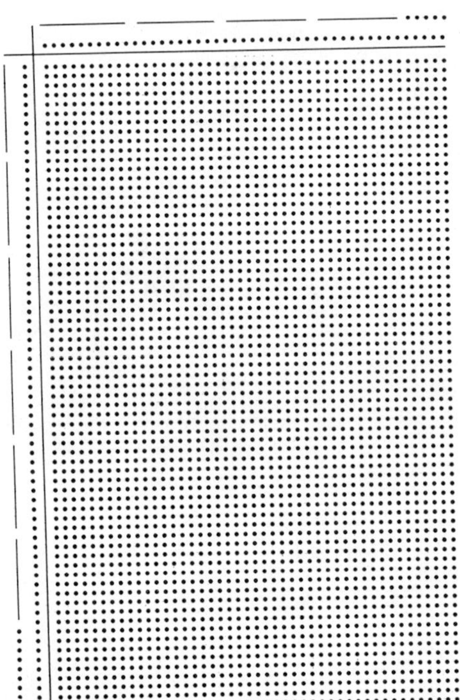

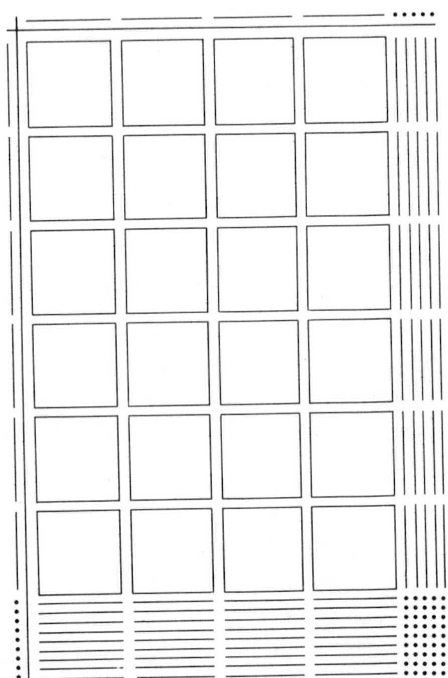

The large array is separated into four subarrays.

1.	a 9 × 5 array of dots	45 dots
2.	a 9 × 4 array of horizontal dashes containing	360 dots
3.	a 6 × 5 array of vertical dashes containing	300 dots
4.	a 6 × 4 array of squares containing	2400 dots

$$\begin{array}{r} 45 \\ \underline{69} \\ 45 \\ 360 \\ 300 \\ \underline{2400} \\ 3105 \end{array}$$

We add, getting 3105 dots all together. We make a layout as in the margin. The two factors are above the first line. The next four terms are numbers of points in the four subarrays. Can you see how to find the number of points in the subarrays merely from the names of the factors without having to draw the array?

We call this layout a "basic facts with zeros" or "partial products with zeros" layout. The 100 products of single-digit factors are the basic facts.

If we leave off the zeros from the dashes and squares, we get the layout called "basic facts without zeros" or "partial products without zeros."

```
   45
   69
   45
  36
  30
 24
 3105
```

Writing the sums of the first two and last two addends, we get

```
  45      45
  69      69
 405     405
2700     270
3105    3105
```

These layouts show how, if we start with the array, count the number of dots in the obvious subarrays, and add these, we are led to the standard layout. The standard algorithm leads to the same layout using the same single-digit multiplications, and a little mental arithmetic.

We show the fine detail by matching single digits in a layout with sets of points in a skeleton array for 34 × 62.

There are 18 digits in the computation. Each digit represents something in the array. Can you identify it? For example, the single 3, indicated by letter *a*, represents the 3 vertical dashes in the left border. There are three 8s. Two of them represent the same set of points, shown by letter *b*. These arise from 4 × 2. The third 8, indicated by *c*, is shown by 8 squares. Any 8 squares will do; there is no distinction among them. These are just 8 of the 18 squares arising from 6 × 3. Can you match the digit marked *d*? The other digits? The zeros are hard to identify, since they do not represent quantity, but are part of the name or show empty places. The 0 in 60, indicated by *e*, can be matched with dashness, the fact that the members in the 3 × 2 array in the upper right are dashes, not dots nor squares.

Before looking at more algorithms, or discussing the standard algorithm, we look at some multiplication attitudes.

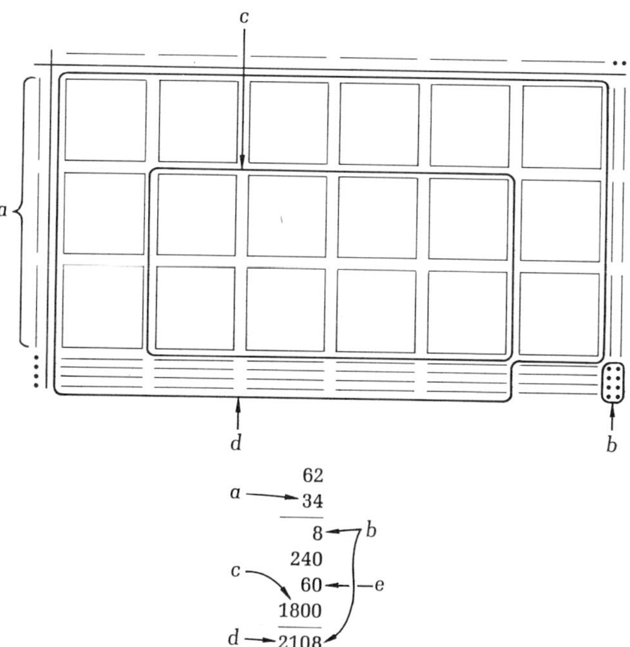

10.6 CARTESIAN PRODUCT The Cartesian product of two sets A,B is defined as the set C of all possible ordered pairs (s, t), where the first element is a member of A and the second a member of B.

EXAMPLE John has two pairs of shoes, black and tan. He has three pairs of clean socks, white, orange, and yellow.

Let $A = \{b,t\}$ be the set of shoes
$B = \{w,o,y\}$ be the set of clean socks

Then $C = (b, w), (b, o), (b, y), (t, w), (t, o), (t, y)$ is the Cartesian product, written $C = A \times B$. C represents the ways John can dress his feet. (b, y) means black shoes and yellow socks; (t, o) means tan shoes and orange socks.

The Cartesian-product pattern can be seen in many situations. For example: possible couples for blind dates from a boys club and a girls club; possible skirt-and-blouse outfits for a girl with set A of skirts and set B of blouses; the set of entrees, A, and the set of desserts, B, in a restaurant.

The name Cartesian product comes from Cartesian coordinates. The set of coordinates of points in the plane (x, y) is the Cartesian product of the coordinates of points on the x axis and the coordinates of points on the y axis. This is discussed more fully under number line and number plane. Cartesian comes from the name of the mathematician Descartes, or des Cartes.

The Cartesian-product definition of multiplication is as follows: By $a \times b$ we mean the number of members in the Cartesian product $C = A \times B$, where A has a members and B has b members. In the shoes-and-socks example, A has 2 members, B has 3 members, and C, the Cartesian product, has 6 members. Therefore, $2 \times 3 = 6$.

The Cartesian-product definition is equivalent to the array definition. The members of the Cartesian product can be arranged in an array by sorting into rows by the first member and into columns by the second member. For our example,

$$\begin{matrix} (b, w) & (b, o) & (b, y) \\ (t, w) & (t, o) & (t, y) \end{matrix}$$

is such an array.

It is more subtle to look on an array as a Cartesian product. The elements may not be ordered pairs (they may not even be pairs at all), or the pairs may not match the rows and columns exactly, but each *position* in the array can be labeled with an ordered pair of numbers, the numbers of the row and column of the position. Since the position labels are then a Cartesian product and since there are as many members in the array as there are positions, the number in the array is given by the number in the Cartesian product.

The Cartesian-product definition is especially liked by people who wish to base numbers and their operations completely on sets.

10.7 TREE A tree pattern is useful in listing possible outcomes when action is taken in stages. For example, suppose we want all the two-digit numerals that can be made with the digits 4, 5, 6, 7, using no digit twice. The numerals, such as 57, will each have a tens digit (5) and a units digit (7). We can write them as fast as we think of them: 57, 46, 67, 54, 65, ... When we finish, we wonder if we have overlooked any numbers, or repeated any. We can be sure if we list the numbers in a pattern, in stages. In stage 1, we list all the first, or tens-digit, possibilities in a fan diagram. The lines merely help focus our attention.

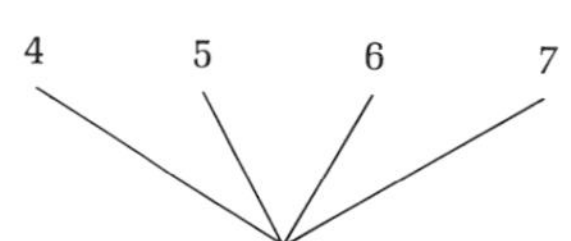

In stage 2, above each digit we draw another fan with the possible second digits:

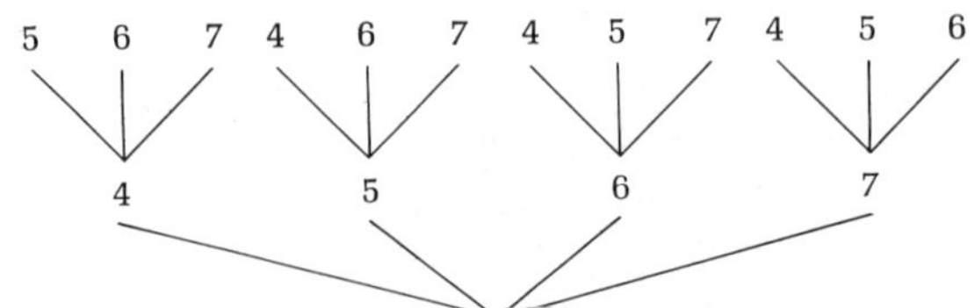

Reading from left to right, we get 45, 46, 47, 54, 56, 57, 64, 65, 67, 74, 75, 76 as the desired numbers. The figure with its branching lines is supposed to remind you of a tree.

This leads to the tree definition of multiplication: By $a \times b$ we mean the number of tips in a two-stage tree with a branches in the first stage. and each fan in the second stage having b branches. In our example, $a = 4$ and $b = 3$. There are 12 (count them) two-digit numbers. Hence, $4 \times 3 = 12$.

The tree definition of product gives the same product as the array definition, since the tips in the last stage could be formed into an array with a rows (one for each of the branches in the first stage) having b in each row. The two-digit numbers in our example form the array

45	46	47
54	56	57
64	65	67
74	75	76

Cartesian products can be formed in a tree pattern, the first element in the first stage, the second element at the second stage.

The tree definition extends nicely to products of three or more factors. In this case, the tree has three or more stages. Besides being a basis for multiplication, the tree is very useful in management and decision making as an aid to listing alternatives.

10.8 SEGMENT-SEGMENT-REGION Multiplication attitudes where the numbers represent segments or operators do not have nice names such as *array* or *tree*. In these cases, we use words associated with the numbers when we name $a \times b = c$.

A rectangle has segments for sides and encloses a region. The area, A, of the region is given by the famous formula $A = l \cdot w$, or area equals length times width. By this we mean if there are l units in the length segment and w units in the width segment, there are $l \cdot w$ rectangles of one length unit long and one width unit wide in the rectangular region.

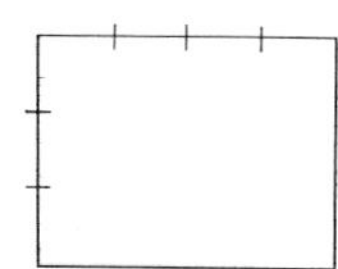

This leads to the following definition: By 3×4 from the segment-segment-region attitude, we mean the area of a rectangular region whose vertical side is 3 long and whose horizontal side is 4 long. This definition is similar to the array definition and is particularly good for generalizing to the multiplication of fractions.

Many people assert that numbers and their operations are pure ideas and do not depend on physical reasoning. However, a great value of arithmetic is that it reflects patterns in life. For example, multiplication comes up as a relation between measures, as in this case. Hence, we recognize an attitude where the factors represent continuous segmentlike quantities. Here are similar examples: speed $\times$ time $=$ distance, rate $\times$ principal $=$ interest, force $\times$ distance $=$ work, and price $\times$ quantity $=$ cost. These ideas are discussed more fully in the chapter on measurement. We note that the nature of the quantities (factors and product) fitting the multiplication pattern are different, whereas in addition the addends and the sum are the same kinds of objects.

10.9 SEGMENT-SEGMENT-SEGMENT The Greeks thought of number as the length of a segment, or as the ratio between a segment and a unit segment. They wanted some scheme whereby each factor is given by a segment and the product also by a segment. The segment-segment-region attitude above almost achieves this. The factors are segments, but the product is a region.

The Greeks solved this problem nicely with the theory of similar triangles. Consider the figure in the margin. The lines OUA and OBC meet at any convenient angle, but AC is drawn parallel to UB. By elementary geometry of similar triangles, we have the ratios $\frac{AO}{UO} = \frac{CO}{BO}$, or $AO \cdot BO = CO \cdot UO$. Now if $AO = a$, $BO = b$, $CO = c$, and $UO = 1$, we have $ab = c$. This construction can be the basis for a definition of multiplication.

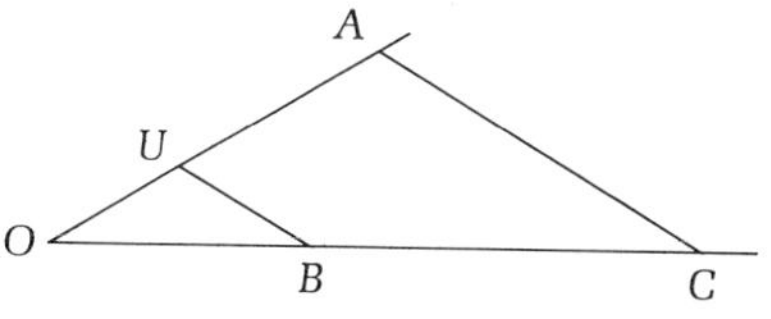

By $a \times b$ we mean the length of OC in the figure, where UO, AO, BO are constructed of length 1, a, b, and AC is constructed parallel to UB.

10.10 OPERATOR-SEGMENT-SEGMENT. OPERATOR-SET-SET In repeated addition, say 5 × 7 = 35, the first factor or multiplier, 5, is an operator. It tells how many times the second factor is to be used as an addend. Some authors call the first factor a "stretcher," a 5-stretcher in the example. A 5-stretcher stretches whatever is put into it to 5 times its size. A segment is stretched to 5 times its length.

Some like to compare an operator to a machine, like a grinder or wringer. If a 7 goes into a 5-stretcher, a 35 comes out. The stretcher also operates on sets. If 7 apples go in, 35 apples come out. This boogies the imagination a bit. We can pull rubber bands and roll dough, but there is no apple machine like this one. Sometimes we see a diagram for a horizontal machine:

7 35

5-operator

Then we think of the second factor as being the operator. The act of taking a 7-segment and stretching it as above we say is 7 multiplied by 5 instead of 5 times 7. We write the process 7 × 5. Computationally there is no difference. There are small advantages to each attitude. In conversation both people need to know how the other thinks to keep things straight.

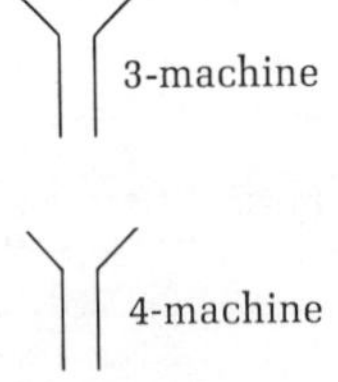

10.11 OPERATOR-OPERATOR-OPERATOR If two operator machines (as in chapter 7) are hooked up, they act together like a single operator machine. This leads to a definition of multiplication. By $a \times b$ we mean an operator equivalent to an a operator followed by a b operator. To determine the name of an operator machine, we imagine putting in a single element and counting what comes out. An upside-down tree diagram is convenient for showing what happens for the example of a 3-machine followed by a 4-machine:

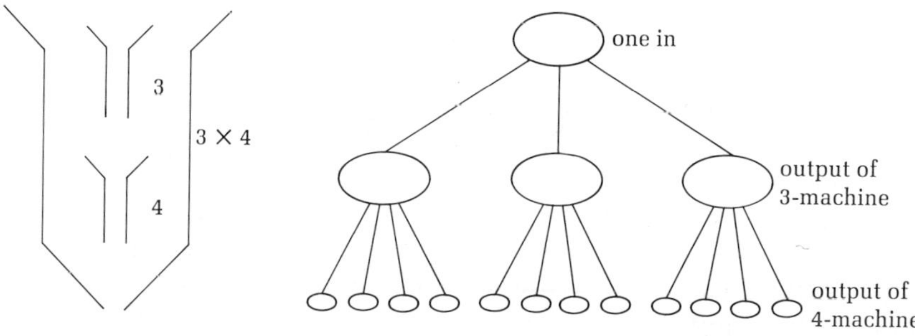

The operator-operator-operator attitude will be helpful later in thinking about multiplying fractions.

10.12 AUTHORITY For many students $a \times b$ means what their teacher or some other authority said it meant: $6 \times 8 = 48$ because the teacher said so. They memorize this statement. They would memorize $6 \times 8 = 23$ if they were told to. One of the purposes of this book is to help people who learn things on the basis of authority realize the broad human need behind the authority. Also, if people see these more human multiplication patterns, the work of memorizing may be reduced as well as the errors in computation.

10.13 THE CAD RULES. COMMUTATIVE PROPERTY We know that addition satisfies the rules (has the properties) $a + b = b + a$ (commutative rule) and $(a + b) + c = a + (b + c)$ (associative rule). Does multiplication satisfy similar rules? That is, "Is it true that $a \times b = b \times a$ and $(a \times b) \times c = a \times (b \times c)$?"

It is true that $3 \times 4 = 12$, and $4 \times 3 = 12$. Therefore, $4 \times 3 = 3 \times 4$ because they give the same answer. Because we will want to use this commutative property in *finding* the standard name of a product, we want to convince ourselves that, for any two whole numbers, $a \times b = b \times a$ *before* we know a standard, recognizable name for either product.

DIGRESSION An interesting question is the following. Using paper and pencil, multiply $875637984397 \times 379843978756$. Then multiply $379843978756 \times 875637984397$. Did you get the same product? Suppose you did not. Which of the following reasons would you accept?

1. For these two specific numbers, the order of multiplication makes a difference.

2. A mistake was made in the computation.

Why do you agree with one reason and not the other? Consider the following. You have never multiplied these two numbers before. Probably no one has ever multiplied these two numbers. Is it satisfactory to claim a mistake because $3 \times 4 = 4 \times 3$, and because in the other multiplications you have done you received the same product in both orders? How many times have you actually multiplied two numbers both ways?

It is certainly plausible that based on experience with other numbers you should guess you should get the same product. There are in mathematics many situations (computations, statements) which are true for all the numbers that have been tried, but which are known not to be true for all numbers.

Many of us are convinced that $a \times b = b \times a$ because "it works," or "my teacher told me so," or "this fact is postulated." It is quite appropriate that, when people are to be convinced of a true statement, any argument that succeeds is a useful argument. Let us look at these more closely.

"It works." This is shorthand for "It has worked in all the cases I know." It is plausible to believe that it will work in new cases not yet met. This is a highly intelligent way for people to act. Most of the time it is wise to act on the assumption that if it worked before it will work again. However, this is not true for most human experience over the long run. Consider a flower. It is blooming now. Will it be blooming a year from now? Will an automobile that works today be working a century from now? Experiences like these make some people hesitate to believe something will always work merely because it works sometimes.

"My teacher told me so." To accept the word of one's teacher is certainly wise. Most of what we know and believe we believe because we have been told. However, some people may feel that occasionally a teacher may be in error, or that it may be a good thing for free human beings not to accept this reason except when no other guide is available.

"It is postulated." Postulates are like rules of a game. If you are to play checkers, say, you must play according to the rules. You may get another very excellent game with the same pieces by changing the rules, but the new game is not checkers. By enforcing the rule $a \times b = b \times a$ we may get a very interesting game, but how do we know that this rule really belongs to the multiplication game? Accepting that it is postulated that multiplication is commutative, the question is then, Why?

To meet these demands we look at the process of multiplication itself. Is there something in the structure of multiplication, the way it arises and occurs in actual situations, that tells us whether or not $a \times b = b \times a$ for all counting numbers? We look at the definitions, of which we have several equivalent forms. The thought process will be discussed in terms of examples. The reader should repeat the argument with different examples, and then in general form.

From the repeated-addition definition of multiplication we get the question whether $3 + 3 + 3 + 3 = 4 + 4 + 4$. It is essential to answer *without* using the fact that both sides are names for 12. It is difficult to do this. Can you make an argument for the large-factor multiplications above?

Can we determine that $4 \times 3 = 3 \times 4$ from the array definition?

This last is almost trivial. If we look at the array

o o o
o o o
o o o
o o o

from the side, it appears as

o o o o
o o o o
o o o o

4×3 describes the number in the set as seen from the end. 3×4 describes the number in the same set as seen from the side. Since both expressions name the number of members in the same set, they name the same number. That is $4 \times 3 = 3 \times 4$. The argument generalizes to the product of any two counting numbers.

10.14 ASSOCIATIVE PROPERTY The associative property is easily seen from the array definition of multiplication. We illustrate with $3 \times 4 \times 5$ to show that $(3 \times 4) \times 5 = 3 \times (4 \times 5)$.

Consider the objects to be little cubes stacked together.

It is easy to imagine, if not to draw, such a brick for any three whole numbers. We conclude that $(a \times b) \times c = a \times (b \times c)$. Special attention should be paid to the cases involving 0 and 1, because the brick appears a bit different. We leave these cases to the reader.

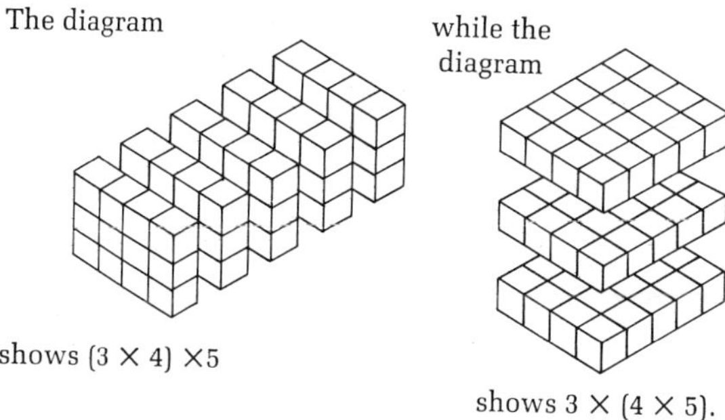

The diagram shows $(3 \times 4) \times 5$ while the diagram shows $3 \times (4 \times 5)$.

Both are expanded versions of

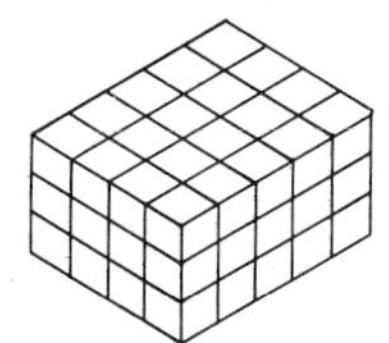

There is a danger in identifying the product of three numbers completely with the volume of a block. What would the product of four numbers be? During the Middle Ages while these ideas were being thrashed out, some people said it was impossible to multiply four numbers together, because it was impossible to have a block with four mutually perpendicular edges.

Let us try to prove the associative property from other multiplication attitudes. Repeated addition is almost useless. It would take a very keen mind to suspect that $20 + 20 + 20 = 5 + 5 + 5 + 5 + 5 + 5 + 5 + 5 + 5 + 5 + 5 + 5$.

The tree definition helps us visualize the associative property. We illustrate with $2 \times (3 \times 4)$ and $(2 \times 3) \times 4$, using small factors to keep the diagram small. The diagram can be easily imagined for any three factors.

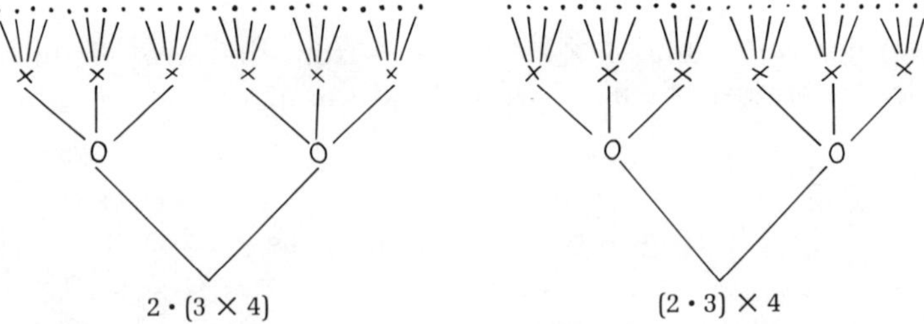

In the left figure we imagine the X connections made first, then the O connections. In the right figure the O connections are made first, then the X connections. Since the final figures are identical, the tips at the top are equivalent.

Owing to the associative property, the expression $3 \times 4 \times 5$ is well defined because it makes no difference which multiplication is done first.

10.15 DISTRIBUTIVE PROPERTY Does it make a difference in evaluating $3 + 4 \times 5$ whether the addition or multiplication is done first? Using parentheses to indicate which is performed first, we have $(3 + 4) \times 5 = 35$ and $3 + (4 \times 5) = 23$.

It does make a difference. So $3 + 4 \times 5$ is not clear, or well defined, without some further information. Since there are two possibilities, we use parentheses as above whenever there is doubt. In a written expression people agree that with no parentheses, the multiplications are performed first. By this convention, $3 + 4 \times 5$ is interpreted as 23.

Suppose we are presented with $6 \times (5 + 3)$. The parentheses say to do the addition first. Suppose that for some reason, such as convenience, we wished to multiply first. Is there any way (formula) that would enable us to multiply first, then add? This problem of interchanging the order of doing two operations is important and arises in many places. We look at an array showing the problem:

```
X X X X X    X X X
X X X X X    X X X
X X X X X    X X X
X X X X X    X X X
X X X X X    X X X
X X X X X    X X X
```

$6 \times (5 + 3)$ is the number of members in an array with 6 rows and $5 + 3$ members in each row. The above array is such, with a slight gap separating each row into 5 and 3.

This array is clearly the union of two arrays as circled below:

```
X X X X X    . . .
X X X X X    . . .
. . . . .    . . .
. . . . .    . . .
. . . . .    . . .
. . . . .    . . .
```

The numbers in these arrays are 6×5 and 6×3. So the number in the united array is $(6 \times 5) + (6 \times 3)$. Hence

$$6 \times (5 + 3) = (6 \times 5) + (6 \times 3)$$

The same argument will apply to any three whole numbers. Hence, $a \times (b + c) = (a \times b) + (a \times c)$. This property is called the *distributive* property of whole numbers. It refers to interchanging the order of performing the two operations, multiplication and addition. You may wish to refer to chapter 7. We say "Multiplication is distributive over addition."

10.16 THE MULTIPLICATION ALGORITHM We next take another look at the dance of the digits whereby we conclude such things as $4 \times 23 = 92$.

By the array definition, we count the numbers in the array.

```
————   ————   x x x
————   ————   x x x
————   ————   x x x
————   ————   x x x
```

To save writer's cramp, we indicate a set of ten by ————. Note that each row in the array has 2 tens and 3 singles. The grouping shows how the digits 2 and 3 enter the discussion.

Clearly there are 8 tens and 12 singles. We regroup the singles shown by a ring:

```
————   ————   (x x x)
————   ————   (x x x)
————   ————   (x x)x
————   ————   (x x)x
```

There are 9 tens and 2 ones. Hence, $4 \times 23 = 92$.

By the repeated addition definition, we consider

23	
23	
23	
23	
12	Adding the 3s gives a sum of 12.
80	Adding the 2s gives a sum of 80.
92	Adding these sums gives 92.

(*a*)		(*b*)		(*c*)		(*d*)
23		23		$^{1}23$		23
4		4		4		4
12	or	12	or	92	or	92
80		8				
92		92				

This numerical layout should be compared with the array pattern above. On the basis of these two patterns, we give the multiplication layouts in the margin.

For (*a*) we say to ourselves (1) "4×3 is 12" and write 12 in a convenient place—here with the units and tens digits in the same columns as the units and tens digits of the factors. (2) Then "4×2 tens is 80" and write 80 in a place convenient with the 12 as shown. (3) Then "Add 12 and 80" and write 92.

In (*b*), step 1 is the same as in (*a*). In step 2 the tens are "understood" and not specifically mentioned. We say "$4 \times 2 = 8$" and put 8 in the tens column.

In (*c*), step 1 is the same as in (*a*) and (*b*). We expect to add the one ten later to the result of the tens multiplication. Instead of writing 12, we write 2 in the units column and write a crutch number 1 in the tens column as a reminder. Sometimes we say "Write 2 and carry 1." In step 3 we say "$4 \times 2 = 8$ and 1 is 9" and write 9 in the tens column.

In (*d*) the argument is the same as in (*c*). We trust our memory and do not write the crutch number.

The reader should identify the various steps here with the array and repeated-addition layout.

The algorithm can also be justified by using the CAD properties of addition and multiplication. Place-value notation is awkward when applying the CAD rules, because the addition symbol is unwritten. Hence, we will use expanded (polynomial) notation, which has no place value; all additions are marked +. Ten is represented by t. We use a double-column presentation with reasons at the right justifying why the phrase on that line equals the one above.

Single-digit multiplications, such as $4 \times 5 = 20$, are called basic facts. Steps 4, 6, and 9 use basic facts. The rest of the steps are manipulating the numbers to get the numerals in the proper position. These positioning steps are taken care of by the positions in the standard algorithm. Since we use polynomial notation, we say $4 \times 3 = t + 2$ rather than 12. In expanded notation we would write $4 \times 5 = 2t$. We assume the 100 basic facts are known.

1.	4×23	statement of problem
2.	$= 4 \times (2t + 3)$	expanded notation
3.	$= 4 \times (2t) + (4 \times 3)$	distributive property
4.	$= 4 \times (2t) + (t + 2)$	basic fact of multiplication
5.	$= (4 \times 2)t + (t + 2)$	associative properties of multiplication
6.	$= 8t + (t + 2)$	basic fact of multiplication
7.	$= (8t + t) + 2$	associative properties of addition
8.	$= (8 + 1)t + 2$	distributive property

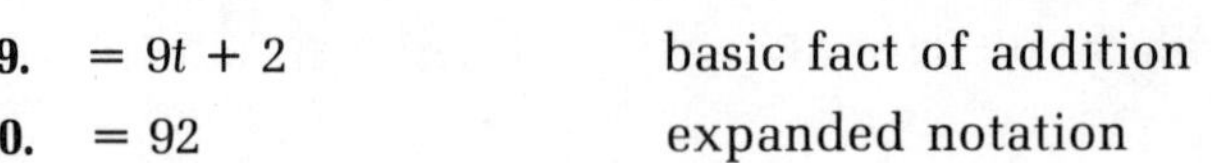

9.	$= 9t + 2$	basic fact of addition
10.	$= 92$	expanded notation

We can justify the algorithm for a larger problem, 69×45, by CAD properties. A very large number of steps are taken. The most interesting steps are uses of the distributive property and basic facts. To show these better we may combine several uses of the CA properties into single steps without showing an order for them. Further, we may stop after showing the partial products with zeros layout, as in the margin, leaving it to the student to continue to the final product.

```
  45
  69
----
  45
 360
 300
2400
----
```

1.	69×45	statement of problem
2.	$= (6t + 9) \times (4t + 5)$	polynomial (expanded) notation
3.	$= 6t(4t + 5) + 9(4t + 5)$	distributive property
4.	$= 6t \cdot 4t + 6t \cdot 5 + 9 \cdot 4t + 9 \cdot 5$	distributive property—leaving out parentheses uses CA for addition
5.	$= 6 \cdot 4t^2 + 6 \cdot 5t + 9 \cdot 4t + 9 \cdot 5$	CA for multiplication
6.	$= 24t^2 + 30t + 36t + 45$	basic multiplication facts
7.	$= 2400 + 300 + 360 + 45$	polynomial notation

These four numbers are the four addends whose sum, 3105, equals 69×45.

10.17 FRONT-END MULTIPLICATION Some people prefer front-end multiplication, doing the largest multiplications first. Consider this layout for 78×35:

```
  35
  78
----
2100
 350
 240
  40
----
2830
 7
```

This layout is quite similar to the standard layout, except that the partial products are written in different order, and the addition is done from the front.

Front-end multiplication is useful for someone wishing only an approximate value, and as a learning exercise for someone who wishes to be sure he understands the dance of the digits. Consider this example: About how much is paid all together if each of 584 people pays $376 for a TV set?

```
   584
   376
------
150000
 24000
 35000
------
209000
```

We know the answer is somewhat more than this because of the terms left out. If we next include those products which have the two right-hand digits zero, we get the layout

```
   584
   376
------
150000
 24000
 35000
  1200
  5600
  3000
------
218800
```

The correct product is somewhat more. We might estimate about 219,000 or 220,000.

10.18 CHECKING MULTIPLICATION Checking can be done by repeating the steps taken originally, interchanging the order of the factors, using more or fewer partial products, doing front-end multiplication, rounding the factors and comparing the product, dividing the product by one of the factors, using the lattice algorithm, changing to a different numeral system and multiplying, casting out nines, using odd and even, and by other ways as well. We discuss a few of these.

Repeating the steps as done originally can be done quickly, with no new writing. However, one is likely to repeat a mistake made the first time. Interchanging the order of the factors and multiplying is excellent, as is dividing the product by one of the factors. You should get the same product and the other factor in these checks. These checks, while accurate, are time-consuming. Casting out nines is discussed later under modular numbers. The odd and even checks to see if the product is of the proper sort. Front-end multiplication checks to make sure the product is about the right size. Such estimates can be done quickly and they catch many errors even if they won't catch them all.

10.19 THE LATTICE METHOD OF MULTIPLICATION This algorithm is similar to the traditional one, yet has a different format. Many people feel it should be taught before the traditional algorithm discussed earlier. We will describe the algorithm by example.

Example 1 Draw a pattern of squares with the factors as shown to find 35×42:

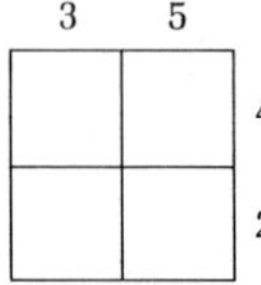

Draw the rising diagonals of each square:

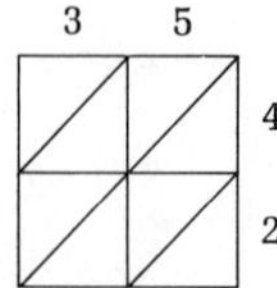

In each square, write the product of the single-digit factors at top and right, as shown:

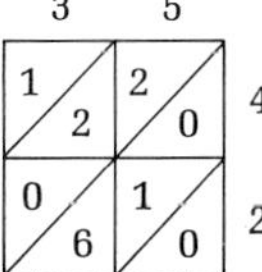

Add diagonally, with sum at the left. The product, 1470, appears to the left and below:

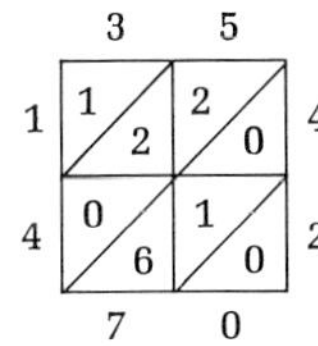

We compare with a basic-facts-without-zeros layout:

$$\begin{array}{cccc}
 & & 3 & 5 \\
 & & 4 & 2 \\ \hline
1 & 2 & & \\
 & 2 & 0 & \\
 & 0 & 6 & \\
 & & 1 & 0 \\ \hline
1 & 4 & 7 & 0
\end{array}$$

In the basic-facts-without-zeros layout, and even more so in the standard layout, it may be difficult to identify the single-digit factors with their specific product. In the lattice layout, this product appears in the column and row shown by the single-digit factors. This positioning is a help. In the lattice, we write the two-digit products on a slant, putting a zero on the left of single-digit products. That is, ${}^{1}{}_{2}$ instead of 12, and ${}^{0}{}_{6}$ instead of 6.

The places showing the same power of ten are on a slant rather than vertical $\begin{array}{ccc} & & 2 \\ & 2 & \\ 0 & & \end{array}$ instead of $\begin{array}{c} 2 \\ 2. \\ 0 \end{array}$

The order of digits appears different in the layouts, but this has no effect. The digits in the sum are placed at the end of the column (slant), with carrying as needed.

EXAMPLE 2 807 × 49 = 39543

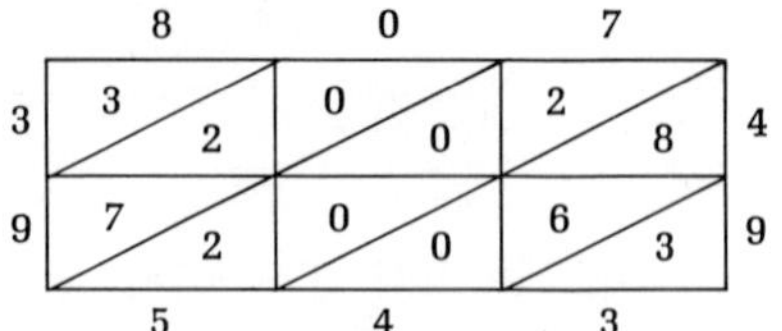

Note carrying 8 + 6 = 14; 1 + 2 + 2 = 5.

10.20 MULTIPLICATION IN ROMAN NUMERALS The array model makes this easy to do. We explain by illustration. 42 × 35 = (XXXXII) × (XXXV).

	X	X	X	V
X	C	C	C	L
X	C	C	C	L
X	C	C	C	L
X	C	C	C	L
I	X	X	X	V
I	X	X	X	V

The two factors are written down the side and across the top of an array. In the body of the array are written the products of individual terms such as X × V = L. The product is the sum of all the terms in the array.

CCCCCCCCCCCC LLLL XXXXXX VV

Grouping and converting:

(CCCCCCCCCC)	CC	(LL)	(LL)	(XXXXX)	X	(VV)
M	CC	C	C	L	X	X

The product is MCCCCLXX.

ANOTHER EXAMPLE We compare the steps in multiplying 205 × 65 in Roman numerals with the basic-facts-with-zeros algorithm in Hindu-Arabic:

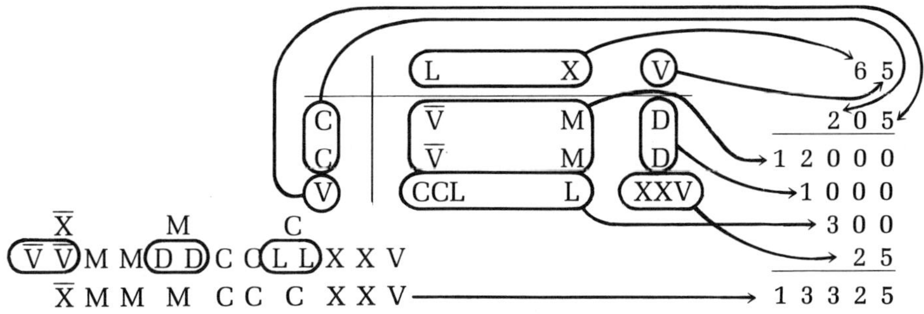

We recall that a bar above a letter multiplies its value by 1000. Hence, $\overline{\text{V}} = 5000$, $\overline{\text{X}} = 10{,}000$. Lines connect corresponding terms in the two layouts.

The layouts in Hindu-Arabic numerals are made clearer when they are compared with the layouts in Roman numerals.

10.21 MULTIPLICATION IN DIFFERENT BASES A multiplication algorithm in a base other than ten is very similar to the base-ten algorithm. We illustrate with base five.

Multiplying by 5 puts a zero on the right of the numeral. $3 \times 5 = 3 \times 10_{\text{five}} = 30_{\text{five}}$. Multiplying by $5^2 = 25 = 100_{\text{five}}$ and $5^3 = 125 = 1000_{\text{five}}$ puts 2 and 3 zeros, respectively, on the right of the numeral. This follows from the definition of the numeral. $3 \times 5^2 = 3 \times 100_{\text{five}} = 300_{\text{five}}$. $3 \times 5^3 = 3 \times 1000_{\text{five}} = 3000_{\text{five}}$.

Base-five facts

×	0	1	2	3	4
0	0	0	0	0	0
1	0	1	2	3	4
2	0	2	4	11	13
3	0	3	11	14	22
4	0	4	13	22	31

EXAMPLE $23_{\text{five}} \times 43_{\text{five}}$. Using a basic-facts–partial-products-with-zeros layout, we have

$$\begin{array}{r} 43_{\text{five}} \\ \underline{23_{\text{five}}} \\ 14_{\text{five}} \\ 220_{\text{five}} \\ 110_{\text{five}} \\ \underline{1300_{\text{five}}} \\ 2144_{\text{five}} \end{array}$$

Indenting from the right rather than writing the zeros implied by the place values of the factors, we have

$$\begin{array}{r}
43_{\text{five}} \\
\underline{23_{\text{five}}} \\
14_{\text{five}} \\
22_{\text{five}} \\
11_{\text{five}} \\
\underline{13_{\text{five}}} \\
2144_{\text{five}}
\end{array}$$

Using crutch numbers and partial addition to write only one partial product for each digit in the lower factor, we have

$$\begin{array}{r}
{}^{1}43_{\text{five}} \\
\underline{23_{\text{five}}} \\
234_{\text{five}} \\
\underline{141_{\text{five}}} \\
2144_{\text{five}}
\end{array}$$

Similarly, in base two, $17 \times 12 = 10001_{\text{two}} \times 1100_{\text{two}}$.

$$\begin{array}{r}
10001_{\text{two}} \\
\underline{1100_{\text{two}}} \\
0_{\text{two}} \\
00_{\text{two}} \\
1000100_{\text{two}} \\
\underline{10001000_{\text{two}}} \\
11001100_{\text{two}}
\end{array}
\quad \text{or} \quad
\begin{array}{r}
10001_{\text{two}} \\
\underline{1100_{\text{two}}} \\
1000100_{\text{two}} \\
\underline{10001_{\text{two}}} \\
11001100_{\text{two}}
\end{array}$$

$$\begin{aligned} \text{The product is } 2^7 + 2^6 + 2^3 + 2^2 &= 128 + 64 + 8 + 4 \\ &= 204 \end{aligned}$$

Some people find the so-called Russian-peasant algorithm for multiplication amusing. The peasant is supposed to be able to add any numbers, multiply by two, or divide by two. Using only these tools he multiplies any two whole numbers as in the next example, 23×21.

Write the two factors, 23 and 21, as heads of two columns. Get numbers in the left column by repeatedly multiplying by 2 and in the right column by repeatedly dividing by two. Throw away any remainders. The columns are extended until 1 appears on the right.

The numbers in the left column opposite even numbers in the right column are thrown away, or crossed off. The sum of the remaining numbers in the left column is the product of the original factors.

23 ×	21	23 ×	21
46	10	~~46~~	~~10~~
92	5	92	5
184	2	~~184~~	~~2~~
368	1	368	1
		483	

The algorithm is made clear by renaming 21 in expanded notation in base two, and comparing the layouts

$$\begin{aligned}
23 \times (2^4 + 2^2 + 1) &= 23 \times 2\,(2^3 + 2) + 23 = \\
46 \times (2^3 + 2) + 23 &= 46 \times 2\,(2^2 + 1) + 23 = \\
92 \times (2^2 + 1) + 23 &= 92 \times 2 \times 2 + 92 + 23 = \\
184 \times 2 + 92 + 23 &= 184 \times 2 + 92 + 23 = \\
368 + 92 + 23 &
\end{aligned}$$

This algorithm has been attributed to several parts of the world besides Russia. Folk genius is everywhere.

EXERCISES

In exercises 1–9, name the multiplication problem shown by the layout. What is the attitude used?

1. $7 + 7 + 7 + 7 + 7 + 7 =$

2. 

3. 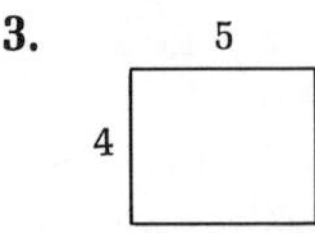

4. 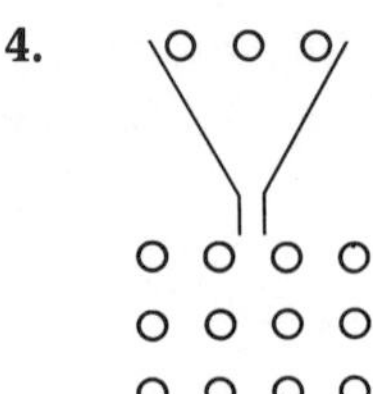

5. 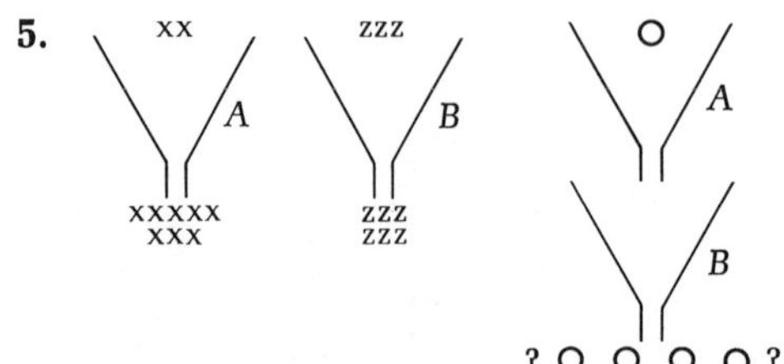

6.
o o o o o o o o
o o o o o o o o
o o o o o o o o
o o o o o o o o
o o o o o o o o

7. $\{a, e, i, o\} \times \{b, r, t\} = \{ab, er, at, ob, ir, it, ar, et, ot, eb, or, ib\}$

8. $7 \times 13 =$

9. A real-estate salesman says, "Buy a desert lot. Its value will triple the first year, and quadruple again in the next two years." What does he promise the price will do in three years?

In each of exercises 10–13 show the given attitude by using its natural algorithm to find 5×3.

10. Repeated addition
11. Array
12. Tree
13. Operator-operator-operator

14. Multiply 26 by 43 three ways using (*a*) the standard algorithm (*b*) the partial-products-with-zeros layout (*c*) the partial-products-without-zeros layout. Identify the step $6 \times 4 = 24$ in each layout.

15. Identify the step $6 + 4 + 1 = 11$ in each layout in exercise 14.

16. Multiply 23 by 23 using the standard algorithm and an array layout. Use streaks and squares where convenient. In each layout, point out what represents the underlined 2s in $\underline{2}3 \times \underline{\underline{2}}3 = 5\underline{\underline{\underline{2}}}9$.

17. Multiply 26 by 43 using an array layout. Point out what corresponds to $6 \times 4 = 24$ and $6 + 4 + 1 = 11$ as in exercises 14 and 15.

18. Multiply XXVI by XXXXIII in Roman numerals. Point out what corresponds to $6 \times 4 = 24$ and $6 + 4 + 1 = 11$.

19. Multiply 7 by 96 using the standard algorithm and partial products with zeros.

20. Expand 96 in polynomial notation. Multiply 7 by 96 in steps where each step uses one of the following types of actions:

(*a*) conversion from Hindu-Arabic numerals to polynomial notation or back
(*b*) use of a basic fact ($4 \times 7 = 2t + 8$ and $t^2 \cdot t = t^3$ we call basic facts.)
(*c*) use of distributive property
(*d*) use of commutative and/or associative properties of addition
(*e*) use of commutative and/or associative properties of multiplication

21. Multiply 26 by 43 in steps as in exercise 20.

22. Use front-end multiplication to find 7×96. Compare with exercise 19.

23. Use front-end multiplication to find 26×43. Compare with the partial-products-with-zeros algorithm and the array layout of exercises 14 and 17.

24. Use front-end multiplication to get the first three digits in the product $463{,}872 \times 756{,}408$. You may need to multiply more than the three front digits in the factors to get the three front digits in the product.

25. Use the results in exercise 24 to write $463{,}872 \times 756{,}408$ as *abc*xxx . . . x, where *abc* are the three digits from exercise 24 and x's are digits still to be determined. How many x's do you have?

26. What is the first front digit in $42{,}774 \times 23{,}380$?

27. Multiply 7 by 48 using the lattice layout and the partial-products-with-zeros layout. Identify five pairs of corresponding terms.

28. Multiply 56 by 384 using the lattice layout and the partial-products-with zeros layout. Identify corresponding terms. What in the lattice layout corresponds to the extra zeros you put into the partial-products-with-zeros layout?

29. For each attitude (repeated addition, array, tree, Cartesian product), make a short teaching plan for showing that multiplication is commutative. If your plan cannot be based on a particular attitude, so state.

30. Rank the attitudes in exercise 29 as to their usefulness as the starting attitude for showing the commutative property of multiplication. Assign grades 1, 3, 5, 7 to the attitudes according to their rank, averaging the scores for tied attitudes. The total of awarded scores is 16. Different people will grade differently, according to their taste. There is no "right" ranking.

31. As in exercise 29, make four short teaching plans for teaching that multiplication is associative.

32. As in exercise 30, score the four attitudes as starting attitudes for teaching the associative property.

33. As in exercises 29 and 30, make four short plans for teaching the partial-products-with-zeros algorithm. Grade the four attitudes as the starting attitude for teaching this algorithm.

34. As in exercises 29 and 30, make four short plans for teaching the lattice algorithm. Grade the four attitudes as the starting attitude for teaching the lattice algorithm.

35. As in exercises 29 and 30, make four short plans for teaching the standard algorithm. Grade the four attitudes as the starting attitude for teaching the standard algorithm.

36. Make a table of the grades you assigned. Add the scores for each attitude. Which has the highest score? Compare your evaluations with those of your neighbors. Can you understand the differences?

37. Locate errors in the following layouts by saying, "Instead of ______, the solver should have ______." If more than one error might have produced the layout, list the simpler ones.

(a)
```
   14
×   7
   88
```

(b)
```
   43
× 21
   43
  86
  803
```

(c)
```
    67
×  48
   566
  288
  3446
```

38. Make up and work an exercise requiring the ability to match steps in an array layout with those in a partial-products-with-zeros layout.

39. Make up and work an exercise showing ability to match steps in multiplying in Roman numerals with the partial products in the standard layout.

40. Show how to find 3×5 using the natural algorithms of the segment-segment-region attitude.

41. Repeat exercise 40 for the segment-segment-segment attitude.

42. Repeat exercise 40 for the operator-operator-operator attitude.

---

36. A possible table:

	TEACHING OBJECTIVE	Comm. Prop.	Assoc. prop.	Partial prod.	Lattice	Stnd.	Total
Mult. attitude	Repeated add.	1	1	5	1	3	11
	Array	7	5	7	7	7	33
	Tree	3	7	3	3	1	17
	Cartes. prod.	5	3	1	5	5	19

Answers to selected questions: **1.** 6×7 repeated addition; **3.** 4×5 segment-segment-region; **5.** 4×2 operator-operator-operator; **7.** 4×3 Cartesian product.

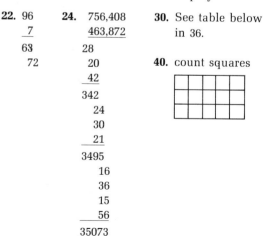

11.
```
. . .
. . .
. . .
. . .
. . .
```

20.

$96 \times 7 = (9t + 6) \cdot 7$	poly. not.
$= 9t \cdot 7 + 6 \cdot 7$	distrib. prop.
$= 7 \cdot 9t + 6 \cdot 7$	CA mult.
$= (6t + 3)t + 4t + 2$	basic fact
$= 6t \cdot t + 3t + 4t + 2$	distrib.
$= 6t^2 + 3t + 4t + 2$	basic fact
$= 6t^2 + (3 + 4)t + 2$	distrib.
$= 6t^2 + 7t + 2$	basic fact
$= 672$	poly. not.

22.
```
 96
  7
 63
  72
```

24.
```
756,408
463,872
```
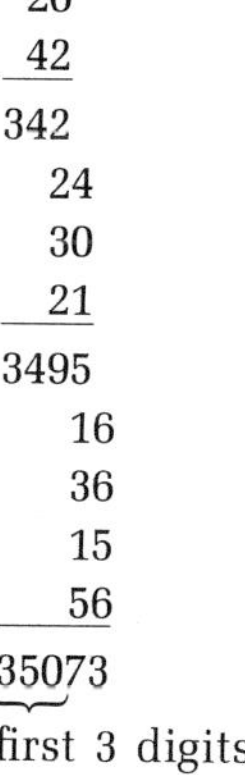
```
  28
   20
   42
 342
    24
    30
    21
 3495
     16
     36
     15
     56
35073
```
first 3 digits

30. See table below in 36.

40. count squares

11

Division

11.1 INTRODUCTION Division is the binary operation whose standard layout looks like $12\overline{)36}$ with the quotient 3 written above. Traditionally, division was regarded as a distinct operation, but today we emphasize its close relation to multiplication. We discuss the traditional attitudes as well as the modern attitudes so that you will (1) recognize when division will be a useful thing to do and (2) be confident and competent in using the division algorithm of your choice.

You do have a choice. Most people in their private lives will use the standard algorithm, the one they learned in school. However, some will change to another which appeals to them more. You will see several algorithms, including one for division in Roman numerals, so that you can understand their advantages and disadvantages.

11.2 DIVISION ATTITUDES We name the common attitudes and illustrate with $36 \div 12 = 3$.

1. *Sharing.* By $36 \div 12$ from the sharing attitude, we mean the number of objects (say apples) each person will have if 36 objects are distributed equally among 12 people.

2. *Measurement or partition.* By $36 \div 12$ from the measurement or partition attitude, we mean the number of sets of 12 objects each that can be formed from 36 objects. The name *partition* arises because the set of 36 is partitioned into parts of 12 each. The name *measurement* arises from the similarity to measuring in feet rather than in inches. We count off the number of segments of 12 inches.

3. *Missing factor.* By $36 \div 12$ we mean the number which multiplied by 12 gives 36. It is the number z satisfying $12 \cdot z = 36$. Slightly different, but equivalent, is to say, "the number which multiplies 12 to get 36, the number y satisfying $y \cdot 12 = 36$." Looking more deeply, we see that every multiplication attitude gives two missing-factor division attitudes, as the missing factor is a or b in $a \times b = c$. We leave it as an exercise to phrase these two division attitudes in terms of repeated addition, array, and tree-multiplication attitudes.

As multiplication is distributive over addition, so is right-hand division. For example,

$$\begin{aligned} 112 \div 4 &= (100 + 12) \div 4 \\ &= 100 \div 4 + 12 \div 4 \\ &= 25 + 3 = 28 \end{aligned}$$

This is useful in thinking about algorithms.

4. *Repeated subtraction.* By $36 \div 12$ from the repeated-subtraction attitude, we mean the number of times 12 can be subtracted from 36.

All these attitudes give the same result, as can be seen by looking at the following array in different ways:

```
x x x x x x x x x x x x
x x x x x x x x x x x x
x x x x x x x x x x x x
```

(*a*) Sorting the 36 objects into 12 columns to find the number in each column shows the sharing attitude.

(*b*) Asking "How many rows of 12 are there?" shows the measurement attitude.

(*c*) The array is a standard layout for multiplication, $y \cdot 12 = 36$, showing the missing-factor attitude.

(*d*) Repeated subtraction is shown by asking, "How many times can a row of 12 be taken away?"

Since these attitudes give the same answer, we may use any one of them to explain any properties or algorithms.

11.3 TERMINOLOGY AND NOTATION In the layout $12\overline{)36}$ (with the quotient 3 written above), 12 is called the *divisor*, 36 the *dividend*, and 3 the *quotient*. The problem is also written $36 \div 12 = 3$, $\frac{36}{12} = 3$, $36/12 = 3$.

From the commutative property of multiplication, or by looking at the array, the relation between the three numbers, 3, 12, 36, can be expressed in three ways: $36 \div 12 = 3$, $36 \div 3 = 12$, and $12 \times 3 = 36$. Each of these relations serves as a possible checking formula for the other two.

EXERCISE Express the relation between 28, 4, and 7 in three different ways. Similarly, show the relation between 6, 5, and 30 in three ways.

11.4 NATURAL ALGORITHMS Each attitude has corresponding natural algortihms. We illustrate some with $221 \div 13$. Note that each algorithm is successful with any numeral system.

1. *Sharing*. Form a set of 221 counters. Sort them into 13 equal piles. The quotient is the number in any pile.

2. *Measurement*. Form a set of 221 counters. Make piles of 13. The quotient is the number of piles.

Sharing and measurement attitudes are quite similar. A set is separated into subsets. In sharing, there are 13 subsets; in measurement, the number of subsets is the unknown. In sharing, the number in each subset is the unknown; in measurement, 13 is the number in each subset. Years ago people made a big thing of the differences. Experience shows, however, that while people see that the attitudes are different, they have difficulty remembering which name goes with which attitude. Often a physical situation calling for division can be looked at easily from either attitude. Sometimes, too, a person may be certain that division is called for, and that a set is to be separated, but he doesn't bother to be sure whether the divisor is to be the number of subsets or the number of members in a subset.

3. *Missing factor*. We use trial and error to find the number that multiplied by 13 gives 221. Try 7: $7 \times 13 = 91$. Too small. Try 32: $32 \times 13 = 416$. Too large. Try 20: $20 \times 13 = 260$. Too large. Try, try again. With luck we may hit on 17: $17 \times 13 = 221$. Hurrah! $221 \div 13 = 17$. This one problem might take a year or two.

4. *Repeated subtraction.* We count the number of times we can subtract 13 from 221. In the layout at the right, there are 17 subtractions.

$$\begin{array}{rl} 221 & \\ \underline{13} & \checkmark \\ 208 & \\ \underline{13} & \checkmark \\ 195 & \\ \vdots & \end{array}$$

11.5 REMAINDERS Division does not always come out even. Suppose 14 candies are to be shared by 3 children. Clearly, after each child is given 4 candies, there are 2 remaining. There are several ways of treating this problem.

1. *Deny that it exists.* We allow division only when the divisor goes into the dividend evenly. We say, "You can't divide 14 by 3.

2. *Accept the fact that there is a remainder.* We say, "14 ÷ 3 = 4 with 2 remainder," which is sometimes abbreviated to "14 ÷ 3 = 4R2."

Some may not like the expression 4R2. It is not an expression for a number. Its meaning is not clear unless we see the left side of the equation.

3. Express the relation between 14 and 3 in multiplication form: 14 = 4 × 3 + 2. One reads this equation in any of three ways:

"14 equals 4 times 3 plus 2."

"14 divided by 3 is 4 with remainder 2."

"14 divided by 4 is 3 with remainder 2."

In letters we have $n = d \cdot q + r$, where n, d, q, r are the dividend, divisor, quotient, and remainder, respectively. The remainder is less than the divisor, as is seen from the sharing process.

An expression such as $17 = 5 \cdot 3 + 2$ can be related to two division problems, 17 ÷ 5 and 17 ÷ 3. However, $41 = 7 \cdot 5 + 6$ relates only to 41 ÷ 7. Can you see why?

4. *Introduce fractions.* Then we say, "$14 \div 3 = \frac{14}{3} = 4\frac{2}{3}$." We consider this further in the discussion of fractions.

We use the particular attitude for a particular problem that suits our purposes best; that is, communicates best at that time. Other phrases equivalent to these are also used.

The word *quotient* is used ambiguously. In $\frac{14}{3}$ we sometimes say, "The quotient is 3 with 2 remainder," and at other times we say, "The quotient is $3\frac{2}{3}$." Usually, we can tell from the rest of the problem which meaning is intended. Sometimes we say "3 is the *partial* quotient."

11.6 SOME EASY DIVISION ALGORITHMS We now look at the question "Given a problem requiring division, and knowing the names of the dividend and the divisor, how do we get the name of the quotient easily?" There are several processes, some of which work in some cases but not in all, and some of which work for any two whole numbers.

1. *Consult your memory.* Everyone has memorized the multiplication facts, and knows the related division statements. From $3 \times 5 = 15$ we have $15 \div 3 = 5$ and $15 \div 5 = 3$, and so on.

A CAUTION From $0 \times 4 = 0$ and $0 \times 5 = 0$, we can conclude that $0 \div 4 = 0$, and $0 \div 5 = 0$, but not $0 \div 0 = 4$ or $0 \div 0 = 5$. We discuss these situations later.

Since not everyone remembers $17 \times 13 = 221$, this method will not work for $221 \div 13 = 17$.

2. *Consult known multiplications.* Multiplication by the base of the numeral system is particularly easy. Multiplying a base-ten numeral by 10 merely joins a zero on the right: $2735 \times 10 = 27350$. Clearly, $27350 \div 10 = 2735$. Division by 10 removes a zero from the right of a numeral.

The author recalls with sadness a student producing the layout in the margin for this problem. This is perfectly correct, but not very efficient.

```
     2735
10)27350
   20
    73
    70
     35
     30
      50
      50
       0
```

3. *Repeated subtraction.* Of the natural algorithms, this appears to be straightforward and to need little equipment or guessing. Consider $35 \div 5$. The quotient can be looked at as the number of times 5 can be subtracted from 35. That is, the number of times 5 "gozinto" 35. We give two possible layouts.

In the first layout a check is placed beside each 5, which is subtracted. The checks are counted when zero is reached. There are 7 checks, so $35 \div 5 = 7$.

```
35
 5  √
30
 5  √
25
 5  √
20
 5  √
15
 5  √
10
 5  √
 5
 5  √  7
 0
```

```
35
 5  √
30
 5  √
25
 5  √
20
20  4  7
 0
```

The second layout starts the same way. However, when the dividend is reduced to 20 we happen to recall that 4 5s are 20. Hence we subtract

20, and place a 4 instead of a single check. The 3 checks and 4 make 7, so 35 ÷ 5 = 7.

EXAMPLE 3695 ÷ 5. Subtracting 5s one at a time will be lengthy. We can subtract 10 5s = 50, or 100 5s = 500, but not 1000 5s = 5000. 200 5s = 1000 appears even easier. We keep subtracting 200 5s until the remainder is less than 1000. We make appropriate changes. We conclude 3695 ÷ 5 = 739.

3695	
1000	200
2695	
1000	200
1695	
1000	200
695	
500	100
195	
100	20
95	
50	10
45	
25	5
20	
5	1
15	
5	1
10	
10	2
0	739

The number of 5s subtracted at any stage is flexible. Someone else might have started by subtracting 100 5s, or 400 5s.

There are fewer steps if the 5s are subtracted in larger sets than in smaller sets or singly. There are eight steps in this layout. How many steps would there be if the 5s were subtracted one at a time? What is the fewest number of steps?

The smallest number of steps is one—for the person who can recognize at a glance that 5 × 739 = 3695. Probably there are few such persons.

This method of dividing is called the *scaffold* algorithm, perhaps because the layout suggests a scaffold beside a building.

4. *Repeated subtraction in a small number of steps.* It is easy to subtract in 10s, 100s, or 1000s. We can see at a glance if that multiple of the divisor is less than the current dividend—the remainder from the previous subtraction. It is almost as easy to subtract in 20s, 200s, or 2000s . . . , 30s, 300s, . . . if we think ahead a bit. In the example 3695 ÷ 5 we saw that 200 × 5 = 1000, and subtracted by 200s.

Even a casual glance at 3695 suggests we subtract 3000 or 600 5s. Indeed, the 3600 in 3695 tells us we could subtract 3500 = 5 × 7 × 100, or seven 100 5s. We could not subtract eight 100 5s. In terms of hundreds of 5s, 700 is the most. We take as a first step

3695	
3500	700
195	

In multiplication form, 3695 = 5 × 700 + 195.

The remainder, 195, is greater than 50, so it contains tens of 5s. The tens (and greater) part of 195 shows that 5 10s, or 50, are contained in 195 three times, but not four. So in the next step we subtract 30 5s.

3695	
3500	700
195	
150	30
45	

```
3695
3500    700
----
 195
 150     30
----
  45
  45      9
----    ---
   0    739
```

Recalling that $5 \times 9 = 45$, we subtract 9 5s at the last step.

11.7 THE STANDARD DIVISION ALGORITHM The standard division algorithm can be regarded as a contraction of the efficient repeated-subtraction algorithm. We put the layouts side by side to show this.

```
    645
 7)4516      7)4516
   42          4200    600
   --          ----
   31           316
   28           280     40
   --           ---
    36           36
    35           35      5
    --           --    ---
     1            1    645
```

The place-value system of numeration automatically keeps track of the multiples of 10 being used. It is customary to think, in the standard algorithm, as follows. "7 won't go into 4. 7 goes into 45 6 times. Write 6 over the 5." This compares with the right-hand layout: "$4516 < 7000$. We cannot subtract a 1000 7s. We can only subtract hundreds of 7s. There are six 100s of 7s, since $600 \times 7 = 4200$ and $4200 < 4516$." The standard algorithm has us write the 6 over the 5 in 4516. The 5 is in the hundreds place in 4516, and the 6 is in the hundreds place in "600 7s." Placing the 6 here keeps track of its place value.

The standard algorithm continues. "$45 - 42 = 3$." This is really short for "$4500 - 4200 = 300$," with the extra zeros not mentioned, but their value implied by the position of 42 and 3 in the layout. "Bring down the 1." There is still a remainder of 316. In the next stage we are subtracting tens of 7s. We are not interested in the units. Bringing down 1 causes us to look only at the 10s: 31. 7 goes into this 4 10s but not 5 10s times. The position shows that we are working with 10s. We need look only at $31 \div 7$, putting 4 in the tens place in the quotient, above the tens place in the dividend, where 1 is. Writing the remainder 3 in the tens position and bringing down the 6 we look at the complete remainder: 36. 7 is contained in 36 5 times. We write 5 in the units place in the quotient, subtract $5 \times 7 = 35$, and have 1 remainder.

With this algorithm it is important to have the maximum multiple of the divisor. For example, if at the second step we estimated "3" rather than "4," we would have

```
   63
7)4516
  42
  31
  21
  10
```

We can still subtract 10 7s from the 10–(or 106) remaining. The tens spot in the quotient is filled by the 3. We erase, or cross out, the 3, and try over with 4:

    4
   6~~3~~5
7)4516
  42
  31
  ~~21~~
  28
  ~~10~~
   36
   35
    1

The resulting layout is messy. Some people prefer the previous algorithm for this reason.

EXAMPLE We compare the layouts for 9642 ÷ 23.

```
    419
23)9642        23)9642
   92              9200    400
   44               442
   23               230     10
   212              212
   207              207      9
     5                5    419
```

Trial quotients To avoid messing up the layout, we sometimes multiply the divisor by a trial quotient at the side, or write the product lightly.

EXAMPLE $\quad 48\overline{)3964}$

What is the tens digit in the quotient? It is 396 ÷ 48. We don't recognize this digit, so we guess. We get help by seeing what we'd have if he dropped the last digits: 39 ÷ 4 = 9. We try 9, which is called the *trial quotient.*

```
     9
48)3964
   432
```

9 × 48 = 432. 432 > 396. 9 is clearly too big. We choose a new number to try—a new trial quotient. 48 is almost 50. 396 ÷ 50 = 7 + (that is, 7 and some more). We try 7 as a new trial quotient.

```
     7
48)3964
   336
    60
```

Since 60 > 48, we could subtract another 48 from 396. So we try 8.

```
     8
48)3964
   384
    124
```

384 < 396, and 12 < 48, so 8 is the tens digit in the quotient.

For the units digit we need 124 ÷ 48. This is approximately 12 10s divided by 4 10s, or about 3. We check this trial quotient by multiplying in our heads or at the side: 48 × 3 = 144. 144 > 124, so 3 is too big. 2 looks reasonable.

```
     82
48)3964
   384
    124
     96
     28
```

Hence, 3964 ÷ 48 = 82 with remainder 28. In multiplication form, 3964 = 48 × 82 + 28.

The scaffold layout is easier to explain, while the standard layout uses fewer digits. Some people shift to the scaffold layout after meeting it for the first time late in life.

11.8 DIVISION IN OTHER BASES Division algorithms in other bases are very similar to those in base ten. We leave it as an exercise to develop such an algorithm for, say, $3412_{\text{five}} \div 4$ and $10110_{\text{two}} \div 11_{\text{two}}$.

11.9 DIVISION IN ROMAN NUMERALS The scaffold algorithm is successful for division in Roman numerals. We subtract convenient multiples of the divisor from the dividend, then add all the multiples. In the algorithm we use the addition, subtraction, and multiplication facts for Roman numerals. A small × means multiplication, while large X means ten. A few multiplication facts are 1 × X = X, V × X = L, X × C = M, X × M = $\overline{\text{X}}$, V × L = CCL, V × D = MMD. We use all numerals in additive form, with no subtractive forms like IV for 4.

CCCLXXXXVI ÷ XXIII

Since there are 3 C's in the dividend, only 2 X's in the divisor, and 10 X's = C, it appears that the divisor can be subtracted X times.

```
XXIII)CCC  L  XXXX  V
        CC        XXX                          X       We shall not
      C    L X         V                               try to be
      XXXXXXI I I I I I I I                    I I I   efficient.
    L  XXXXXX  XXXX  V  V  I I I I   I                 Some
       XXXXXX  I I I I I I I I I                       renaming is
    L      XXXX       V    I                           necessary.
              XX    I I I                      I
    L      XX  I I I
           XX  I I I                           I
    L    or                                            Renaming
    XXXX  I I I I I I I I I I
    XXXX  I I I I I I                          I I
              I I I I                          XI I I I I I I = XVII
```

Hence, CCCLXXXXV ÷ XXIII = XVII with IIII remainder. In multiplication pattern, CCCLXXXXV = XXIII × XVII + IIII.

11.10 MISSING-FACTOR ALGORITHM We can also look at the algorithm from a missing-factor point of view rather than one that emphasizes division as repeated subtraction. In $3964 \div 48$, we are looking for the missing factor $\square$ and remainder $\triangle$ in the sentence

$$3964 = 48 \times \square + \triangle$$

The number $\square$ is between 10 and 100, since $480 < 3964 < 4800$. If a is the number of tens and b the number of units, $\square = a \times 10 + b$. The algorithm gives us a way of determining a and b.

$$\begin{array}{r} 8 \\ 48\overline{)3964} \\ \underline{384} \\ 12 \end{array}$$

We wish to split 3964 into three parts: $3964 = 48 \cdot a \cdot t + 48 \cdot b + \triangle$. The layout in the margin tells us

$$\begin{aligned} 3964 &= 3840 + 124 \\ &= 48 \cdot 8 \cdot t + 124 \\ &= 48 \cdot a \cdot t + 48 \cdot b + \triangle \end{aligned}$$

Hence, $a = 8$. Subtracting shows we wish to split 124:

$$124 = 48b + \triangle$$

We use either

$$\begin{array}{r} 2 \\ 48\overline{)124} \\ \underline{96} \\ 28 \end{array} \quad \text{or} \quad \begin{array}{r} 82 \\ 48\overline{)3964} \\ \underline{384} \\ 124 \\ \underline{96} \\ 28 \end{array}$$

finding $124 = 48 \cdot 2 + 28$. Hence

$$\begin{aligned} 3964 &= 3840 + 124 \\ &= 3840 + 96 + 28 \\ &= 48 \cdot 8 \cdot t + 48 \cdot 2 + 28 \\ &= 48(8t + 2) + 28 \\ &= 48 \cdot 82 + 28 \end{aligned}$$

The standard layout can bc regarded as showing all these steps—several by the layout position and pattern.

11.11 CHECKING DIVISION As in other operations, the first and most important check is to see if the answer appears about right. One way this can be done is by rounding the dividend and divisor. 3764 ÷ 72 is about 4000 ÷ 70 and should be about 50. Perhaps the most accurate check is to multiply quotient and divisor. To check whether 3635 ÷ 49 = 74 with 9 remainder, one thinks of the multiplication form 3635 = 49 × 74 + 9, expands the right-hand side, and compares with 3635. Doing the problem over is a good check. Usually an error will be a simple manipulative error, so switching to something different will be useful if the above checks don't satisfy you. Switching to the scaffold layout, or converting and dividing in another numeral system, is good. Separating 3635 into two addends makes a good check. That is, since 3635 = 3145 + 490, then 3635 ÷ 49 = 3145 ÷ 49 + 490 ÷ 49. But 3145 ÷ 49 = 64 remainder 9, and 490 ÷ 49 = 10. Therefore, 3635 ÷ 49 = 64 remainder 9 + 10 = 74 remainder 9.

Sometimes people can remember some numbers, simplifying the layout to *short division*. For example, one mutters something like "6 goes into 8 once with 2 remainder. Write 1 above the 8. 6 goes into 29, 4 times with 5 remainder. 6 goes into 52, 8 times with 4 left over. 6 goes into 42, 7 times." The standard layout, in contrast, is called *long division*. Can you match the steps in the long division layout with the mutterings?

```
   1487         1487
6)8922       6)8922
               6
               29
               24
                52
                48
                 42
                 42
```

Most people can learn to do short division with single-digit divisors, a few with small double-digit divisors. Can you see $11\overline{)8932}$ with quotient 812 written above?

Short division makes a good check after rounding the divisor to one digit.

EXERCISES

1. Compute (*a*) 24 ÷ 6 = (*b*) 108 ÷ 9 =

2. Draw layouts that would show 24 ÷ 6 = 4 from the following attitudes:

(*a*) repeated subtraction
(*b*) sharing (find number in subset)
(*c*) measurement (find number of subsets)

3. State two division problems (such as 6 ÷ 2 = 3) associated with 37 × 23 = 851.

4. In bridge, 52 cards are dealt to 4 players. How many does each player hold? What division problem is shown here? What division attitude is illustrated?

5. What division problem is suggested by 7 × □ = 56?

6. Jennifer wishes to show 18 ÷ 3 using missing factor. Show two array diagrams that may be used.

7. Show how 15 ÷ 5 can be illustrated using missing factor and a tree diagram.

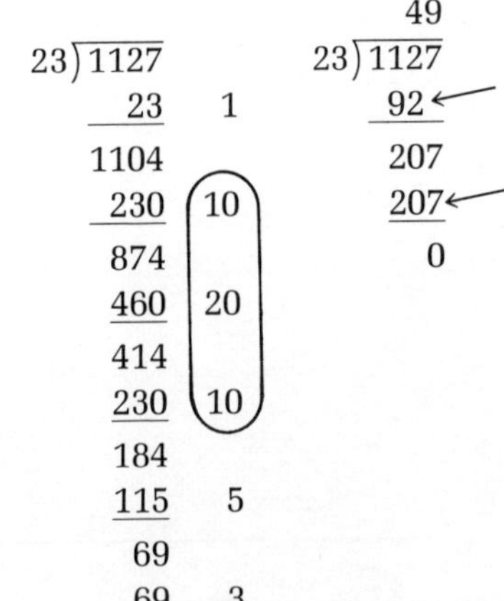

8. Match the terms *given factor, missing factor,* and *product* with the terms *dividend, divisor,* and *quotient* and the numbers in 10 ÷ 2 = 5.

9. Candy lost her glasses at a picnic. She promised 3 nickels to each cub scout who would help look for them. If she has 60 cents, how many scouts can she hire? What division attitude are you using?

10. Divide LVI by VII using repeated subtraction.

11. Divide LVI by VII using missing-factor and array multiplication.

12. (*a*) What in the standard layout matches the numbers circled in the scaffold layout? (See margin.)
(*b*) What in the scaffold layout matches 92 in the standard layout?
(*c*) What in the scaffold layout matches 207 in the standard layout?

13. (*a*) What in the standard layout matches the circled 160 and 80 in the scaffold layout? (See margin.)
(*b*) What in the scaffold layout matches the upper 32 in the standard layout? The lower 32?

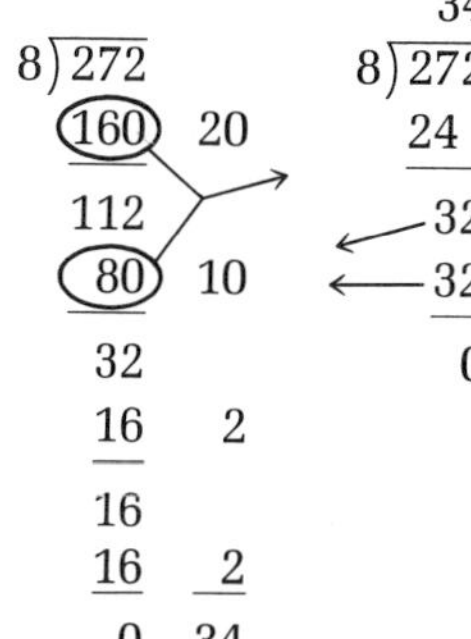

14. Show that the steps in the standard layout for 272 ÷ 8 can be matched with finding x and y in $272 = 8(10x + y)$. This is explaining the standard algorithm from the missing-factor attitude.

15. Explain the standard algorithm for 899 ÷ 29 from the missing-factor attitude as in exercise 14.

16. Divide MMCCLXV by XV in Roman numerals. What division attitude are you using?

17. Divide 2265 by 15. Match two of the steps with corresponding steps in exercise 16.

18. Match digits in the standard division algorithm with x, y, z in $855 = 5(100x + 10y + z)$.

19. Divide 47 by 8 to get a quotient and remainder.

20. Write the result of problem 19 in multiplication form $n = d \times q + r$.

21. Write two division problems from $87 = 12 \times 7 + 3$.

22. Write a division problem from $395 = 9 \times 42 + 17$.

23. Why can two division problems be written from the multiplication in exercise 21, but only one from exercise 22?

24. Write the results of 693 ÷ 37 in multiplication form.

25. Divide MCL by XV in Roman numerals.

26. Divide 1150 by 15. Match two figures in this layout with those in the layout for exercise 25.

27. How many 5-card poker hands can be dealt from a deck of 52 cards if at least 13 cards must be left for drawing?

28. How many cards will each person have if as many as possible are dealt evenly to 5 people, and at least 13 cards must remain undealt from the pack of 52?

Answers to selected questions:

2. (*a*)

```
24
 6 √
18
 6 √
12
 6 √
 6
 6 √
 0
```

(*b*) 0000 / 0000 / 0000 / 0000 / 0000 / 0000

(*c*) four columns of o o o o o o

3. 851 ÷ 37 = 23; 851 ÷ 23 = 37; **5.** 56 ÷ 7; **8.** given factor: divisor 2; missing factor: quotient 5; product: dividend 10.

10.

```
 LVI
 VI I
 XXXXVI III
 VI I
 XXXVVII
 VII
 XXVVIIIII
 VII
 XXVI II
 VI I
 XVI I IIII
 VI I
 VVI I II
 VI I
 VI I
 VI I
```

12. (*a*) 4 in 49

(*b*)

```
  230
  460
+ 230
```

(*c*) 23 + 115 + 69

29. How many yards are there in 60 feet? Is this problem more like exercise 28 or exercise 29? Why?

30. How many celery sticks were at the party if each of 14 people ate 7, and 5 were left over?

31. Starting with 6132 beans Joseph sorts them into 84 equal piles, with maybe a few left over. Sandra starts with 6132 beans and makes little piles with 84 in each. Do you believe that any beans they have left over will show the same number? Do you believe the number of piles Sandra gets equals the number of beans in each of Joseph's piles? Why?

32. If Joseph and Sandra started with a different number, say 8377, would your conclusions be the same? Do you need to divide?

---

14. What number multiplied by 8 gives a product less than 272? (30). What number multiplied by 8 gives a product 32? (4).

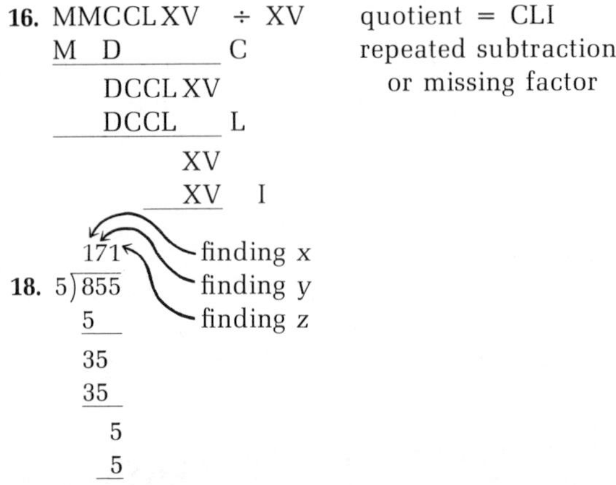

20. 47 = 8 × 5 + 7; **23.** remainder > 9; **27.** 7; **29.** 20; 27; measurement or partition not sharing; **31.** yes; yes; both can be matched with the same array.

12

Measurement and Denominate Numbers

12.1 INTRODUCTION Phrases like "10 feet," "4 hours," "15 kilometers per hour," "20 pounds" are called *denominate numbers,* or numbers with names of units. They result from measurements. The number part is called the *measure,* and the name gives the unit. In "15 pounds," "15" is the measure and "pound" is the unit.

While measuring is a physical process, we discuss it because measurements of length are the foundation of several theories of numbers and number operations. Many of the most important applications of arithmetic come in measurement. Approximation and rounding off play much more important roles in measurement than they do in counting.

There are various levels of knowledge about measurement. Most people have a rote knowledge of basic measuring. They count the number of teaspoons of sugar. They read the dial on the scale at the grocery and the markings on a ruler. Deeper knowledge comes from answering the following questions:

1. Why do people measure?

2. How big is ______? Can you imagine a region of 30 square feet, a length of 8 centimeters, a speed of 10 feet per second? To have a vivid personal understanding of such denominate numbers, it helps to know how measurements are made (or at least might be made).

3. How many feet long is 48 inches? Conversion problems like this come up when units are changed.

4. How do the operations on whole numbers generalize to denominate numbers? Can we add (multiply) 5 feet to (by) 18 inches or 18 minutes?

5. How exact are measurements? How can one discuss errors and approximations reasonably?

6. What is the difference between metric units and the traditional English units?

You will find some answers to these questions in this chapter.

12.2 THE PURPOSE AND NATURE OF MEASUREMENT A measurement serves three purposes. It enables people to reproduce a twin of the same size as the measured object without direct comparison. It gives people a sense of "how big" the object is without ever seeing it. It furnishes a useful standard for comparing value (the apples are 39 cents per pound, 100°, 60 kilometers per hour).

Measurement answers the question "How long is a mystery segment?" or "How big. . . . ?" or "How much . . . ?" Such a measurement is a matching with a known familiar quantity. We say, "The segment is as long as my hand is wide," "About as far as from my home to school," "6 inches," "8 minutes," "5 kilograms." We see a direct comparison between measurement and counting when a mystery set is matched 1:1 with a known familiar set–usually a counting set such as 1, 2, 3, 4.

While there are standard units in every field of life, these units may have little personal impact. Consider the little old lady who regularly drank a Gold Elixir of Life because it was guaranteed that if she drank it for only 1200 months, she would live to be 100 years old. We strongly recommend that everyone have his own personal common object to think about as a match for the common units. Some possibilities are mentioned in later sections of this chapter.

We also think of measurement as the ratio of the mystery object to a standard unit object of the same nature. Something weighs 5 pounds if the ratio of its weight to a standard 1-pound object is 5. Ratio is discussed again in chapters 17 and 23.

Some people like to include counting discrete objects as a measurement operation when they get answers like "7 cars" and "3 dozen eggs." This is a matter of taste. Certainly these are denominate numbers, and have many of the properties of measurement. However, we will discuss measurement of continuous objects, where the units can be chosen at the convenience of the measurer. We will leave it as an excellent exercise to determine which statements also apply to the counting of discrete objects.

While we instinctively believe there is a true value, we learn from experience and theoretical physics that all measurements are approximate. Except in imagination we cannot have a stick exactly 3 feet long. If two sticks, each measuring 3 feet long, are put side by side, they will not match exactly. Whenever this or a corresponding experiment has been tried, careful observation has detected a negligible, but nonzero mismatch. This contrasts with counting sets. We can have 6 guests for a party, and never, never have 5.999 guests.

Accurate measurement is the basis of modern construction of buildings, refrigerators, and the like. It may be a surprise that this is not true everywhere. In many places of the world today, as in former times, people build by eye, putting parts in so they fit. The man who first discovered how to make two wheels about the same size, with axles close to the center, must have made a fortune in the chariot business.

Buying, selling, and trading in general are swap situations. Each person tries to make his offering more attractive than the others. Measurement is a magnificent tool for foiling the packaging specialists, who are expert at making a little seem like a lot. Measurement helps people make better judgments and get what is really wanted, undistracted by clever camouflage.

12.3 MEASURING SEGMENTS We measure a segment when we say it is "as long as ———," naming a known familiar segment. Usually the familiar segment is a multiple of a standard unit, like "6 inches," but it need not be.

A scale or ruler is often used. A scale may be thought of as many familiar segments on top of each other with the same starting point and different ending points. A scale is a segment with points labeled, usually with numbers. The segment being measured is placed with one end at the starting point. The number on the scale closest to the other end is then proclaimed as the length of the segment. In the following diagram we say $\overline{AB}$ is 4 long.

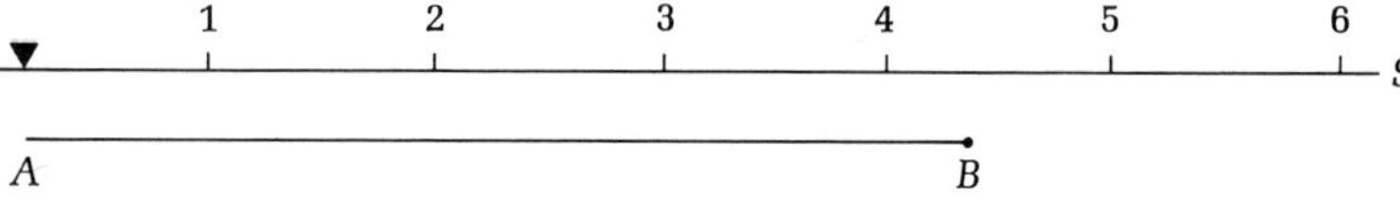

Usually the scale is marked in equal intervals. With experience we often measure closer than to the nearest labeled point; for example, we might say $\overline{AB}$ is 4.4 long. We do this by eye. This is equivalent to pretending we have a finer net of labeled marks and are reading to the closest of these. Clearly there is a limit to how fine we can make such visual estimates. Since this closer reading adds nothing to the theory, we will assume in what follows that all marks used are labeled, that we read to the closest mark, and that with the instruments and skill available it is impossible to read closer.

The scale need not be uniform (with equal intervals) and the numbers need not increase in standard order. On scale T, we would say $\overline{AB}$ is 8 long.

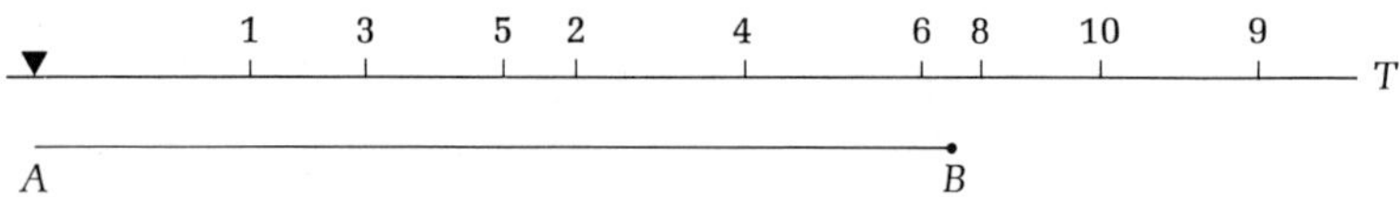

Measuring on scale T is quite offensive to us. We like the scales we are used to, because (1) the intervals are equal, (2) the labels are in standard order, and (3) the scale is a model of the number line.

Scale T is actually more accurate than scale S, because the marked points are closer together on the average. The following experiment will show that T is better than S when it comes to giving the information "how long." Prepare a mystery segment such as $\overline{AB}$. Determine its length by scale S (say 4). Draw the segment showing the error (4 B). Similarly measure $\overline{AB}$ by scale T (say 8) and draw the segment showing the error (B 8). For a single segment there is no strong reason to insist that the error from one scale is larger or smaller than the error from the other. But if we measure many segments of various lengths, we will see that the errors from scale T are less on the average than on scale S. In particular, if we put the error segments from S end to end, the segment so formed will be longer than the segment formed by putting the error segments from T end to end.

Various tools are used to measure the length of segments. We list some of them:

1. *Blocks.* High-precision machinists use steel blocks with thickness very carefully made to prescribed dimensions—1 inch, $\frac{1}{2}$ inch, 1 centimeter.

Several of these blocks are stacked together to make a precise segment to compare with an unknown segment.

2. *Feeler gauges*—thin strips of metal with carefully made thickness. One or more of these can be used to measure a gap between two points, as in a spark plug, or the width of a crack at the edge of a drawer or door. Sheets of paper make good feeler gauges. The number that can be put into a crack gives the size of the crack. The thickness of the paper is measured by, say, 1/500 the thickness of a pad of 500 sheets.

3. *Rulers.* Yardsticks and meter sticks have scales marked on straight sticks of wood, plastic, or metal. Sometimes the zero on the scale is at the end of the stick, sometimes just in from the end, and rarely in the middle. If the zero is at the end, the sticks can be put end to end. If not, the sticks must overlap if two sticks are combined in making a measurement. On worn sticks the first marked centimeter from the end is sometimes less than a centimeter long. Hence, for more accurate work, people do not use the end of the stick for measurement.

4. *Tapes and chains.* Steel tapes that roll into convenient packets are useful. Often the first few inches have finer graduations to make accurate measurement easier. Surveyors use very long tapes called chains, because years ago they used actual chains.

5. *Light.* Measurements of the most extreme accuracy are made by the use of light of one color. Standards are expressed as so many wavelengths of a light of a certain color. Various laser and radar beams can be used to measure distance by measuring the time it takes the light to travel the segment being measured. Distances to the moon are measured this way.

6. *Personal units.* Some people find that the length of a finger joint is about one inch, the width of a fingernail is about one centimeter, their step is about 30 inches long, the distance from their nose to the end of their outstretched arm is about one yard, the distance they can distinguish the two eyes in a face is about 100 yards. You may wish to use some such personal units that you have checked.

A standard scale is a uniform scale. Some standard segment (meter, inch, foot, . . .) is copied many times. These copies are laid end to end. The endpoints of each copy are marked and then numbered in sequence to form a scale. Clearly, a scale is very similar to a number line. In many ways we regard the terms as the same. Here are portions of an inch scale and a centimeter scale.

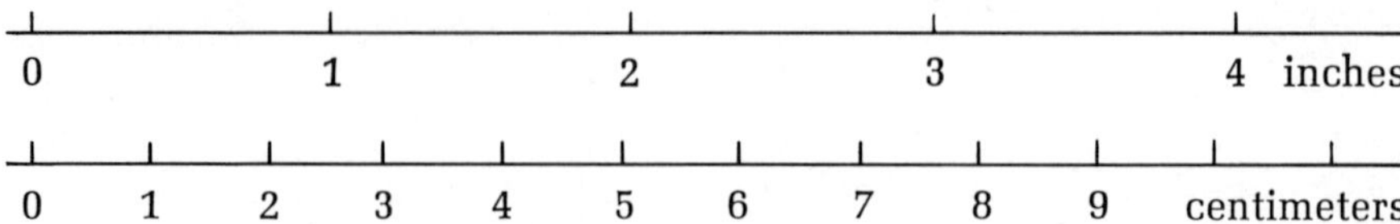

If a segment being measured is not negligibly close to an exact number of units long, we usually use a scale with smaller units, such as halves, quarters, eighths, tenths of the unit. This problem was probably a major reason for inventing fractions. We will not use fractions in this chapter, and will assume that all measurements are close enough when given in whole units.

Equivalent to reading off a scale, and perhaps even a more basic idea, is counting the number of units "in" the segment measured. The width of a floor is measured by counting the number of tiles across. The distance between two places in a field is measured by counting the number of steps. An automobile odometer counts the number of revolutions of the wheels. For many purposes, defining length as the number of unit segments is more satisfactory than reading from a scale.

Measurements obtained by directly comparing the object with a scale or unit are called *direct* measurements. Measurements using a formula combining measures of other things are called *indirect*. For example, we can measure the diameter of a circle directly with a ruler. We can measure the circumference indirectly by multiplying the diameter by π (about 3.14). How would you measure the circumference directly?

Sometimes distance is measured indirectly from speed and time measurements. How far does he go if he travels 3 hours at 40 miles per hour?

If no standard scale or tape measure is available, people may measure distance in unusual units. For example, several children together can measure the width of the room in "books." Each child holds his book next to the book next to his until the width of the room is spanned by books. The children then count the books. This shows that the unit one uses can be chosen by the measurer, and does not depend on the object measured.

In every case, measurements are approximations or estimates. Different measures of the same segment are not exactly the same. Some people call measuring by eye "estimating" and take "measuring" to mean measuring with an instrument. There are extensive discussions beyond the scope of this book on ways of finding the length of segments with accuracy and low cost. Usually a very good measurement is costly, so in practice some trade-off is made between accuracy and expense.

12.4 UNITS AND CONVERSION Clearly, the answer to "how long" depends not only on the segment but on the unit used. We will wish to express length in different units. For example, a segment 3 feet long is also 36 inches long, 1 yard long, and 91 centimeters long. (91.44 centimeters is closer, but we report to the nearest centimeter.) 3 feet = 36 inches in the sense that both are measures of the length of the same segment.

A common problem is converting lengths from one unit to another—that is, given the length in one unit, determining the length in another.

Perhaps a scale marked in two units is available:

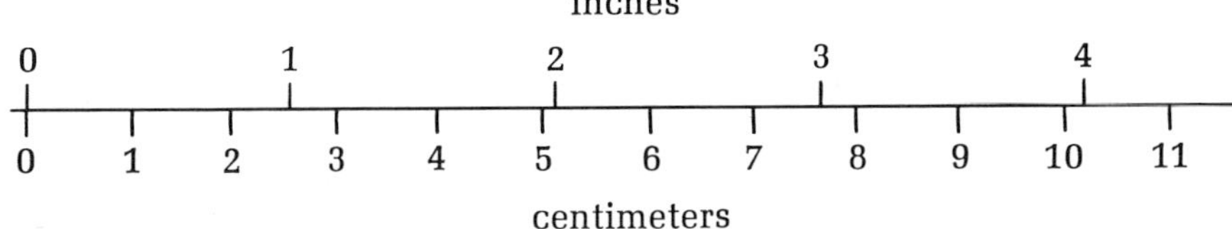

From this scale we can read the number of centimeters in 3 inches (8), or the number of inches in 10 centimeters (4).

Lacking such a scale, we can convert as in the following examples. (The size of one unit in terms of the other is a key idea.)

EXAMPLE How many yards in 36 feet? Since 3 feet form a yard, we ask how many sets of 3 are there in 36? There are 12 sets; hence there are 12 yards in 36 feet.

EXAMPLE How many inches in 4 feet? We can imagine 4 one-foot segments, end to end. Each foot is replaced by 12 one-inch segments. Hence there are $4 \times 12 = 48$ inches.

Good notation in mathematics suggests correct steps, even though the correct reasons may be quite different. The phrase "4 feet" suggests "4 times 1 foot" just as "4 sevens" means "4 times seven." Since 1 foot = 12 inches, we substitute 12 inches for 1 foot and get 4 feet = 4×12 inches. Similarly, since 1 foot = $\frac{1}{3}$ yard, 36 feet = $36 \times \frac{1}{3}$ yard = 12 yards.

EXAMPLE How many centimeters in 5 yards? We imagine each yard separated into 3 feet, each foot into 12 inches, and each inch into 2.54 centimeters. In 5 yards there are $5 \times 3 = 15$ feet; $15 \times 12 = 180$ inches; $180 \times 2.54 = 457.20$ centimeters.

12.5 MEASUREMENT OF FORCE AND WEIGHT Pushes, pulls, and weights are examples of forces that are measured by comparison with a standard force, such as a pound or a gram. These standard forces are usually the weights of certain standard masses or quantities of material. When we say a boy weighs 98 pounds, we mean the earth pulls on him with the same force that it pulls on a set of 98 one-pound masses.

It turns out that it is a tricky problem to compare forces directly. Usually some sort of lever and a measurement of distance are also involved. Here are four common ways of measuring weight:

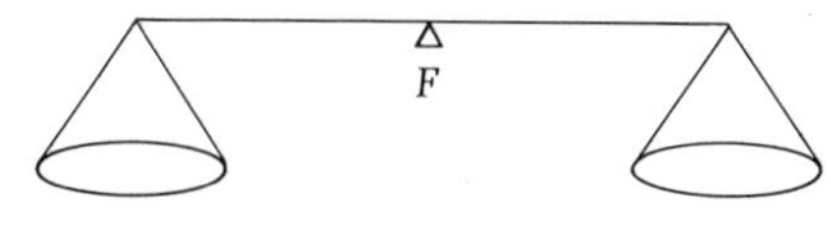

1. *Double-pan balance.* Two similar pans are suspended from a bar. The pans balance at a fulcrum F, or knife edge, at the center of the bar. If two masses are placed one in each dish, and the pans still balance, we say the weights are equal. If one of the masses consists of, say, 3 one-pound weights, the other mass weighs 3 pounds. The equality principle says that the product force times distance from F must be the same on both sides to balance.

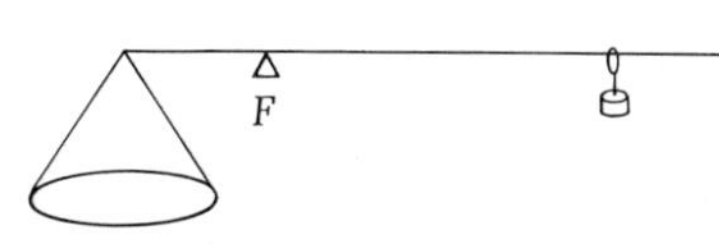

2. *Variable-arm balance.* The quantity to be weighed is placed in a pan balanced by a fixed weight which slides along the bar. The equality principle is the same as before, force times distance from F must be the same on both sides. Here the variable quantity in the pan is compensated for by the variable distance to a constant weight on the other side. Many platform scales work on this principle.

3. *Pointer variable-arm balance.* This device uses the same principle as the one in method 2, but has a pointer showing the weight. It is used in many grocery stores.

4. *A force can be measured by the amount it stretches a spring.* Kitchen and bathroom scales work on this principle.

The most common units of weight are ounce, pound, ton, gram, kilogram. When we say something weighs 3 pounds, we mean (1) it will balance 3 one-pound weights in a double-pan balance, (2) it will balance when the sliding weight or pointer indicates 3 pounds, or (3) it will stretch a spring as much as 3 pounds will stretch it.

We learn in physics that mass and weight are different. Mass represents an amount of material; weight is a force. Under standard conditions, a 1-pound mass is pulled by gravity with 1 pound of force. The same mass weighs only 1/6 as much on the moon. Further discussion is left to physics classes.

Some useful personal weights are a pound of margarine, a 5 cent coin weighs 5 grams, your own weight, a gallon jug of water weighs a little less than 10 pounds, a full-size car weighs about 2 tons.

12.6 MEASUREMENT OF AREA Area is the size of a region. Terms such as "47 square inches," "6 acres," "17 square centimeters" are common. Such denominate numbers are comparisons. They say, "This region is the same size as 47 squares, 1 inch on a side," or "It is as big as 6 plots, each 1 acre in size." Every area measure involves comparing with some standard region.

How does one compare two regions? How does the number 2 come up when one says that the area of ⌓ is 2 times the area of □ ?

Directly, it is done by tiling, tesselating, or cutting and pasting. We imagine the standard or unit region made of paper, with many copies. We suppose that we can cut and paste two of the regions □ so that the result exactly covers the given region.

The rectangular region ⊞ is exactly covered by 6 unit regions—no cutting necessary. The rectangle is said to be tesselated, or tiled, by the 6 squares. The rectangle is completely covered, with no gaps or overlaps.

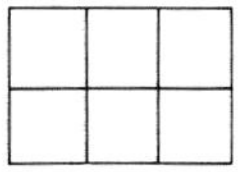

The triangle ◺ can be covered by one unit by means of an appropriate cut and paste.

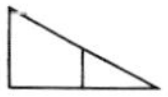

What sort of region makes a good standard region? Many different types have been used. While politics and personal pique (I want a yard twice as big as that man's; I want a field that will raise as much hay as his does) played some role in the past, we now want standard regions that will tesselate or tile conveniently. That is, for standard regions we like pieces which, like pieces in a jigsaw puzzle, will just fit together to cover the region being measured. The square is usually the unit region.

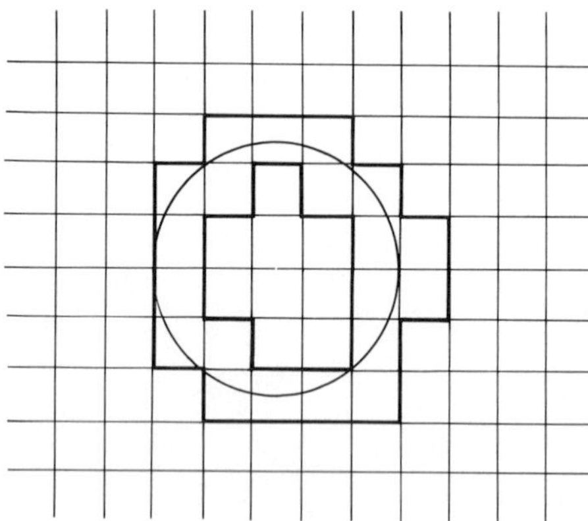

In measuring a segment with unit segments we very quickly found the particular number of unit segments closest to the given segment. This may not happen so easily with regions. If we cover the region with squares we may find that the edges don't fit at all well.

For example, if we cover a circular table top with tiles, we find the edges don't fit anywhere. We show what to do by example. In the figure in the margin, a circle is drawn on a piece of graph paper. The area of the circle is the number of squares it contains, where we suppose that one square is our unit region and that one square is one tile. The outer polygon shows 29 squares that have at least some part inside the circle. Of these, 9 are completely inside the circle and are shown by the inner polygon. As far as whole squares are concerned, we say that $9 < A < 29$; 9 is called a lower bound and 29 an upper bound.

The next thing is to split the 20 squares that are partly inside the circle into smaller regions of known area—smaller squares or triangles. In this way we get more refined upper and lower bounds for A. We continue until the difference is negligible. We note that we can never get an exact value for A in this manner.

You may want to copy a figure like this to check the answers you get by taking smaller and smaller squares. We may also estimate what portion of each of the 20 partly-in squares is actually within the circle. Adding these values gives an estimate of the area, but there is no way of knowing whether the figure is too large or too small, or by how much. Many polygons can be measured by dissection—that is, we cut standard squares into pieces that will exactly cover the polygon. Without going into details, we give some formulas that result from such "cutting and pasting."

$$\text{area of rectangle} = \text{length} \times \text{width}$$

Clearly, the area is the number of squares, and the length and width are measured in units, one of which makes a side of the standard square.

This formula is easily seen when the sides are an integral number of units. When the sides are fractional or irrational lengths, the formula still holds but the argument is more subtle.

The following diagrams suggest how $A = bh$ for parallelograms and $A = \frac{1}{2}bh$ for triangles.

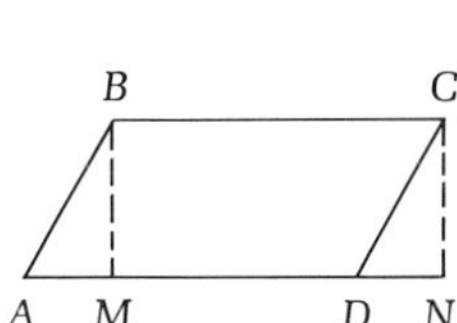

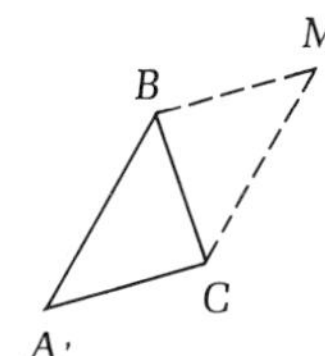

The sectors of a circle look like little triangles, each of whose heights is r, the radius. The sum of the bases is the circumference C. Hence, we are led to the formula for the area of a circle:

$$
\begin{aligned}
A &= \tfrac{1}{2}rb_1 + \tfrac{1}{2}rb_2 + \cdots \\
&= \tfrac{1}{2}r(b_1 + b_2 + \cdots) = \\
A &= \tfrac{1}{2}rC
\end{aligned}
$$

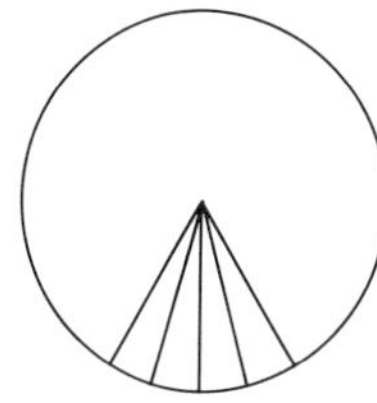

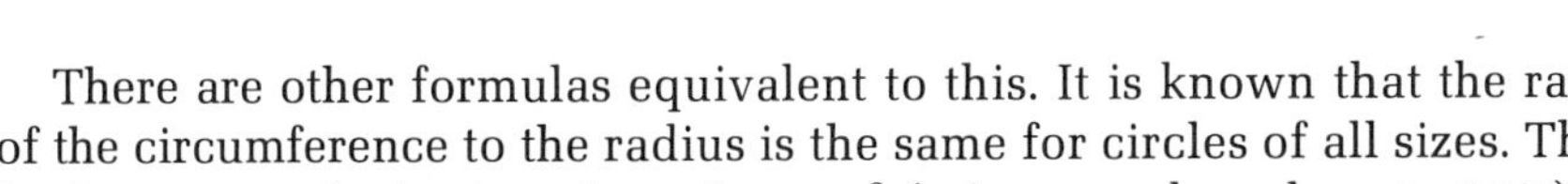

There are other formulas equivalent to this. It is known that the ratio of the circumference to the radius is the same for circles of all sizes. That is, $C = 2\pi r$; substituting gives $A = \pi r^2$ (π is a number close to 22/7).

Area formulas are available for only a few regions. A common trick is to cut the region up into regions for which formulas are available. Then add these areas. Often we approximate a region by a rectangle, especially if the error involved is small. Some people use small stones or seeds as approximate tiles. One popcorn-loving family measures the size of its pans by the number of kernels of popcorn necessary to cover the bottom one kernel deep.

Several mechanical ways of measuring areas of irregular figures are available. For example, the figure can be drawn on graph paper like the circle on page 220. We measure the area as the number of intersections of graph lines inside the figure, and say that this is the number of squares. In our example there are 18 points inside the circle, so we might estimate the area as 18 squares. If a point falls exactly on the circle it is usually counted as half a point.

The area may be measured by weighing. The figure is cut from a uni-

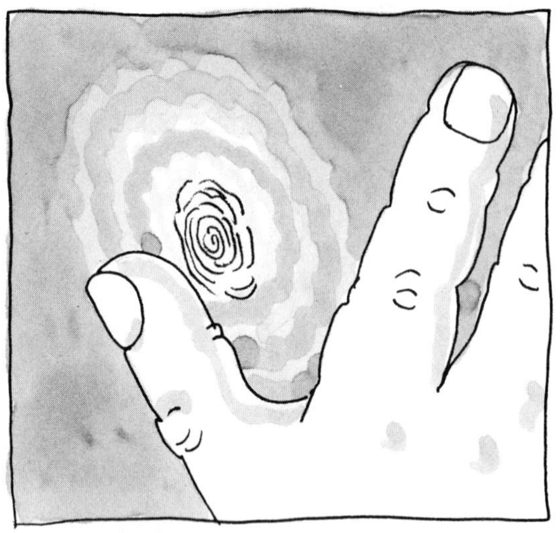

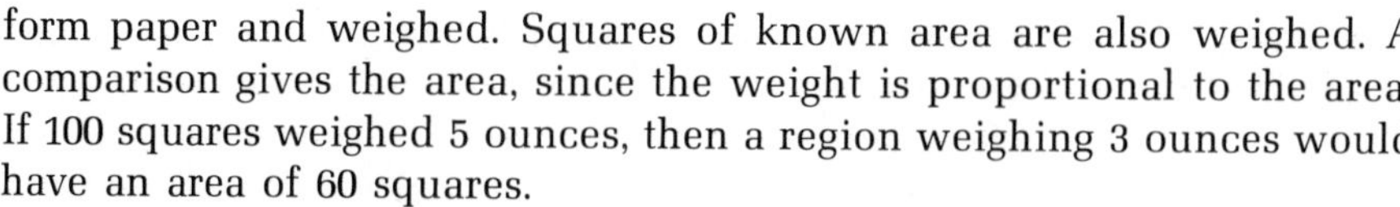

form paper and weighed. Squares of known area are also weighed. A comparison gives the area, since the weight is proportional to the area. If 100 squares weighed 5 ounces, then a region weighing 3 ounces would have an area of 60 squares.

A satisfactory spring scale for such weighings can be made by fastening several rubber bands together to make a soft spring. One end is fastened to the wall and the other end to a paper clip. Marks are placed on paper on the wall showing where the rubber is stretched by figures of known size held by the clip. The area of a figure of unknown size is shown by how far this piece stretches the rubber.

It is customary to say "square inches" and "square feet" if the standard regions are squares one inch and one foot on a side.

There are mechanical tools, such as the planimeter, that measure areas, but these are beyond the scope of this book.

Personal units of area are probably not as useful as others. Commonly a square inch or square centimeter is used. A thumbprint is about a square inch, a little finger print is about a square centimeter. Usually in estimation we "square up" the region, approximating it by a rectangle. Then we estimate the sides and apply the formula $A = lw$.

12.7 MEASUREMENT OF VOLUME Volume is measured by fluids, or by tiling. A jar, glass, or bottle has its volume measured in quarts, pints, liters, or other similar units. To say that a bottle contains 3 pints means that the water in three containers of 1 pint each just fills the bottle. In this approach one thinks of pouring, and not in terms of piling blocks together. The tiling approach asks how many little blocks can just be fitted together to equal the three-dimensional region. We speak of units like cubic inches, cubic meters, square foot-inches.

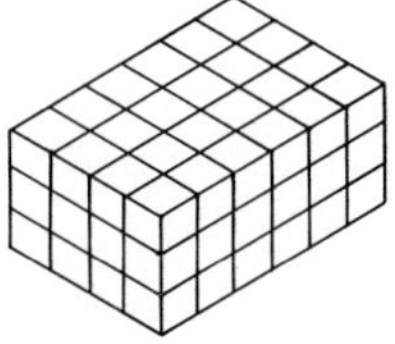

Suppose we have a box 3 inches by 4 inches by 6 inches. Cubes 1 inch on a side can be packed smoothly in the box. 72 cubes go in, so we say the volume of the box is 72 cubic inches.

Very seldom does one try to estimate the number of pints or gallons directly from linear measures; but frequently one estimates the volume in cubic inches by multiplying the length, width, and depth of a block of about the size of the object being measured. The volume of a pancake, for example, is estimated as the area of one face times the thickness.

Fluid measure can be expressed in cubic units, even though one seldom stacks blocks of water or gasoline. Clearly, a can 10 inches by 10 inches by 10 inches would hold 1000 cubic inches of water.

12.8 DIRECT AND INDIRECT MEASUREMENT. SPEED As mentioned above, direct measures occur when the unit is compared directly with the object measured. Measurements with rulers and by counting tiles are examples. Indirect measures occur when something else, often a distance, is measured directly, and this result converted. Using formulas such as $A = \pi r^2$, $A = l \cdot w$, volume $= lwh$ are examples of measuring indirectly. The speedometer of a car shows that speed can be converted to distance (the scale reading on the odometer), which is "read" in terms of speed, such as miles per hour.

Some things, like speed, are almost always measured indirectly. Think what a task it would be to measure directly. How would you arrange 30 objects, each going 1 mile per hour, to compare with a car to see if the car was or was not going 30 miles per hour?

Some people think of speed as derived from distance and time. They say average speed is the distance traveled divided by the time taken. This extends the concept of division from division of numbers (repeated subtraction or missing factor).

12.9 OPERATIONS WITH DENOMINATE NUMBERS Do the following expressions make sense? 3 hours + 10 hours; 7 feet + 3 inches; 4 pounds − 8 ounces; 5 feet × 3 feet; 12 feet × 2 inches; 18 pounds ÷ 6 pounds; 6 feet ÷ 3 minutes; 13 degrees × 20 days; 4 pounds × 2 feet. If they make sense, is there some simpler name, some simpler denominate number, some better-known unit, in each case?

For such expressions to make sense, we must describe some human activity where it makes sense, and then determine the appropriate denominate number.

3 HOURS + 10 HOURS Mr. Peters studied mathematics for 3 hours, then gardened for 10 hours. His total time was 13 hours. We note that the units are the same: hours. When we think of an actual situation, we see it is appropriate to add the measures.

7 FEET + 3 INCHES The reader can make up an example. It is not legitimate, of course, to add the measures, 7 + 3. Your example should show, however, that it is possible to write 7 feet as 84 inches, and, hence, 7 feet + 3 inches = 87 inches.

4 POUNDS − 8 OUNCES You may make up a story about a girl and a box of candy. By using 16 ounces = 1 pound, we get 4 pounds − 8 ounces = 56 ounces, or $3\frac{1}{2}$ pounds.

From these examples, we appear to get the rule: If two denominate numbers describe things that can be put together, it makes sense to add (subtract) the numbers. If the units are the same, the sum has the same unit, and the measure (the pure number part) of the sum is found by adding the measures of the two denominate numbers.

This rule is true most of the time, but not always. One always has to think of the situation being described to see if the quantities involved act this way. Perhaps the most obvious counterexample comes from mixing things of different sizes. If a cup of salt and a cup of large beans are mixed, the mixture occupies less than two cups. Careful measurements show that mixing fluids of different substances can act the same way. The smaller particles fill in the spaces between the bigger ones. Here is another example: The theory of relativity tells us that velocities do not add in the "normal" way. A velocity of 0.6 the speed of light added to another similar velocity gives a velocity of 0.88 the speed of light, not 1.2. The reasons are beyond this book.

5 FEET × 3 FEET People use this to describe the area of a rectangle 5 feet long and 3 feet wide. From a sketch we see that squares, 1 foot on a side, will tile the rectangle, and that there are 5 × 3 = 15 of them. We see that 5 feet × 3 feet = 15 square feet. By analogy with exponents for numbers, we often write 15 feet^2.

12 FEET × 2 INCHES Following the same meaning as above, this could mean the area of a rectangle 12 feet long by 2 inches wide. Multiplying the measures we get 24—but 24 what? Here the unit would be a rectangle 1 foot long and 1 inch wide. We might call it a "foot-inch." It is customary, however, to convert to feet, getting 12 feet × $\frac{1}{6}$ feet = 2 square feet, or to inches, getting 144 inches × 2 inches = 288 inches^2. Note this looks like 288 inch × inch.

It appears that when we multiply denominate numbers, we multiply the measures, and get a unit whose name is the "product" of the factor units. At least these names are written side by side. Amazingly, with appropriate interpretation, this makes good sense in many cases. But one must check this out.

18 POUNDS ÷ 6 POUNDS How many 6 pound bean bags could you fill with 18 pounds of beans? The answer, 3, is a reasonable interpretation of

$$\frac{18 \text{ pounds}}{6 \text{ pounds}}$$

It appears that the measures divide and the units cancel, just as if they were numbers. This is the sort of human situation we would like division to represent.

6 FEET ÷ 3 MINUTES We might ask "What could we do that involves a segment of 6 feet and a time interval of 3 minutes?" Some judicious guessing may suggest something could move smoothly 6 feet in 3 minutes. It would move 2 feet in 1 minute. That is, the object would have a speed of 2 feet per minute. This suggests that division of denominate numbers yields a rate whose measure is the quotient of the measures and whose unit is "units per unit."

13 DEGREES × 20 DAYS Heating contractors use this product to describe the weather. If the temperature is 13 degrees below freezing for 20 days, the figure is 13 degrees × 20 days = 260 degree-days. The number of degree-days in a month measures how severe the weather is and relates closely to the demands for fuel.

4 POUNDS × 2 FEET Here we are rich indeed. There are two common interpretations for this. If a 4-pound weight is lifted 2 feet, we say we have done 8 foot-pounds of work. If a force of 4 pounds pushes on a lever 2 feet from the point P, we say that it makes a torque of 8 pounds-feet about P.

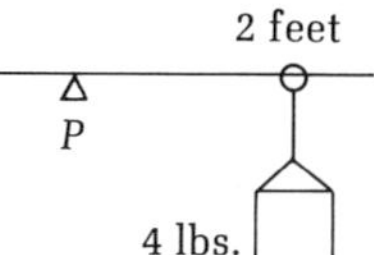

25 ÷ 3 SECONDS If 25 events occur in 3 seconds, 25 ÷ 3 is called the frequency. The unit is *per second* or *events per second.* In wave motion, cycles per second are used.

Frequently the units are named after people who played an important role in their use. For example, in radio the unit *cycles per second* is called Hertz, after a leading pioneer in the discovery of radio waves.

12.10 ERROR IN MEASUREMENT Whenever we make a measurement or count a large number, it is always safest to assume we don't get the exact true value. Indeed, our best experience says that no two measurements are ever exactly the same. We say that two quantities have equal size if the difference is negligible.

We do not wish to be inaccurate, nor do we wish to become neurotic, so we want a precise way of discussing such unavoidable differences in good measurements.

Suppose we measure the volume of a tank as 103 cups. It is convenient to imagine a mythical true value, v, as found by some "perfect measurer." Then we define an error, e, by $103 = v + e$. If the true value were 101, then the error is 2. If v were 104, the error is $^{-}1$. Looked at another way, if we knew our reading and the true value, v, did not differ by more than 3, then we would know $100 < v < 106$. We recall that $a < b < c$ means "a is less than b, and b is less than c." We also say, "the absolute value of the error, written $|e|$, is less than 3."

In general, when making a measurement or reading, r, we imagine a true value t and an error e so that $r = t + e$. We never know t or e exactly, but we can often make very good statements about a bound for $|e|$, such as $|e| < 2$.

Alternatively we may report a measurement with three numbers. In measuring the length of a rectangular field, one might say, "My best measurement is 73 yards, and I'm certain it is longer than 71 and shorter than 76 yards." In symbols, $71 < L < 76$ and $L = 73$ is the best estimate. Most often the best estimate is the center of the interval of certainty, but this is not necessary nor always true.

What we have called *error* is often called *tolerance*. You can be sure when purchasing 5 gallons of gas you did not receive exactly 5 gallons. The government bureau of weights and measures will have established an acceptable tolerance or error. You can only be sure of receiving an amount within that tolerance—perhaps one cup—of the measured 5 gallons.

What is an acceptable error in one place may be intolerable in another.

Overestimating the weight of a beauty queen by 30 pounds is disastrous, but is probably negligible for an elephant. It is convenient to consider *relative error,* or e/t. An error of 30 pounds in a person's weight whose true weight is 120 pounds would give a relative error of $\frac{1}{4}$, or 25 percent. In a 6-ton elephant, the relative error is 30/12000, or 1/400, or .25 percent. We can never know relative error exactly. However, we can frequently estimate it by estimating e and using the measurement for t. Estimating a person's weight as 125 and being convinced the error is less than 5 pounds, we estimate the relative error as less than $5/125 = 1/25 = 4$ percent. Relative error is somewhat related to the idea of significant figures, which are discussed in the next chapter.

This error, the difference between the measured value and the true value, is often called *absolute error* in contrast with *relative error.* Some might use error for relative error, saying "The poll taker estimated the vote with an error of 2 percent, and the absolute error was 13,500." What is meant is "relative error of 2 percent, and absolute error of 13,500."

Since error is always involved in a measurement, many people believe measurements should be made as accurately as possible. Unfortunately, there appears to be a law of nature which says that accurate measurements are expensive and very accurate measurements are very expensive indeed. This cost may be in time and psychic irritation as well as in dollars spent for measuring devices.

Using devices readily available, we may measure a weight w and conclude that $120 < w < 150$. We may well believe that by spending some time and \$10 we could locate w in an interval only 5 wide rather than 30 wide. Spending more time and \$10,000, we can locate w in an interval only .05 wide. What should be done?

Sophisticated measurers ask, "Is the extra information given by shortening the interval of certainty worth the extra money, time, and anxiety?" Often it is, and the effort is spent for the shorter interval—the more accurate reading. Often it is not, so no further work is done. The measurer says the measurement was "close enough" and the error was "negligible."

12.11 METRIC AND ENGLISH UNITS Common English units such as foot, yard, pound, and ton developed as reasonable units in the fairly isolated community of medieval England. The inch and foot were convenient units for small measures, the mile for large measures. At first there was little attention paid to the relation between them. Then later they were standardized at 12 inches = 1 foot, 5280 feet = 1 mile—rather peculiar figures, but convenient because they have many factors. Similar systems developed in other societies.

During the Age of Enlightenment, the eighteenth century, the French developed the metric system, a decimal system based on the meter, supposedly one ten-millionth the arc of a quarter of a great circle of the earth. Slowly the metric system has been adopted for common use almost everywhere. In the United States it is widely used for scientific and technical purposes, and is spreading to other fields as well.

A short table of conversion values follows.

1 yard = 0.9144 meters	1 meter = 1.093613 yards
1 pound = 0.45359237 kilograms	1 kilogram = 2.2046230 pounds
1 gallon = 3.785306 liters	1 liter = 0.264179 gallons
1 inch = 2.54 centimeters	1 centimeter = 0.3937008 inches

The units in the metric system vary by multiples of ten. For example,

1 meter = 10 decimeters = 100 centimeters = 1000 millimeters

1 kilogram = 10 hectograms = 100 decagrams = 1000 grams

Metric measurements fit base-ten numerals very well.

Periodically the leading nations will sponsor a conference to review the measurement systems in use. Accurate measurement has shown that the earth is not a perfect sphere, so the meter is no longer defined in terms of the arc of a great circle. It is now defined in terms of the wavelength of light, a length undisturbed by gravity, temperature, or pressure.

Most people agree that the world would be better off if everyone used the same units. However, changing from one system to another involves many problems. Some people think countries adopting the metric system should do so gradually—perhaps using both English and metric at the same time. Others compare a dual system with driving on both sides of the road. They say that a fixed date should be set after which everyone would use metric.

The problems are not so much getting people used to different units, as in getting agreement on which of several reasonable and possible sizes to make. For example, certain kinds of bread are made in standard one-pound loaves. When changed to the metric system, should the bread be made in .45359237 kilogram loaves? .45 kilo loaves? 0.4 kilos? or 0.5 kilo loaves? How much variation is allowed before we say the loaf is too light? How much new equipment must be purchased to make the change? Should the taxpayer buy this equipment for the baker? This agreement must be reached by busy people in all industries.

12.12 CONVERTING MEASUREMENTS (*Continued*) Converting measurements involving compound units, as in speed, or changing from English to metric units is quite similar to what is done in changing length units.

EXAMPLE What is the speed in feet per second of a runner covering a mile in 4 minutes? We know there are 5280 feet in 1 mile and 60 seconds in 1 minute. The speed is the generalized quotient of distance in feet divided by time in seconds. The distance is 5280 feet and the time is $4 \times 60 = 240$ seconds. Hence, speed $=$ v feet per second $=$ 5280 feet/240 seconds $=$ 5280/240 feet/second $=$ 22 feet/second. Instead of *per*, we write /, which suggests division.

A good notation suggests operations that yield correct answers and serve as a memory device, even if the operations are subtle and perhaps hard to justify. We can look at this operation slightly differently. If speed $=$ v feet/second $=$ 1 mile/4 minutes, and if we multiply the fraction on the right by 1 $=$ 5280 feet/1 mile and by 1 $=$ 1 minute/60 seconds, we think

$$v \text{ feet/second} = (1 \text{ mile}/4 \text{ minutes})(5280 \text{ feet}/1 \text{ mile})(1 \text{ minute}/60 \text{ seconds})$$

$$= \frac{1 \cdot 5280 \cdot 1}{4 \cdot \quad 1 \cdot 60} \cdot \frac{\text{mile}}{\text{minutes}} \cdot \frac{\text{feet}}{\text{mile}} \cdot \frac{\text{minute}}{\text{seconds}}$$

We may think of 4 and 60 dividing into 5280, the mile in the numerator and the mile in the denominator forming a ratio of 1, and the 2 minutes similarly canceling to yield

$$\text{speed} = v \text{ feet/second} = 22 \text{ feet/second}$$

Similarly, we ask the speed in miles per hour of a dash man who runs 100 yards in 9.4 seconds. Speed $= \frac{100}{9.4}\frac{\text{yards}}{\text{seconds}} = x\frac{\text{miles}}{\text{hour}}$. Let us suppose we know 3 feet = 1 yard, 5280 feet = 1 mile, 1 hour = 60 minutes, and 60 seconds = 1 minute. Then,

$$x\frac{\text{miles}}{\text{hour}} = \frac{1000}{94}\cdot\frac{\text{yards}}{\text{seconds}}\cdot\frac{3\text{ feet}}{1\text{ yard}}\cdot\frac{1\text{ mile}}{5280\text{ feet}}\cdot\frac{60\text{ seconds}}{1\text{ minute}}\cdot\frac{60\text{ minutes}}{1\text{ hour}}$$

We are led to the extra factors on the right because—

1. the unwanted units are canceled, leaving the desired units;
2. each factor represents a relation between units we know or can find in a conversion table. We note each factor equals one. The right side of the above equation telescopes to

$$x\frac{\text{miles}}{\text{hour}} = \frac{1000}{94}\cdot\frac{3}{1}\cdot\frac{1}{5280}\cdot\frac{60}{1}\cdot\frac{60\text{ miles}}{1\text{ hour}} = 21.8\frac{\text{miles}}{\text{hour}}$$

If we had realized that 1760 yards = 1 mile, we could have used the ratio 1 mile/1760 yards rather than the two ratios 3 feet/1 yard and 1 mile/5280 feet. The example shows that while this ratio would have made the problem slightly easier, it was not necessary.

The same process makes conversions from English to metric units easy. We ask how many square meters are in a recreation homestead of 5 acres. Only one figure beyond those we have had is needed. There are 640 acres in 1 square mile.

$$M\text{ square meters} = M\text{ meters} \times \text{meters} = 5\text{ acres} =$$

$$5\text{ acres}\frac{1\text{ mile}\cdot 1\text{ mile}}{640\text{ acres}}\cdot\frac{5280\text{ feet}\cdot 5280\text{ feet}}{1\text{ mile}\cdot 1\text{ mile}}\cdot\frac{12\text{ in.}\cdot 12\text{ in.}}{1\text{ foot}\cdot 1\text{ foot}}$$

$$\cdot\frac{2.54\text{ cm}\cdot 2.54\text{ cm}}{1\text{ in.}\cdot 1\text{ in.}}\cdot\frac{1\text{ meter}\cdot 1\text{ meter}}{100\text{ cm}\cdot 100\text{ cm}}$$

Hence, we get

$$M\text{ meters}^2 = \frac{5}{1}\cdot\frac{1}{640}\cdot\frac{5280}{1}\cdot\frac{5280}{1}\cdot\frac{12}{1}\cdot\frac{12}{1}\cdot\frac{254}{100}\cdot\frac{254}{100}\frac{1}{100}\frac{1}{100}m^2$$

$$= 20{,}234\text{ meters}^2 = 2.0234\text{ hectares.}$$

A hectare is 10,000 meters^2 and is the metric unit of area used the way acre is used in English units.

Of course, if one had a more complete conversion table which said that 1 hectare = 2.471 acres, one would use the shorter computation:

$$\begin{aligned} \text{area} = M \text{ meters}^2 &= 5 \text{ acres} \frac{1 \text{ hectare}}{2.471 \text{ acres}} \cdot \frac{10{,}000 \text{ meters}^2}{1 \text{ hectare}} \\ &= 20{,}235 \text{ meters}^2 \end{aligned}$$

The difference in results of the two methods reflects round-off errors in the conversion factors, 2.54 and 2.471.

EXERCISES

1. Describe what you look at, listen to, feel, or whatever and what you see, hear, feel, or whatever when you become aware of the measurement of these things:

(*a*) the contents of a 13-ounce can of nuts
(*b*) the contents of a carton of milk
(*c*) your weight
(*d*) your age
(*e*) the length of a class period
(*f*) the length of a book
(*g*) 36 ounces of fresh bananas at a market.

2. Why is the distance from Los Angeles to San Francisco given in miles rather than inches? Could it be given as correctly in inches?

3. Enid measures a stick as 6 inches long. Frank draws a line 6 inches long with a ruler. Will Frank's line be exactly as long as Enid's stick? How much difference might there be between the lengths?

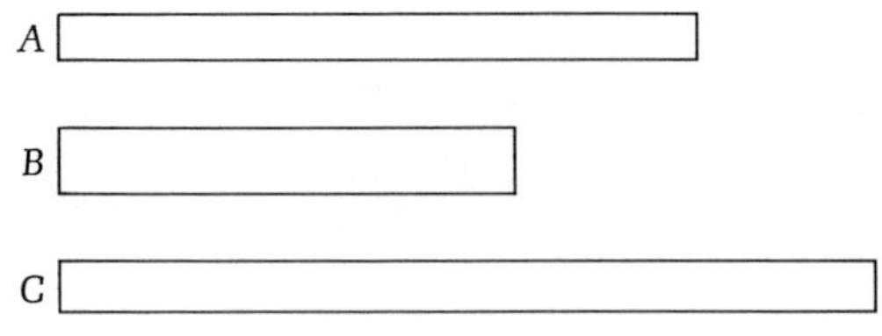

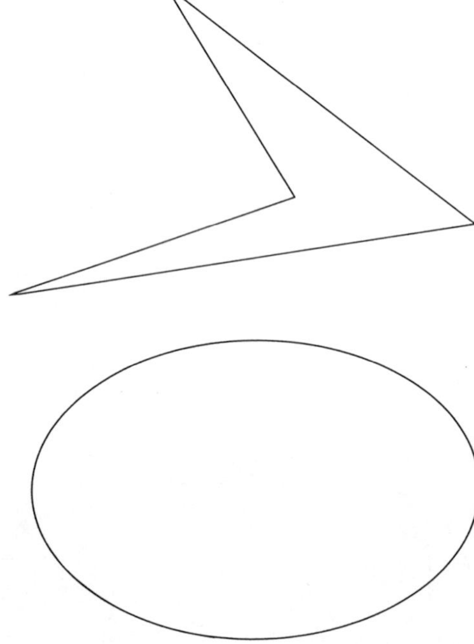

4. Copy the scale on page 213.

(*a*) Using this scale, measure the length of rectangle *A* to the nearest division.

(*b*) Using the scale, but without looking at rectangle *A*, draw a segment as long as the measurement. Compare rectangle and segment.

(*c*) Do the same as *a* and *b* using rectangle *B*.

(*d*) Copy the scale on page 214. Using this scale, make the same constructions and comparisons of rectangles *A* and *B*.

(*e*) Imagine repeating these constructions and comparisons with many rectangles. Do you think the segments drawn from the scale in *a* reproduce the long sides of the rectangles as well as, better than, or worse than the segments drawn from the scale in *d*?

5. Name objects about your body, or objects commonly available for use in comparison, that show (*a*) 1 inch, (*b*) 1 centimeter, (*c*) 1 foot, (*d*) 1 pound, (*e*) 10 grams.

6. Estimate the perimeter of the figure in the margin in inches. Give an upper and lower bound, saying, "The perimeter is between ______ and ______ inches long."

7. Compare your estimates with your neighbors'. Are there any numbers that would lie in all your intervals?

8. Use a ruler to get a better estimate or measurement.

9. Find the area in square inches of the oval, using graph paper or drawing convenient squares. Copy the oval on a sheet of paper first.

10. Calibrate beans or popcorn: Draw a 1-inch square. Cover the interior with beans (or grains of popcorn), *n* being the number required. Check by covering a 2-inch square. Now cover the oval with beans. What do you get for the area of the oval?

11. For each of the following, name a situation where the operation makes sense. State what the product represents, and give the denominate number for the product. If you can think of no such situation, so state.

(*a*) 6 miles ÷ 10 minutes

(*b*) 4 pounds × 3 feet

(*c*) 10 feet + 6 inches

(d) 8 feet × 4 hours
(e) 4 square feet × 50 inches

12. Convert the following measurements as indicated:
(a) 11 feet to inches
(b) 4 yards to meters
(c) 30 centimeters to inches
(d) 15 pounds to grams
(e) 440 yards in 58 seconds to miles per hour
(f) 60 miles per hour to kilometers per hour

13. In a double-pan balance, a quart carton of milk and a box of margarine balance a sack of nuts. How much do the nuts weigh?

14. Where should a 100-pound boy sit on a teeter-totter to balance an 80-pound girl who sits 200 centimeters from the center?

15. A 100-pound boy sitting 150 centimeters from the center balances a girl sitting 2 meters from the center on a teeter-totter. How much does the girl weigh?

16. Show how to cut a right triangle into two pieces and put them together to form a rectangle. Can this be done in more than one way?

17. Show how to cut a parallelogram into two pieces and reassemble them to form a rectangle.

18. Show how any triangle can be cut into no more than four pieces that can be put together to form a rectangle.

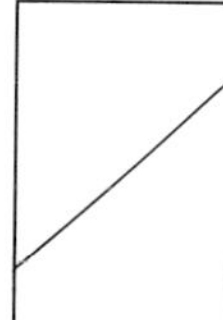

19. Show how a rectangle can be cut into four pieces that can be put together to form a square. (HINT: Try cutting the rectangle first as in the margin.)

20. The speed at which an egg dropped from a building h feet tall hits the ground is given by $s = 8\sqrt{h}$ feet per second. How many miles per hour is the egg moving when it hits the ground if dropped from a 25-foot building?

Exercises 21 and 22 are good class games.

21. The leader displays a segment whose length each person estimates in centimeters (for example, "It is between 11 and 14 centimeters"). The score is $10/w$ if the true length lies in the interval, where w is the width of the interval of estimation. The total score on 6 segments is called a hexad. Some people like to score each correct estimate $20 - w$ rather than $10/w$. Thus an excellent score on a hexad is 100.

22. The leader displays a segment that shows the length of some object visible to everyone in the room. People guess what the object is.

23. Sheila gets 25 miles per gallon of gasoline with her small Ford, while Frédérique gets 7 kilometers per liter in her Renault. Who gets the better mileage?

24. Jerome buys 5 pounds of sugar for 83 cents. Hans buys 3 kilograms for 10 marks. If 3 marks equals one dollar, who pays more? What is the cost in cents per kilogram in each case?

Answers to selected questions: **1.** (*b*) label; (*f*) last page number; (g) scale; **3.** no; as much as 1/2 inch; **4.** (*a*) 3; (*d*) 4, 5;

8. $7\frac{9}{16}$; **11.** (*a*) speed of car, 6 miles per minute; (*c*) Add 6 more inches of dirt, result is total thickness of dirt. $10\frac{1}{2}$ feet or 126 inches; (*d*) no meaning; (e) volume of box, 200 square feet-inches, or 28,800 cubic inches, or $16\frac{2}{3}$ cubic feet; **12.** (*c*) 11.81; (e) 15.52.

16. 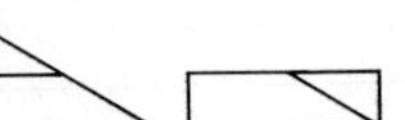 yes

23. Sheila

13

Approximations

13.1 INTRODUCTION The hairs of one's head are numbered, but only God and bald men know what the number is. Practically speaking, only when we are counting a relatively few objects can we be sure of the exact number. Most of the time we are working with numbers that are only close to the numbers of interest.

In this chapter we discuss how approximate numbers arise, how we may talk about them and compute with them. We use inequalities, ideas of absolute and relative error, and significant digits. As a final fillip, we discover a nonassociative arithmetic.

We realize that many people restrict the theory of arithmetic to a theory of exact numbers. They say there are no such things as approximate numbers. What we will consider here is a part of statistical inference–how to express inexact knowledge with exact numbers.

13.2 HOW APPROXIMATE NUMBERS ARISE Approximate numbers arise from at least six processes: (*a*) measuring, described earlier, (*b*) inexact counting, (*c*) rounding off, (*d*) computing with approximate numbers, (*e*) using inexact formulas, (*f*) blundering.

13.3 MEASURING In chapter 12 we saw that measurements are never exact. We supposed that human measurement is an approximation of what some perfect measurer would get–the "true" value. Although we actually never know perfectly, we give an interval of certainty in which we know the true value lies. We might say, based on a measurement of 100 inches and our knowledge of our measuring, that $96 < s < 106$. That is, the true measure is between 96 and 106, and our best estimate is 100. We also might say, $s = 100 \pm 6$, meaning that our best estimate is 100 and the error is not more than 6. However, our known interval is shorter. We might say $s = 101 \pm 5$, which reproduces the interval correctly, but the best estimate is off.

Measurements are taken for a purpose, as a guide for action. If one action would be taken if $90 < s < 110$, and another if s is outside this interval, the measurement and any of the reports based on it are good enough to tell us which action to take.

On the other hand, if one of three different actions is to be taken as $s < 98$ or $98 < s < 102$ or $102 < s$, then the measurement above was not good enough to guide our choice. One needs to repeat the measurement with increased care and accuracy.

More careful measurements, resulting in narrower intervals of certainty, cost something to make. If the cost is low in time, money, inconvenience, and so on, we remeasure. In general, if the cost is less than the value to be gained by more accurate measurement, we remeasure.

13.4 INEXACT COUNTING How many letters are on this page? How many people in this class? How many at the last play or concert you attended? Most people find it difficult to count exactly beyond 500 objects. At the close of a guessing contest, a little girl was asked to count the number of beans in a jar. After some time she announced, "I've counted them ten times and here are the ten answers." On the other hand, a reporter counted the people watching Old Faithful's five-minute eruption in Yellowstone Park, and announced, "2987 people were watching."

Now, it is possible to count large numbers of things exactly. Where this is important, as in tallying election ballots, people spend large sums to make sure the figure is exact.

In many cases the counter states an upper bound for the error in his counting. He knows this by the way he did his counting. He announces his result by giving a best estimate, and a sure interval. You might say you counted 48 people in a room. You might have missed 3 or 4, or counted 1 or 2 people twice. In this case you say that p, the number of people, satisfies $46 \leq p \leq 52$, and that your best guess is 48. The Old Faithful counter counted in conservative blocks and felt sure he wasn't more than 200 over, but might have been 700 under, so he said the number g satisfied $2787 < g < 3687$, with 2987 as his best guess.

Note that in inexact counting your best guess need not come in the center of the interval of certainty. The answer is similar in form to the approximate numbers arising from measurement. Counters frequently juggle their figures to make their "best guess" lie at the center of the "interval of certainty." The people-counter might say $p = 48 \pm 4$ or $p = 49 \pm 3$, depending on whether he extended his interval of certainty to have 48 at the center or whether he changed his best guess to the middle of the interval. If the whole discussion is merely casual conversation, one doesn't worry much anyway; on the other hand, if important decisions are at stake, then great care is taken in reporting the count.

EXAMPLE The foreman in a diamond-cutting plant, after a minute's counting, said there were about 1130 diamonds in a pile, but certainly between 1110 and 1155. It required an hour for a workman to count them carefully, getting exactly 1149. Was the foreman right? Do you think it worth the workman's time to count? What if it had been rubber bands? If the workman had reported 1105 diamonds would the foreman have suspected him of carelessness? Why?

Since the careful count of 1149 was within the foreman's bounds, the foreman's figures and the workman's figures support each other.

13.5 ROUNDING OFF Ever since Mark Twain took a dime from the collection plate rather than put in a dollar because the preacher talked too long, preachers have been careful not to tell their parishioners more than they wanted to know. Similarly, with expensive equipment and careful procedures, correct statements can be made like, "There were 34,785 paid admissions," "The national debt is $704,896,372,025.38." These numbers contain all available information, but it is more information than the reader may want to know. We consider rounding off.

Suppose you wished to express the information in 34,785 as well as possible with one nonzero digit, and maybe several zeros. What number would you give? In particular, which of the following would you prefer: 100,000; 10,000; 40,000; 50,000; 30,000; 700?

Actually the number you would prefer would depend on the use you would make of the information. Most of the time, however, we choose a number so that the difference between it and the exact number is as small as possible. In this case the number is 30,000. The difference, or error, is 4785. The next-best number is 40,000, with an error of 5215. We say we

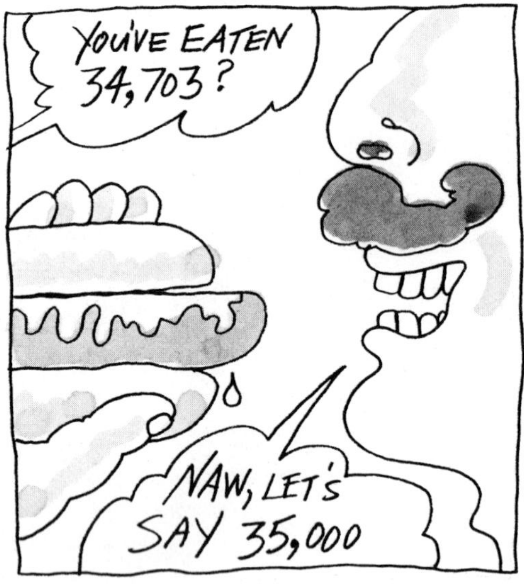

have rounded off 34,785 to 30,000. We say we have rounded off to the nearest 10,000. We have rounded 34,785 to one significant digit.

Perhaps we have starved our reader. He is willing to take a bigger bite of the hot dog to get some weenie with the bun. In that case we can round to two significant digits or to the nearest thousand. The closest such number is 35,000. The error is 215. The next-closest is 34,000, with an error of 785. Hence, we round 34,785 to 35,000, in rounding to two significant digits. The two significant digits are 3 and 5. The zeros in 35,000 are placeholders. If they were left off, we would be grossly misleading. They serve to put 3 and 5 in the correct places so that they have the correct place value.

Similarly, if we round 34,785 to three significant figures we get 34,800. The error is 15. The next-closest number is 34,700 with an error of 85. Since $15 < 85$, we choose 34,800. This is rounding to the nearest hundred.

Similarly, rounding 94,500 to the nearest 10,000 gives us 90,000. Rounding to the nearest 1000, or two significant digits, presents a problem. Two candidates present themselves—94,000 and 95,000. The error in each is 500. Which is preferable? Actually you may choose either number. You choose the more convenient one. If you are going to do further calculations, one number may be more convenient than the other. As a rough rule, lacking any other guide, people round to the nearest *even* digit. Hence, they choose 94,000. In this way we round down about as often as we round up. Even numbers are slightly easier to handle; for example, you may wish later to divide by 2. Rounding to the even digit when rounding away from 5 is called the *computer's rule.*

Many electronic calculators follow a $\frac{5}{4}$ rule. That is, they round up if the rounding digit is 5, and down if it is 4. This is almost the same as the computer's rule and is cheaper.

It is not necessary to actually compute the error or difference between the given figure and its rounded approximation to choose the closer of the two candidates. If the digit in the first place to be made zero is less than 5, just drop these digits. If the figure is greater than 5, raise the preceding significant digit one. If the figure is 5, and all others to the right are zero, round to the nearest even digit.

Another question arises in rounding the debt figure. To one significant figure, 704,896,372,025.38 rounds to 700,000,000,000. To two significant figures, it rounds to 700,000,000,000; to three, 705,000,000,000. How can we tell whether 700,000,000,000 has one or two significant figures?

We can't with this notation. Fortunately people seldom worry about this. If they do, they use a different notation, called *scientific notation,* which shows which digits are significant and which are placeholder zeros. In scientific notation the three rounded numbers above are written 7×10^{11}, 7.0×10^{11}, and 7.05×10^{11}. We write the number as a power of 10 times a number between 1 and 10. The power of 10 gives the magnitude, while each digit of the number between 1 and 10 is a significant digit.

$3 \times 10^4 = 30{,}000$ and is rounded to one significant digit

$3.0 \times 10^4 = 30{,}000$ and is rounded to two significant digits

$3.00 \times 10^4 = 30{,}000$ and is rounded to three significant digits

$9.45 \times 10^4 = 94{,}500$ and is rounded to three significant digits

$5.0 \times 10^{-3} = .0050$ and is rounded to two significant digits

Suppose we are given a number rounded by the computer's rule. What can we say about the number before it was rounded? Naturally we don't expect to get all the digits. Again, the situation is clearer from examples.

EXAMPLE 2300 is rounded to two figures. The original number n lies in the interval, $2250 < n < 2350$. Note that the sign is $<$ and not $\leq$.

EXAMPLE 54,000 has two significant figures. The original number m lies in the interval $53{,}500 \leq m \leq 54{,}500$.

EXAMPLE 4.80×10^4 has three significant digits. Our best guess for the number z is 48,000, but we know it lies in the interval $47{,}950 \leq z \leq 48{,}050$.

In each case, to find the limits we add and subtract 5 in the first non-significant place. You can try several numbers in and out of the interval and test to see how they round off. Can you check the equal signs on the boundaries of the intervals?

Sometimes we state the intervals this way: $n = 2300 \pm 50$, $m = 54000 \pm 500$, $z = 48{,}000 \pm 50$. In the first case, $n = 2300 \pm 50$ means that n is some number greater than what we get with the minus sign and less than what we get with the plus sign. Similarly with the others.

13.6 COMPUTING WITH APPROXIMATE NUMBERS When two approximate numbers are added (subtracted), the sum (difference) is an approximate number. The problem is to decide what one can say about the error in the sum (difference), and how to say it. You may think of joining two sets or segments whose size is known only approximately.

Suppose we want $C = A + B$, where $A = 20 \pm 2$ and $B = 40 \pm 1$. Is $C = 60 \pm 2$ (2 is the largest possible error in A or B), $C = 60$ (errors cancel in addition), or $C = 60 \pm 3$? To simplify discussion, we assume that A and B are whole numbers. Then A is one of $\{18,19,20,21,22\}$ and B is one of $\{39,40,41\}$. The array in the margin has these five values of A listed along the bottom and the three values of B along the side. At the points of the array are the sums of the corresponding values. Not all sums are printed.

41	59	60		62	63
40	58				62
39	57	58			61
	18	19	20	21	22

It is clear that the largest value is in the upper right corner and the smallest value is in the lower left. Since the sum C is one of these fifteen numbers, clearly $57 \leq C \leq 63$, or $C = 60 \pm 3$.

A similar argument, which we leave you to make, will show the following examples of $A + B = C$.

A	B	C
26 ± 3	48 ± 5	74 ± 8
$10 < A < 19$	$3 < B < 5$	$13 < C < 24$

Can you make up a rule for the relation between A, B, and C?

The process extends nicely to adding three or more numbers. For example, for $A = 20 \pm 5$, $B = 50 \pm 6$, $C = 30 \pm 7$, then

$$D = A + B + C = 100 \pm 18$$

CAUTION It is not clearly agreed whether $B = 50 \pm 6$ means $44 < B < 56$ or $44 \leq B \leq 56$. It is usually clear from the context of the problem which meaning is intended. Hence, 50 ± 6 can be used for either without catastrophe, thus avoiding the burden of two notations. If the meaning is not clear from the problem and the difference is not important, then one uses the more convenient meaning.

Similar discussions are needed for the other operations. Consider subtraction and $Z = Y - X$, where $Y = 50 \pm 3$ and $X = 17 \pm 2$. We draw a grid with Y values on the side, X values along the bottom, and the differences at the grid points. Not all values are printed. You may put in more.

53	38	37		35	34
52	37				33
51			34		
50					
49					
48	33				29
47	32	31		29	28
	15	16	17	18	19

Clearly, the largest value for $Y - X$ is 38 and the smallest is 28, at the upper left and lower right corners, respectively. Hence, $28 \leq Z \leq 38$, or $Z = 33 \pm 5$. Similar arguments will give the following table of subtraction.

Y	X	Y − X = Z
40 ± 5	10 ± 5	30 ± 10
60 ± 6	23 ± 8	37 ± 14
$35 < Y < 43$	$18 < X < 28$	$7 < Z < 25$

Can you make up rules for subtracting approximate numbers?

The grid diagram below is for multiplication, where $W = U \times V$ with $U = 20 \pm 3$ and $V = 30 \pm 2$.

32	544					704	736
31							713
30	510	540		600			
29	493	522					
28	476	504					644
	17	18	19	20	21	22	23

Is it clear that $476 < W < 736$?

We can also say $W = 606 \pm 130$. We may not like this last expression, even though it is correct. For U approximately equal to 20 and V approximately equal to 30, we may want to say that W is approximately 600 and write $W = 600 \pm e$, where e is the maximum possible error. If we insist on this, then $W = 600 \pm 136$ is the appropriate statement. We have placed 600 at the center of the interval, but we've had to enlarge the interval.

Grid diagrams will give the following table of products of approximate numbers.

U	V	W	W
20 ± 3	30 ± 2	606 ± 130	600 ± 136
40 ± 6	50 ± 3	2018 ± 420	2000 ± 438
50 ± 2	80 ± 7	4014 ± 510	4000 ± 524
$20 < U < 25$	$37 < V < 40$	$740 < W < 1000$	

It appears it is easiest to note that the product W lies between the product of the lower bounds of U and V and the product of the upper bounds.

While the absolute error in the sum of two approximate numbers is the sum of the absolute errors of the addends, the absolute error in a product is not described so easily. It turns out, however, that the relative error in a product (quotient) is approximately the sum (difference) of the relative errors of the terms.

Let $u = 20 + 3$ and $v = 30 + 2$. Then we think of u,v as being 20 and 30, with relative errors of 3/20 and 2/30, respectively. The product is 600 + 136, with relative error 136/600. Anticipating fractions, we see that this is close to $3/20 + 2/30 = 130/600$.

Again, if we think of 17 as $20 - 3$, and 32 as $30 + 2$, we can think of $17 \times 32 = 544$ as $20 \times 30 = 600$, with a relative error of $-56/600$. We note that this is very close to $-3/20 + 2/30 = -50/600$, the sum of the relative errors of the addends.

12	3	$2\frac{2}{5}$	2
11	$2\frac{3}{4}$	$2\frac{1}{5}$	$1\frac{5}{6}$
10	$2\frac{2}{4}$	2	$1\frac{4}{6}$
9	$2\frac{1}{4}$	$1\frac{4}{5}$	$1\frac{3}{6}$
8	2	$1\frac{3}{5}$	$1\frac{2}{6}$
	4	5	6

A grid analysis for division appears in the margin for $R = P \div Q$, where $8 \leq P \leq 12$ and $4 \leq Q \leq 6$. Is it clear that the highest and lowest values occur at the upper left and the lower right respectively? Are you convinced that the largest quotient comes from dividing the largest dividend by the smallest divisor? Does the smallest quotient come from dividing the smallest dividend by the smallest divisor or by the largest?

The values in the grid are common fractions. Since fractions are discussed more fully later, you may wish to postpone detailed discussion until you have studied them.

The example $\frac{122}{13} = 9\frac{5}{13}$ will illustrate that the relative error in a quotient is approximately the relative error in the dividend minus the relative error in the divisor. If we think of 122 as 120 with relative error $\frac{2}{120}$, and 13 as 12 with relative error 1/12, then 122/13 is thought of as 120/12 with relative error $-\frac{8}{130}$. We note that $-\frac{8}{130}$ is close to $\frac{2}{120} - \frac{1}{12} = -\frac{8}{120}$.

The following condensed argument uses algebra to show these results for multiplication and division. Readers uneasy with algebra may skip this, or substitute easy numbers for the letters to see what happens. Let $u = U(1 + x)$ and $v = V(1 + y)$, then x and y are the relative errors in u and v. Then

$$uv = UV(1 + x)(1 + y) = UV(1 + x + y + xy)$$

This shows the relative error in uv is $x + y + xy$, which is close to $x + y$ when xy is small.
Similarly

$$\frac{u}{v} = \frac{U}{V} \cdot (1 + x) \cdot \frac{1}{1 + y}$$

Now we may write

$$\frac{1}{1 + y} = \frac{1 + y - y}{1 + y} = \frac{1 + y}{1 + y} - \frac{y}{1 + y} = 1 - \frac{y}{1 + y}$$

and

$$(1 + x)\left(\frac{1}{1 + y}\right) = (1 + x)\left(1 - \frac{y}{1 + y}\right) = 1 + x - \frac{y}{1 + y} - \frac{xy}{1 + y}$$

which is about $1 + x - y$ when x and y are small. This shows that the relative error in u/v is about $x - y$, the relative error in u minus the relative error in v.

Many authors limit their discussion of approximate numbers to significant figures and rounding off, and do not discuss inequalities and intervals. They suggest rules such as this: "When adding rounded numbers, round the sum to the same place as the first significant digit in the least accurate of the addends." For example, in adding 340 + 970 + 136 where 340 and 970 have two significant figures and 136 has three, this rule would round 1446 to 1450 as the sum. One might be tempted to say the sum 1450 has 3 significant figures and hence the sum S satisfies $1445 < S < 1455$. A more accurate analysis shows that

$$335 + 965 + 135.5 < S < 345 + 975 + 136.5$$

or

$$1435.5 < S < 1456.5$$

Serious computers, when combining approximate numbers, use interval analysis as above to find the appropriate interval for the result. They do not use significant figures except in rough approximations.

Statistical treatment The grid diagram showed that $(20 \pm 2) + (40 \pm 1) = (60 \pm 3)$. That is, the error in the sum can be as large as 3. In most cases the error is closer to 0 than it is to 3. In courses in statistics you may discuss how small errors are much more likely to occur in the sum than large errors.

13.7 USING INEXACT FORMULAS Many formulas do not give exact results, yet they are used because they are easy to apply and the results are good enough for the purposes at hand. Legally, people may agree to settle affairs on the basis of an accepted formula—even though the formula is inaccurate. Here are some inexact formulas:

1. Area of a circle $= \sqrt{10}\, r^2$. Many formulas use a convenient approximation for an irrational number. Here we approximate π in $A = \pi r^2$ by $\sqrt{10}$. Other approximations often used for π are 3, 3.14, $\frac{22}{7}$, 3.1416.

2. The perimeter (distance around) of a simple convex closed curve is twice the greatest distance across plus the shortest distance across. Here are some examples of such curves:

3. The area of a parallelogram is the product of two adjacent sides.

4. The square root of a number:

$$\sqrt{N} = k + \frac{N - k^2}{2k}$$

where k^2 is the largest square contained in N. This formula would estimate $\sqrt{12} = 3 + \frac{3}{6} = 3.5$ and $\sqrt{17} = 4 + \frac{1}{8} = 4.125$.

In frontier days many such rules of thumb were used. In advanced mathematics, statistics, physics, and engineering, many other approximate formulas are also used—especially in checking and using computers.

13.8 BLUNDERING Blunders do occur. Unfortunately there appears to be no good theory of handling blunders. In all our discussion of approximate numbers, we assume they are free of blunders. Some people use the word *error* or *mistake* for "blunder." We have used the word *error* for unavoidable differences, as in measurements and rounding off.

Since everyone makes blunders, the best advice appears to be the following for avoiding their evil consequences:

1. Admit you are going to make them.

2. Do your work clearly and neatly so blunders will be easily seen.

3. Budget a substantial amount of energy, time, and money to checking for blunders.

4. Check, check, check.

Is there this side of heaven
A lass older than 'leven
When given the chore
To add two and four
Has never written down seven?

13.9 A NONASSOCIATIVE ARITHMETIC In earlier chapters we saw that addition and multiplication of whole numbers are associative. This property plays an important role in understanding, in computation, and later in algebra.

The examples of nonassociative operations were not surprising. Indeed, what was surprising was the idea that anyone would even bother to test subtraction or division for associativity. We present here an arithmetic that probably appears reasonable and yet is nonassociative.

We call it "Round Off to Two Significant Digits" Arithmetic. We always round our numbers to two significant digits. For example, $12 \times 12 = 140$ (144 rounds to 140). Similarly, $12 \times 13 = 160$ (156 rounds to 160). If multiplication is associative, then $(a \times b) \times c = a \times (b \times c)$ for all numbers. In particular

$$(12 \times 12) \times 13 = 12 \times (12 \times 13)$$

But $$(12 \times 12) \times 13 = 140 \times 13 = 1800$$

and $$12 \times (12 \times 13) = 12 \times 160 = 1900$$

The arithmetic is not associative!

Perhaps this is another reason why one must be very cautious in interpreting results of operations with rounded numbers.

EXERCISES

1. A path is more than 80 yards and less than 90 yards long. Give a reasonable estimate of the length of the path, and an upper bound to the error.

2. A scale is guaranteed to be off by no more than 3 ounces. For a box measuring 45 ounces on this scale, give the interval in which the true weight lies.

3. Mamie tells time with a sundial that may be off by as much as 7 minutes. She wishes to be listening to the radio for an announcement that will be given at noon, exactly. According to the sundial, what time should she start listening to the radio, and how long should she be prepared to wait for the announcement?

4. What relative error does a postman make when he is off by 2 pounds in estimating the weight of a $12\frac{1}{2}$-pound package?

5. If Gloria estimates distances with a relative error less than $\frac{1}{10}$, give an interval for the true width of a room she estimates as 34 feet wide.

6. Take a handful of beans. Think of something you ought to do but haven't done. Estimate the number of beans in the handful and then count them. If the relative error of your estimate is less than 5 percent, put off until tomorrow what you would have done today if the relative error were greater than 5 percent.

7. A length which is within 3 of 300 is multiplied by a weight which is within 2 of 80. What is the possible error in estimating the product as 24,000?

8. Round off 3,645,500 to (*a*) one significant digit, (*b*) two, (*c*) three, (*d*) four, (*e*) six significant digits.

9. A number rounded to the nearest thousand is 36,000. Write the interval in which the exact number must fall.

10. Rounded to the nearest hundred, $z = 40{,}000$. State the interval for z.

11. The measured length of side A of a 5-sided polygon is 364 feet, and the error may be as big as 40 feet. Give upper and lower bounds for the length.

12. The other four sides of the polygon in exercise 11 satisfy

$$\begin{aligned} 200 \leq B &\leq 230 \\ 122 \leq C &\leq 140 \\ 110 \leq D &\leq 120 \\ 5 \leq E &\leq 7 \end{aligned}$$

(*a*) Give upper and lower bounds for the perimeter.
(*b*) What is your best estimate of the perimeter?

13. The measured length of a rectangle is 250 feet, with a possible error as large as 5 feet. The measured width is 100 feet, with a possible error up to 3 feet. Give the best (narrowest) bounds for the area of the rectangle.

14. Suppose you are allowed to remeasure one side of the rectangle in exercise 13. Your remeasure will be within one foot of being correct. Should you remeasure the length or the width? Why? Can you use relative-error arguments as a guide to your decision?

15. The measured area of a triangle is 1440 square feet, with a possible error up to 80 square feet. The measured base is 25 feet with a possible error up to 5 feet. What is your best estimate of the height of the rectangle? Give the best possible interval for the height.

16. If "$a*b$" means "multiply a by b and round the product to two significant figures," then 12, 12, 13 are three numbers, a, b, c such that $a*(b*c) \neq (a*b)*c$. Can you find another triple of such numbers? Note that * is an operation which is commutative but not associative. Simple operations with these properties are not easy to find.

Answers to selected questions: **1.** 85 yards, 5 yards; **5.** (30.6, 37.4) feet; **7.** (−834, 846); **12.** (*a*) 761–901 feet; (*b*) 831 feet; **13.** 23765, 26265; **14.** width; sum of new relative errors is least.

14

Number Theory

14.1 INTRODUCTION The counting numbers {1,2,3,...} are used so widely, have such a regular pattern, and are so numerous that many interesting patterns and relations exist among them. People find these patterns challenging, rich sources of conjecture, and whetstones for sharpening thought and logical principles. The teasing questions appear to be "Do you see a pattern?"; "In how many ways can you . . . ?"; and "How do you know?" While some of the richest and most profound problems arise in number theory, we will limit ourselves to famous problems in the Additive Structure of Numbers and the Multiplicative Structure of Numbers. In this chapter, by "number" we mean "counting number."

14.2 THE ADDITIVE STRUCTURE OF NUMBERS In a certain auto race of 10 laps the rules require each driver to make one stop at the end of some lap. Before the race, the driver plans where he will stop. Clearly, he thinks of expressions for 10: $1 + 9$, $2 + 8$, $4 + 6$, $8 + 2$, $3 + 7$, ... The first addend gives the laps before the stop and the second gives the number after.

The driver must make a decision. Five possible plans are suggested by the sums above. Are there others? Before reading further, see if you can identify nine possible plans for the driver.

Do you see that any number n can be written as the sum of two numbers in $(n - 1)$ ways, where different orders of addends are regarded as different ways? Note that the pairs to be added can be listed (say, for $n = 10$) as $1 + 9$, $2 + 8$, $3 + 7$, . . . , $8 + 2$, $9 + 1$, where 1 is the first member of the first pair and $(n - 1)$ is the first member of the last pair.

People like to think of 10 as being separated into parts, or partitioned. $1 + 9$ and $7 + 3$ are called *partitions* of 10. If order is important and 1 + 9 is considered different from 9 + 1, they are called *ordered partitions* of 10. Partitioning a number and addition correspond to factoring a number and multiplication. 2 × 5 is a factored form of 10.

Sometimes people do not like to consider 4 + 3 and 3 + 4 as different, since the same addends are involved. Then they say that 7 can be written in three ways, 9 can be written in four ways. What we have is not an argument about which is the correct answer, but two problems.

How many ways can a number be written as a sum of two addends (1) if the same addends in different order form different ways, and (2) if ways are different only if the addends are different? We can rephrase these problems: (1) How many ordered partitions are there? (2) How many unordered partitions are there? For example, there are six ordered and three unordered partitions of 7.

Try the following to determine the number of unordered partitions: For several numbers, list the unordered partitions. (For 7 they are 1 + 6, 2 + 5, and 4 + 3.) Then make a table showing, for each number, the number of partitions (pairs of addends). See if you can see a pattern and successfully predict the number of pairs of addends for some numbers you have not tried yet.

EXAMPLE 1 A red die and a green die are rolled. Each die shows a number from 1 to 6. What sums can occur and how can they occur? This is a problem dear to the hearts of backgammon and craps players. Clearly, the sum will be a number from 2 to 12. If both show 1 (or 6) the sum is 2 (or 12). We can get 4 in three ways, just as the theorem above states. But we can get 8 in only five ways—2 + 6, 3 + 5, 4 + 4, 5 + 3, and 6 + 2—since we do not have 1 + 7 or 7 + 1. This is an example of the additive structure of numbers, but with restricted addends.

We leave as an exercise the listing of the various ways sums can be found rolling two dice.

14.3 MULTIPLE ADDENDS

EXAMPLE 1 A red die, a white die, and a green die are rolled. Since each die shows a number from 1 to 6, the sum of the numbers shown by the three is a number from 3 to 18. We ask for the different ways we can get a total of 5.

It is helpful to have a formal, organized way of listing these ways. We

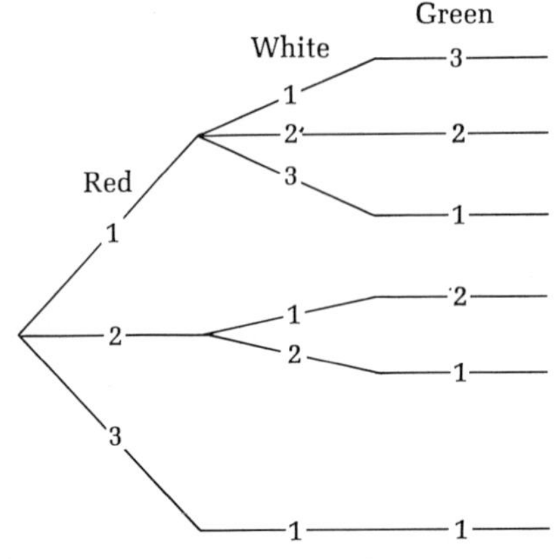

use a tree diagram, giving first the number on the red die, then on the white, and finally on the green.

Counting shows that there are six ways of getting 5. We can make a small table, filling in the blanks in the following sentence:

With □ on red, we get △ on white and green together in ○ ways.

□	△	○
1	4	3
2	3	2
3	2	1
		6

We make the table by putting 1, 2, 3, . . . in □, then (5 − □) in △, and the number of ways of getting △ with two dice in ○. Adding the values in ○ gives the total number of ways.

EXAMPLE 2 We use the above pattern to find the number of ways of getting 14 with three dice. First we use a tree to make a list (we leave out one branch as an exercise for the reader). We fill in □, △, and ○ in the sentence.

With □ on red, we get △ on white and green together in ○ ways.

□	△	○
1	13	0
2	12	1
3	11	2
4	10	3
5	9	4
6	8	5
		15

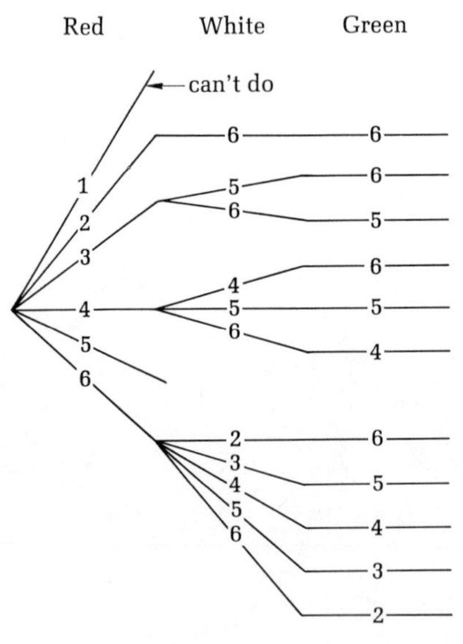

Hence, there are fifteen ways the value 14 can be obtained on rolling three dice.

If we wished to leave the dice out, we could write this problem as follows: "In how many ways can 14 be written as the sum of three addends, none greater than 6? Different orders of addends are regarded as different ways." We did not need to actually list the ways for this question, but listing was a good help to see the additive structure involved.

EXAMPLE 3. We remove the restriction "none greater than 6" and ask, "In how many ways can n be written as the sum of three addends?" For

$n = 5$, the same pattern as with the dice occurs. For $n = 14$, the tree and table would be extended. Let us look at the table. Now red, white and green refer to the order of the addends in $R + W + G = 14$. We use the earlier result that m can be written as the sum of *two* addends in $m - 1$ ways.

□	△	○
1	13	12
2	12	11
3	11	10
⋮	⋮	⋮
10	4	3
11	3	2
12	2	1
		78

The sum of the numbers in ○, $1 + 2 + 3 + \ldots + 11 + 12$, is 78. Hence, 14 can be written as the sum of three numbers in seventy-eight ways.

In more advanced work on permutations and combinations, it is shown that $\frac{(n-1)(n-2)}{2}$ gives the number of ways n can be written as the sum of three numbers. For $n = 5$, the formula gives $\frac{4 \cdot 3}{2} = 6$, and for $n = 14$, it gives $\frac{12 \cdot 13}{2} = 78$, which checks our results.

People interested in statistics have developed formulas for many problems in the additive structure of numbers. All these problems can be solved by careful listing and counting, as is shown here. If you are going to do this sort of thing frequently, the formulas may be great timesavers.

14.4 MULTIPLICATIVE STRUCTURE A set of twelve points forms a rectangular array in two ways:

$$\begin{pmatrix} \cdot & \cdot & \cdot & \cdot & \cdot & \cdot \\ \cdot & \cdot & \cdot & \cdot & \cdot & \cdot \end{pmatrix} \quad \text{and} \quad \begin{pmatrix} \cdot & \cdot & \cdot & \cdot \\ \cdot & \cdot & \cdot & \cdot \\ \cdot & \cdot & \cdot & \cdot \end{pmatrix}.$$

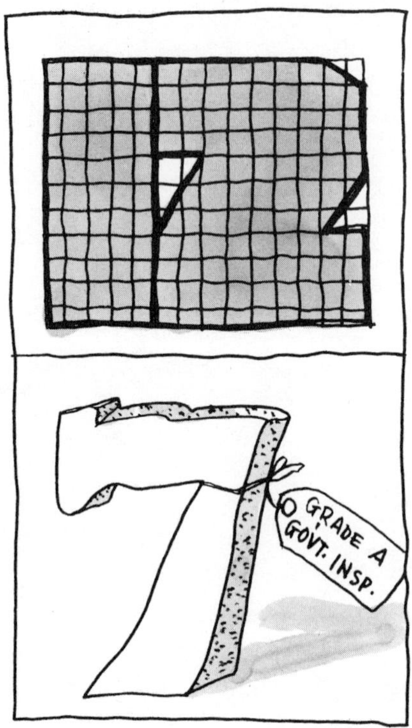

A set of seven points does not. Fractions such as $\frac{3}{12}$ can be reduced to $\frac{1}{4}$. Fractions such as $\frac{3}{7}$ cannot be reduced. Numbers like 12 that form rectangles were called *rectangular numbers* by the ancients, and *composite* (composed of two numbers) by moderns who wish to keep geometric ideas and words out of discussions about numbers. Numbers like 7 which don't form rectangles are called *prime.* People who set out orchards or chairs in classrooms, or who want to simplify fractions, are interested in which numbers are composite and which are prime.

Clearly, multiplication is involved in this concept. Indeed we can define a prime number as one divisible by two factors only, itself and one. Composite numbers have more than two factors. Twelve has six factors—1, 2, 3, 4, 6, 12. Nine has three factors—1, 3, 9. Seven has two factors—1, 7. Hence, 12 and 9 are composite numbers, while 7 is a prime number.

Is the number 1 prime or composite? The idea of forming or not forming an array doesn't apply very well to one point. Further, while 1 is divisible by itself and 1, there is only one factor.

As happens so often in mathematics, a nice idea works fine in most cases but breaks down for some. The ideas of prime and composite are clear for larger numbers and we would like to extend them to 1. Such extensions of a concept are usually made in the convenient way. Alas! It turns out to be inconvenient to list the number 1 as either composite or prime. We shall want composites to have at least three factors. Counting 1 as a prime would make listing the prime factors of a composite number awkward. Would the prime factors of 4 be 2, 2, or 2, 2, 1, or 2, 2, 1, 1, 1? For these and similar reasons, we say that 1 is neither prime nor composite.

A common distinction is that a composite is divisible by some number other than itself and 1. A prime is not. To see whether a number is composite, we merely divide it by all the numbers smaller than itself and see if the division comes out even.

We make a partial list of numbers with their divisors, and classify them as prime or composite (see top of opposite page).

Make such a list for numbers from 1 to 60. For example, you will find for 13 that 2, 3, 4, 5, 6, 7, 8, 9, 10, 11, 12 do not divide evenly. Hence, the only divisors of 13 are 1 and 13. As you develop your list, you may find some shortcuts. You should be able to find all the divisors of 48, for instance, without trying all the numbers 1 through 48.

In the additive structure of numbers we found that any number greater

n	Divisors	Prime	Composite
1	1		
2	1, 2	✓	
3	1, 3	✓	
4	1, 2, 4		✓
8	1, 2, 4, 8		✓
15	1, 3, 5, 15		✓
17	1, 17	✓	
30	1, 2, 3, 5, 6, 10, 15, 30		✓

than 1 could be written as the sum of smaller numbers. In fact, any smaller number could be an addend. No such simple rule holds for the multiplicative structure. We look for patterns. Do you see any? Can you write a composite number greater than 100? Do you see any patterns for writing a great many composite numbers? Do you think there are more composites or more primes among numbers under 1,000,000?

Consider the following array as a pattern guide. Write the numbers 2 through 61 in six columns and ten rows in this pattern:

2	3	4	5	6	7
8	9	10	11	12	13
14	15	16	17	18	19
20	21	22	23	24	25
26	27	28	29	30	31
...					

Now cross off each number you found to be composite. Do some rows or columns seem richer in primes (or composites) than other rows or columns? Do you think this will continue if the list is extended?

You may note that, except for the first row, primes appear only in the columns headed 5 and 7, and most numbers in these columns are primes. Every row has at least one prime, some have two. Do you think these properties will hold if the list is extended? Actually some do, some don't. All the larger primes appear in the columns headed 5 and 7, but they become scarcer. Many rows, such as the one beginning with 116, have no primes in them. The two rows beginning with 30032 have no primes.

You may have noticed, as you developed the table, that if A is a divisor of C, then C/A is also a divisor. For example, 3 is a divisor of 36, and so is $36/3 = 12$. Divisors come in pairs. One is smaller than or equal to the other. Hence, we need find only the smaller divisors by trial. For example, with 36, as we test 1, 2, 3, 4, 5, 6 we get quotients of 36, 18, 12, 9, 7^+, 6. ($7^+ = 7$ + a remainder.) As the trial numbers increase, the quotients decrease. When these two sequences meet or pass each other you have gone far enough. Hence, we know the divisors of 36 are 1, 2, 3, 4, 6, 9, 12, 18, 36.

When you test 39 with 1, 2, 3, 4, 5, 6, 7, you get quotients of 39, 19^+, 13, 9^+, 7^+, 6^+, 5^+. The numbers in these sequences have passed each other. Any divisor larger than 7 will have a companion smaller than 5^+. But these possible companions have already been tested, so you need go no further. The divisors of 39 are 1, 3, 13, 39. You may have noticed that the smaller divisor of C is $\leq$ (less than or equal to) $\sqrt{C}$ (square root of C). This is another guide to the amount of testing needed.

EXAMPLE 1 In testing for divisors of 230, Peter tested 2, 3, 4, . . . , 15 and found 1, 2, 5, 10, 23, 46, 115, and 230 as the factors. Marie said, "You've worked too hard. Clearly, $230 = 23 \times 10$, and $10 = 2 \times 5$. If I take the eight subsets of {2,5,23}, I will get all the factors." Can you show that the products of the members of Marie's subsets are indeed the factors that Peter found?

What Marie did was first find two factors of 230, just as Peter did. Then, instead of looking for more factors of 230, she looked for further factors of the two, 23 and 10. 23 is prime and had no interesting factors, while $10 = 2 \times 5$.

Now, is it possible that Peter, who tested all numbers, might find some factors that Marie would miss? You might try their methods on 45. Peter writes $45 = 2 \times 22^+, 3 \times 15, 4 \times 11^+, 5 \times 9, 6 \times 7^+, 7 \times 6^+$. Marie writes $45 = 3 \times 15 = 3 \times 5 \times 3$. The different subsets of {3,3,5} are $\emptyset$, {3}, {5}, {3,3}, {3,5}, {3,3,5}. Do they get the same factors?

There is a famous theorem called the Unique Prime Factorization Theorem by some, the Fundamental Theorem of Arithmetic by others, and less pompous names by the irreverent. It says, "A number can be written as a product of primes in only one way (except for the order of writing the primes)." This means that while 230 can be written differently with composite factors (23×10, 46×5, 115×2, 230×1), it can be written in only one way with prime factors ($2 \times 5 \times 23$). ($5 \times 2 \times 23$, $23 \times 2 \times 5$, . . . are regarded as the same as, or equivalent to, $2 \times 5 \times 23$.) While we will not prove this theorem, let us check it on 225.

EXAMPLE 2 One person notices $225 = 5 \times 45$. Another sees 9×25. A third recalls that $15 \times 15 = 225$. Factoring these expressions further, we get

$$225 = 5 \times 45 = 5 \times (9 \times 5) \quad = 5 \times 3 \times 3 \times 5$$
$$225 = 9 \times 25 = (3 \times 3) \times 25 \quad = 3 \times 3 \times 5 \times 5$$
$$225 = 15 \times 15 = (3 \times 5) \times 15 = 3 \times 5 \times 3 \times 5$$

Hence, we see that there is only one set of prime factors of 225, {3,3,5,5}. We note that the same factor can appear more than once in the set of prime factors. We regard the repeated factors as somehow "different" if we wish to use the convenience of set language. The different subsets are ∅, {3}, {5}, {3,3}, {3,5}, {5,5}, {3,3,5}, {3,5,5}, {3,3,5,5}. So the factors of 225 are 1, 3, 5, 9, 15, 25, 45, 75, and 225.

The unique factorization theorem guarantees that Peter will not find factors Marie will miss. Whatever factors Peter finds can be written as a product of primes. These primes will be a subset of the prime factors of the number. Hence, his factor will be one of Marie's as well.

The unique factorization theorem tells us there is no need to test composite numbers as possible factors. Hence, our simplest rule for determining whether a number is prime or composite is as follows:

1. Test whether the number is divisible by 2, 3, 5, 7, 11, . . . , the prime numbers in increasing order. If any divide evenly, the number is composite.

2. Continue step 1 until you have tried the largest prime less than or equal to the square root of the number. Equivalently, continue until the quotient on dividing by a prime is less than or equal to that divisor.

14.5 LEAST COMMON MULTIPLE You will see later that in adding fractions it is good to rewrite the fractions so that they all have the same denominator. When the fractions are reduced to lowest terms, this common denominator is a common multiple of the denominators. Often we want the *least* such common multiple, primarily because smaller numbers are easier to work with. Let's consider some examples of common multiples.

EXAMPLE 1 24 is a multiple of 4, because $24 = 6 \times 4$. Any number of the form $k \times 4$ is a multiple of 4. The set of multiples of 4 is {4,8,12,16,20, . . .}. Similarly, the set of multiples of 6 is {6,12,18,24, . . .}. The intersection of these sets is the set of common multiples, the numbers that are multiples of both 4 and 6. Looking closely we can guess that the common multiples are {12,24,36, . . .}. Do you agree? One starts at the low ends of the sets, looks at both sets simultaneously, and notes those numbers he sees in both.

The smallest common multiple is also called the *least common multiple* (LCM). It is 12 in this example.

Will every pair of numbers have a least common multiple? We look at two more examples.

EXAMPLE 2 (12, 15). The multiples of 12 are {12,24,36,48,60,72,84,96,108, 120,132,144, . . .}. The multiples of 15 are {15,30,45,60,75,90,105,120,135, . . .}. We see that there are common multiples; 60 and 120 are two of them. Can you find two more? The least common multiple is 60.

EXAMPLE 3 (5, 7). The multiples of 5 are {5,10,15,20, . . .}. The multiples of 7 are {7,14,21,28,35,42,49,56,63,70,77, . . .}. Common multiples are 35 and 70. Can you find a third one?

We conjecture that every pair of numbers has a least common multiple. From (5, 7) we see that the product of two numbers is a multiple of each. Since there are common multiples, there must be a smallest one. As we look at our examples we see that for one pair, (5, 7), the LCM is the product. For the other two pairs, (4, 6) and (12, 15), while the product is a common multiple, the LCM is smaller. Is this just luck, or is there some way we could tell before we found it whether or not the LCM is the product of the numbers? We notice that 5 and 7 are prime. Could this be the reason? You may want to use the idea of common factors to help you answer.

We note that the set of common multiples is the intersection of the two sets of multiples.

14.6 GREATEST COMMON FACTOR (DIVISOR)

If we look over the list of numbers and their factors or divisors, we see that every pair of numbers has a common factor, 1, and some pairs have larger common factors—2, 3, 5, 6, ...

EXAMPLE 1 GCF (4, 6). The factors of 4 are {1,2,4}. The factors of 6 are {1,2,3,6}. The common factors, the intersection of these two sets, are {1,2} and the greatest one is 2.

EXAMPLE 2 GCF (12, 15). The factors of 12 are {1,2,3,4,6,12}. The factors of 15 are {1,3,5,15}. The greatest common factor is 3.

EXAMPLE 3 GCF (5, 7). The factors of 5 are {1,5}. The factors of 7 are {1,7}. The greatest common factor is 1.

EXAMPLE 4 GCF (200, 300). The factors of 200 are {1,2,4,5,8,10,20,25,40,50, 100,200}. The factors of 300 are {1,2,3,4,5,6,10,12,15,20,25,30,50,60,75,100, 150,300}. The common factors are {1,2,4,5,10,20,25,50,100}. The greatest is 100.

These greatest common factors were found by finding all the factors of both numbers, finding the intersection of these sets, and picking out the greatest. There is a shorter way.

We write the two numbers as products of prime factors. The intersection of the sets of prime factors gives the common prime factors. Multiplying these together gives the greatest common factor.

EXAMPLE 5 GCF (200, 300). The prime factors of 200 are {2,2,2,5,5} and of 300 are {2,2,3,5,5}. The intersection of these sets is {2,2,5,5}. Their product is $2 \times 2 \times 5 \times 5 = 100$, the greatest common factor.

EXAMPLE 6 GCF (252, 30). The factors of 252 are {2,2,3,3,7}. The factors of 30 are {2,3,5}. The intersection of these sets is {2,3}. Their product is 6, the greatest common factor of 252 and 30.

Sometimes it is quite a chore to find all the prime factors of two numbers. Here are some examples of Euclid's method of finding the GCF of two numbers without factoring. We use the fact that the difference of two numbers, say $42 - 27$, is divisible by any number that divides both. We don't have to compute the difference to know it is divisible by 3. Similarly, the sum of two numbers, say $44 + 66$, is divisible by any factor that divides both. We know the sum is divisible by 11 and also by 2.

EXAMPLE 7 Find the GCF of 4 and 6. Since $6 = 4 + 2$, we know any number that divides both 4 and 2 divides 6. Clearly, 2 divides both 2 and 4, and hence divides 6. Since $6 - 4 = 2$, we know that any number that divides both 6 and 4 must divide 2. No common factor of 4 and 6 can be larger than 2, since it must divide 2. Hence, 2 is the GCF.

EXAMPLE 8 GCF (12, 15). Since $15 = 12 + 3$, any factor of 12 and 3 is a factor of 15; 3 is a factor of 3 and 12, and hence of 15. Since $15 - 12 = 3$, any factor of 15 and 12 is a factor of 3, and thus can be no larger than 3. Hence, 3 is the GCF (12, 15).

EXAMPLE 9 GCF (5, 7). Since $7 = 5 + 2$, we look for the GCF,of 5 and 2. Since $5 = 2 \times 2 + 1$, and $5 - 2 \times 2 = 1$, any common factor of 5 and 2 is a factor of 1. Hence, 1 is the greatest common factor. 5 and 7 are prime, and hence have no common factor other than 1.

EXAMPLE 10 GCF (252, 30). Instead of subtracting 30 at a time, we use division. We get $252 = 8 \times 30 + 12$ and $252 - 8 \times 30 = 12$. Any factor of 252 and 30 is a factor of 12. Any factor of 30 and 12 is a factor of 252. We use the fact than any number dividing 30 also divides 8×30.

The problem is simplified to finding GCF of 30 and 12. But $30 = 2 \times 12 + 6$ and $30 - 2 \times 12 = 6$. Any factor of 12 and 6 is a factor of 30. The GCF of 30 and 12 is a factor of 6. Since 6 is a factor of 6 and 12 and hence 30, 6 is the GCF of 12 and 30, and also the GCF of 30 and 252.

The Euclidean algorithm, as this method is called, is particularly useful with large numbers. Suppose we wish the GCF of 1121 and 4189. Division, or repeated subtraction, gives $4189 = 3 \times 1121 + 826$ and $4189 - 3 \times 1121 = 826$. Hence, the required number is also the GCF of 1121 and 826. Again

$1121 = 826 + 295$	so we seek the GCF of 826 and 295.
$826 = 2 \times 295 + 236$	so we seek the GCF of 295 and 236.
$295 = 236 + 59$	so we seek the GCF of 236 and 59.
$236 = 4 \times 59$	hence, 59 is the GCF of 1121 and 4189.

It would be quite a chore finding the prime factors of 1121 and 4189.

Two numbers are called *relatively prime* if the GCF is 1, even though one or both may be composite. 34 and 27 are relatively prime.

The distributive property shows that if $A + B = C$, then any common factor of A and B is a factor of C, and any common factor of B and C is a factor of A. Suppose that d is a factor of A and B—that is, $A = ad$, $B = bd$. Then $A + B = ad + bd = (a + b)d = C$, which says that d is a factor of C. Suppose that e is a factor of B and C—that is, $B = fe$, $C = ge$. Then $B = C - A = fe - ge = (f - g)e$, which says e is a factor of B.

14.7 PRIME FACTORS, LCM, AND GCF As suggested earlier, there is an easy, short way of finding LCM using prime factors. This also shows a relation between LCM and GCF.

EXAMPLE 1 Consider $72 = 2 \times 2 \times 2 \times 3 \times 3$ and $54 = 2 \times 3 \times 3 \times 3$. Any multiple of 72 has at least three 2s and two 3s. Any multiple of 54 has at least one 2 and three 3s. Any common multiple must have at least three 2s and three 3s, and hence is of the form $k \cdot 2 \cdot 2 \cdot 2 \cdot 3 \cdot 3 \cdot 3$. The least multiple has $k = 1$ and is $2 \cdot 2 \cdot 2 \cdot 3 \cdot 3 \cdot 3 = 216$. 216 is the least common multiple of 72 and 54.

Similarly, any divisor of 72 has no more than three 2s or two 3s. Any divisor of 54 has no more than one 2 or three 3s. Any common divisor has no more than one 2 and two 3s. Hence, 18 is the greatest common factor of 72 and 54.

EXAMPLE 2 Consider 4 and 10. $4 = 2 \cdot 2$, $10 = 2 \cdot 5$.

Any multiple of 4 has to have at least two 2s.

Any multiple of 10 has to have at least one 2 and one 5.

Any number with two 2s and one 5 will be a common multiple, and is of the form $k \cdot 2 \cdot 2 \cdot 5$. The least one has $k = 1$ and is 20.

So 20 is the least common multiple of 4 and 10.

Any factor of 4 has no more than two 2s.

Any factor of 10 has no more than one 2 and one 5.

Only a number with no more than one 2 is a common factor.

So the greatest common factor of 4 and 10 is 2.

We make a table of some pairs we have discussed.

Do you see any relation between the last three columns? Will this relation hold for all pairs? Try another pair to check.

A	B	LCM	GCF	$A \times B$
4	6	12	2	24
12	15	60	3	180
5	7	35	1	35
200	300	600	100	60,000
4	10	20	2	40

EXAMPLE 3 Consider the union of all the prime factors of 12 with those of 15.

$$\left\{\begin{matrix} 2 \cdot 2 \cdot 3 & \\ & 3 \cdot 5 \end{matrix}\right\}.$$

Let us take out one of each pair of the common factors. This gives

$$\left\{\begin{matrix} 2 \cdot 2 \cdot 3 & \\ & 5 \end{matrix}\right\} \text{ and } \{3\}.$$

We see the product of the factors remaining is the least common multiple 60, and the product of the factors removed is the greatest common factor. But the product of all the factors is $12 \times 15 = 180$.

EXAMPLE 4 Consider (72, 54). The union of all the prime factors of each is

$$\left\{\begin{matrix} 2,2,2,3,3 & \\ & 2,3,3,3 \end{matrix}\right\}.$$

We remove one of each pair of duplicated factors, getting

$$\left\{\begin{matrix} 2,2,2,3,3 & \\ & ,3 \end{matrix}\right\} \qquad \text{and} \qquad \{2,3,3\}$$

The LCM is $2 \cdot 2 \cdot 2 \cdot 3 \cdot 3 \cdot 3 = 216$, the GCF is 18, and the product is $72 \times 54 = 3888$.

Since this process can be copied for any pair of numbers, we always have, for any two numbers A,B

$$(\text{LCM}) \times (\text{GCF}) = A \times B$$

You might use this relation to find the LCM for numbers hard to factor. For example, to find the LCM of 72 and 54, using Euclid's algorithm:

$$72 = 54 + 18 \qquad 54 = 18 \cdot 3 \qquad \therefore 18 \text{ is the GCF}$$

$$\text{LCM} = \frac{72 \times 54}{18} = 72 \times 3 = 216$$

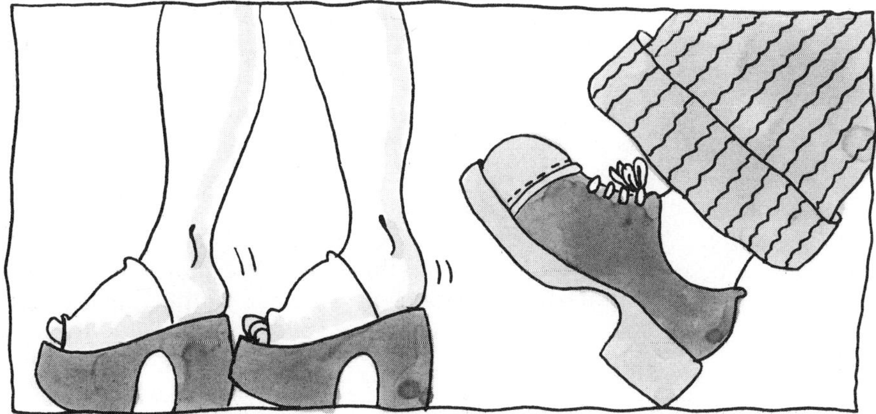

EXAMPLE 5 A tall man who takes steps 36 inches long goes for a walk with a girl who takes steps 28 inches long. She obviously takes more steps than he does. If they start out exactly together moving their left feet at the same time, will they ever be exactly together again? If so, how far have they gone? How many steps did each take? Which foot did each one land on when they were exactly together again for the first time?

Clearly, they will be exactly together again after he takes 28 steps and she takes 36 steps. They will have gone 28 × 36 inches. Both took an even number of steps, so both landed on their right feet. Can you see this as a "common multiple" problem? Now, when were they *first* together?

14.8 SIEVES FOR PRIMES Factoring a number is easy, as explained above. We merely try the primes in order. While the method is straightforward, we wish a more mass-production method if we wish to find a lot of primes and test a lot of numbers.

One of the best ways was developed by the Greeks, and is called the sieve of Eratosthenes. We give a brief discussion of the idea.

We have all seen sieves with some of the holes plugged. Most sieves are round. We want to make a linear sieve. Imagine a long pipe with equally spaced holes big enough to let a ball fall through. We number the holes 1, 2, 3, . . . and imagine they go on forever. The pipe with holes is our sieve.

We roll a ball down the sieve. It falls out hole 1. We plug hole 1 with a white plug. We roll the ball again. It rolls past hole 1 because it's plugged up and falls in through hole 2. We plug hole 2 with a red plug, and plug all the holes that are multiples of 2 with black plugs. These are (4, 6, 8, 10, . . .). This is every other hole.

We roll the ball again. It falls through hole 3. We plug with a red plug, then plug every multiple of 3 with a black plug—6, 9, 12, 15, . . . , every third plug. Some holes are already plugged, we just check the plug to be sure it is tight. We roll the ball. It falls through 5. We plug 5 with a red plug and every fifth hole thereafter—10, 15, 20, . . . —with black plugs, merely checking those already plugged.

And so on. We roll the ball. It goes through the first unplugged hole. We plug this hole red, and all multiples of this hole black. For example, after a while the ball will fall through 101. We black-plug every 101th hole thereafter—202, 303, 404, 505, 606, ...

When we have finished, all the red holes will be primes and all the black holes will be composite numbers. This concept has been used to form extensive tables of primes going into the millions.

You can make a paper sieve as follows. List the numbers, say to 100 (1, 2, 3, 4, 5, . . . , 100). Then draw a line under each number as you would put in a black plug. When you finish, some numbers will have several lines under them. Those not underlined are the primes. You may want to arrange the numbers in rows of 6 as was done on page 253 and compare the results.

2	3	$\underline{4}$	5	$\underline{\underline{6}}$	7
$\underline{8}$	$\underline{9}$	$\underline{\underline{10}}$	11	$\underline{\underline{12}}$	13
$\underline{\underline{14}}$	$\underline{\underline{15}}$	$\underline{16}$	17	$\underline{\underline{18}}$	19

We know that there are an infinite number of composite numbers. Anyone can give a very large even number and guarantee that it is composite. The opposite is hard. Indeed if you were to write a number with 100 digits it would likely be very hard to tell if it were or were not prime.

In 1964 two Americans proved that $(2^{11213} - 1)$ is a prime. This number in base ten has almost 3400 digits. At that time this was the largest number known for sure to be prime. Such knowledge about primes is of little use to engineers, chemists, or physicists, but it adds insight to the structure of numbers.

Are there larger prime numbers? Maybe someday you, a student of yours, or someone you know will name a larger prime.

14.9 INFINITY OF PRIMES A look at the six-column table shows that all the numbers in four of the columns (after the first row) are composite, and many in the other two columns. Clearly, in one sense, there are more composites than primes.

It might seem reasonable to argue that after some very large number all the remaining numbers can be constructed by multiplying smaller ones together. Certainly for the additive structure, any number after 1 can be formed by adding smaller ones together. If this happens, we would say the number of primes is finite, although doubtless very large. Are the primes finite?

Actually the Greeks answered this question. The primes never run out. The number of primes is infinite. No matter how large a number you take, there is a prime larger. We give in modern notation the Greek proof from Euclid.

Some preliminaries first. From the definition of division, we know that if we divide $7 \times 483 + 1$ by 7 we'll get a remainder of 1. The number $2 \times 3 \times 5 \times 7 + 1 = 211$ is not divisible by 2, or 3, or 5, or 7, because there is a remainder of 1 in each case. Similarly, if P_1, P_2, P_3, P_4 are any four primes, then $P_1P_2P_3P_4 + 1$ is not divisible by any of them, because there is a remainder of 1 in each case.

Now suppose there were a largest prime—that is, the number of primes is finite, and there is a number such that all larger numbers are composite. We could list the primes $P_1, P_2, \ldots, P_n$, where P_n is the largest prime. Consider the number $Q = P_1P_2 \ldots P_n + 1$. None of the primes divides Q, since each division leaves a remainder of 1. But since every number > 1 has a prime factor, Q must have a prime factor. Hence, there must be a prime other than $P_1, P_2, \ldots P_n$. Since this must be true for any n, there must be more primes than any number we care to name—an infinitude of primes.

As we look at the six-column number table, we see that some rows have two primes, others only one. Since the primes become very scarce, we might suspect that after far enough along in the table no lines with two primes will appear. It is not known whether this is true or not. This is called the "Twin Primes" problem.

The fact that there is no largest prime can be demonstrated on the six-column sieve (table of counting numbers greater than 1). Consider the following:

1. Every multiple of 2 and/or 3 appears in the columns headed 2, 3, 4, or 6. Also, every number in these columns is a multiple of 2 or 3 or both.

2. Every prime (except 2 and 3) and every product of primes not including 2 or 3 falls in the columns headed 5 or 7.

3. No prime except 2, 3, and 5 is a factor of two numbers in the same row, and 5 only if the numbers are ends of the row.

Suppose P is a large prime.

Let $N = 5 \cdot 7 \cdot 11 \cdot 13 \ldots P$, the product of all primes less than or equal to P, except 2 and 3. By statement 2, N falls in the column headed 5 or the column headed 7 in some row. Let x be the number in that same row in the column headed 7 or 5. Then, from statement 1, x does not have 2 or 3 as a factor. From statement 3, x has none of the other primes as factors. Since all numbers have at least one prime factor, x must have a prime factor greater than P.

This argument shows that no matter how large a prime P is, there is always a larger prime P_1. Hence, we may get as many primes as we please. There are an infinite number of primes.

EXERCISES

All numbers are counting numbers unless otherwise stated.

1. List three ways 12 can be written as the sum of two numbers with the larger number first.

2. Three children of the same weight wish to ride a teeter-totter. If one child sits 10 feet from the center on one side, where must the other two sit on the other side?

3. In how many ordered ways can 8 be written as the sum of three numbers?

4. List 5 as the sum of three numbers, largest addends first, in as many ways as possible. Addends may be the same.

5. In how many ways can a total of 10 be rolled with a red, a green, and a white die?

6. In how many ways can 37 cents be paid in U.S. coins having values of 1, 5, 10, and 25 cents?

7. Write 33 as the sum of four numbers where the largest addend is the largest possible.

8. Write 33 as the sum of four numbers where the largest addend is as small as possible.

9. Write 33 as the sum of four numbers where the smallest addend is (*a*) as small as possible, and (*b*) as large as possible.

10. List all the factors of 18. Which factors are prime?

11. Write 144 as the product of prime factors.

12. Pyxie wants a box containing 84 cubic inches. Give her the dimensions for eight possible different-shaped boxes. (Use counting numbers for edges.)

13. List the factors of 93.

14. What is the largest prime number less than 100?

15. Make a 6×10 sieve of Eratosthenes (prime-number sieve) for the numbers 2–61. Which number is underlined most? How many twin prime pairs are in this range?

16. Find the least common multiple (LCM) of
(*a*) [4, 6] (*b*) [16, 10] (*c*) [5, 8, 12]

17. Find the greatest common factors (GCF) of
(*a*) [6, 15] (*c*) [144, 360] (*e*) [84, 147, 372]
(*b*) [12, 54] (*d*) [1161, 1188]

18. Find (*a*) *A*, (*b*) *B*, (*c*) LCM (*A*, *B*), and (*d*) GCF (*A*, *B*) if

$$A = 2 \times 2 \times 3 \times 3 \times 3 \times 5 \times 7 \times 7 \times 11 \text{ and}$$
$$B = 2 \times 3 \times 5 \times 5 \times 11 \times 11$$

19. In exercise 18, how many factors of 5 are there all together in LCM (*A*, *B*) and GCF (*A*, *B*)? How many factors of 5 are there all together in *A* and *B*?

20. Answer the same questions as in exercise 19 about factors of 3.

21. In a race, *L* makes a lap in 48 seconds and *M* makes a lap in 40 seconds. At what time after starting does *M* first pass *L*?

22. Give two other times after starting that *M* passes *L*.

23. Write a number greater than 1,000,000 that has 23 and 7 as factors.

24. Write the smallest number greater than 1,000,000 that has 3, 5, and 7 as factors. (HINT: Which numbers have 3 as a factor?)

25. Are there any primes greater than 1700? Why do you think so? If there are any, how would you find the smallest one?

26. How many primes are there between 2 and 55? Between 56 and 110? Which interval has more?

27. Guess which interval, (*a*) between 1 and 1,000,000 or (*b*) between 1,000,001 and 2,000,000, has more primes? Give reasons why your guess is plausible.

28. Make the same guess as in exercise 27 for the intervals (*a*) between 99,000,000 and 100,000,000 and (*b*) between 100,000,000 and 101,000,000. Do you feel as sure of your guess here as in exercise 27?

Answers to selected questions: **1.** 11 + 1, 10 + 2, 8 + 4, and others; **2.** 5 feet from center; **3.** 21; **5.** 27; **8.** 9 + 8 + 8 + 8 and others; **9.** (*b*) 9 + 8 + 8 + 8; **12.** 10 possible: (1, 1, 84), (1, 2, 42) . . . . (2, 2, 21), (2, 6, 7), (3, 4, 7); **17.** (*b*) 6; (*d*) 27; (*e*) 3; **21.** 240 seconds; **23.** 1,000,132 + *n* (161); **26.** 15; 13; between 2 and 55.

15

Number Patterns

15.1 INTRODUCTION "Checkerboards, backgammon boards, go boards are good for games, but the best board to play on is the counting numbers," according to one famous mathematician. Number games explore patterns and sequences that are often useful as well as interesting. First we consider arithmetic progressions and geometric progressions. These sequences move like a parade down the avenue of numbers. Then we discuss figurate numbers, whose sets form geometric figures. These numbers are a link between arithmetic and geometry. We close the chapter with numerology and the ways patterns in numbers are sometimes linked with choices in life, for lucky numbers have an enchanting mystique for many people. Most mathematicians believe the use of numbers in this way is highly unreliable.

15.2 ARITHMETIC PROGRESSIONS The most fundamental pattern is that of the counting numbers themselves, $\{1,2,3,\ldots\}$, sometimes called the "one more than" pattern. The even numbers $\{2,4,6,\ldots\}$ and the odd numbers $\{1,3,5,\ldots\}$ show the "two more than" pattern. Any sequence that follows a "d more than" pattern is called an *arithmetic progression*. Usually only a few terms are wanted. For example, (7, 10, 13, 16) is an arithmetic progression with $d = 3$, first term 7, last term 16, and 4 terms. This fits the following story. Sharon tosses cards on the floor and picks them up one at a time as a figure-improvement exercise. She starts with 7 cards and increases the number by 3 each day. How many cards does she pick up on the first, second, third, fourth day? How many cards on the tenth day? The twentieth day?

It is easy to merely list the 10 numbers to get the number of cards on the tenth day. But can you see a pattern in the numbers from which you could figure out the tenth number after, say, only 8 numbers? 5? Let us make a table:

Today	1	2	3	4	5	6 . . . 10 . . . 20
Number of cards today	7	10	13	16		
Cards more than yesterday		3	3	3	3	
Cards more than first day		3	6	9	12	

Can you see that the entries for day 10 are 10, 34, 3, 27? Which of these four entries did you get first? Next? Third? Last? Can you get the four entries for day 20?

Can you see that for day y, the number of cards is $7 + 3(y - 1)$?

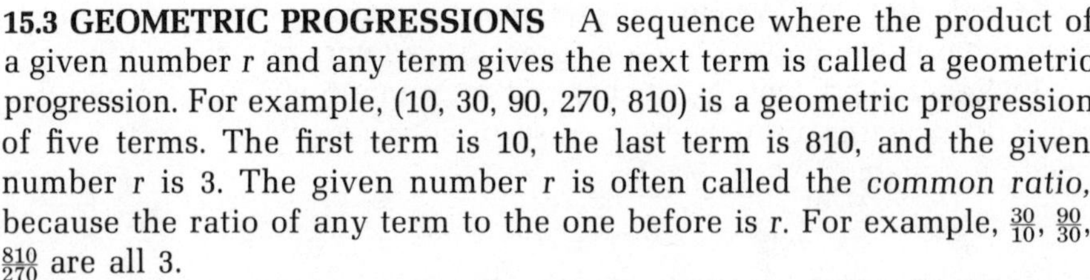

15.3 GEOMETRIC PROGRESSIONS A sequence where the product of a given number r and any term gives the next term is called a geometric progression. For example, (10, 30, 90, 270, 810) is a geometric progression of five terms. The first term is 10, the last term is 810, and the given number r is 3. The given number r is often called the *common ratio,* because the ratio of any term to the one before is r. For example, $\frac{30}{10}$, $\frac{90}{30}$, $\frac{810}{270}$ are all 3.

It is said that a collection of guinea pigs will triple in one year. If there are 10 at the end of year 1, there will be 30 at the end of year 2. How many will there be at the end of year 3? 4? 6?

Members of a club form a telephone chain. To call a meeting, the president calls two members. Each of these calls two members, who then each call two members, and so on. Let us call the president the first generation, those he calls the second generation, those these call the third generation, and so on. Members in the ith generation call those in the $i + 1$ generation. How many people are in the third generation? The fourth? The fifth? The tenth?

If the fourth generation just joined the membership, how many members are there in the club?

15.4 FIGURATE NUMBERS If a set of n chips can be arranged in a geometric figure, n is given the name of the figure. Here are some examples.

Rectangular numbers If the chips can bc arranged to form a rectangle, the number is a rectangular number. For example,

show that 12, 8, and 30 are rectangular numbers. Usually one wishes the sides to be at least 2 long. Some people think there is a relation between this idea and the idea of prime and composite numbers. Do you?

Square numbers If the rectangle is a square, the number is a square number.

and

show that 9 and 25 are square numbers. Do you see any others? Sometimes we call these numbers *squares* rather than *square numbers.*

Triangular numbers The diagrams

show that 3, 6, 10 are triangular numbers. What would be the next number? The next? It is convenient to call 1 a triangular number. Then when the triangular numbers are arranged in increasing order, the third number's triangle has 3 spots on a side. What is the number with 4 spots on a side? (Actually a number is not a leopard with spots. We mean the triangle associated with the fourth number has four spots on a side.) What is the seventh number?

Some people think they see a relation between triangular numbers and arithmetic progression. Do you see any relation?

Some people like to rotate the triangle through 180° and put two triangles together, as in the margin.

o x x x x
o o x x x
o o o x x
o o o o x

Some people think this figure helps them determine what the fourth triangular number is. They say the figure shows that twice the fourth number is $4 \times 5 = 20$. If you do the same thing for the second triangular number, what would you get?

Hexagonal numbers The figures

show that 1, 7, 19 are hexagonal numbers associated with 1, 2, 3, respectively. What are the hexagonal numbers for 4 and 5? Do you see a pattern in the numbers? How much do you add to number 1 to get number 2? To number 2 to get number 3? To number 3 to get number 4?

This way of exploring a sequence of numbers by subtracting each from the next is called the *method of differences,* because you look at the differences between the terms. It is sometimes helpful to look at the differences between the differences also. For example, Z = (1, 6, 16, 31, 51, . . .) is a sequence whose differences are (5, 10, 15, 20, . . .). The differences in this sequence are (5, 5, 5, . . .). These last two sequences are called *first differences* and *second differences,* respectively, of the sequence 1, 6, 16, 31, 51, ... What is the next term in Z? What would be the second differences of (5, 10, 15, 20, . . .)?

Pentagonal numbers We leave it as an exercise to draw the diagrams to show that 1, 6, 16 are the first three pentagonal numbers. Can you find the next three as well?

Sometimes several numbers lead to figures that are similar but of different size. Can you find such sets or sequences of numbers from the earlier examples?

15.5 NUMEROLOGY

People are always looking for easy answers to hard problems. Since mathematics, and especially arithmetic, has been the most successful tool for this, people often ask questions that are inappropriate and force easy answers. Questions such as "Should I marry this girl?" "Should I ask the boss for a raise today?" "What is my lucky number?" are typical.

There is a great wealth of literature containing a large number of computational rules purporting to answer such questions. Often the rules are said to be handed down from antiquity; the reasons for them are lost, are proclaimed in vague terms, or are said to be secrets. We cannot go into all of these computations, but here are a few.

Do you have a lucky number, say one of the digits 1, . . . , 9? If so, associate numbers with what you may do, and do those things tagged with your lucky number. In particular, (1) bet on this number at roulette, (2) bet on the horse with this number at the races, (3) buy this many gallons of gasoline at the service station, (4) invite this many guests to your next party, and (5) speak admiringly to this many people every day. If you follow these rules you will find people will like you, you won't have fatal automobile accidents, and you will win more often at gambling.

Here is one highly recommended way of finding your lucky number. Note that in any part of the computation where a two- or more digit number arises, you add the digits. Give each letter in your name the number associated with its position in the alphabet; i.e., a = 1, b = 2, c = 3, . . . , y = 25 = 7, z = 26 = 8. Add the numbers for your names. This is your lucky number. Thomas Jefferson's lucky number was 20 + 8 + 15 + 13 + 1 + 19 + 10 + 5 + 6 + 6 + 5 + 18 + 19 + 15 + 14 = 3. Some people call this number your "name lucky number." Other lucky numbers come from your birthdate and your address.

Some people claim that every number has a certain degree of luck for you—some more than others. Your problem is to find the luckiest. People who have the same number for their name, birthdate, and address are thrice endowed.

Name numbers found by multiplying the letter numbers together rather than adding them are universally regarded as powerful lucky numbers. Sometimes these are called P numbers, for powerful product numbers. T. Jefferson's P number was 9.

The luckiest numbers are 1, 3, 7. The least lucky are 4 and 8. If zero turns up as your lucky number you are doomed. Lengthy penance is required to escape the "curse of zero." Many expert numerologists suggest that extra exercises in arithmetic, frequently done for six months, will go a long way toward giving protection.

You may use different names in different places—such as Hourice,

Hourice Jones, and Miss Jones. Hence you may have different lucky numbers for different places. If you don't like the number you get, you can change your name, use an initial, or put in a title such as Mister.

You may like to use a different alphabet. Some like to use a Chaldean or Hebrew alphabet where (A, I, J, Q, Y) = 1, (B, K, R) = 2, (C, G, L, S) = 3, (D, M) = 4, (E, H, N, T, X) = 5, (U, V, W) = 6, (O, Z) = 7, (F, P) = 8. Such numbering ties the theory to the mystic past. Nine is not used because this is the number for a nine-lettered name of God and is considered sacrilegious.

Do you think numerology is a reliable guide? Most mathematicians think it is nonsense. If you go about pinning numbers to events, you'll find some nice, some bitter. Some coincidences will occur that you will particularly remember and you may believe that the number is lucky. Careful record keeping will doubtless show no correspondence between numbers and the pleasantness of events.

On the other hand, people do like to figure their name number and their friend's name number. It makes a useful grouping game at parties. Finding name, birthdate, and address numbers is a good exercise in counting, addition, and multiplication. Timid people may be more willing to try something new if they believe it is lucky for them.

You may wish to make up some rules of your own for identifying lucky numbers. The author is glad to guarantee that if you do the five things mentioned at the start of this section you will have the outcomes promised. Indeed, if you speak admiringly to many people and give parties you will soon have many friends. If you bet regularly and frequently on your lucky number you will win more frequently than you have in the past. You will also lose more frequently. The chances of your having a fatal accident are small, and if you do, you can't hold the author to the guarantee.

EXERCISES

1. Zelda ate 3 cookies on Monday, 5 on Tuesday, and so on, increasing by two cookies each day. How many cookies did she eat on Saturday? How many did she eat all together during the week (6 days)? Name the number pattern formed by these cookies.

2. Suppose the population increases by $\frac{1}{3}$ every 10 years. If the population is 243,000,000 in 1975, what will it be in 1985? 1995? 2005? 2015? Name the number pattern formed by these population figures.

3. The inventory at a used car lot contained 23, 27, and 31 cars on June 1, 2, and 3 respectively. If the inventory continues to grow at the same rate, how many cars will there be on June 5? June 6? June 7?

4. When will the inventory in exercise 3 be greater than 50 cars?

5. Write the first four triangular numbers.

6. What is the sum of the first five triangular numbers?

7. What is your name lucky number?

8. Do you have a name, nickname, home name, birthplace, or . . . whose number is 7? What is it?

9. Suppose in 1000 races you bet on the horse with your lucky number. Do you think you will win many times? What else do you expect will happen?

10. Draw a diagram to show that a triangular number is half a rectangular number.

11. A snake eats every 4 days. If it ate on July 3, name the other days in July on which it ate.

12. What is the lucky number for Jane? The powerful *P* number?

Answers to selected questions: **1.** 13; 48; arithmetic progression; **2.** 324,000,000; 432,000,000; 576,000,000; 768,000,000; geometric progression; **5.** 1; 3; 6; 10; **9.** yes; lose many times; **12.** 3; 7.

16

Extending Ideas

16.1 INTRODUCTION Perhaps even more important than numbers, the objects studied in arithmetic, are the methods of thinking. We look in particular at the ways ideas grow and are extended.

What is a dog? What is a set? What is the color green? Try answering these questions. Then ask these questions: Is your idea of a dog the same today as it was when you were a small child? Do you think you have the same idea as dog breeders and biologists? Ideally the author would like you to wrestle with these questions a bit before going on.

We hope you will agree that your concepts start as small and simple ideas, then are extended as you become more experienced. You will see the question changes slightly from "What is a dog?" to "How can we sort things into dogs and not-dogs?" Sometimes we give an example of such a classification rather than a definition in words. We say "A, X, Z are dogs, B, P, U are not dogs," where A, X, Z, B, P, U are known. We define by example and classification.

Surely to many of the very young, *dog* is a proper name, like Spot, and applies to a particular animal—the family pet. The collie from down the street is no more a *dog* than the kitchen chair, but a fuzzy doll may be a dog. As time goes on, the child enlarges and restricts his acceptable set quite drastically as he finds it more convenient to include some things and exclude others. In maturity he seeks a way of classifying any creature he has met or will meet, and calls this classifying criterion his definition of *dog*. Sometimes the border between *dog* and *not-dog* is not very clear. He may not be quite sure whether wolves, coyotes, hyenas, wild dogs, chihuahuas, Saint Bernards, and foxes are or are not also *dogs*.

As one becomes more familiar with the word and the concept, he may extend it to quite different objects. Many pictures and sculptures are called dogs. An asset that loses value is called a dog. Some people say they ride from town to town on a dog—a phrase suggested by the name of a popular bus company.

We note that an idea like "dog" or "green" will seem quite clear at first. With more experience, we wish to consider some new objects. Our ideas of "dog" are extended in order to classify the new objects satisfactorily.

This process of extending or generalizing our ideas with greater experience is one of the most important and significant human activities. We give some examples from mathematics and urge you to recognize this process elsewhere.

16.2 EXTENDING SETS The first intuitive sets are collections of several objects such as a set of books, a set of chairs, a set of dishes. Such sets have at least 3 or 4 members and not more than 200. They have certain definite characteristics. All the members have a characteristic common property, such as being a dish or a book. There are more than a few of them. There are so many that the individual nature of the members tends to be lost and one thinks of them as a collective; yet there are not so many that the individual nature is obscured. Grains of sand at the beach are too small and too numerous to be considered a set. In common language, people do not speak of the set of grains of sand, or the set of noses on a person's face, or the set of eyes. The individuality of each eye is so strong that one says "pair of eyes."

Many of these facets are not spelled out, but are reactions based on common use. Clearly, as one matures he finds some points need greater clarification. Among these are—

1. What kind and how dominant must the common property of members be for the members to form a set?

2. What is the largest number of members a set may have?

3. What is the smallest number of members?

4. How individual, how separable, or how different from each other must the members of the set be?

Mathematicians in particular wish these questions to be clearly answered so that the answers can be precisely applied in borderline cases in ways agreeable to everyone.

Now it turns out that the concept of common property is a Gordian knot as far as leading to a clear, reliable definition is concerned. After a hard struggle, the mathematicians threw up their hands saying, "This is not the central concept in sets. The idea of a collection is important, but the reason for the collection is not important. We shall extend the concept of set by no longer requiring any common property. Anyone who forms a three-member set from a blade of grass in London, a loaf of bread in Yellowstone, and a stone wall in Timbuctoo is an idiot, but he probably won't hurt anyone, so we'll allow him to do this. Anyone may put together any objects whatsoever and proclaim the result a set."

Some people like to say that belonging to the set is a common property. This is a specious reason, for if the idea of common property is to have any value, the objects must have the property before they are classified in the same set.

We next seek to name a largest number of members for a set. The whole concept is so intuitive that it is impossible to give any good maximum. So mathematicians make no upper limit. A set can have as many members as one desires to put in it.

What is the fewest number of members a set can have? Here the situation is different. As explained in chapter 4 on sets, we find it extremely convenient to extend the set idea to include sets with only one member, and even to the empty set with no members.

There is no logical reason why the idea could not be extended to sets with negative numbers of members. Physicists have discovered antimatter. Perhaps three books made from antimatter would be a set of $^{-}3$ books. But the concept of sets with negative members has not been useful and ideas like multiplication and Cartesian product do not extend to negative numbers as we would like. Hence mathematicians have agreed to stop extending the idea of sets with the empty set. The smallest number of members in a set is zero.

Books and dishes are completely separate. Fingers are fastened together at the palm. How distinct must the members be to be different? Again, this is such an awkward idea that we extend the concept of set so that the set-maker need be the only one who can identify the different members. The members can be quite varied, such as ideas, objects formed by imaginary cuts along imaginary lines, or sets themselves.

We see that the set of books or the set of dishes is the same for the mathematician as for the layman. The mathematician who seeks to explore the set idea finds it most useful to extend the concept from its intuitive beginnings.

16.3 EXTENDING THE IDEA OF NUMBER A sneaking question to ask a group of intelligent, sophisticated adults is, "What is a number?" Everyone knows what a number is. But a close look shows that most people's ideas are a goulash formed from many sources joined together in a shotgun marriage. These ideas are actually a chain of generalizations or extensions. One starts with counting numbers, which are wonderful and completely satisfying for many purposes, but they are not enough. We find certain situations, such as lengths of segments and points on a line, that are not named by counting numbers but are nevertheless much like those that are. So we find it greatly convenient to extend the idea of number to include the rationals, whole numbers, negatives, roots, and reals. We will look at these extensions in later chapters.

In advanced algebra, a detailed study is made of the trail of ideas starting with counting numbers and leading to concepts that stretch one's intuitive ideas of number drastically. This is called "constructing the number system." This detailed formal study is beyond our purposes.

16.4 GENERALIZING MULTIPLICATION Multiplication is an example of an operation that started with one attitude and was then extended and generalized to other attitudes. The first intuitive attitude is usually repeated addition. This attitude is excellent for multiplying counting numbers, but fails miserably when attempted with fractions, negatives, and radicals. So, either multiplication had to be thrown away when we extended numbers or we had to invent appropriate extensions of multiplication. The latter was done and now, as was seen in earlier chapters, we have several extensions of multiplication that are quite useful. We have called these attitudes array, segment-segment-region, operator-operator-operator, tree, and similar names.

16.5 ILLUSTRATIONS You may find it interesting to trace the development of other mathematical and nonmathematical ideas. In each case, look at the simplest, most rudimentary examples first. Describe the people for whom these are the whole story, such as the baby who thinks *dog* is

the proper name of the family pet, or the child who thinks "1, 2, 3, 4, 5" are the numbers and *all* the numbers. Then describe how the idea may be extended to your current idea. Here are some ideas to try this on: addition, length, subtraction, equals, division, is the same as, is greater than, angle, area, point, line.

16.6 SUMMARY The form or process of generalization can be described in steps:

1. Define a concept in a simple, intuitive, obvious case by means of a sorting rule. This is often called an *operational* definition.

2. Consider some objects that should be sorted but that the sorting rule doesn't fit.

3. Change the sorting rule so that the old objects can be sorted as before, while the new objects can also be sorted.

The new sorting rule is the extended or generalized concept. It is useful to express the sorting rule in words as well as by actions.

EXERCISES

1. The idea of set in "the set of dishes" is extended when we say "the set of people in the world." What property does the latter set have which the former lacks and which might cause the beginner to have trouble with the concept?

2. Similarly to exercise 1, point out the extension to "the set of theories of God in history."

3. Similarly to exercise 1, point out the extension to "the set of orchids growing wild in Antarctica."

4. Discuss how the idea of subtraction can be generalized beyond "take away." Name a subtraction situation in which nothing is taken away and we must stretch our minds to identify something taken away.

5. Describe a situation in which multiplication is called for, but repeated addition is senseless. Describe how repeated addition is generalized and extended to an attitude that does make sense in your situation.

Answers to selected questions: **1.** cannot be seen at one time; **3.** no objects or ideas; the null set,

17

Fractions and Rational Numbers

17.1 INTRODUCTION To conquer and control, Caesar thought of Gaul as divided into three parts. For similar reasons we think of the ideas of fractions and rational numbers as separated into three parts: the number, the names, and the human experiences that lead to rational numbers and make them worth talking about. More specifically, fractions are symbols like $\frac{2}{3}$, $\frac{1}{4}$, $\frac{7}{5}$, .234. These symbols are the names for numbers. We will see that $\frac{1}{2}$, $\frac{2}{4}$, $\frac{3}{6}$, . . . , while different fractions, are all names of the same rational number. In this chapter we think of a rational number as an idea with different names that arise in several types of situations. In chapter 20 we will take a slightly different approach and define rational numbers as an extension of whole numbers without regard to these physical situations.

We will discuss various physical happenings from which the ideas of fractions and rational numbers might arise, and which illustrate their meanings. In the next chapters we discuss operations such as addition.

While it is clear that fractions are symbols and rational numbers are ideas, common usage often allows either "fraction" or "number" for either the symbol or the idea. For example, it is impossible to add the fractions $\frac{1}{5}$ and $\frac{2}{5}$—we add the numbers whose names are the fractions $\frac{1}{5}$ and $\frac{2}{5}$. Nobody ever wrote the number $\frac{3}{4}$ on a sheet of paper. We write the fraction $\frac{3}{4}$ instead. Or, if we like, the name of the number whose name is $\frac{3}{4}$. Yet people do speak of adding fractions and writing numbers.

We are mainly interested in ideas. Sometimes the ideas are hidden by the many words used in exact language. When it is clear from the context whether the name or the idea is being used, we may use either *fraction* or *rational number* to simplify the language.

The name *rational* comes from the ratio use, explained in section 17.4. These numbers are not particularly "wise" or "unemotional."

17.2 TYPICAL ATTITUDES TOWARD FRACTIONS AND RATIONAL NUMBERS People often think and feel about fractions according to their main uses. This is, of course, true of most concepts. Among the attitudes people have toward them are

1. names of objects with a history
2. ratios of sizes of objects
3. results of division, or missing factors
4. operators or instructions
5. points on a number line
6. symbols to be used as suits our fancy

The first four are especially important as the physical foundations or illustrations for the operations with rational numbers. The number line is vivid, suggestive, and highly recommended. The nominal use of fractions as an identifying or decorative symbol is painful to mathematicians. It is mentioned only to remind us that just because something might be called a rose doesn't mean it smells sweet. We consider these attitudes and uses in detail.

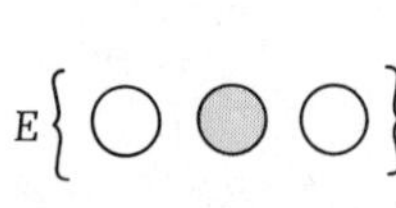

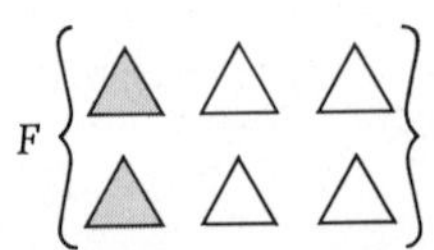

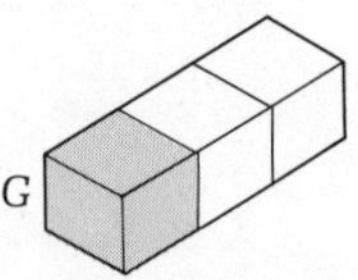

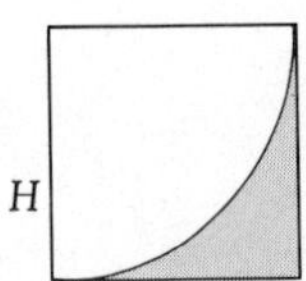

17.3 FRACTIONS AS THE NAMES OF OBJECTS WITH A HISTORY It is customary to illustrate $\frac{1}{3}$ with pictures like this:

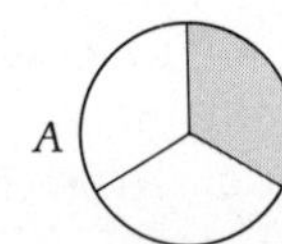

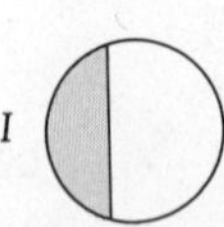

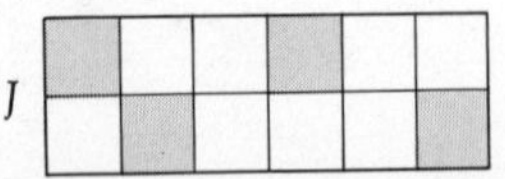

Something called a whole is separated into 3 parts of the same size. One of these parts is tagged with the label $\frac{1}{3}$. The "whole" may be any size or shape, in one or several pieces. The key is that the history of the portion tagged is the same in both cases. By history we mean what happened to a whole to produce the objects, or how the object arose. We picture some other objects that show $\frac{1}{3}$.

A is a circle, with the shaded part showing $\frac{1}{3}$. The 3 angles at the center have the same size. These pie-shaped figures show fractions vividly.

We may imagine *B* as a long thin rod with the middle of the 3 parts of the same size being labeled $\frac{1}{3}$. Of course, each of the other 2 parts also show $\frac{1}{3}$, and can be so labeled. The separation may be physical or merely imagined.

C is a rectangle, D is a line segment with the portions showing $\frac{1}{3}$ marked. E is a set of 3 objects, F a set of six. In these examples, the different objects in the set are equivalent to each other. We say they are the same size.

G is a brick, or block, H is a square with certain off-center curve. It is not so clear that H can be separated into 3 parts, each having the same size as the shaded part. Certainly the 3 parts won't be congruent—that is, duplicates of each other. However, their areas will be the same. In this sense, they will be the same size. Similarly, in I a portion of the circle is cut off. The circle cannot be separated into 3 duplicate parts, but the areas will be the same.

In J a rectangle has been separated into 12 small congruent squares that are duplicates of each other. Any set of 4, such as those shaded, can be regarded as one of 3 parts of the rectangle. The remaining 8 squares can be collected into two other sets of 4 each.

The piece of a circle as in A is perhaps the best beginning illustration of $\frac{1}{3}$. Its shape gives not only itself, but we can easily imagine the whole, the entire circle, and the fact that this piece is one of 3 parts.

The other illustrations are better when it comes to putting parts together. However, when we look at the object showing $\frac{1}{3}$ in C by itself, we cannot tell what the whole is. Hence, we cannot say whether this object shows $\frac{1}{3}$, or $\frac{1}{2}$, or some other fraction. It is perhaps more accurate to say that "$\frac{1}{3}$ is the name of the history of the object," rather than of the object itself. The objects are all different. It is their histories that are similar. We use the shorter "name of object" for convenience.

The fraction $\frac{2}{3}$ can be shown in two ways as an object with a history. First, $\frac{2}{3}$ labels an object made of 2 of 3 equal parts of a whole. Examples of such objects are shaded in the following:

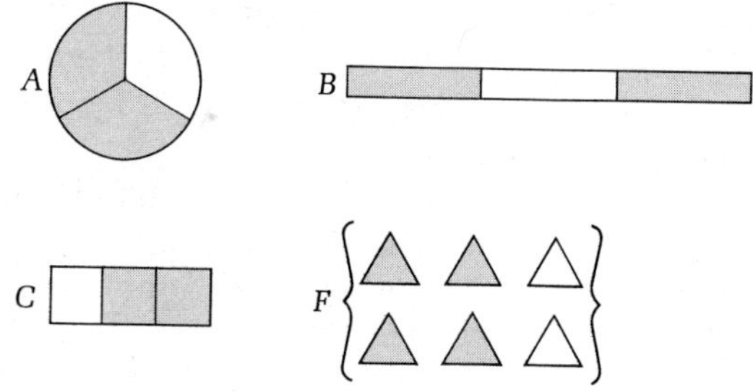

We leave it as an exercise for the reader to draw diagrams showing $\frac{2}{3}$ for the wholes shown in the earlier examples.

Second, $\frac{2}{3}$ labels one part when two wholes are separated into 3 parts of the same size. Consider these diagrams:

B

In *A* the whole is a rectangle. In the first diagram, a whole is separated into 3 parts by dotted lines, and 2 are shaded showing $\frac{2}{3}$ as before. In the second diagram two wholes are placed end to end. Their union is separated into 3 parts by dotted lines. (The middle part is half from one of the wholes and half from the other.) One of these parts is shaded. A glance shows that the shaded parts of the two diagrams are the same size. Both show $\frac{2}{3}$, even though their histories are slightly different.

In *B* the whole is a set of 3 members. The shaded subset of 2 shows $\frac{2}{3}$ in the left diagram. In the right, we have two wholes—a set of 6 members. Separating into 3 subsets and taking one yields a set of 2. This also shows $\frac{2}{3}$.

We see that "whole" is determined at the start and is the same in the entire discussion. Two wholes put together remain two wholes. They are not a "new whole."

In a fraction such as $\frac{2}{3}$, the lower number—3—is called the *denominator*. The upper number—2—is called the *numerator*. Some people also call them "downstairs" and "upstairs" in the fraction. *Denominator* comes from the Latin word meaning to name, or to identify. Denomination in money comes from the same word. The denominator names the size of the individual pieces. *Numerator* comes from the Latin word meaning to count. Hence the numerator counts the number of pieces.

In the early days people were steeped in the physical background of their numbers. A fraction like $\frac{3}{4}$ had to represent 3 little bits, such as 3 apples. A fraction like $\frac{5}{4}$ didn't make sense, because there were only 4 parts available from the whole. As we will see, the use of fractions was later extended so that $\frac{5}{4}$ did make sense. Even so, the "little-bits" die-hards called fractions with numerators bigger than or equal to the denominators

"improper," while those with the numerator less than the denominator were "proper." These terms are still used, but with no moral meaning.

$\frac{3}{4}$ does not have to be 3 little bits. It can be one of 4 parts of 3 wholes.

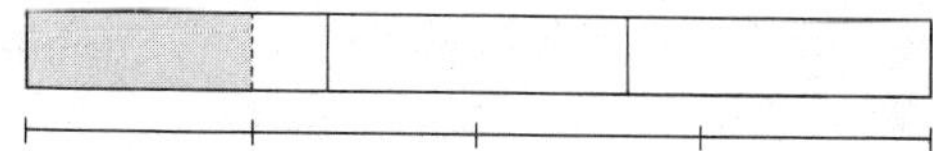

Hence, $\frac{3}{4}$ can be the name of a single thing rather than 3 things. Other attitudes support the idea that $\frac{3}{4}$ is one number, not 3 little bits.

In discussing fractions as names of objects, some authors say that the whole is divided into "equal" parts, meaning parts of the same size. We may do the same. However, it is currently popular to use "equal" as meaning "the same" rather than "the same size." "$A = B$" means A and B are different names of the same piece.

Improper fractions, such as $\frac{5}{4}$, have two interpretations. In the first, we think of a whole as separated into 4 parts. Then from some extra source we get another piece just like one of these 4. The 5 parts show the fraction $\frac{5}{4}$. In the second, we think of separating 5 wholes into 4 parts. One part shows $\frac{5}{4}$.

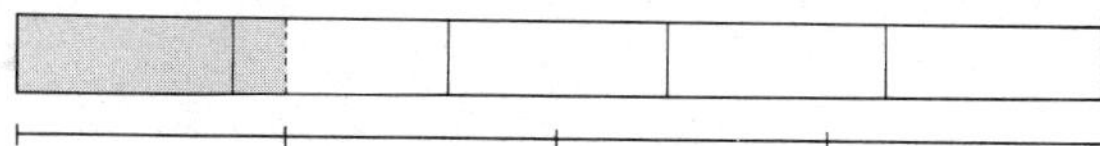

17.4 FRACTIONS AS RATIOS OF SIZES OF OBJECTS Two objects of the same nature (segments, regions, times) are often compared by giving their ratio. For example, people make statements like–

1. The small box of candy is only $\frac{1}{3}$ as heavy as the large box.
2. The olive jar holds $\frac{3}{5}$ cup.
3. The desk is $\frac{29}{12}$ feet wide.
4. The width of the sheet of paper is $\frac{4}{5}$ the length.
5. The ratio of men to women at the PTA meeting was $\frac{2}{3}$; or, There were $\frac{2}{3}$ as many men as women at the PTA meeting.
6. Corn cooks in $\frac{3}{4}$ the time that beans do.
7. In four games, Joe's batting average was $\frac{5}{18}$.

Let us look at what these statements mean. Can we describe the common thread in all these events that makes fractions a useful part in each description?

Each of these statements shows a comparison between the sizes of two objects of the same type. The size may be a count, a length, a time interval, or a weight. We consider each statement in turn. We point out two ways of showing a fraction, one by counting subparts, the other by counting multiples.

The subparts method If two objects A,B can be separated into subparts of the same size and if there are a parts in A and b parts in B, we say the ratio of the size of A to the size of B is $\frac{a}{b}$. Frequently we say, "The ratio of A to B is $\frac{a}{b}$." Note that A and B are objects, while a and b are counting numbers.

The multiples method Suppose we have many copies of A and B. If we find that s copies of A are the same size as r copies of B, then we say the ratio of A to B is $\frac{r}{s}$. We note that in s copies of A there are $s \times a$ subparts and in r copies of B there are $r \times b$ subparts. Can you see that $s \times a = r \times b$? Later we will see that from $sa = rb$, we conclude $\frac{a}{b} = \frac{r}{s}$ on dividing both sides by s and b.

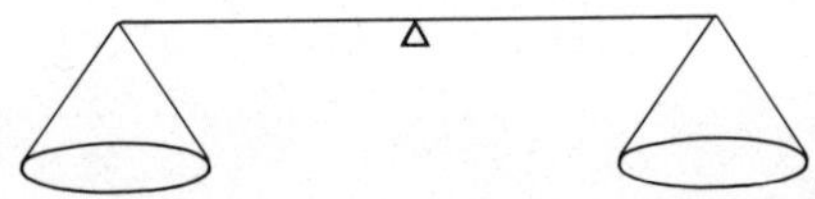

1. The small box of candy is only $\frac{1}{3}$ as heavy as the large box. Think of an equal-arm pan balance. We put the candy from the small box in one pan and find we can separate the candy in the large box into 3 parts, each of which just balances the candy in the small box. There is one subpart in the small box, 3 in the large. The ratio of the small box to the large box is $\frac{a}{b} = \frac{1}{3}$.

The multiples method is shown by noticing that 3 of the small boxes just balances one of the large. The ratio of the small box to the large is $\frac{r}{s} = \frac{1}{3}$.

Closely related to the subparts method is the following. Suppose the pieces of candy all weigh the same. Then, if for each piece we take from the small box, we take 3 pieces from the large, the candy will run out in both boxes at the same time. We call this the *nibble* method.

2. The olive jar holds $\frac{3}{5}$ cup. We suppose that we have a measuring cup marked off in fifths. Filling the olive jar with water (or some other fluid or rice) and pouring the contents into the measuring cup, we find the level at the third mark. A fifth of a cup is a subpart that fits the cup 5 times and the olive jar 3 times. The ratio of the olive jar to the cup is $\frac{3}{5}$.

We may have a cup, but not one marked off in fifths. If we are lucky we may find a small spoon and observe that 24 spoonfuls fill the olive jar and 40 fill the cup. The ratio is $\frac{24}{40}$. We may think of 8 spoonfuls as a subpart contained 3 times in the jar and 5 times in the cup. The ratio is $\frac{3}{5}$.

We needed special equipment—a measuring cup or a lucky spoon—for the subparts method. The multiples method requires nothing but the jar, a cup, and some rice. We use rice because it sweeps up easily. If you have web feet or are at the beach, use water.

Fill the cup. Fill the jar from the cup. When the jar is full, dump it and fill again from the cup. When the cup runs out, fill it again and continue. When the cup runs out just as the jar is full, stop. In this case, $r = 3$ cups filled the jar $s = 5$ times. The ratio of the jar to the cup is $\dfrac{r}{s} = \dfrac{3}{5}$.

3. The desk is $\frac{29}{12}$ feet wide. Measuring with an inch ruler, we find that the desk is 29 inches wide. There are 12 inches in one foot. Using multiples, we would find that 12 desks placed side by side are 29 feet long. This might be awkward.

4. The width of a sheet of paper is $\frac{4}{5}$ the length. Perhaps we measured the paper with a scale having a unit that fitted 4 times into the width and 5 times into the length. Perhaps we measured with a small unit, say a millimeter, and found that dividing the number of millimeters in the width by 4 gave the same quotient as dividing the length in millimeters by 5. Can you identify the subparts used here?

The multiples method says that if we line up 4 sheets lengthwise, we get a segment as long as we would by lining up 5 sheets widthwise. The diagram shows the ratio of width to length is $\dfrac{4}{5}$. We may also have noticed that 5 times the width in millimeters is equal to 4 times the length in millimeters.

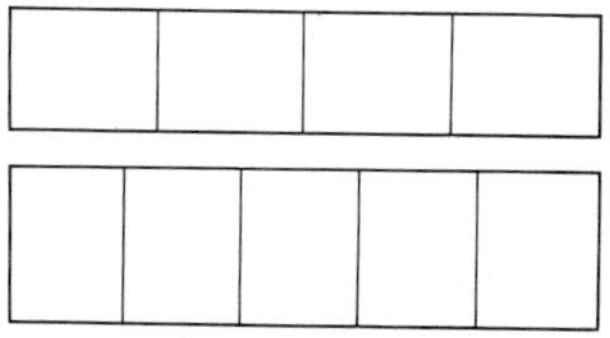

5. The ratio of men to women at the PTA meeting was $\frac{2}{3}$. We regard sets with the same number of men or women as being equal subparts. In this case there were 3 such sets of women and 2 of men. Suppose there were 12 men and 18 women. The ratio is

$\frac{12}{18}$ if each person is a subpart

$\frac{6}{9}$ if 2 people are a subpart

$\frac{4}{6}$ if 3 people are a subpart

$\frac{2}{3}$ if 6 people are a subpart

The multiples method indicates that 2 times the number of men equals 3 times the number of women.

NOTE We do not know how many men there are. The ratio tells us the comparative sizes of the groups, but not the absolute sizes.

The nibble method would show that if every time 2 men left the meeting 3 women also left, the group would shrink until neither men nor women remained.

6. Corn cooks in $\frac{3}{4}$ the time that beans do. If we multiply the time to cook beans by 3, the product equals the time to cook corn multiplied by 4. If we have no clock, we could start two cooks, one cooking corn, the other cooking beans. When the corn cook has finished 4 batches, the bean cook will have finished 3.

7. In several games, Joe's batting average is $\frac{5}{18}$. Joe made 5 hits in 18 times at bat, or 10 hits in 36 times at bat, or some multiple of 5 (say $5k$) hits in the same multiple of 18 ($18k$) times at bat.

From these discussions, we see the following ways of finding the ratio between two segments $\overline{AB}$ and $\overline{CD}$.

1. Find a subsegment that fits evenly into both segments. Count the number of times it is contained in each, say a in $\overline{AB}$ and c in $\overline{CD}$. The ratio of these segments $\overline{AB}$ to $\overline{CD}$ is $\frac{a}{c}$.

2. Measure $\overline{AB}$ and $\overline{CD}$ in terms of a short unit. If the short unit goes into $\overline{AB}$ about a times and into $\overline{CD}$ about c times, the ratio of $\overline{AB}$ to $\overline{CD}$ is about $\frac{a}{c}$. More precisely, if $a < AB < a + 1$ and $c < CD < c + 1$, the ratio $\frac{\overline{AB}}{\overline{CD}}$ satisfies

$$\frac{a}{c+1} < \frac{\overline{AB}}{\overline{CD}} < \frac{a+1}{c}$$

Instead of $\frac{a}{c}$ as the estimate of the ratio, some people may prefer a fraction with a smaller denominator, if there is one in the interval $(\frac{a}{c+1}, \frac{a+1}{c})$. For example, if the first segment is slightly more than 29 millimeters long and the other slightly more than 49 millimeters long, we might give their ratio as $\frac{3}{5}$ rather than $\frac{29}{49}$. Do you see that $\frac{3}{5}$ falls in the interval $(\frac{29}{50}, \frac{30}{49})$?

3. Lay several segments equal to $\overline{AB}$ end to end alongside several segments of $\overline{CD}$ laid end to end. By observation, then by counting, we see that b of the $\overline{AB}$ segments are as long as d of the $\overline{CD}$ segments. Then the ratio $\frac{\overline{AB}}{\overline{CD}}$ is $\frac{d}{b}$.

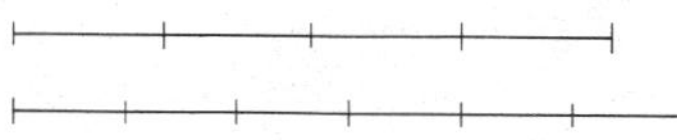

Such a union of segments is called a *train*.

Suppose a train of b $\overline{AB}$ segments is just a bit shorter than a train of d $\overline{CD}$ segments. Then $b(AB) < d(CD)$, where AB and CD are the lengths of the segments. Then the ratio $\frac{AB}{CD}$ is less than $\frac{d}{b}$.

At another point of our lined-up segments, we may notice that p $\overline{AB}$

segments are slightly longer than $q\ \overline{CD}$ segments. Then the ratio $\frac{AB}{CD}$ is greater than $\frac{q}{p}$. Hence, we can say that the ratio is in the interval $(\frac{q}{p}, \frac{d}{b})$. We may reasonably use any convenient fraction in the interval as our estimate of the ratio.

4. It may be that the two segments are "incommensurable"—that is, there is no segment evenly fitting both. There are no counting numbers b, d such that a train of $b\ \overline{AB}$ segments is exactly as long as a train of $d\ \overline{CD}$ segments. Then we may only approximate the ratio.

17.5 RATIONAL NUMBERS AS RESULTS OF DIVISION—MISSING FACTORS The young lady was saving her money for a trip to Slobovia. She explained her interest in Slobovia as follows: "Families there have $3\frac{3}{4}$ children apiece. I've always wanted to see what a $\frac{3}{4}$ child looked like."

It is quite annoying to people who have learned to divide whole numbers that they cannot divide 3 by 4. There is no whole number that satisfies $4\ \square = 3$. The missing factor which multiplied by 4 gives 3 does not exist.

We look at this idea again. Mathematicians are unique people. If they want something badly, they (1) say they want it, (2) describe what they want, (3) devise a notation for what they want, (4) name it, (5) state emphatically that they have it, and (6) act like they have it. This attitude is called wishful thinking by ordinary mortals and is usually considered lunacy. In mathematics it is called generalization, or extending the number system, and is considered brilliant.

So we say we want all such missing factors to exist, so that we can divide any two natural numbers. What would we name a number that satisfied $4\ \square = 3$ and would be the quotient of 3 divided by 4?

Certain candidates present themselves. When we divide 32 by 4 we can write 32/4 or $\frac{32}{4}$ or $32 \div 4$ or $4\overline{)32}$ or 32:4. The colon in the last expression is $\div$ with the bar dropped. Some people even use (32, 4). We have chosen the single-bar notation, 3/4 or $\frac{3}{4}$, as the name of our newly proclaimed number. The $3 \div 4$ and $4\overline{)3}$ were rejected as being too suggestive of the division process and being more complicated to write. The (3, 4) was rejected as not distinctive enough. In reference to ratios, 3:4 was once used extensively but is not used much today.

Henceforth we act like we have a new number, which we join to the set $\{0,1,2,\ldots\}$. These ideas make sense for $\frac{a}{b}$ where a, b are any two whole numbers, provided $b \neq 0$.

This process may seem a bit foreign, indeed weird. However, it will be satisfactory if we can show that it is consistent with the other attitudes. Many people (actually, most mathematicians) prefer this attitude toward rational numbers. The notion is completely defined in terms of the previously discussed whole numbers. Clearly, the idea does not depend on physical objects or operations. This satisfies many people who want to be sure that mathematical ideas are independent of the physical universe. See chapter 20 for more details.

17.6 FRACTIONS AS OPERATORS OR INSTRUCTIONS "$\frac{2}{3}$" or "$\frac{2}{3}$ of" is the instruction "Separate into 3 equal parts and take 2." The instruction exists by itself and is understandable, even if it doesn't have any effect until it operates on something. Such an instruction is like the phrases "Down with" and "Viva" so loved by political rioters.

Other instructions that can be labeled $\frac{2}{3}$ are "Stretch to twice its size, separate into 3 equal parts, and take one part." "Take 2 of these, separate into 3 equal parts, and take one."

An instruction for an improper fraction might be "$\frac{5}{2}$," which means "Stretch to 5 times its size, separate into 2 parts, and take one."

When we consider how often we see phrases like "$\frac{2}{3}$ of," we realize this is a widely used attitude, although many people have not learned to think of numbers as operators. The operator idea generalizes and is used widely in advanced mathematics and science.

Fractions are sometimes described as hooked up "stretchers" and "shrinkers." For example, $\frac{3}{4}$ is the name of a machine or operator formed by hooking up a 3-stretcher with a 4-shrinker. Anything going into the 3-stretcher is stretched to 3 times its size, and then the 4-shrinker shrinks it to $\frac{1}{4}$ its size. Would it make any difference if the object went into the shrinker first and then the stretcher?

17.7 COMPARING ATTITUDES TOWARD RATIONAL NUMBERS We have given four different attitudes toward fractions. That is, four kinds of situations that fractions help to describe and to explain. Are the implied ideas consistent? It would be a shame if people used the same name in different situations with contradictory implications.

Compare $\frac{2}{3}$ as an object with a history, with $\frac{2}{3}$ as a ratio. True, the object with a history is thought of as part of the whole, while objects being compared by ratio are separate. However, a look at the process of naming the object shows its ratio to the whole is $\frac{2}{3}$.

Object with a History	**Ratio**
Join 2 of 3 equal parts of a whole.	The whole is separated into 3 subparts, the object into 2. The subparts are equal. The ratio is $\frac{2}{3}$.
The object is one of 3 equal parts of 2 wholes. The object shows $\frac{2}{3}$.	Three of the objects are the same size as 2 of the wholes. The ratio is $\frac{2}{3}$.

The operator is closely related to the object with a history because the operator may be regarded as giving the history to a whole.

The missing-factor attitude dovetails nicely with the ratio attitude. The rational number $\frac{2}{3}$ is the number that multiplied by 3 gives 2. The object whose ratio to the standard or whole is $\frac{2}{3}$ is the object that when tripled equals 2 of the standards or wholes.

You may wish to compare the attitudes in other ways as well. At any rate, we can be confident that we may use any attitude we please, or even switch attitudes in the middle of the problem, and still be confident of the outcome.

17.8 FRACTIONS AS NAMES OR SYMBOLS A fraction is a convenient name for anything. Just as we have seen that the symbols . . . 4, 5, 6, . . . are useful as names for athletes, houses on the street, and so forth, we can use fractions for names. An example is, "He lives at $834\frac{1}{2}$ Maple Street." Decimal fractions are used for books in libraries. An arithmetic book may be named 513.67. Fractions are useful as names because they allow for many different names within a category already given a number name, such as 513. In particular, fractions name points on the number line between the counting numbers. When fractions, rather than words such as Sally, are used to name objects or as a design motive in a decoration, they have no relation to rational numbers.

17.9 FRACTIONS NAMING POINTS ON A NUMBER LINE Consider the figure

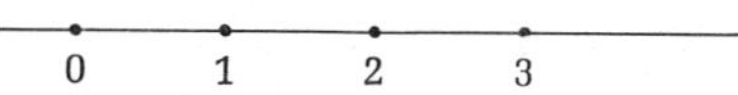

This is part of a number line, with four points labeled. It is customary to think of the line as extended to the right, with other points labeled with the names of other counting numbers. We usually think of the points so labeled that the distances between 0 and 1, 1 and 2, and so on are all equal. When this is done, we use the number line as an aid in understanding numbers and relations between them.

Strictly speaking, a number line does not have to have equal distances between points labeled with counting numbers. A logarithmic scale on a slide rule is an example. Numbers are independent of geometry. However, with these distances all equal the number line furnishes a good tie between geometry and arithmetic. Many arithmetic operations have simple, vivid, illuminating geometric analogues. Similarly, many geometric properties can be interpreted in terms of arithmetic. Hence, in what follows we assume the distances to be equal. The question of extending the above figure to the left as well as to the right we leave till later.

While the figure can be extended by showing more counting numbers, it can also be extended by showing points between counting numbers. We could label these points with letters like A, B, C or with names like George and Mabel. However, we want the names to fit into the structure of the counting numbers in some fashion. We use fractions, and the object-with-a-history point of view. We first consider the segment (0, 1). Calling this segment a whole, we get other segments for various fractions as illustrated:

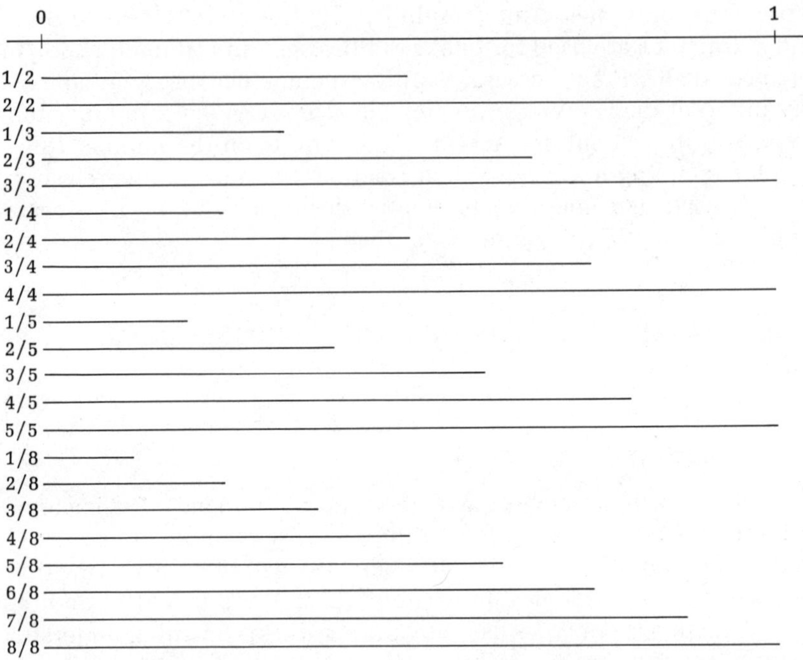

We place each of these segments on the segment (0, 1) with the left end at zero, and label the point at the right end with the name of the segment (see top of opposite page). We note that some points have more than one tag. For example $\frac{1}{2}$, $\frac{2}{4}$, and $\frac{4}{8}$ all name the same point. Can you find other names for these points, using fractions with larger denominators? On the next number line we illustrate some other points between counting numbers and give some of their fractional labels. Can you locate and name some other points?

17.10 THE UNIQUENESS PROBLEM If $\frac{2}{3}$ is the name of a point on the number line, then some other names for the same point are $\frac{4}{6} = (2 \times 2)/(3 \times 2)$, $(2 \times 3)/(3 \times 3) = \frac{6}{9}$, and in general, $(2 \times n)/(3 \times n)$, where n is any counting number. If a whole segment is separated equally into $3 \times n$ subsegments, then n of these subsegments make one segment, 3 of which make the whole. Two of these show $\frac{2}{3}$; these two contain $2 \times n$ subseg-

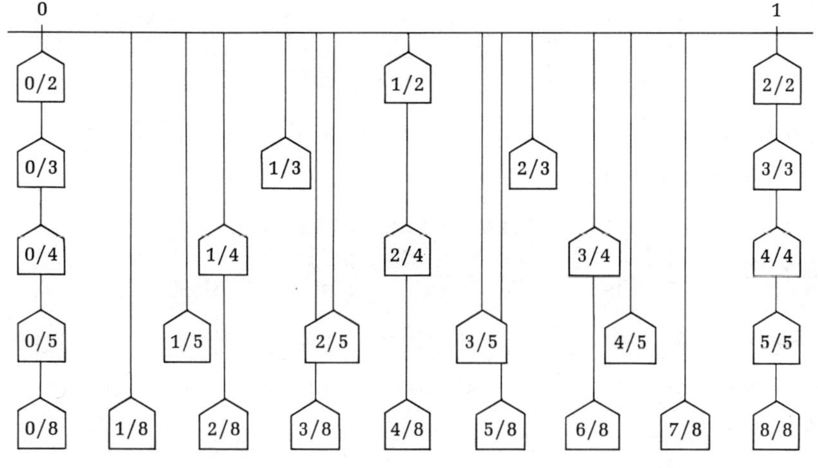

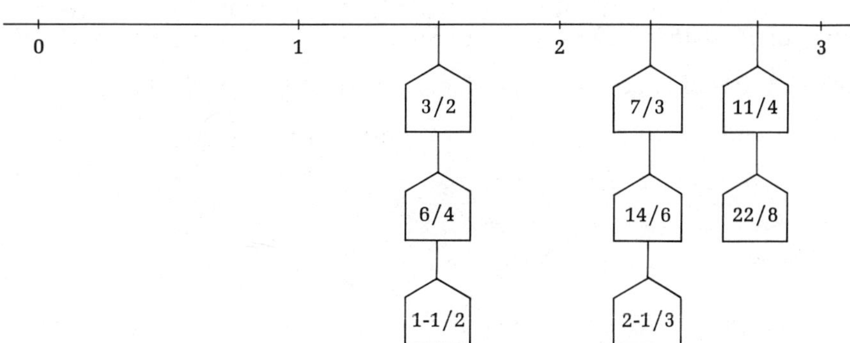

ments. Hence, the same point has the names $\frac{2}{3}$ and $(2 \times n)/(3 \times n)$. The figure in the margin illustrates the case for $n = 4$.

The same discussion applies to any fraction $\frac{a}{b}$. Even though $\frac{2}{3}, \frac{4}{6}, \ldots, \frac{2 \times n}{3 \times n}$ are names of the same point, do we wish to go so far as to say they are names of the same number? We check the various situations that led to fractions to see if these different names describe them:

1. *Objects-with-a-history.* The above discussion for segments can be applied to other types of objects to show that the same object can have different histories. Of course, in saying an object could have had either the history $\frac{2}{3}$ or the history $\frac{8}{12}$, we assume it didn't hurt the object to be a combination of 8 pieces rather than 2.

2. *Instructions and operations.* Arguments like the one above say that the instructions "Separate equally into 3 parts and take 2" and "Separate equally into 12 parts and take 8" have the same result. Hence we regard the instructions as the same. $\frac{2}{3}$ and $\frac{8}{12}$ are names of the same operation. Some people prefer to say "equivalent" rather than "same."

3. *To indicate ratio, or comparison of size.* The statement "The ratio of A to B is $\frac{2}{3}$" means that if object A is separated equally into 2 parts and B is separated equally into 3 parts, the parts of A are the same size as those of B. If these parts are separated again into 4 smaller parts, the smaller parts of A are the same size as those of B. We say the ratio is $\frac{8}{12}$. Since A and B are unchanged, their relative size is unchanged; so we say that $\frac{2}{3}$ and $\frac{8}{12}$ name the same ratio.

4. *The result of division.* We say that $2 \div 3 = \frac{2}{3}$, and $8 \div 12 = \frac{8}{12}$. But is $2 \div 3$ also $\frac{8}{12}$? We check by multiplying by 3 to see if we get 2 as a result. As will be shown later, $\frac{8}{12} \times 3 = \frac{8 \times 3}{12} = \frac{24}{12} = 2$. Hence, $\frac{8}{12}$ is a result of $2 \div 3$, since by the missing-factor attitude of division, $2 \div 3$ names anything that multiplied by 3 gives 2.

5. *The solution to a linear equation.* $\frac{2}{3}$ is a solution to $3y = 2$. Is $\frac{8}{12}$? The argument above shows yes. $3 \times \frac{8}{12} = \frac{24}{12} = 2$.

6. *Nominal uses.* If someone wishes to name his dog $\frac{2}{3}$, he might or might not feel that $\frac{8}{12}$ is also the dog's name. Nominal uses, however, are independent of other uses, and we may not wish to require $\frac{2}{3} = \frac{8}{12}$ here.

With this unanimity we make the assertion that the number whose name is $\frac{2}{3}$ also has the name $(2 \times n)/(3 \times n)$. In general, the number $\frac{a}{b}$ is also the number $\frac{an}{bn}$. Conversely, if one name is $\frac{an}{bn}$, then another name is $\frac{a}{b}$.

Does this tell the whole story about allowable names? If $\frac{2}{3}$ is one name, must every other name be of the form $\frac{2n}{3n}$? How about $.666 = \frac{666}{1000}$? Using some advanced knowledge that may not seem unreasonable, we prove every name is of the form $\frac{2n}{3n}$.

Suppose $\frac{2}{3} = \frac{p}{q}$, where p and q are counting numbers. Then from what has been said, $\frac{2}{3} = \frac{2q}{3q}, \frac{p}{q} = \frac{3p}{3q}$. Hence, $\frac{2q}{3q} = \frac{3p}{3q}$. Since the denominators are equal, the numerators must be equal. Hence, $2q = 3p$. Since 2 is a factor on the left, it must also be a factor on the right. Since 2 is not a factor of 3 it must be of p. That is, $p = 2h$. Similarly 3 is a factor of q. Say $q = 3k$. Then, by substituting, we have $2 \times 3k = 3 \times 2h$, or $6k = 6h$. Then if we call h and k both n, we have $p = 2n$ and $q = 3n$. Hence we conclude that two fractions cannot be names of the same rational number unless both are of the form $\frac{an}{bn}$, with perhaps different n. $\frac{4}{6}$ and $\frac{10}{15}$ are both names of the same rational number by this standard, since the first is $\frac{2 \times 2}{3 \times 2}$ and the second is $\frac{2 \times 5}{3 \times 5}$.

We conclude that $\frac{2}{3} \neq .666 = \frac{666}{1000}$. Can you explain why? Recall that many people use .666 instead of $\frac{2}{3}$.

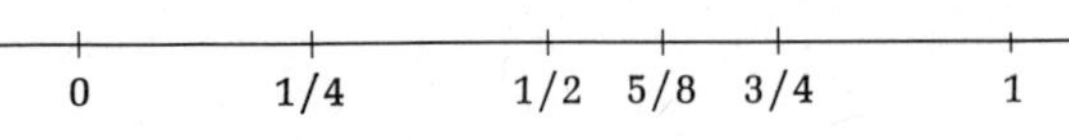

17.11 NATURAL ORDER OF RATIONAL NUMBERS BY SIZE Rational numbers can be ordered by size of segments similarly to the way counting numbers are ordered. $\frac{5}{8} < \frac{3}{4}$, because a $\frac{5}{8}$ segment is shorter than a $\frac{3}{4}$ segment. On a number line increasing from left to right, the point $\frac{5}{8}$ is to the left of the point $\frac{3}{4}$. Assuming this type of number line, we say that a number to the left of $\frac{3}{4}$ is less than $\frac{3}{4}$,

It is tedious to construct segments, or plot points, to see which of two fractions is the larger. Further, we might make a mistake, as, say, in asking whether $\frac{4}{7}$ or $\frac{9}{16}$ is the larger. A way less open to error is to rename the fractions with the same denominator. We compare $\frac{64}{112}$ and $\frac{63}{112}$. Clearly, $\frac{63}{112} < \frac{64}{112}$, as we can see if we think of 63 and 64 segments, each segment $\frac{1}{112}$ of a whole. Hence, $\frac{9}{16} < \frac{4}{7}$.

Can you see how this thinking leads to the rule $\frac{a}{b} < \frac{c}{d}$ if and only if $ad < bc$?

With this definition of order, the fractions are dense. That is, between any two fractions is a third. Actually there are many. For example, adding the numerators and adding the denominators makes a new fraction between the two. Can you see that between $\frac{2}{3}$ and $\frac{3}{4}$ is $\frac{(2+3)}{(3+4)} = \frac{5}{7}$? So are $\frac{7}{10}$ and $\frac{8}{11}$.

Geometrically this means that between any two points on the number line there is another point (or fraction naming the point). From this we see that between any two fractions we have an infinite number of fractions. Can you show that there is no smallest positive fraction? What is the smallest positive integer? (HINT Suppose there were a smallest positive fraction. Could you find a fraction between it and zero?)

17.12 ORDERING FRACTIONS BY HEIGHT AND NUMERATOR While size is by far the most common criterion for ordering fractions, we may also order them by height and numerator. The height, by definition, is the sum of numerator and denominator. Here is a table listing fractions in this way:

Height	Fractions
1	$\frac{0}{1}$
2	$\frac{0}{2}, \frac{1}{1}$
3	$\frac{0}{3}, \frac{1}{2}, \frac{2}{1}$
4	$\frac{0}{4}, \frac{1}{3}, \frac{2}{2}, \frac{3}{1}$
⋮	
9	$\frac{0}{9}, \frac{1}{8}, \frac{2}{7}, \frac{3}{6}, \frac{4}{5}, \frac{5}{4}, \frac{6}{3}, \frac{7}{2}, \frac{8}{1}$
⋮	

All the fractions can be arranged in a row, which goes on indefinitely $\frac{0}{1}, \frac{0}{2}, \frac{1}{1}, \frac{0}{3}, \frac{1}{2}, \frac{2}{1}, \frac{0}{4}, \frac{1}{3}, \frac{2}{2}, \frac{3}{1}, \frac{0}{5}, \frac{1}{4}, \frac{2}{3}, \frac{3}{2}, \frac{4}{1}, \frac{0}{6}, \frac{1}{5}, \frac{2}{4}, \frac{3}{3}, \frac{4}{2}, \frac{5}{1}, \ldots$

We see that it is the fraction names that are ordered, and not the rational numbers. $\frac{1}{2}$ and $\frac{2}{4}$ are names of the same rational number, but these fractions are at different places in the list.

This order is introduced here primarily to show an order different from the order by size, in line with our desire to show at least two ways of doing something. However, this ordering has some interest. According to this order the fractions are not dense. There is no fraction between $\frac{1}{4}$ and $\frac{2}{3}$, for example. This order allows setting up a 1:1 correspondence between fractions and counting numbers. In this list $\frac{1}{2}$ is the fifth fraction.

17.13 MIXED NUMBERS The point Q on the number line in the margin is $\frac{1}{3}$ the way between *B* and *C*. The fraction name is $\frac{10}{3}$. We also say $3 + \frac{1}{3}$ and $3\frac{1}{3}$.

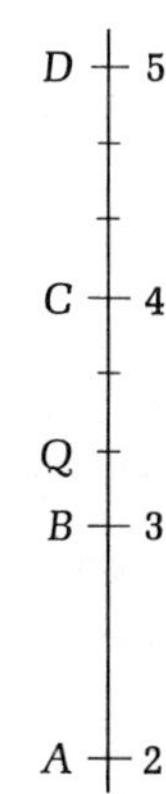

Numerals like $3\frac{1}{3}$ and $4\frac{3}{4}$ are sometimes called *mixed numbers,* because the name appears to be a mixture of a counting number and a fraction. Needless to say $4\frac{3}{4}$ is the name of a single number, just as Paul Johnson is the name of one person—not a mixture of one person named Paul and another named Johnson.

17.14 SUMMARY Fractions and rational numbers are perhaps the first really big generalization people make. We showed how some of the ideas of counting numbers generalized to rational numbers (segment and operator, but not set or sequence). We see that many fraction attitudes are in active use. These include fractions as the name of an object with a history (or the history of an object), ratio, result of division or missing factor, and operator.

The number line is an excellent visual model of fractions. It helps to show that every rational number has many fraction names and to appreciate the natural order of fractions by size.

That order may be imposed by the worker and is not necessarily a property intrinsic to the fractions as shown by ordering fractions by height and numerator. This ordering is of little practical use.

Fractions greater than one can be written as mixed numbers, showing another way some rational numbers have many names.

EXERCISES

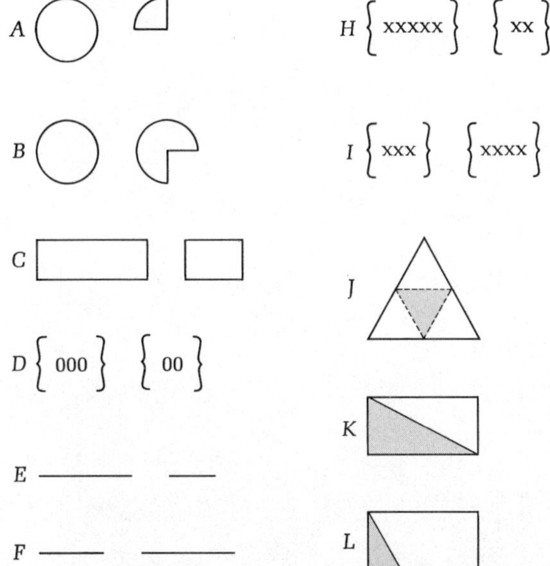

1. Pairs of objects, such as figures, sets, and segments, are shown in the margin. For each of the pairs of objects, name the fraction shown by the second object, where the first is the whole. Some objects are shaded parts of wholes.

2. For each of the following pairs,

(*a*) Determine a subunit that fits both members evenly.
(*b*) Give the ratio of the first to the second as determined by this subunit.
(*c*) Is this subunit the largest possible? If not, give the largest possible subunit. If so, give a smaller subunit.
(*d*) Give the ratio as determined by the new subunit *c*.

A. 3 pounds, 7 pounds
B. 5 inches, 2 inches
C. 3 feet, 6 feet
D. 2 yards, 4 feet
E. 350 people, 400 people at political meetings
F. 20 couples, 5 people at a party

3. If one out of 3 people camping were a girl, how many girls would there be among 12 campers?

4. If $\frac{2}{3}$ of the class voted, and there were 30 children in the class, how many voted?

5. If the ratio of those who voted to those who did not vote was $\frac{2}{3}$, and there were 30 children in the class, how many voted?

6. In each of the following, two segments are labeled with letters. Further multiples of each segment are shown of equal length. Determine the ratio of the labeled segments, first to second.

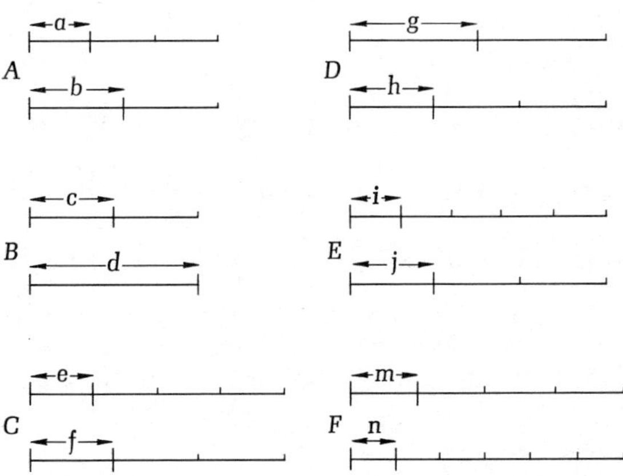

7. Can you predict when the ratio is greater than 1? Less than 1?

8. Roy and Ian start walking together. Roy says, "Ian, I've taken 24 steps and you've taken 20 steps, yet we're at the same place. How can this be?" Can you answer? What is the ratio of the length of Roy's step to Ian's? Who has the longer step?

9. In exercise 8 were Roy and Ian at the same place before Roy's 24th step? If so, at which step? If Roy and Ian continue to walk, give 3 more places they will be together again.

10. Here is a set of rational number names:

$$\{12\tfrac{6}{9}, \tfrac{2}{3}, \tfrac{23}{15}, \tfrac{7}{4}, 17\tfrac{18}{13}, \tfrac{5}{8}\}$$

(a) Give two numerators.
(b) Give two denominators.
(c) Give two proper fractions.
(d) Give two improper fractions.
(e) Give two mixed numbers.

11. In a–j state which fraction attitude best describes the fraction use in that sentence. Restrict yourself to

A. history of an object
B. ratio
C. operator
D. result of division
E. symbol
F. none of these

Check your answers with a neighbor. If your answers differ, discuss the differences. You may both be right because you look at the sentence differently.

(a) He is $\frac{5}{3}$ as tall as he was five years ago.
(b) The Democrats hope to get 60 percent of the votes.
(c) Half a loaf is better than none.
(d) Less than $\frac{1}{4}$ the Ganges water flows out of the main mouth of the river.
(e) He pole-vaulted $17\frac{2}{3}$ feet.
(f) Since the perimeter of the square is 3 feet, each side is $\frac{3}{4}$ feet.
(g) Divide 7 by 8 to get $\frac{7}{8}$.
(h) Since 5 boxes hold 12 quarts, each holds $\frac{12}{5}$ quart.
(i) They invited only $\frac{2}{5}$ as many girls as boys to their party.
(j) Millicent bought $\frac{3}{4}$ kilograms of bread.

12. Here is a portion of a number line with several points labeled:

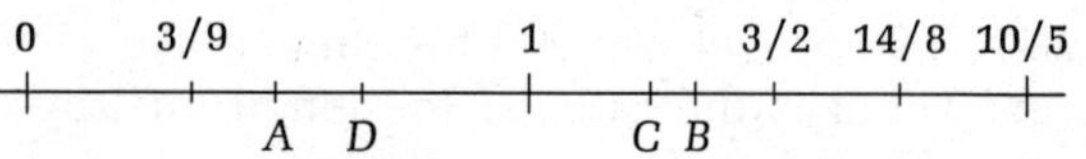

(a) Write fraction names for A, B, C, D.
(b) Write another fraction name for each of these 4 points.
(c) Name the point halfway between 1 and C.

13. Draw a line segment 10 centimeters long with the left end labeled "0" and the right end labeled "1." What fraction name do you give to

(a) P, 4 cm from the left end?
(b) Q, 2.5 cm from the right?
(c) R, 1 cm from the right?
(d) S, $3\frac{1}{3}$ cm from left?

14. Give three different fraction names for each of the following numbers: (a) $\frac{6}{8}$ (b) $\frac{3}{5}$ (c) $\frac{14}{20}$

15. Does every rational number have a name where at least one of the numerator or denominator are prime numbers? Give examples.

16. How many names does a rational number have where numerator and denominator are relatively prime? (Relatively prime means having no common factor other than 1.)

17. List these fractions in increasing order by size:
$\frac{2}{5}, \frac{3}{7}, \frac{4}{3}, \frac{1}{3}, \frac{1}{2}$.

18. Find a fraction between $\frac{3}{5}$ and $\frac{2}{3}$.

19. Order the fractions in exercise 17 by height and numerator (see section 12).

20. Arrange two sets of beans in three ways to show the three names you gave to $\frac{6}{8}$ in exercise 14.

21. You have become the proud holder of a bag of newly minted gold Centenario coins in Mexico. Assume that a U.S. nickel weighs 5 grams. Using a double-pan balance, how would you determine the weight of a Centenario? Assume that you have a bag of nickels, each weighing 5 grams, and that a Centenario is not a multiple of 5 grams.

22. Choose two rods from a set of colored rods, say a yellow rod and a blue rod. If the yellow rod is unit length, what is the length of the blue? Show in two ways.

23. Michael and Jennifer wish to determine the ratio of the width of Jennifer's hand to the width of Michael's. How can they do this without a ruler? Can they compare the lengths of their hands to determine the ratio of the lengths?

24. P is a cup holding exactly one pint, while V is a vase you just made from clay. Tell how you could determine the capacity of V using P and water. Marking P or V is not allowed.

Answers to selected questions: **2.** (*A*) 1 lb.; 3/7; yes—8 oz.; 6/14; (*C*) 1 foot; 3/6; no—3 feet; 1/2; (*F*) 1 person; 40/5; no—5 people; 8/1; **5.** 12; **6.** (*A*) 2/3; (*B*) 1/2; (*E*) 3/5; **9.** yes; when Roy had taken 6 steps and Ian 5; when Roy has taken 12, 18, 32, 40, . . . steps; **10.** (*a*) operator; ratio; (*d*) ratio; history of object; (*i*) operator; **11.** (*a*) any two of 6, 2, 23, 7, 18, 5; **12.** (*a*) 1/2, 4/3, 5/4, 2/3; **15.** no; 4/9, 10/21; **20.** 45 beans and 60 beans; arrange in sets of 15 to show 3/4, in sets of 5 to show 9/12, in sets of 3 to show 15/20; **24.** If V of the vases exactly fill the cup P times, the vase contains $\frac{P}{V}$ pints.

18

Adding and Subtracting Fractions

18.1 INTRODUCTION What leads us to say $\frac{1}{3} + \frac{1}{4} = \frac{7}{12}$ rather than $\frac{1}{3} + \frac{1}{4} = \frac{2}{7}$? There are situations where when we combine $\frac{1}{3}$ and $\frac{1}{4}$ we get $\frac{7}{12}$ and others where we get $\frac{1+1}{3+4} = \frac{2}{7}$. Why do we wish to call the first case *addition,* and not even give a name to the second?

We do this because we want addition of fractions to be a kissing cousin, or at least a nodding acquaintance, of addition of whole numbers. We will explore the attitudes toward addition, and the attitudes toward fractions to see if they can be easily extended and combined to give an attitude toward addition of fractions. We shall not be surprised if some can and some cannot. Subtraction will be discussed in this chapter, because it is so close to addition.

We will discuss the algorithms—the addition and subtraction dances of the digits. Some people prefer to discuss multiplication of fractions (chapter 19) before addition because the algorithm is easier. We prefer to follow the order used with counting numbers.

18.2 ADDITION DEFINED Addition of whole numbers arises from joining segments, putting sets together, or continued counting. The object-with-a-history attitude gives segments that can be put together nicely. If $\overline{OU}$ is a unit, then $\overline{OT}$ shows $\frac{1}{3}$ and $\overline{OF}$ shows $\frac{1}{4}$. If we place $\overline{OF}$ at the end of $\overline{OT}$, we get $\overline{OS}$, which shows $\frac{1}{3} + \frac{1}{4}$.

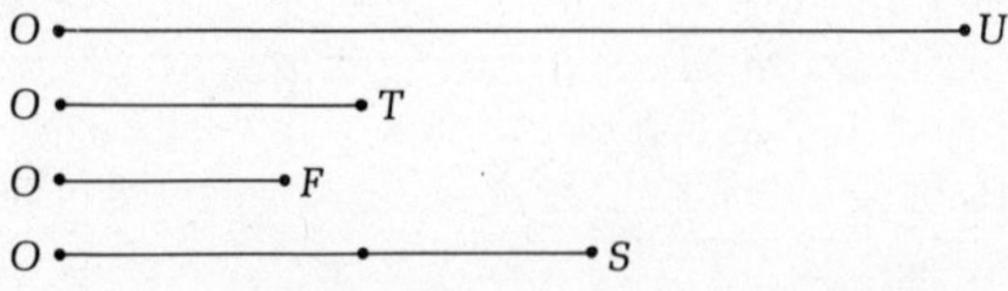

Similarly, we show $\frac{1}{3}$ and $\frac{1}{4}$ in a pie diagram (gourmets love mathematics).

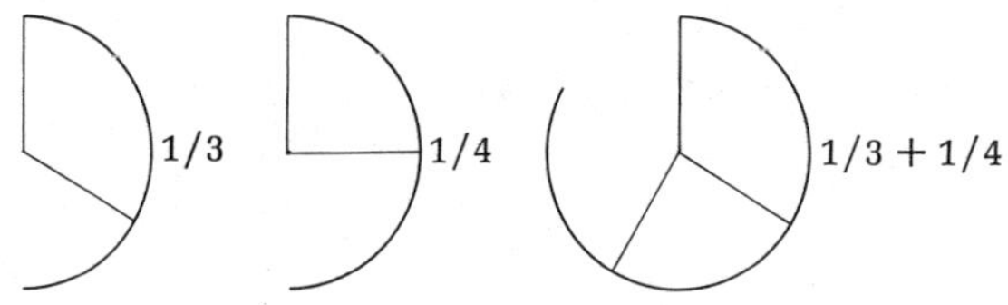

So the addition operation extends nicely in this attitude to yield a sum $\frac{1}{3} + \frac{1}{4}$. It is not clear whether this sum has a single-fraction name.

Careful measurement as well as mental theoretical argument shows that, if the pieces showing $\frac{1}{3}$ are properly separated into 4 parts and the pieces showing $\frac{1}{4}$ are properly separated into 3 parts, all these small parts will be of the same size, and 12 of them will make up a whole. Hence, if we separate a whole into 12 parts the same size and put 7 of them together, we get a piece the same size as $\frac{1}{3} + \frac{1}{4}$. Hence we say $\frac{1}{3} + \frac{1}{4} = \frac{7}{12}$.

To prove this we note that if $\overline{OU}$ is separated into 3 parts, and each of these including $\overline{OT}$ is separated into 4, we have $\overline{OU}$ separated into 12 parts. Similarly, if $\overline{OU}$ is separated into 4 parts and each of these including $\overline{OF}$ is separated into 3, $\overline{OU}$ is again separated into 12 parts. These parts are of the same size (can you see why?) and the union of $\overline{OT}$ and $\overline{OF}$, put end to end, contains 7 of them. We leave it as an exercise to discuss the pie graphs.

Extending putting sets together to show fraction addition is not so easy. We seldom label a set $\frac{1}{3}$, but instead label it with a counting number. We shall not try to extend joining sets to addition of fractions as objects.

Similarly, people do not count except with counting numbers. We do not try to define addition of fractions by continued counting. We could probably force some way of continued counting with fractions, but this would not be a natural thing to do, as it was with putting segments end to end or putting $\frac{1}{4}$ gallon of water in a bucket containing $\frac{1}{3}$ gallon.

In these examples it was natural to combine objects named by fractions. Is there anything natural to do to combine operators, or instructions named by fractions? To see what might happen, is there a natural way of combining a 4-operator with a 5-operator, a machine that stretches things to 4 times their length with one that stretches things to 5 times their length?

The natural way is to hook them up, one behind the other. But the result is a 20-machine, not a 9-machine. There appears to be no natural, obvious way of combining operators that shows addition. Hence we will not try to define addition of fractions from the operator attitude.

The ratio attitude toward fractions is delicate as a foundation for addition. In general, it fails, although in certain special cases it works. Consider the following: "The ratio of boys to girls in the Latin class is $\frac{1}{4}$, and in the speech class is $\frac{2}{3}$." There appears to be nothing reasonable to do with these classes that shows $\frac{1}{4} + \frac{2}{3}$. Suppose the Latin class had 5 boys and 20 girls, and the speech class had 4 boys and 6 girls with the ratios $\frac{5}{20}$ and $\frac{4}{6}$. We might reasonably ask, "What would the ratio be if we combined the classes?" Assuming no overlap, the ratio would be $\frac{9}{26}$. We might be tempted to say $\frac{5}{20} + \frac{4}{6} \stackrel{?}{=} \frac{9}{26}$. But since we have already said we could use $\frac{1}{4}$ and $\frac{2}{3}$, we would also say $\frac{1}{4} + \frac{2}{3} \stackrel{?}{=} \frac{3}{7}$. But $\frac{3}{7} \neq \frac{9}{26}$. Since we want to get the same "sum" regardless of the names used in the addends, we do not like calling $\frac{9}{26}$ the "sum." In general, there is no natural way of combining things named by ratios that leads to addition.

However, if two ratios compare two things to the same standard unit thing, addition comes easy. If the ratio of the amount of water in a blue pail to the water in a red pail is $\frac{1}{3}$ and the ratio of the water in a green pail to that in the red pail is $\frac{1}{4}$, then it is easy to see that the ratio of the blue and green pail together to the red pail is $\frac{1}{3} + \frac{1}{4}$, in the sense of putting segments together.

In fact, all measurements are ratios. $\frac{4}{3}$ feet + $\frac{1}{2}$ foot = $(\frac{4}{3} + \frac{1}{2})$ feet. We can add these ratios, which compare two segments with the same standard segment–one foot.

18.3 ADDITION ALGORITHMS We develop the famous addition algorithm, or rule, from two points of view, on the grounds that if something is worth doing, it's worth doing twice. We put segments end to end to get $\frac{a}{b} + \frac{c}{d} = \frac{(ad + bc)}{bd}$.

We show the processes with $\frac{3}{4} + \frac{2}{3}$, leaving it as an exercise to show the processes with other fractions and in general.

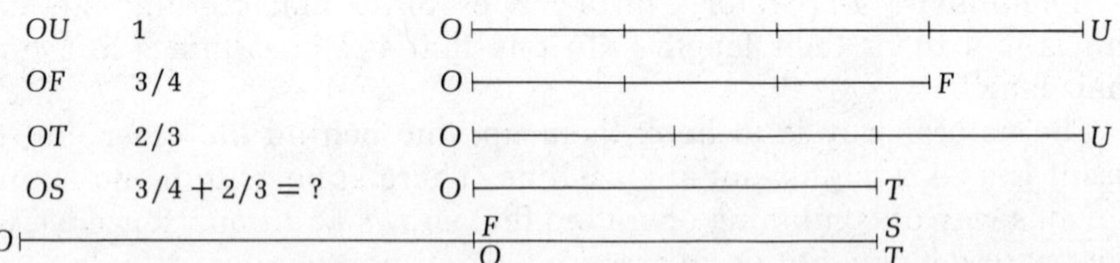

We wish to separate $\overline{OF}$, $\overline{OT}$, and the unit segment $\overline{OU}$ into small segments, all the same size, which fit evenly into all 3 segments. Clearly, if the three parts forming $\overline{OF}$ are each separated into 3 segments, and the two parts of $\overline{OT}$ are each separated into 4 segments, these segments are the same size, and 12 fit into $\overline{OU}$. There are $3 \cdot 3 = 9$ in $\overline{OF}$, $2 \cdot 4 = 8$ in $\overline{OT}$, and $8 + 9 = 17$ in $\overline{OS}$. Hence

$$\frac{3}{4} + \frac{2}{3} = \frac{3 \cdot 3}{4 \cdot 3} + \frac{2 \cdot 4}{3 \cdot 4} = \frac{9}{12} + \frac{8}{12} = \frac{17}{12}$$

We recall the definition of a $\frac{3}{4}$ segment as one such that 4 of them placed end to end make 3 unit segments. We find a name for $(\frac{3}{4} + \frac{2}{3})$ by asking how many such segments placed end to end will equal an exact number of segments.

It may not be clear whether or not 7 of these segments will form an exact number of units. But it is clear that 12 segments do. For suppose we connected 12 segments $(\frac{3}{4} + \frac{2}{3})$ long. We can imagine shifting the $\frac{3}{4}$ and $\frac{2}{3}$ segments around until we had 12 $\frac{3}{4}$ segments followed by 12 $\frac{2}{3}$ segments. The 12 $\frac{3}{4}$ segments make 9 units. The 12 $\frac{2}{3}$ segments make 8 units. Hence the 12 $(\frac{3}{4} + \frac{2}{3})$ segments make 17 units. We say $\frac{3}{4} + \frac{2}{3} = \frac{17}{12}$.

We would like to describe the dance without thinking about the segments. We do this in two steps. First, we see we can add two fractions having the same denominator by adding the numerators. $\frac{2}{5} + \frac{4}{5} = \frac{6}{5}$.

Second, we change $\frac{3}{4} + \frac{2}{3}$ to adding two fractions having the same denominator. Some other names for $\frac{3}{4}$ are $\frac{6}{8}, \frac{9}{12}, \frac{12}{16}, \frac{18}{24}, \ldots$. Other names for $\frac{2}{3}$ are $\frac{4}{6}, \frac{8}{12}, \frac{10}{15}, \frac{16}{24}, \ldots$. We run our eyes up and down for names with the same denominator. Hence,

$$\begin{aligned} \frac{3}{4} + \frac{2}{3} = \frac{9}{12} + \frac{8}{12} &= \frac{18}{24} + \frac{16}{24} \\ = \frac{17}{12} \qquad &= \frac{34}{24} \end{aligned}$$

In general, to add $\frac{a}{b} + \frac{c}{d}$, we change the names of the fractions to fractions with the same denominators. What are candidates for the common denominators? Clearly bd and $2bd$ are two such candidates. Hence,

$$\frac{a}{b} + \frac{c}{d} = \frac{ad}{bd} + \frac{bc}{bd} = \frac{2ad}{2bd} + \frac{2bc}{2bd}$$

$$= \frac{ad + bc}{bd} \quad \text{or} \quad \frac{2ad + 2bc}{2bd}$$

Usually $\frac{ad + bc}{bd}$ is satisfactory.

EXAMPLES $\frac{1}{2} + \frac{1}{5} = \frac{5}{10} + \frac{2}{10} = \frac{7}{10}$

$$\frac{3}{4} + \frac{1}{6} = \frac{3 \cdot 6}{4 \cdot 6} + \frac{4 \cdot 1}{4 \cdot 6} = \frac{18}{24} + \frac{4}{24} = \frac{22}{24} = \frac{2 \cdot 11}{2 \cdot 12} = \frac{11}{12}$$

Replacing $\frac{22}{24}$ by $\frac{11}{12}$ is called *reducing the fraction*. In this case, we divide numerator and denominator, or top and bottom, by 2. When we finish a problem, we usually leave the fractions in *lowest terms*—that is, with the smallest numbers—unless there is a special reason.

$$\frac{5}{6} + \frac{4}{9} = \frac{45}{54} + \frac{24}{54} = \frac{69}{54} = \frac{23}{18}$$

We note that we could have used 18 as a common denominator at the first step.

$$\frac{5}{6} + \frac{4}{9} = \frac{15}{18} + \frac{8}{18} = \frac{23}{18}$$

This is probably easier, because the numbers used are smaller. On the other hand, people who don't like looking for smallest denominators prefer using any common denominator, such as the product of the denominators, even at the cost of larger numbers and more reduction at the end.

In chapter 14 on number theory we discussed ways of finding small common denominators of fractions as common multiples, and of finding common factors.

In adding three or more fractions it may be useful to find a common denominator of all. We build the LCM of the denominators from the prime factors. Every prime factor of every denominator must appear in the LCM as many times as it appears in that denominator. A convenient layout for determining this is shown by the example in finding the LCM of 12, 25, 10, 18.

2	12	25	10	18
	6	25	5	9

Using an L-shaped division symbol, we write the numbers as shown, using 2 as the first required factor. Under the line goes the quotient if the number is divisible by 2, or the number itself otherwise. Since 2 again divides one of the terms we repeat the process.

2	12	25	10	18
2	6	25	5	9
	3	25	5	9

Two is not a factor of any of the new terms, but 3 is. We repeat with 3 as divisor twice. Then we continue with 5 as a divisor twice. When the numbers on the bottom row are all 1s, we are finished. The LCM is the product of all the divisors on the left. In this case, the LCM of 12, 25, 10, and 18 is $2 \times 2 \times 3 \times 3 \times 5 \times 5 = 900$.

We check by showing that each of 12, 25, 10, 18 does divide 900 and that two 2s are required by 12, two 3s by 18, and two 5s by 25.

2	12	25	10	18
2	6	25	5	9
3	3	25	5	9
3	1	25	5	3
5	1	25	5	1
5	1	5	1	1
	1	1	1	1

18.4 SUBTRACTING FRACTIONS The take-away and missing-addend attitudes toward subtraction of whole numbers extend quite easily to subtracting fractions. We leave it as an exercise for the student to draw segment and pie diagrams showing these two subtraction attitudes.

The algorithm comes easily from the addition algorithm. We write fractions having the same denominator and subtract the numerators. For example, $\frac{3}{4} - \frac{1}{3}$ can be written $\frac{9}{12} - \frac{4}{12}$, which is $\frac{5}{12}$. In general we have the formula

$$\frac{a}{b} - \frac{c}{d} = \frac{ad}{bd} - \frac{bc}{bd} = \frac{ad - bc}{bd},$$

$$\frac{5}{6} - \frac{2}{9} = \frac{45}{54} - \frac{12}{54} = \frac{33}{54}.$$

$\frac{33}{54}$ can be reduced to $\frac{11}{18}$. We note that we could have written $\frac{5}{6}$ and $\frac{2}{9}$ with 18 as denominator instead of using the formula. We have to decide for each problem whether it is easier to look for the smallest common denominator and work with smaller numbers, or to substitute in the general formula, work with larger numbers, and reduce the fraction at the end. Sometimes one is easier, sometimes the other. Here are some examples.

$$\frac{23}{4} - \frac{7}{2} = \frac{23}{4} - \frac{14}{4} = \frac{9}{4}$$

$$\frac{1}{3} - \frac{1}{4} = \frac{4}{12} - \frac{3}{12} = \frac{1}{12}$$

$$\frac{2}{3} - \frac{1}{12} = \frac{24}{36} - \frac{3}{36} = \frac{21}{36} = \frac{7}{12}$$

18.5 SITUATIONS CALLING FOR SUBTRACTING FRACTIONS Most subtraction situations described under subtracting counting numbers extend to subtraction when the terms involve fractions. We give a few:

1. Mary has $\frac{1}{3}$ gallon of juice but wants $\frac{2}{5}$ gallons. How much more does she need? Missing addend.

2. Peter ate $4\frac{1}{2}$ inches off one end of a hot dog $11\frac{3}{4}$ inches long. How much was left? Take away.

3. How much taller is Carl at $5\frac{1}{6}$ feet than Sandra at $4\frac{3}{4}$ feet? Comparison.

18.6 SUMMARY Addition and subtraction of fractions are direct generalizations of the joining-segments attitude for addition of whole numbers. Union of sets and continued-counting addition attitudes do not generalize so well.

People who concentrate on learning formulas have the simple formula

$$\frac{a}{b} \pm \frac{c}{d} = \frac{ad \pm bc}{bd}$$

which always works. If b and d have a common factor, a smaller common multiple will give a smaller numerator and denominator than that given by the formula.

EXERCISES

1. Draw segments showing (*a*) $\frac{1}{2}$ (*b*) $\frac{3}{4}$ (*c*) $\frac{1}{2} + \frac{3}{4}$.

2. Compute using common denominators:

(*a*) $\frac{2}{3} + \frac{1}{3}$ (*d*) $\frac{2}{3} + \frac{3}{4}$
(*b*) $\frac{1}{2} + \frac{1}{4}$ (*e*) $\frac{3}{5} - \frac{1}{2}$
(*c*) $\frac{3}{7} - \frac{1}{7}$ (*f*) $\frac{2}{3} - \frac{1}{2}$

3. Form 3 sets of beans, one which shows a "whole," the second shows $\frac{3}{5}$, and the third shows $\frac{1}{4}$. (HINT: Can you use just any set of beans for the whole?)

4. Using the sets in exercise 3, show $\frac{3}{5} + \frac{1}{4}$.

5. Using the sets in exercise 3, show $\frac{3}{5} - \frac{1}{4}$.

6. Using segments show it is possible to add $\frac{2}{3} + \frac{1}{4}$ without finding a common denominator.

7. With colored rods, find a unit rod, a $\frac{1}{3}$ rod, a $\frac{1}{2}$ rod. Show $\frac{1}{3} + \frac{1}{2}$ without finding a common denominator.

8. Find the standard name for the rod in exercise 7 showing $\frac{1}{3} + \frac{1}{2}$.

9. Describe a way (algorithm) of finding the standard name for $\frac{2}{3} + \frac{1}{4}$ that does not require finding a common denominator. Apply the algorithm to get the sum without finding the common denominator and adding numerators.

10. Can you explain the layout at right for finding the least common multiple of 6, 12, 10, and 15? What is the LCM?

2	6	12	10	15
3	3	6	5	15
2	1	2	5	5
5	1	1	5	5
	1	1	1	1

11. Use the results of exercise 10 to find

$$\frac{1}{6} + \frac{5}{12} + \frac{3}{10} + \frac{4}{15}$$

12. Let a one-inch square be a whole. Draw $\frac{1}{4} + \frac{3}{5}$.

13. Compute $4\frac{2}{3} + 7\frac{4}{5}$.

14. Compute $8\frac{1}{3} - 3\frac{1}{2}$.

15. Why is the operator attitude poor for explaining addition? In answering this, answer "If $\frac{3}{4}$ and $\frac{1}{5}$ are (objects, ratios, operators), how do we get the (object, ratio, operator) $\frac{3}{4} + \frac{1}{5}$?"

16. Do addition attitudes for counting numbers generalize to fractions? In each of the following three questions, (*a*) name the addition attitude. Then (*b*) determine if the problem makes sense when 4 is replaced by $\frac{1}{2}$, and 7 by $\frac{2}{3}$.

A. Joseph drank 4 glasses of milk in the morning and 7 in the afternoon. How many did he drink all together?

B. Jeannette tutored 4 children whose ages ranged from 8 to 9, and 7 children aged 9 to 10. How many children did she tutor all together?

C. Crossing the stream, Myron stepped as far as stepping-stone 4, and then stepped 7 more. On what stone did he end?

Answers to selected questions:

1. $\frac{1}{2}$ $\frac{3}{4}$ $\frac{1}{2} + \frac{3}{4}$

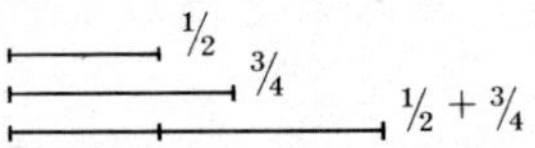

2. (*a*) 3/3; (e) 6/10 − 5/10 = 1/10.

6. 2/3 1/4

2/3 + 1/4

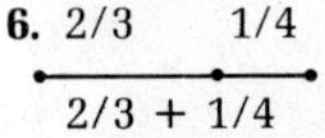

10. 60; **13.** $12\frac{7}{15}$; **16.** (*A*) uniting sets; yes; (*B*) uniting sets; no; (*C*) continued counting; no.

19

Multiplication and Division of Rational Numbers

19.1 INTRODUCTION The author once noticed a beautiful woman seated at the next table in a restaurant. Feeling she might be willing to discuss a topic of universal interest, he said, "Excuse me, will you tell me what multiplication is?" She smiled in a friendly way and said, "I've always thought of it as repeated addition." The conversation continued, more and more fascinating, until her husband arrived. Beauty fades rapidly under some circumstances, and it was with great sadness we didn't explore how to add $\frac{3}{4}$, $\frac{2}{3}$ times. For the number of times something is done is always a counting number. What would she have said?

For many people there is no relation between what they say multiplication is and what they do with fractions. They instinctively feel that multiplication of fractions is related to multiplication of counting numbers, because (1) every whole number can be written as a fraction and (2) things called multiplication must be related. (Many people believe all the Smiths in America are descended from Captain John Smith and Pocahontas.)

This chapter brings a little sunshine into life. After following it, readers will be able to relate the multiplications in several ways, even using different meanings for fractions. They will recognize situations pleading for multiplication and can point out how $\frac{a}{b} \times \frac{c}{d} = \frac{ac}{bd}$ fits.

We relate division of fractions to different attitudes toward division and toward fractions. This leads to several ways of showing the rule, "To divide fractions, invert the divisor and multiply by the dividend."

19.2 ATTITUDES There is no doubt that multiplication of counting numbers comes before multiplication of fractions. To relate the two, we ask which attitudes toward multiplication of counting numbers $p \times q = r$ make sense when p and q are fractions. We then ask what sense is made to get a fraction for r. We hope that the values of r resulting from different attitudes will agree.

Some examples of fraction multiplication situations may guide our thinking. (*a*) Find the area of a rectangle $\frac{2}{5}$ feet wide and $\frac{3}{4}$ feet long. (*b*) Robert drank $\frac{3}{4}$ of the $\frac{2}{5}$ quart of orange juice. (*c*) Find out how much of a shipment of rice gets to market if the highway passes roadblocks put up by two groups: the VC allows $\frac{2}{5}$ of a shipment through the first and the RVN allows $\frac{3}{4}$ of a shipment through the second. (*d*) What is $\frac{2}{5} \times \frac{3}{4}$?

We compare these problems with counting number multiplication. We see that repeated addition; array, and tree approaches don't apply directly, because these require counting number factors. However, segment-segment-region, operator-segment-segment, operator-operator-operator, and formal rule attitudes do appear to fit these four situations. We write out the details.

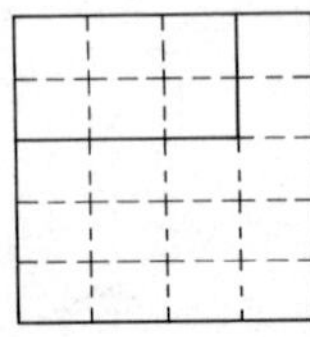

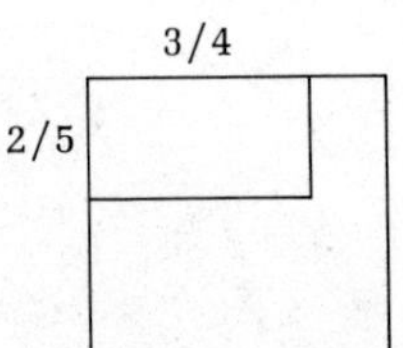

19.3 SEGMENT-SEGMENT-REGION Certainly one can have a rectangle $\frac{2}{5}$ feet by $\frac{3}{4}$ feet. We draw such a rectangle with sides part of the sides of a unit square. To find the area of this rectangle we separate the sides into the number of equal parts given by the denominators and draw parallel lines. The resulting figure shows the unit square separated into 5×4 small congruent rectangles, and the $\frac{2}{5} \times \frac{3}{4}$ rectangle separated into 2×3 small rectangles. The area of the rectangle is the ratio of these numbers, or

$$\frac{2 \times 3}{5 \times 4} = \frac{6}{20} = \frac{3}{10}$$

Many of the multiplications of denominate numbers can be usefully regarded as special cases of this situation. In geometry we have other area and volume problems. The volume of a box is area of base times height. The base is represented by one segment, the height by the other. Force $\times$ distance = work, in physics. Temperature $\times$ time = degree-days, in studying climate. Speed $\times$ time = distance, in travel. These are situations where the factors can be represented by segments of fractional length, and the product represented by a region.

19.4 OPERATOR-SEGMENT-SEGMENT In "$\frac{3}{4}$ of $\frac{2}{5}$ quart of juice" we represent $\frac{2}{5}$ quart of juice by a segment and use $\frac{3}{4}$ as an operator, or instruction. Here are two common interpretations of $\frac{3}{4}$ as an operator: (*a*) separate __________ into 4 parts of the same size, and take 3 of these; and (*b*) join 3 things the same size as __________, separate into 4 parts, and take one. Some people phrase these as (*a*) shrink __________ to $\frac{1}{4}$ its size, then stretch the result to 3 times that size; and (*b*) stretch __________ to 3 times its size and shrink the result to $\frac{1}{4}$ that size. The __________ is for the number or segment operated on. Both these ways show $\frac{3}{4}$ as the combination of two operators described with counting numbers.

We apply the operator $\frac{3}{4}$ to $\frac{2}{5}$ to see what results. It is useful to think of $\frac{2}{5}$ as *AB*, a part of a unit segment *AU*.

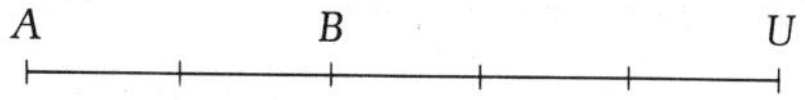

A B U

Separating *AB* into 4 parts and joining 3 leads us to the segment *AZ*.

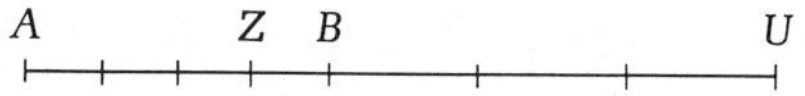

A Z B U

If each of the segments in *BU* is separated into 2 parts, we see that *AU* is separated into 10 equal parts, and that *AZ* has 3 of them. Hence, *AZ* is called $\frac{3}{10}$, and we say $\frac{3}{4} \times \frac{2}{5} = \frac{3}{10}$.

If we hadn't noticed that separating each of the 5 segments forming *AU* into 2 equal segments would separate *AB* into 4 equal segments, we could do the following. Separate each of the 5 segments forming *AU* into 4 equal segments. Then, no matter how many of the fifth segments formed *AB*, the same number of the new short segments will be *AQ* or $\frac{1}{4}$ of *AB*, the result of shrinking *AB* by 4.

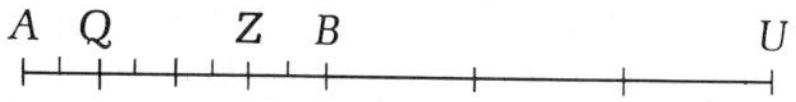

A Q Z B U

AZ, formed from 3*AQ*, will contain 3×2 of the short segments. *AU* contains 4×5 short segments, since there are 4 in each of 5 longer segments.

Hence, $AZ = \frac{3 \times 2}{4 \times 5} = \frac{6}{20} = \frac{3}{10}$, or $AZ = \frac{3}{10} AU$.

Alternatively, to triple AB, then shrink by 4, we first get $AT = 3AB$.

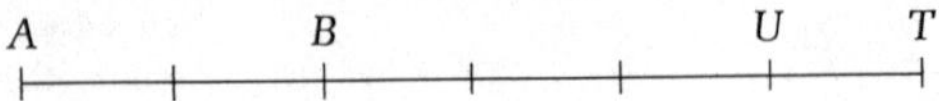

AT contains $3 \times 2 = 6$ segments, of which 5 form AU. We separate AT into 4 equal parts, of which one forms AZ.

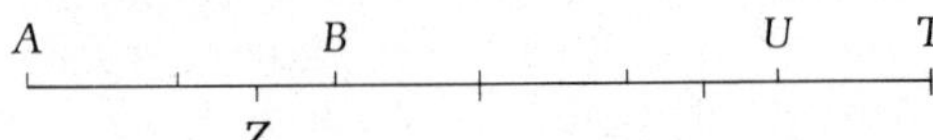

Since none of the separating points for AT falls on U, it is not clear what the ratio of AZ to AU is. If, however, each of the 6 segments forming AT is separated into 4 parts, then U must be one of the separating points. AT contains $4 \times 6 = 24$ short segments. Separating AT into 4 parts and taking one yields AZ of 6 segments. AU has $4 \times 5 = 20$ short segments. $AZ = \frac{6}{20} = \frac{3}{10}$.

19.5 OPERATOR-OPERATOR-OPERATOR The rules by which the VC, the RVN, business discounters, and many tax collectors operate is by using fractions as operators. For certain manufactured items the organizer allows $\frac{2}{5}$ of the selling price to the manufacturer for making them. (The rest goes for distribution.) The manufacturer spends $\frac{3}{4}$ of what he gets for labor (the rest for materials, overhead, etc.) What fraction of the selling price goes for labor? We do not know how much is made, or what the selling price is. We are working solely with operators, or instructions. We have the pattern of an operator formed by one operator followed by a second. In counting numbers, we said that the name of the combined operator was the product of the names of the operators hooked up together. If we use the same attitude, what short name would we give $\frac{3}{4} \times \frac{2}{5}$, the operator formed by a $\frac{2}{5}$ operator followed by a $\frac{3}{4}$ operator?

We can't work well in a vacuum. We ask, say, how much labor would get for building something that sold for 1. The selling price is represented by a unit segment. The manufacturer's money is shown by a segment $\frac{2}{5}$ long. Our problem reduces to the previous one of finding the product of a $\frac{3}{4}$ operator operating on a $\frac{2}{5}$ segment. The result was a $\frac{3}{10}$ segment. Since a $\frac{3}{10}$ segment is the result of a $\frac{3}{10}$ operator working on a unit segment, we say the short name of $\frac{3}{4} \times \frac{2}{5} = \frac{3}{10}$.

19.6 FORMAL MANIPULATION BY CAD Suppose we thought of $\frac{3}{4}$ and $\frac{2}{5}$ as just symbols—not standing for anything, neither sets nor segments nor operators. What short symbol would we want to equal $\frac{3}{4} \times \frac{2}{5}$?

Actually we are free to put down any symbol, such as $\frac{3}{4} \times \frac{2}{5} = 7$ because 7 is our lucky number, or $\frac{3}{4} \times \frac{2}{5} = 0$ because 0 is so easy to add and multiply, or maybe $\frac{3}{4} \times \frac{2}{5} =$ @ because @ is a symbol on the typewriter we never understood. But we might say that (1) we can "multiply them like counting numbers," in the sense that we want the CAD properties to hold, and (2) we want $4 \cdot \frac{3}{4}$ to be 3 and $5 \cdot \frac{2}{5}$ to be 2. Let's see what that gives us.

From $y = \frac{3}{4} \times \frac{2}{5}$ we get

$$4y = 4(\tfrac{3}{4} \times \tfrac{2}{5}) = (4 \times \tfrac{3}{4}) \times \tfrac{2}{5} = 3 \times \tfrac{2}{5}$$

where we have used the associative property and property (2).

Next, multiplying both members by 5, we get

$$\begin{aligned} 5 \times 4 \times y &= 5 \times (3 \times \tfrac{2}{5}) = (5 \times 3) \times \tfrac{2}{5} = (3 \times 5) \times \tfrac{2}{5} \\ &= 3 \times (5 \times \tfrac{2}{5}) = 3 \times 2 \end{aligned}$$

where we have used associative and commutative properties as well as property (2). Writing $5 \times 4 \times y = 4 \times 5 \times y = 3 \times 2$ by our rule for making fractions, we see that $y = \frac{3 \times 2}{4 \times 5} = \frac{3}{10}$. Note that our property (2) can be stated this way: if $az = b$, then another name for z is $\frac{b}{a}$.

19.7 MULTIPLICATION PROPERTIES We have been pleased to see that each of the most obvious generalizations of multiplication of counting numbers that made sense when applied to fractions yielded the name $\frac{3}{10}$ for $\frac{3}{4} \times \frac{2}{5}$. So we feel secure in saying $\frac{3}{10} = \frac{3}{4} \times \frac{2}{5}$. The processes used are clear and we see that if we took any two fractions $\frac{a}{b}$ and $\frac{c}{d}$, each of these analyses would give $\frac{a}{b} \times \frac{c}{d} = \frac{ac}{bd}$. So we take this as the formula for the product of two fractions.

We should work out the details with some other pairs of fractions to be sure we understand the steps.

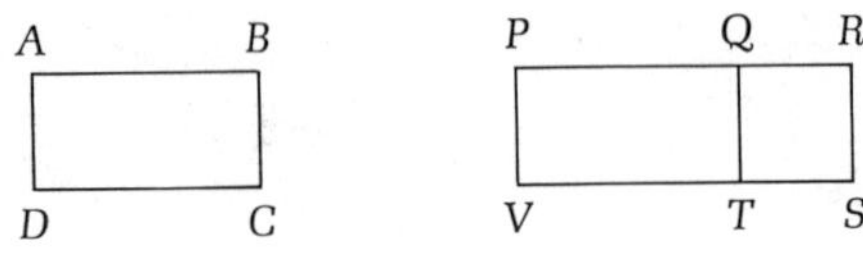

There is a bonus. Because multiplication (and addition) of fractions is defined according to the segment-segment-region attitude, we know that the commutative and distributive properties hold. Looking at the rectangle *ABCD* from two directions shows the commutative property, while the figure PQRSTV shows the distributive property. We leave it as an exercise to give directions for looking at these figures to see the CD properties.

Further, since the product follows the operator-operator-operator pattern, the associative property holds. For, clearly, if 3 operators or machines are hooked up one after another, the names we give will not affect what the machines do. That is, if the combination of $A \times B$ is called L and $B \times C$ is called R, then $L \times C$ and $A \times R$ are both names for $A \times B \times C = (A \times B) \times C = A \times (B \times C)$. These ideas hold for fractions as well as for counting numbers.

19.8 DIVISION OF RATIONAL NUMBERS Here are some situations where dividing fractions comes up: (1) A rectangular tile $\frac{1}{2}$ foot wide has an area of $\frac{1}{3}$ square foot. How long is the tile? (2) The circumference of a circle is about $\frac{22}{7}D$. How big a diameter D is required for a circumference of $\frac{1}{3}$ foot? (3) A person travels $\frac{3}{5}$ mile in $\frac{1}{6}$ hours. How fast is he going in miles per hour? (4) A $\frac{2}{5}$-scale statue ($\frac{2}{5}$ as tall as the subject) is $\frac{3}{4}$ yards tall. How tall is the subject? (5) $7\frac{2}{3}$ ounces of sweet-smelling water cost $6\frac{1}{2}$ dollars. What is the cost per ounce?

Arithmetically, each of these is a special case of the more general problem: Given the names of two rational numbers, how do we find the name of their quotient?

We shall look at different ways of manipulating fractions based on various attitudes toward division and fractions. Some lead to convenient names for the quotient, others support the rule "Invert the denominator and multiply."

19.9 WHOLE-NUMBER ATTITUDES THAT GENERALIZE We recall that division in whole numbers arises from (1) partitioning of sets into sets of a given size (measurement) or a given number of equal sets (sharing), (2) repeated subtraction, (3) forming a ratio of two quantities, and (4) finding a missing factor. Fractions arose as (1) objects, (2) operators, (3) ratios, (4) solutions of equations, and (5) symbols showing division.

Multiplication attitudes to use with missing factor approaches include (1) repeated addition, (2) Cartesian product, (3) array, (4) tree, (5) segment-segment-region, (6) segment-operator-segment, (7) operator-operator-operator, (8) segment-segment-segment, and (9) formal manipulation.

Even a brief look at this list shows that some fraction attitudes won't fit some division attitudes. Division as repeated subtraction or as sharing is awkward with fractions. Division as finding the ratio or missing factor makes good sense. Using $\frac{2}{3} \div \frac{5}{7}$ we will look at some of the situations where division of fractions makes sense, leave others as exercises, and ignore the cases that are meaningless. We can follow the action better with this example than with $\frac{1}{2} \div \frac{1}{3}$. The reader will get help by taking these steps with $\frac{1}{2} \div \frac{1}{3}$, $\frac{2}{3} \div \frac{1}{4}$, and other easy examples.

19.10 MISSING FACTOR–SYMBOL By $\frac{2}{3} \div \frac{5}{7}$ we mean a number, y, for which $\frac{5}{7}y = \frac{2}{3}$. The fractions are number names that we can multiply by any whole number or fraction, and that show associative and commutative properties. Recalling that $7 \times \frac{5}{7} = 5$ and $3 \times \frac{2}{3} = 2$, we expect a simpler equation if we multiply both sides by 7 and 3.

$$7 \times \tfrac{5}{7}y = 7 \times \tfrac{2}{3} \quad \text{or} \quad 5y = \tfrac{14}{3} \quad \text{then}$$
$$3 \times 5y = 3 \times \tfrac{14}{3} \quad \text{or} \quad 15y = 14 \quad \text{hence, } y = \tfrac{2}{3} \div \tfrac{5}{7} = \tfrac{14}{15}$$

We might have renamed the fractions with the same denominator, 21. That is, we seek y such that

$$\tfrac{5}{7}y = \tfrac{2}{3} \quad \text{or} \quad \tfrac{15}{21}y = \tfrac{14}{21}$$

Multiplying both members by 21 gives

$$15y = 14 \quad \text{or} \quad y = \tfrac{14}{15}$$

Since we can multiply by fractions, and we know that $\frac{7}{5} \times \frac{5}{7} = 1$, it seems good to multiply both members of $\frac{5}{7}y = \frac{2}{3}$ by $\frac{7}{5}$, getting

$$\tfrac{7}{5} \times \tfrac{5}{7}y = \tfrac{7}{5} \times \tfrac{2}{3} \quad \text{or} \quad 1 \cdot y = y = \tfrac{7}{5} \times \tfrac{2}{3} = \tfrac{14}{15}$$

19.11 INDICATED DIVISION–SYMBOL We indicate the division by a compound fraction, and suppose that the same kinds of things can be done to a compound fraction as can be done to a simple fraction. A simple fraction has whole numbers above and below the bar. A compound fraction has a fraction either above or below the bar or in both places.

$$\frac{2}{3} \div \frac{5}{7} = \frac{\frac{2}{3}}{\frac{5}{7}}$$

We suppose that we can multiply both numerator and denominator of the compound fraction by the same counting number without changing the value of the fraction, just as we can a simple fraction. 7 and 3 appear to be interesting choices, since they will change the compound to a simple fraction.

$$\frac{\frac{2}{3}}{\frac{5}{7}} \cdot \frac{7}{7} = \frac{\frac{14}{3}}{5} \cdot \frac{3}{3} = \frac{14}{15}$$

It might have been interesting to change to fractions with the same denominator, then multiply by 21.

$$\frac{\frac{2}{3}}{\frac{5}{7}} = \frac{\frac{14}{21}}{\frac{15}{21}} \cdot \frac{21}{21} = \frac{14}{15}$$

Recalling that $\frac{7}{5} \times \frac{5}{7} = 1$,

$$\frac{\frac{2}{3}}{\frac{5}{7}} \times \frac{\frac{7}{5}}{\frac{7}{5}} = \frac{\frac{2}{3} \times \frac{7}{5}}{1} = \frac{\frac{14}{15}}{1} = \frac{14}{15}$$

Some people like to justify or explain these steps by comparing them with the missing-factor attitude (section 19.10). Other people like to proclaim a "Rule of 1." Expressions such as $\frac{7}{7}$ and $\frac{21}{21}$ are other names for 1. These steps reducing the compound fraction to a simple fraction are merely multiplying by 1 (a highly convenient 1, to be sure), which does not change the value of the fraction.

The three approaches of each section support each other. The last approach leads vividly to the famous rule "Invert and multiply"—that is, to get $\frac{a}{b} \div \frac{c}{d}$, form $\frac{a}{b} \times \frac{d}{c}$. Can you identify this in the discussions?

19.12 RATIO-SEGMENT The previous sections gave arguments that manipulated numbers by assuming properties for fractions similar to properties for whole numbers. Some people will want to go deeper in their analysis. Suppose we look at fractions as names of objects, choosing segments to represent such objects. By division we could mean several things, but here we use ratio. That is, by $\frac{2}{3} \div \frac{5}{7}$, we mean the ratio of a segment $\frac{2}{3}$ long to a segment $\frac{5}{7}$ long.

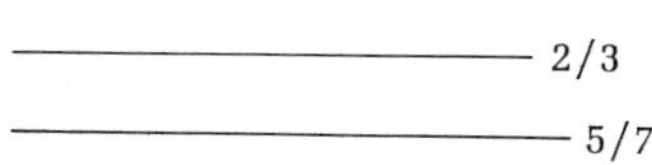

We recall two ways of finding the ratio of segments: through (1) subdivision and (2) multiples. In subdivision, we find a short segment that fits into each of the segments a counting number of times. If this short segment fits into the $\frac{2}{3}$ segment h times and into the $\frac{5}{7}$ segment k times, our desired quotient, or ratio, is $\frac{h}{k}$. We review the multiples method in a later paragraph.

For subdivision to work, it is essential to find this short segment. Euclid, thousands of years ago, outlined a general way, called Euclid's algorithm. However, for certain fractions we can find it by observation. Consider two unit segments with the $\frac{2}{3}$ and $\frac{5}{7}$ segments marked.

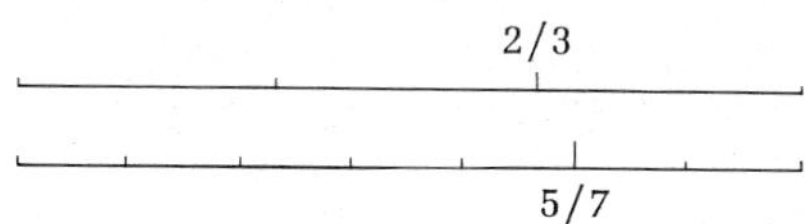

If each of the thirds in the upper diagram is separated into 7 parts, the upper unit is separated into 21. If each of the sevenths in the lower diagram is separated into 3 parts, the lower unit is separated into 21. Hence all the little segments are the same size. The $\frac{2}{3}$ segment contains 14 little ones; the $\frac{5}{7}$ segment contains 15. Hence the ratio is $\frac{14}{15}$. $\frac{2}{3} \div \frac{5}{7} = \frac{14}{15}$.

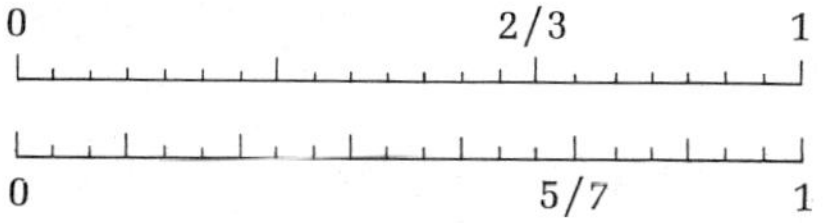

This subdivision and the diagram correspond quite closely to renaming the fractions with the same denominator as done earlier.

The multiples method for finding the ratio is to find a multiple m of the $\frac{2}{3}$ segment that is the same length as some multiple p of the $\frac{5}{7}$ segment. The ratio $\frac{2}{3} \div \frac{5}{7}$ is then $\frac{p}{m}$. This uses the idea that the ratio of A to B is $\frac{p}{m}$ if $mA = pB$. Using smaller segments so that the diagram will fit the page, we mark off multiples of $\frac{2}{3}$ and $\frac{5}{7}$ on two parallel lines.

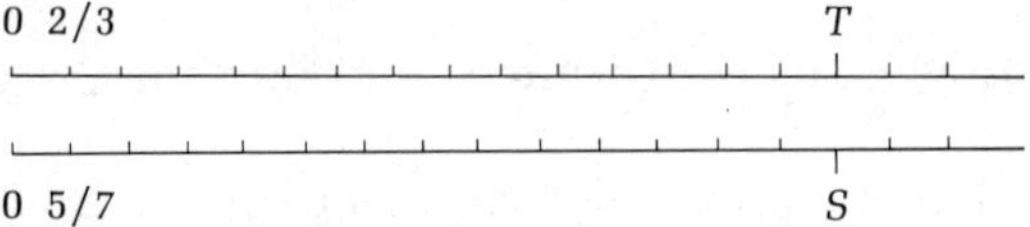

It appears that the points at T and S are the same distance from 0. There are 15 segments in $0T$; 14 in $0S$. Hence the ratio $\frac{2}{3} \div \frac{5}{7} = \frac{14}{15}$.

We suspect we may have been misled by an optical illusion to identify T and S. Perhaps we should have gone one more segment, and the ratio would have been $\frac{15}{16}$. Or one less, with the ratio $\frac{13}{14}$. We check. Fifteen $\frac{2}{3}$ segments make a length of 10 units. Fourteen $\frac{5}{7}$ segments also make a length of 10. Hence $0T$ does equal $0S$.

This second, or multiples, approach to finding the ratio of two segments does not match as closely with the earlier ways of finding the simple fraction name for $\frac{2}{3} \div \frac{5}{7}$. It compares with the ratio-solution of equation approach in section 19.14 below.

19.13 RATIO-SYMBOLS Using ratio as the meaning of division, we wish to find numbers c and d such that $\frac{2}{3} \div \frac{5}{7} = \frac{c}{d}$, where $\frac{2}{3}d = \frac{5}{7}c$. Multiplying both sides first by 7 and then by 3 we get $14d = 15c$. Since $14 \cdot 15 = 15 \cdot 14$, an obvious solution is $d = 15$, $c = 14$. Hence, $\frac{2}{3} \div \frac{5}{7} = \frac{14}{15}$.

19.14 RATIO–SOLUTION OF EQUATION By $\frac{2}{3} \div \frac{5}{7}$ we mean $y \div x$ where $3y = 2$ and $7x = 5$. By division as ratio we wish c and d so that $y \div x = \frac{c}{d}$ or $yd = xc$. Multiplying both sides of $3y = 2$ by 5 gives $15y = 10$. Multiplying both sides of $7x = 5$ by 2 gives $14x = 10$. Hence $15y = 14x$. Comparing with $yd = xc$, we see a solution is $d = 15$, $c = 14$, and $\frac{y}{x} = \frac{14}{15}$.

19.15 MISSING FACTOR–SEGMENT Multiplication attitudes of segment-segment-segment, segment-segment-region, and operator-segment-segment lead to missing-factor division approaches. For segment-segment-segment, we draw the following figure:

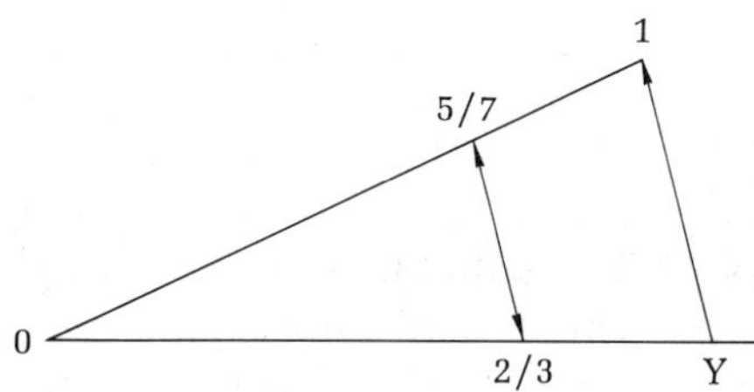

The order of drawing is as follows: (a) draw two lines meeting at 0 at a convenient angle; (b) locate $\frac{5}{7}$, 1, and $\frac{2}{3}$, and draw line $\frac{5}{7}$–$\frac{2}{3}$; (c) draw 1–Y parallel to $\frac{5}{7}$–$\frac{2}{3}$. From similar triangles $0Y = \frac{\frac{2}{3}}{\frac{5}{7}}$. The problem is to determine the fraction name for 0Y. To do this, we mark (d) the seventh points on the upper line. (e) Through these points, draw lines parallel to 1–Y meeting the lower line at points A, B, . . . , E. Clearly, 0Y is separated into 7 segments. We need the length of each segment, which we deduce from $\frac{2}{3}$. 0–$\frac{2}{3}$ is separated into 5 segments. It is not easy, perhaps, to get the size of one segment from this. If we separate each segment into two smaller segments the same length, there will be 10 of these segments 0Q, QA, . . . in 0–$\frac{2}{3}$. Ten is divisible by 2, so we see that 5 make up $\frac{1}{3}$.

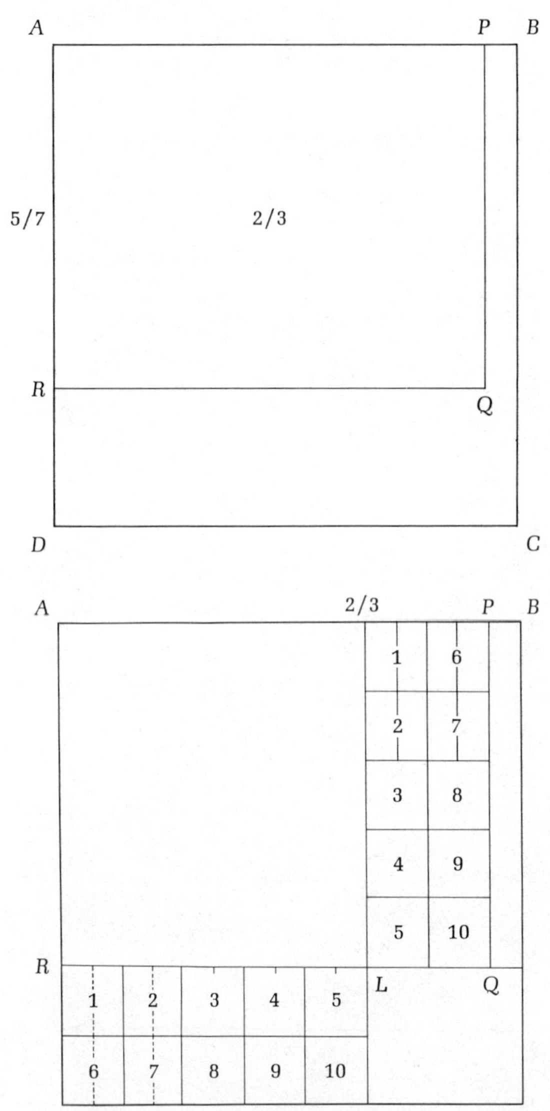

Hence, 15 make a unit on the lower line. Since there are 14 of these segments in 0Y, $0Y = \frac{\frac{2}{3}}{\frac{5}{7}} = \frac{14}{15}$. The complete diagram is as follows:

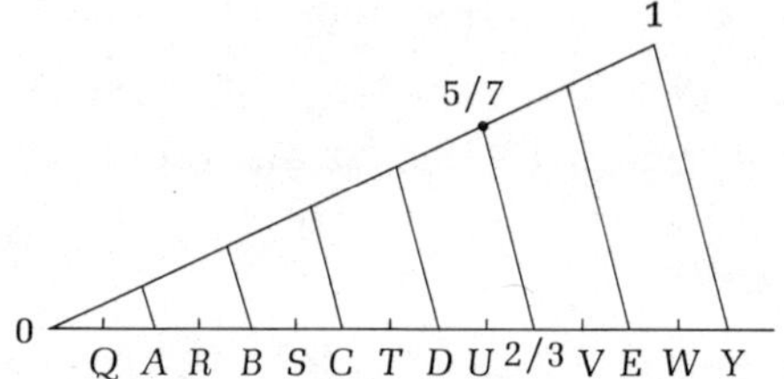

19.16 MISSING FACTOR-SEGMENT (CONTINUED) The segment-segment-region attitude toward multiplication is perhaps one of the most powerful for discussing fraction multiplication. It is sad that it is poor for discussing division. We see why when we consider the figures in the margin.

To find a more convenient name for $\frac{2}{3} \div \frac{5}{7}$ means to find the length of a rectangle whose width is $\frac{5}{7}$ and whose area is $\frac{2}{3}$. The figure in the margin shows a rectangle *APQR* with width $AR = \frac{5}{7}$, area $= \frac{2}{3}$ and *AP* to be found. *ABCD* is a unit square drawn in a standard position. Our challenge is to cut up this figure in such a way that these numbers will be clear. Can we, for example, cut up the unit square into congruent rectangles so that $\frac{2}{3}$ of them form *APQR*, the edges form *AR* and *AP*, and it is easy to determine the length of *AP* by counting? Let's try. One such dissection is shown below. Rectangle $A\frac{2}{3}HD$ has area $\frac{2}{3}$. *RLHD* is cut into 10 rectangles that fit on the right of $\frac{2}{3}L$. These 10 rectangles are cut into 20 so that the segment *RL*, which is $\frac{2}{3}$, will be separated into an even number of segments. Five of these segments make $\frac{1}{3}$; 15 make the unit side of the unit square. Since there are 14 in *AP*, $AP = \frac{14}{15} = \frac{\frac{2}{3}}{\frac{5}{7}}$.

19.17 MISSING FACTOR–OPERATOR Fractions are used extensively as operators, as suggested by the phrase "$\frac{5}{7}$ of." Indeed, many people think that to multiply by a fraction, as in "$\frac{2}{5}$ times," one changes the "times" to "of" and operates—often by mentally cutting and pasting. We have called such an attitude the operator-segment-segment attitude. One uses a $\frac{2}{5}$ operator on something (represented by a segment) with the result represented as a segment.

From this attitude, our problem $\frac{2}{3} \div \frac{5}{7}$ would mean "How long a segment is required so that if I take $\frac{5}{7}$ of it I will have a $\frac{2}{3}$ segment?" This structure is shown also by the questions "How much milk must I start with if $\frac{5}{7}$ of it is $\frac{2}{3}$ gallon?" and "How much sugar do I need if $\frac{5}{7}$ of it is $\frac{2}{3}$ pounds?"

Perhaps the easiest way to work this problem is to use abstract structure, manipulating the numbers as symbols. That is $y = \frac{2}{3} \div \frac{5}{7}$ means $\frac{5}{7}y = \frac{2}{3}$. Then use ideas given above. However, we can use a method common to pioneers, which might be called a *unit-fraction method.* Our pioneer forbears did not trust decisions made on juggling numbers alone. They wanted to think in terms of the actual milk as much as the numbers.

They argued as follows. If $\frac{5}{7}$ of the milk is $\frac{2}{3}$ gallon, then $\frac{1}{7}$ of the milk is $\frac{1}{5}$ of $\frac{2}{3}$ gallon. $\frac{1}{5}$ of $\frac{2}{3}$ gallon is $\frac{2}{15}$ gallon. All the milk will be seven times this or $\frac{14}{15}$ gallon. Hence, $\frac{2}{3} \div \frac{5}{7} = \frac{14}{15}$.

19.18 MISSING FACTOR-OPERATOR (CONTINUED) One instruction or operator followed by another make a third operator. This hook-up leads to multiplication of fractions as an attitude that we call operator-operator-operator. Interpreting $\frac{2}{3} \div \frac{5}{7}$ as missing factor, this attitude gives the question "What instruction should follow 'take $\frac{5}{7}$ of' so that the combined instruction is 'take $\frac{2}{3}$ of'?" Some people like to say "What machine should be hooked up to a $\frac{5}{7}$-machine to make a $\frac{2}{3}$-machine?"

It is almost impossible to think of combining operators and renaming the result without either imagining what happens to something operated on by the operators, or using pure number manipulations with the associative and commutative properties of multiplication. This reduces the problem to one considered earlier. We leave this as an exercise for the reader.

EXERCISES

1. Which of the following multiplication attitudes make sense for $\frac{3}{5} \times \frac{2}{7}$? Repeated addition, array, Cartesian product, tree, segment-segment-region, operator-operator-operator, operator-segment-segment.

2. For two of the attitudes in exercise 1 that make sense, describe what is meant by $\frac{3}{5} \times \frac{2}{7}$.

3. Use the most convenient attitude to use to show $\frac{2}{5} \times \frac{3}{4}$. This demonstration should show clearly why we define

$$\frac{a}{b} \times \frac{c}{d} = \frac{ac}{bd}$$

4. Choose a set of beans, W, for a whole. Can you operate on this set nicely with the operator $\frac{2}{3}$? If not, choose another set for W. Let B be the set produced. Can you operate on B nicely with the operator $\frac{1}{2}$? If not choose a new W.

5. Describe two other sets that would work nicely for W in exercise 4.

6. Describe a set W on which you could operate nicely with operators $\frac{3}{4}$ and $\frac{2}{3}$ in that order. Describe the set C remaining. What number does C show?

7. Can you operate on W in exercise 6 with the operators $\frac{2}{3}$ and $\frac{3}{4}$ in that order? Find three sets for W where (*a*) you can operate in the order $\frac{3}{4}$ then $\frac{2}{3}$ but not in the order $\frac{2}{3}$ then $\frac{3}{4}$, then (*b*) W on which you can operate in the order $\frac{2}{3}$ then $\frac{3}{4}$ but not in the order $\frac{3}{4}$ and $\frac{2}{3}$, and (*c*) W on which you can operate in either order.

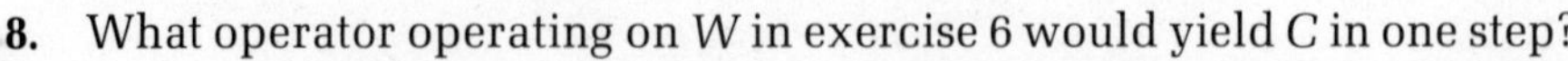

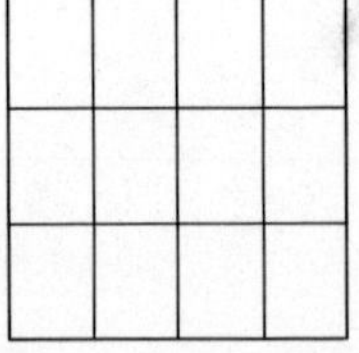

8. What operator operating on W in exercise 6 would yield C in one step?

9. What multiplication sentence is suggested by exercises 6–8? What attitude towards multiplication is illustrated?

10. Copy the diagram at left, which is often used to illustrate $\frac{1}{3} \times \frac{3}{4} = \frac{3}{12}$. Mark what shows $\frac{1}{3}$ in the figure. Mark what shows $\frac{3}{4}$. Mark what shows $\frac{3}{12}$.

11. Point out how the union of 4 of whatever shows $\frac{3}{12}$ will show 1. Hence $4 \times \frac{3}{12} = 1$.

12. The argument in exercise 11 shows that $\frac{3}{12}$ is another name for $\frac{1}{4}$. Which fraction attitude is used in this argument (*a*) object with a history, (*b*) ratio, (*c*) solution to a linear equation, (*d*) operator, (*e*) symbol, (*f*) none of these?

13. Compute (*a*) $\frac{2}{3} \times \frac{4}{5}$ (*b*) $\frac{12}{7} \times \frac{14}{15}$

14. Make a diagram like that in exercise 10 to show $\frac{3}{4} \times \frac{4}{5}$.

15. Make a diagram like that in exercise 10 to show $1\frac{1}{2} \times 2\frac{2}{3}$.

16. Make a diagram to show

$$\tfrac{2}{3}(\tfrac{4}{5} + \tfrac{3}{4}) = \tfrac{2}{3} \times \tfrac{4}{5} + \tfrac{2}{3} \times \tfrac{3}{4}$$

What law or property does your diagram show?

17. Which of the following attitudes towards division (which are fine for $12 \div 3 = y$) make sense for $\frac{2}{3} \div \frac{1}{4} = x$? Write yes or no.

(*a*) x is the number of times $\frac{1}{4}$ goes into $\frac{2}{3}$.
(*b*) x is the number of times $\frac{1}{4}$ can be subtracted from $\frac{2}{3}$.
(*c*) x is the ratio of a $\frac{2}{3}$ segment to a $\frac{1}{4}$ segment.
(*d*) x is the number which multiplied by $\frac{1}{4}$ yields $\frac{2}{3}$.
(*e*) x is the side of a rectangle whose area is $\frac{2}{3}$ and whose other side is $\frac{1}{4}$.
(*f*) x is the number of eggs each person gets if $\frac{2}{3}$ eggs are shared equally among $\frac{1}{4}$ people.
(*g*) x is the number shown when $\frac{2}{3}$ is separated into $\frac{1}{4}$ parts of the same size.
(*h*) x is a fraction whose numerator is $\frac{2}{3}$ and whose denominator is $\frac{1}{4}$. (The same rules for operating on fractions with whole numbers in numerator and denominator hold for x.)

18. Using *h* in exercise 17, multiply $\frac{\frac{2}{5}}{\frac{4}{3}}$ by 1 (disguised as $\frac{b}{b}$) getting $\frac{\frac{2}{5} \times b}{\frac{4}{3} \times b}$. Name two convenient numbers for *b*, either of which would change the numerator and denominator to integers. Show that the result can be written as $\frac{2}{5} \times \frac{3}{4}$.

19. Rewrite the fractions in the numerator and denominator of $\frac{\frac{2}{5}}{\frac{4}{3}}$ so they have the same denominator. Multiply by 1 (or $\frac{c}{c}$). What convenient value do you take for c? Show that the result can be written as $\frac{2}{5} \times \frac{3}{4}$.

20. Find a convenient fraction, q, so

$$\frac{\frac{2}{5} \times q}{\frac{4}{3} \times q} = \frac{\frac{2}{5} \times \frac{3}{4}}{1}$$

21. Compute
(*a*) $\frac{2}{5} \div \frac{1}{2} =$ (*c*) $\frac{3}{8} \div \frac{3}{4} =$
(*b*) $\frac{4}{3} \div \frac{2}{1} =$ (*d*) $5\frac{1}{2} \div 2\frac{3}{4} =$

22. (*a*) Find a segment that will fit evenly into segment A which is $\frac{2}{3}$ long and into segment B which is $\frac{1}{4}$ long.

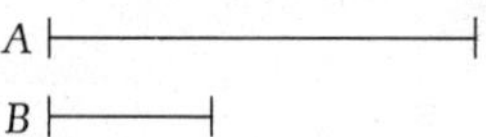

(*b*) Find another segment for *a*.
(*c*) What is the ratio of $\frac{2}{3}$ to $\frac{1}{4}$ from the subunits attitude? Why?
(*d*) Find numbers X, Y such that $X \cdot \frac{2}{3} = Y \cdot \frac{1}{4}$.
(*e*) What is the ratio $\frac{2}{3}$ to $\frac{1}{4}$ from the multiples attitude? Why?

23. By what number would you multiply both sides of $\frac{2}{3}Z = \frac{3}{7}$ to eliminate fractions? What do you get?

24. Draw a unit square. Show how to draw or to cut and paste to form a rectangle of area $\frac{2}{3}$ and width $\frac{1}{2}$ without first knowing the length. Determine the length to show

$$\frac{2}{3} \div \frac{1}{2} = \frac{4}{3}$$

25. A bicycle goes $\frac{3}{4}$ kilometer in 4 minutes. What was the time in hours? What was the speed in kilometers per hour?

26. An engine used $7\frac{1}{2}$ gallons of diesel oil in 2 hours and 5 minutes. What was the rate of oil consumption in gallons per hour?

Answers to selected questions: **1.** segment-segment-region; operator-operator-operator; operator-segment-segment; **6.** $W = 4$ beans; $C = 2$ beans and shows $\frac{3}{4} \times \frac{2}{3} = \frac{1}{2}$; **7.** no; (*a*) W = 4, 8, 16 beans; (*b*) 6, 18, 30; (*c*) 12, 24, 36; **9.** $\frac{3}{4} \times \frac{2}{3} = \frac{1}{2}$; operator-operator-operator; **13.** (*a*) $\frac{8}{15}$; (*b*) $\frac{8}{5}$; **17.** (*a*) no; (*b*) no; (*c*) yes; (*d*) yes; (*e*) yes; (*f*) no; (*g*) no; (*h*) yes.

19. $\frac{\frac{2}{5}}{\frac{4}{3}} = \frac{\frac{2 \times 3}{15}}{\frac{4 \times 5}{15}} \quad \frac{15}{15} = \frac{2 \times 3}{4 \times 5}$

21. (*a*) $\frac{4}{5}$; (*b*) $\frac{2}{3}$; (*c*) $\frac{1}{2}$; (*d*) 2; **23.** 21; $14z = 9$; **25.** $\frac{1}{15}$ hour; $\frac{3}{4} \div \frac{1}{15} = \frac{45}{4} = 11\frac{1}{4}$ kilometers per hour.

20

A Unified Theory of Fractions

20.1 INTRODUCTION The uses and attitudes that lead to fractions and their properties are highly varied and may appear inconsistent. We do not multiply the two pieces of pie that we commonly use to introduce the fraction idea. Nor do we add the operators "$\frac{1}{3}$ of" and "$\frac{3}{4}$ of." None of the intuitive approaches to fractions that show the most common uses of fractions is satisfactory as a basic theory for all the uses of fractions we want. This is in sharp contrast with counting numbers, where either sets or segments furnish a good base for the number ideas and the operations. Because of the limited applicability of the elementary fraction attitudes, people shift from one to another as convenience suggests. Some people wonder if the properties that develop clearly in one situation really apply in others. Can we trust the results of number manipulation?

What is needed is a theory that will apply to all the uses of fractions. The theory should be broad enough so that each of the common attitudes will apply to a part of it. We would like the theory to support the algorithms—the number dances we do to find answers—and to relate well to whole numbers and sets.

This is a tall order. The Greeks came a long way toward this with their number theory based on line segments. Fractions related easily to whole numbers. All operations were clearly defined. However, the algorithms are very obscure, and it is hard to identify an operator or instruction with a segment.

Modern mathematics has a theory that meets these requirements, but at a cost of eliminating intuitive experience from the definitions, even though the definitions are formed to agree with the various intuitive experiences. Everything is expressed solely in terms of the symbols. More accurately, we assume that we know the whole numbers and build everything on the whole numbers. We will suggest two ways this is done.

20.2 FRACTIONS AS SOLUTIONS OF LINEAR EQUATIONS Starting with whole numbers, we define $\frac{2}{3}$ as something that satisfies the equation $3y = 2$. This is all we know about it, or care to know. In particular, we don't know if $\frac{2}{3}$ names a segment, or an operator, or anything.

We assume that the multiplication makes sense and satisfies the CAD (commutative, associative, and distributive) properties just as the multiplication of whole numbers does.

This definition works nicely. We get multiplication quickly. To get $\frac{2}{3} \times \frac{5}{7}$, we want wz where $3w = 2$ and $7z = 5$. Multiplying the left and right members, we get

$$3w \cdot 7z = 2 \cdot 5$$
$$\text{or} \qquad 21wz = 10$$

Hence, $wz = \frac{10}{21}$.

Addition is almost as easy. For $\frac{2}{3} + \frac{5}{7}$ we wish $w + z$. Multiplying the first equation by 7 and the second by 3, we get

$$21w = 14$$
$$\text{and} \quad 21z = 15$$

Now we add left sides and right sides, then use the distributive property, getting

$$21x + 21z = 14 + 15$$
$$21(w + z) = 29$$

Hence, $w + z = \frac{29}{21}$.

To show that this theory of fractions applies to pies, segments, instructions, and operators, we need to identify the multiplication and units in these terms.

For example, to show $3w = 2$ in terms of segments, we wish segments w such that 3 of them put end to end form a 2-segment. This interpretation of fractions is given earlier. It may be a bit bothersome to students who believe that $\frac{2}{3}$ is a segment found by separating a whole into 3 parts and joining 2 of them.

We will leave it as an exercise for the student to identify what is called multiplication in several situations and to show that in these cases where $\frac{2}{3}$ is used as the name of something, then 3 times that something makes sense, and the product is named 2. Such identification has to exist if the fraction language is to tell a useful story. You may also want to define subtraction and division and show that they fit the patterns we expect from intuitive attitudes.

Some people object to this theory. They say, "If the only numbers we have are counting numbers, then $\frac{2}{3}$ does not exist. $3w = 2$ is a rule for identifying $\frac{2}{3}$ if we see the number. Just because we know how to identify a number doesn't mean there is such a number." Some people call definitions such as $3w = 2$ *unicorn definitions*. They know how to define and recognize a unicorn (perhaps as a horse with a long horn on his forehead), but they feel it a complete waste of time to design a saddle and harness or a diet for the beast. They want a definition of fraction that displays the fractions solely from whole numbers, with no make-believe about their existence. We will see how mathematicians have done this.

20.3 FRACTIONS AS ORDERED PAIRS OF NUMBERS A fraction is now defined as an ordered pair (2, 3) of counting numbers. There is no implication of division—just the pair. We now define sum, difference, product, and quotient of these ordered pairs. Of course, we define them by the same formulas that fit fractions we studied in previous chapters. Matching (2, 3) with $\frac{2}{3}$, we define

$$(2, 3) \times (5, 7) = (10, 21)$$
$$(2, 3) + (5, 7) = (29, 21)$$

Can you see why $(a, b) \times (c, d) = (ac, bd)$ and $(a, b) + (c, d) = (ad + bc, bd)$ are chosen for the definitions of product and sum of (a, b) and (c, d)?

$(a, 1)$ is matched with the counting number a.

Is it reasonable to say that (2, 3), which has nothing to do with division, being merely a pair of counting numbers, is properly called $\frac{2}{3}$ or 2 divided by 3?

We apply the unicorn definition. If the product of 3 and (2, 3) is 2, then (2, 3) is appropriately called $\frac{2}{3}$. Of course, we match 3 with (3, 1) and 2 with (2, 1)

$$3 \times (2, 3) = (3, 1) \times (2, 3) = (6, 3)$$

by the multiplication formula.

Alas! The pair (6, 3) is a different pair from (2, 1). So we add another rule: We regard two pairs (p, q) and (r, s) as equal if and only if $ps = rq$. This is another way of saying that (p, q) and (r, s) can be reduced to lowest terms. For example, $\frac{4}{10}$ and $\frac{6}{15}$ are the same rational number, since $4 \cdot 15 = 10 \cdot 6$, and both fractions reduce to $\frac{2}{5}$.

We now conclude that $(6, 3) = (2, 1)$, since $6 \cdot 1 = 3 \cdot 2$. Hence $3 \times (2, 3) = 2$ and (2, 3) is properly named $\frac{2}{3}$.

It is perhaps acceptable that all the properties of fractions we have discussed earlier hold with this new ordered-pair definition. While we will not discuss this further here, we comment that this trick of defining new types of numbers as ordered pairs of old numbers is used several places in mathematical theory. It gives a way of constructing new numbers without appealing to the physical intuition that makes the new numbers

useful. This shows that the number properties do not depend on the physical properties of the things that led us to consider the numbers in the first place. Two + three is five, even if the sticks which we put end to end to show this bend a bit under our enthusiasm.

EXERCISES

1. Give linear equations that are satisfied by
(*a*) $\frac{1}{2}$ (*b*) $\frac{2}{3}$ (*c*) $\frac{5}{7}$

2. Using CAD properties and the solution of linear equation attitude toward fractions, show the following:
(*a*) $\frac{1}{2} \times \frac{1}{3} = \frac{1}{6}$ (*b*) $\frac{2}{3} \times \frac{4}{5} = \frac{8}{15}$

3. As in exercise 2 show

(*a*) $\frac{1}{2} + \frac{1}{3} = \frac{5}{6}$ (*b*) $\frac{2}{3} + \frac{3}{5} = \frac{19}{15}$

4. As in exercise 2, show
(*a*) $\frac{1}{3} \div \frac{1}{4} = \frac{4}{3}$ (*b*) $\frac{2}{3} \div \frac{4}{5} = \frac{5}{6}$

5. Show that $\frac{3}{4} = \frac{6}{8} = \frac{9}{12}$

6. What does "multiply by 5" mean? Give physical interpretations of "5 × 3/5 = 3" for these situations.
- (*a*) *P* is a bag containing 3/5 pounds of almonds.
- (*b*) Amy drank 3/5 pint of orange juice.
- (*c*) The split peas cooked in 3/5 hour.
- (*d*) The ratio of women to men in Alaska is 3/5.
- (*e*) The temperature was 3/5 degree Celsius.
- (*f*) The Chevrolet goes 3/5 as far as a Volkswagen on a gallon of gas.
- (*g*) John Legg ran 3/5 as fast when he was an eighth-grader as he does as a senior.
- (*h*) One franc is worth 3/5 mark.
- (*i*) The width of a card is 3/5 as long as its length.

7. Let (*a*, *b*) represent the fraction $\frac{a}{b}$ where *a* and *b* are whole numbers. Give the standard symbols for (*a*) (3, 5), (*b*) (4, 1), (*c*) (2, 6).

8. Compute:
(*a*) (3, 4) × (2, 3), (*b*) (2, 5) + (3, 2)

Answers to selected questions: **1.** (*a*) $2x = 1$; (*b*) $3y = 2$; **3.** (*a*) $2x = 1$; $3y = 1$; $6x = 3$; $6y = 2$; $6(x + y) = 5$; $x + y = \frac{5}{6}$; (*b*) $3x = 2$; $5y = 3$; $15x = 10$; $15y = 9$; $15(x + y) = 19$; $x + y = \frac{19}{15}$.

5. $\frac{3}{4} = \frac{3 \cdot 2}{4 \cdot 2} = \frac{3 \cdot 3}{4 \cdot 3} = \frac{6}{8} = \frac{9}{12}$

6. (*b*) 5 of Amy's drinks would make 3 pints; (*d*) If there were 5 times as many women, there would be 3 women for every man; (*f*) If the Chevrolet went 5 times as far as it did, it would have gone 3 times as far as the Volkswagen.

21

Integers

21.1 INTRODUCTION In this chapter we extend the concept of numbers from the whole numbers to the integers. The integers are the set $\{\ldots {}^-3, {}^-2, {}^-1, 0, {}^+1, {}^+2, {}^+3, \ldots\}$. We discuss the way integers arise as convenient names, and as answers to problems that have solution in counting numbers. Then we extend the operations of addition, subtraction, multiplication, and division so that they apply to negative integers as well as to whole numbers. We hope to do this so that (1) the obvious number patterns of counting numbers extend to integers, (2) the attitudes that led to, say, addition of counting numbers, extend to addition of integers, (3) the attitudes for the other operations extend similarly. When thinking of counting numbers as integers, we often call them *positive* and write . . . $^+4$, $^+5$, $^+6$, ...

We close with a section showing how the integers can be "constructed" from the counting numbers.

21.2 THINGS NAMED BY NEGATIVE INTEGERS The idea here is to extend counting numbers to negative integers by considering places where we use counting numbers and seeing if the concept extends meaningfully. Sometimes it does and sometimes it doesn't.

1. *Names of points on a number line, or scale.* Consider the figure

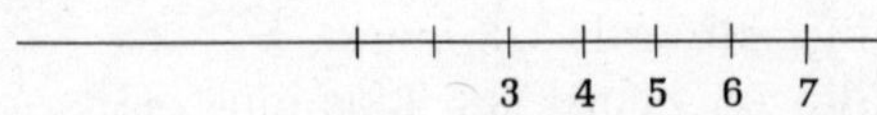

which is a line with points labeled 3, 4, 5, 6, 7 at equal intervals. Can this labeling be extended to the right and left? We continue to the right using counting numbers, increasing by one for each interval. We continue to the left using decreasing counting numbers and zero. We are now out of labels if we want to use decreasing whole numbers. But the line goes on, and points marking off congruent intervals can be placed on the line as far to the left as we please. It is traditional to label these points as shown in the following figure:

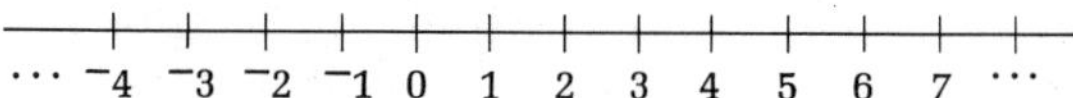

The numeral $^{-}2$ can be read "opposite two," because with respect to 0 the point $^{-}2$ is just opposite the point 2.

It turns out that this quality of being opposite is the quality we look for in other uses of $^{-}2$. The numeral $^{-}2$ is usually read "negative two," and "minus two." The symbol $\overline{2}$ is sometimes used instead of $^{-}2$.

Note that the dash is high in front of the numeral, and is not in the middle. The dash in the middle, such as -3, means "subtract," a command or operation. After people are familiar with the ideas, they can tell from the way the words are used whether the dash means "subtract" or is part of the symbol for "negative two." Then they aren't so careful how high they write the dash.

2. *Temperature readings.* Thermometers show the temperature by locating a point on a scale by a colored line or a pointer. If the weather is below freezing, the temperature on the centigrade scale is read as negative, or below zero. 3° below freezing is called $^{-}3°$.

3. *Elevation* is the height of land above or below a certain point, such as sea level or the first floor of a building. $^{+}200$ feet is the height above sea level, while $^{-}75$ feet marks a position 75 feet below sea level, or the level chosen as zero.

4. *Sets.* Modern mathematics presents counting numbers as properties of sets. This property does not extend to negative numbers. While we have the empty set with zero members, people have not yet found it useful to have a set with a negative number of members. What would it mean to have "a set of $^{-}2$ pencils"?

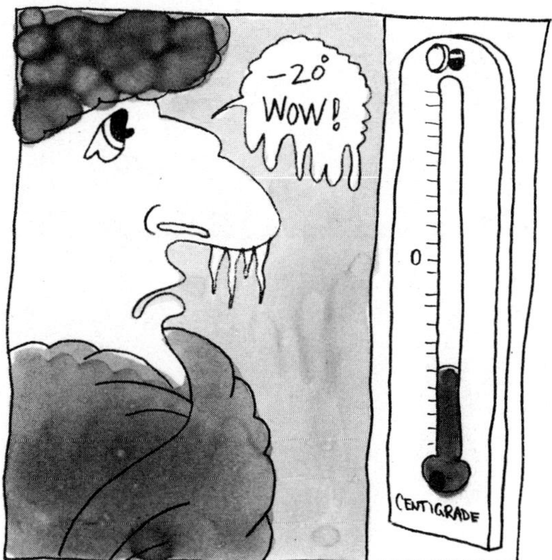

5. *Values of assets, debts, notes, worth.* There is a natural opposite relation in buying, selling, trading, and valuing that can be described with integers. A check for \$6 that can be cashed is labeled $^{+}6$. A dun or bill for \$7 that must be paid is labeled $^{-}7$. Duns are not called bills here, because the word *bill* is commonly used not only for duns to be paid but also for bank notes or money. We may think of a dun as a piece of paper standing for a debt, and a check as a piece of paper standing for an asset. A check for \$6 and a dun for \$7 together have a worth of $^{-}1$. A person with notes of values $^{-}4$, $^{+}7$, $^{-}6$ has worth $^{-}3$.

6. *Giving and receiving.* These are obviously opposite actions. One decides which action he will call positive, and the other is then negative. Suppose we say receiving 3 times is labeled positive three, $^{+}3$. Giving 5 times is then labeled $^{-}5$. We note that the objects labeled are commands or operators rather than things.

7. *Game counters and cards.* Some games have counters of opposite value. For example, blue chips might count $^{+}1$ each, and red chips count $^{-}1$ each. A pot with 8 blue and 5 red chips would be worth $^{+}3$. Similarly, playing cards sometimes have instructions such as "Take 4 jumps forward" and "Jump back 2." Such cards can be labeled $^{+}4$ and $^{-}2$, respectively.

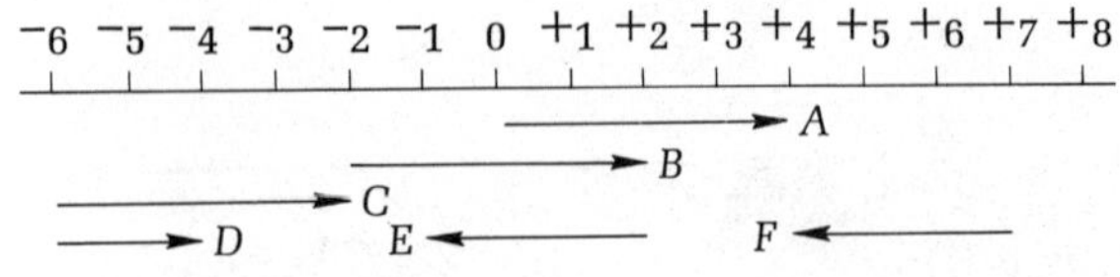

8. *Linear displacement* is directed motion along a line, a number line, a series of stepping stones, or a curve. Game cards have instructions such as "Move 3 steps forward." Such a move is called a linear displacement (linear, because the path is a line—perhaps curved—and displacement is another name for movement). Displacements are often represented by arrows. In the figure in the margin, displacements along a number line are represented. Displacements *A* (from 0 to 4), *B* (from $^{-}2$ to 2), and *C* (from $^{-}6$ to $^{-}2$) are labeled $^{+}4$. They are considered equal, because they would each result from the instruction "Move 4 to the right." *D* is a $^{+}2$ displacement. *E* is a $^{-}3$ displacement.

Displacements could be along a sequence of stepping stones, a curve, or a line in any direction:

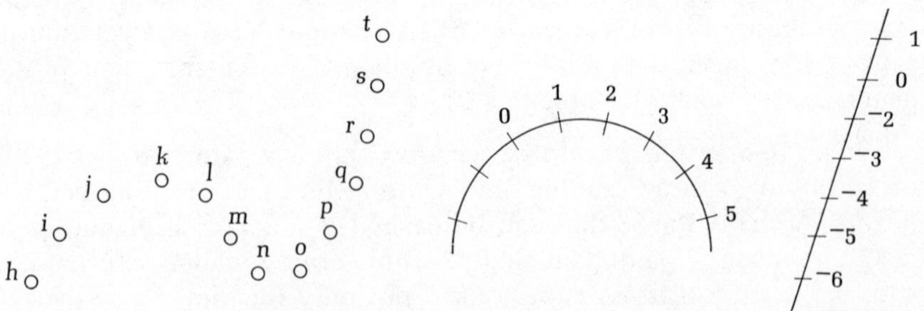

Positions on the paths can be named in any convenient way such as by letters in the stepping stones and integers in the above figure. A $^-3$ displacement on the stepping stones would be a motion, say from q to n, or from k to h.

9. *Velocity.* Velocity is directed speed. Velocity along a line in one direction is labeled positive; in the other direction it is labeled negative. For example, a bus moving east along an east-west street at 3 kilometers per hour might have a velocity labeled as $^+3$. A man (or another bus or car) moving west at 2 kilometers per hour would have a velocity of $^-2$.

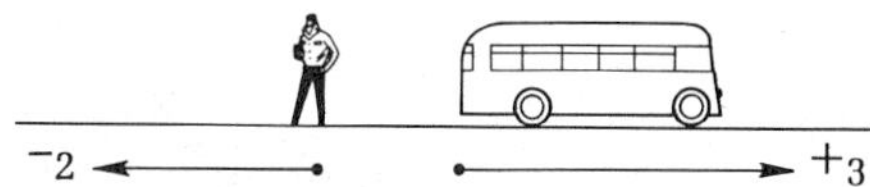

10. *Instants in time.* Just as points on a line are labeled by integers with zero at some central spot, so instants in time are labeled by integers with zero at some important event. The steps taken in launching a space rocket are planned in detail long before the exact moment of blast-off is set. For example, $^-17$ labels the instant 17 hours before the planned firing. Other steps might be started at this instant. $^+2$ labels an instant 2 hours after firing. There are 6 hours from $^-2$ to $^+4$. Seconds, days, or other time units might be used. The oppositeness of before and after are reflected by the oppositeness of negative and positive.

WARNING The calendar dates labeled A.D. and B.C. are not quite like this. The year 2 B.C. means the second year before Christ and refers to an interval, not an instant. Similarly for 3 A.D. There is no year labeled 0.

11. *Directed time intervals.* We label the interval from 7 P.M. to 9 P.M. $^+2$, and the interval from 6 P.M. to 1 P.M. $^-5$. An interval that ends at an instant later than the starting instant is called positive. An interval that "ends" at an instant earlier than the starting instant is called negative.

Just as displacements along a line are represented by vectors or arrows, so are directed time intervals. We draw a "calendar"—a number line with points representing instants, with arrows showing directed time intervals.

Some people have difficulty thinking of an interval that ends earlier than it began. We point out that the direction of an interval depends on the viewer. If we look backward, we may think of the interval from now to the day we were born, or from now to last Christmas, as a negative interval.

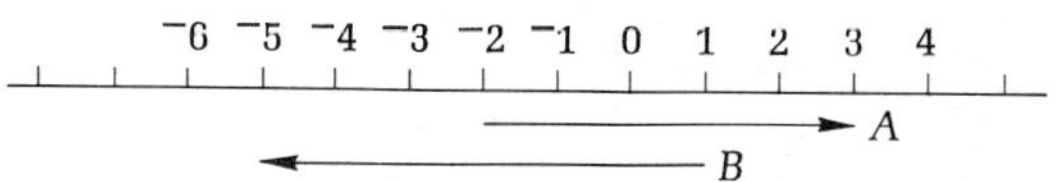

Interval A from $^-2$ to $^+3$ is labeled $^+5$.
Interval B from 1 to $^-5$ is labeled $^-6$.

12. *Forces.* A helium-filled balloon is tied to a small weight. Suppose the balloon pulls up with a force of 3 grams, and the weight pulls down with a force of 3 grams. These opposite forces can be labeled by integers.

If we think of the upward force as positive and the downward force as negative, we label the balloon's force as $^{+}3$ and the weight's force as $^{-}3$. A block of wood under water is pulled down by the force of gravity, and pushed up by the buoyancy, or floating force, of the water. The gravity force might be labeled $^{-}5$ and the buoyancy force labeled $^{+}6$. The units might be grams, or pounds, or other units of force.

Forces along a line may be labeled with integers. A railroad locomotive pulls with a force of $^{+}40{,}000$ pounds. Friction in the railroad cars pulls with a force of $^{-}38{,}000$ pounds. Such forces can be represented by vectors (← →). The figure in the margin shows a box on a road pulled by 3 ropes with forces of $^{+}10$, $^{+}12$, and $^{-}15$ acting on it.

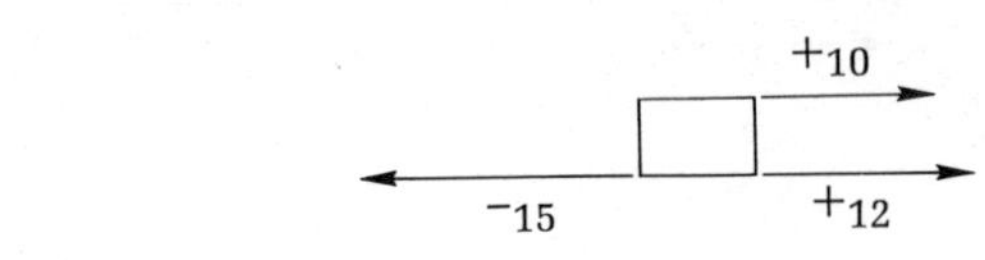

13. *Torques.* Torques, or twists, about an axis or axle can be in opposite directions. The figure shows a bar free to turn about a point 0. The one-pound force at A produces a positive, counterclockwise torque about 0 of $^{+}4$ pounds-feet. The one-pound force at B produces a negative, clockwise torque of $^{-}3$ pounds-feet.

21.3 STRUCTURAL DEFINITION OF NEGATIVE INTEGERS AS SOLUTIONS OF SUBTRACTION PROBLEMS Students soon learn that there is no whole number named $(3 - 5)$, or that satisfies $x + 5 = 3$. Similarly, there is no whole number y that satisfies $y + 4 = 0$. Many mathematicians would like to have solutions to such equations.

When mathematicians want something badly enough, they just say that they have it. They say, "Let us invent some new numbers which will extend the set of counting numbers so that all equations of the sort $x + 5 = 3$ will have solutions," and presto, vito—there you are. We can now solve any equation $x + a = b$, where a and b are whole numbers. Having given birth to these new numbers, we have to name them, look at their other properties, and see if we really want to keep them.

It is no surprise that the number satisfying $x + 4 = 0$ is named $^{-}4$, the same "opposite 4" we used earlier to label points on a line. Later we will justify this use of the same name in different places.

In general the solution to $y + N = 0$ is called ^{-}N. We soon agree that we want the solutions to $y + 5 = 1$ and $z + 7 = 3$ to be the same as the solution to $x + 4 = 0$, because we want to be able to subtract the same number from both sides of an equation and have the differences equal.

21.4 ADDITION AND SUBTRACTION OF INTEGERS USED TO NAME THINGS In section 21.2 we listed the following objects, which are often labeled with integers, sometimes positive, sometimes 0, sometimes negative: points, temperatures, elevations, directed values of sets whose members have directed values, giving-receiving, game counters, linear displacement (directed motion along a line), velocity, instants in time, directed time intervals, forces, and torques.

We lay the foundation for addition of integers in addition of counting numbers. We restrict ourselves to cases where our labels are counting numbers, and where something we do with the objects is helpfully described by addition.

Immediately we see that not all pairs can be put together in such a way that addition makes sense. We list three cases where addition is useful, two where it is not—and leave other possibilities to you for further exploration.

EXAMPLE 1 *Value of sets of objects with directed value.* Suppose we play a game with counters, using white and red beans. Each white bean is worth $^{+}1$; each red bean is worth $^{-}1$. In dish one are 12 white and 5 red beans; the dish contains value $^{+}7$. In dish two are 6 white and 2 red for a value of $^{+}4$. If all the beans are put in dish three, there are 18 white and 7 red with a value of $^{+}11$. We note that $^{+}7 + {}^{+}4 = {}^{+}11$.

Now clearly we could have dishes with more red than white beans, and hence a negative value. We can put all the beans in one dish. Hence it would seem reasonable to define addition of integers in a way that agrees with these values. Let us make a table showing several possibilities.

Dish one			Dish two			Dish three			Number Equation
White	Red	Value	White	Red	Value	White	Red	Value	
12	5	$^{+}7$	6	2	$^{+}4$	18	7	$^{+}11$	$^{+}7 + {}^{+}4 = {}^{+}11$
6	14	$^{-}8$	7	4	$^{+}3$	13	18	$^{-}5$	$^{-}8 + {}^{+}3 = {}^{-}5$
22	7	$^{+}15$	6	17	$^{-}11$	28	24	$^{+}4$	$^{+}15 + {}^{-}11 = {}^{+}4$
7	13	$^{-}6$	4	11	$^{-}7$	11	24	$^{-}13$	$^{-}6 + {}^{-}7 = {}^{-}13$

We note that the number equations are made by writing the equation $\square + \triangle = \bigcirc$, then blindly putting the value of dish one in $\square$, the value of dish two in $\triangle$, and the value of dish 3 in $\bigcirc$. Since the first such equation agrees with our previous agreements, and the other three were formed in ways very similar to the way the first was formed, we define addition of integers to agree with these four equations and others like them.

You will find it interesting to find other sums, and to make up a rule.

EXAMPLE 2 *Checks and duns.* Suppose Mrs. Jingle has a check for $12 and a dun for $5. What is her net worth? Suppose Mr. Jingle has a check for $6 and a dun for $2. What is his net worth? What is their combined net worth? What is the value of their checks? Their duns?

Similar check and dun stories can be devised to fit the other numbers in the above table, with Mrs. Jingle corresponding to dish one and Mr. Jingle to dish two. We get exactly the same number equations as in the bean illustration.

EXAMPLE 3 *Linear displacements.* Imagine a row of dancers, one at each point on a number line. The step "Move 5 places to the right" we call a $^{+}5$ displacement. The dancer at 8 goes to 13. The dancer at $^{-}12$ goes to $^{-}7$, and so on. Similarly, under a $^{+}3$ displacement the dancer at 6 goes to 9, the dancer at $^{-}1$ goes to $^{+}2$. Under a $^{-}6$ displacement, the dancer at 11 goes to 5, the dancer at 2 goes to $^{-}4$, the dancer at 0 goes to $^{-}6$.

In a dance, one step follows another. Is it clear that $^{+}5$ displacement followed by $^{+}3$ displacement has the same effect as $^{+}8$ displacement? We check a few dancers.

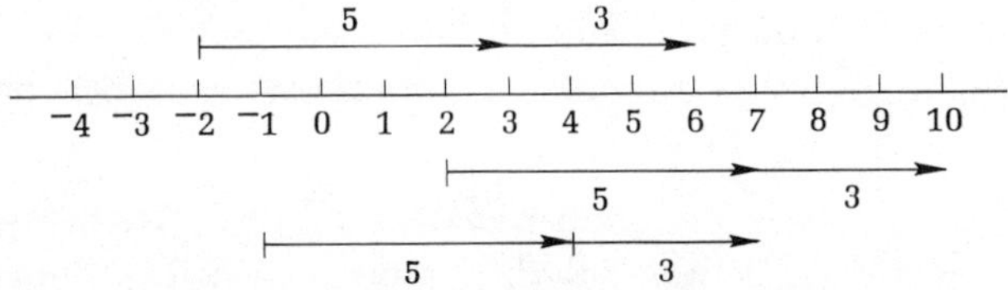

The dancer at $^{-}2$ went to 3 then to 6.

The dancer at 2 went to 7 then to 10.

The dancer at $^{-}1$ went to 4 then to 7.

In each case the dancer has moved 8 to the right. The displacements show the pattern 5, 3, 8, which is the addition pattern for positive numbers, $5 + 3 = 8$.

Since negative steps can follow other steps we ask the effect of $^-8$ displacement followed by $^+3$ displacement. We check a couple of dancers and see that all dancers will show the same results.

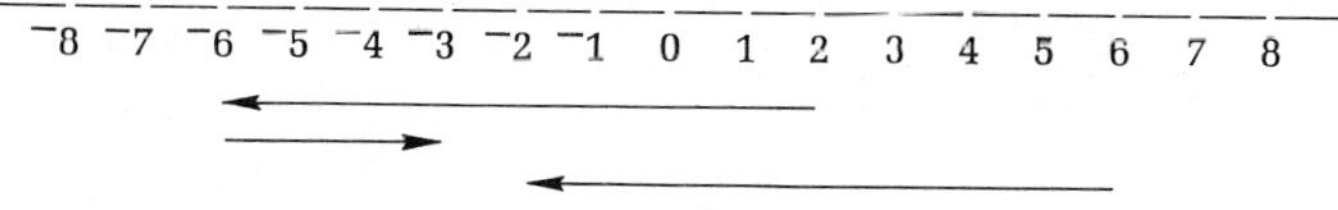

We see that the dancer at 2 goes to $^-6$ then to $^-3$, while the dancer at 6 goes to $^-2$ then to $^+1$. We see the effect of $^-8$ followed by $^+3$ is the same as a displacement of $^-5$. Since the steps follow the same pattern as for positive only steps, we say the names of the displacements show the addition pattern, and hence, $^-8 + {}^+3 = {}^-5$. We are quite pleased that this is the same result as in the two earlier examples. This reinforces us in our decision to define addition in this way.

By checking the steps of several dancers, you will find that displacements following each other yield the addition equations $^+15 + {}^-11 = {}^+4$ and $^-6 + {}^-7 = {}^-13$ as before.

That not all uses of integers support addition is shown by the next two examples. Consider two points labeled $^+6$ and $^+5$. There is nothing simple one would do with these two points that suggests the point $^+11$. Temperatures similarly do not support addition. What would you do with one dish of water at 14° and another dish at 5° to suggest 19°? We leave it as an exercise to show that, while there is nothing one can do with 2 points, one can combine a point and a displacement to yield a point in a way which shows addition and which furnishes a framework for extending addition of positive integers to addition of all integers.

21.5 SUBTRACTION Having addition, we get subtraction easily from the missing-addend attitude.

1. $^+23 - {}^-6 = \square$ can be read as "What number added to $^-6$ yields $^+23$?" and can also be written $^-6 + \square = {}^+23$. Recalling that $^+6 + {}^-6 = 0$, and using the associative property, we get

$$^+6 + (^-6 + \square) = {}^+6 + {}^+23$$
$$(^+6 + {}^-6) + \square = {}^+6 + {}^+23$$
$$\square = {}^+29$$

We leave it as an exercise to do similar subtractions with pairs other than $^{+}23$ and $^{-}6$. These lead to the rule "To subtract, change the sign of the subtrahend and add to the minuend."

The argument in this section is elegant, but some people will also want an argument based on objects. We turn to examples of these in the next section.

21.6 SUBTRACTION (CONTINUED) Sets with directed values, especially sets made of white (+) and red (−) beans, form useful bases for extending subtraction of positive integers to subtraction of all integers. Both take-away and missing-addend subtraction attitudes fit very nicely. Using missing addend, we will explore several examples to see how it applies to any numbers.

1. $^{+}12 - {}^{+}7 = ?$ What do we add to $^{+}7$ to get $^{+}12$? Let 7 white and 0 red be the set A of beans with value $^{+}7$. Recall that each white bean has value $^{+}1$ and each red bean has value $^{-}1$. Choose a set of value $^{+}12$ with more than 7 white and 0 red beans—22 white and 10 red will do nicely as set B. Clearly, we need to join a set C of 15 white and 10 red to set A to get set B. The value of $15w$ and $10r$ is $^{+}5$. (We abbreviate white and red.) The three values (12, 7, 5) show subtraction, $12 - 7 = 5$, as we learned it with counting numbers. We will use the same process with bean sets to define subtraction with integers.

2. $^{+}4 - {}^{+}13 = ?$ What do we add to $^{+}13$ to get $^{+}4$? Set A with value $^{+}13$ is $13w$, $0r$. A set B with more of each kind of bean and value $^{+}4$ is $14w$, $10r$. The set to be joined is $1w$, $10r$. This set has value $^{-}9$. So we say $^{+}4 - {}^{+}13 = {}^{-}9$.

Our next examples we give in a table (see top of opposite page), leaving it to the reader to give the words for the steps.

The number equations are found by putting the value for set C in the circle in the pattern

$$\square - \triangle = \bigcirc$$

where $\square$ is the value for set B and $\triangle$is the value for set A.

Example	Set A w	r	v	Set B w	r	v	Set C w	r	v	Number Equation
a) $^{+}12 - ^{+}7$	15	8	$^{+}7$	22	10	$^{+}12$	7	2	$^{+}5$	$^{+}12 - ^{+}7 = ^{+}5$
b) $^{+}4 - ^{+}13$	15	2	$^{+}13$	20	16	$^{+}4$	5	14	$^{-}9$	$^{+}4 - ^{+}13 = ^{-}9$
c) $^{+}6 - ^{-}2$	0	2	$^{-}2$	8	2	$^{+}6$	8	0	$^{+}8$	$^{+}6 - ^{-}2 = ^{+}8$
d) $^{-}4 - ^{+}5$	8	3	$^{+}5$	8	12	$^{-}4$	0	9	$^{-}9$	$^{-}4 - ^{+}5 = ^{-}9$
e) $^{-}3 - ^{-}8$	0	8	$^{-}8$	6	9	$^{-}3$	6	1	$^{+}5$	$^{-}3 - ^{-}8 = ^{+}5$
f) $^{-}12 - ^{-}9$	2	11	$^{-}9$	5	17	$^{-}12$	3	6	$^{-}3$	$^{-}12 - ^{-}9 = ^{-}3$

You may wish to see if you can find different, and perhaps simpler, sets for these examples. We look at similar examples from a take-away point of view.

g) $^{+}13 - ^{+}4 = ?$ A set D with value $^{+}13$ is 20*w*, 7r. A set E with value $^{+}4$ is 4*w*, 0r. If we take set E from D we get F: 16*w*, 7r with value $^{+}9$. The three values of the three sets are 13, 4, 9, showing subtraction $13 - 4 = 9$.

h) $^{-}4 - ^{-}9$. Some sets D with value $^{-}4$ are 0*w* 4r, 3*w* 7r, 10*w* 14r. We choose the third, because it has enough beans to take away a set E, 1*w* 10r, of value $^{-}9$. The set F that is left is 9*w* 4r of value $^{+}5$. Hence we say $^{-}4 - ^{-}9 = ^{+}5$. We make a table showing the values of sets F that are left when sets E are taken away from sets D.

Example	Set D w	r	v	Set E w	r	v	Set F w	r	v	Number Equation
g) $^{+}13 - ^{+}4$	20	7	$^{+}13$	4	0	$^{+}4$	16	7	$^{+}9$	$^{+}13 - ^{+}4 = ^{+}9$
h) $^{-}4 - ^{-}9$	10	14	$^{-}4$	1	10	$^{-}9$	9	4	$^{+}5$	$^{-}4 - ^{-}9 = ^{+}5$

It may be helpful to have a simple routine for generating the sets D and E. Consider $^{+}5 - ^{-}6$. Perhaps the simplest set of value $^{+}5$ is 5*w* 0r. But we cannot take 6r from a dish with no red beans. However, a set with 6*w* 6r has value zero, and can be joined to 5*w* 0r without changing the value. Now we have a set D, 11*w* 6r. We take away set E, 0*w* 6r, leaving set F, 11*w* 0r of value $^{+}11$. Hence, $^{+}5 - ^{-}6 = ^{+}11$.

Some people like to represent this last operation in numerals as follows:

$$\begin{aligned} {}^{+}5 - {}^{-}6 &= {}^{+}5 + ({}^{+}6 + {}^{-}6) - {}^{-}6 \\ &= {}^{+}5 + {}^{+}6 + {}^{-}6 - {}^{-}6 \\ &= {}^{+}5 + {}^{+}6 \\ &= {}^{+}11 \end{aligned}$$

Do you see where we have used ${}^{+}6 + {}^{-}6 = 0$ and ${}^{-}6 - {}^{-}6 = 0$?

Similarly

$$\begin{aligned} {}^{-}4 - {}^{+}2 &= {}^{-}4 + {}^{-}2 + {}^{+}2 - {}^{+}2 \\ &= {}^{-}4 + {}^{-}2 \\ &= {}^{-}6 \end{aligned}$$

Do you see where we used ${}^{-}2 + {}^{+}2 = 0$ and ${}^{+}2 - {}^{+}2 = 0$?

All these examples lead to the rule "When subtracting signed numbers, change the sign of the subtrahend (second number) and add to the minuend (first number)."

When signed numbers are represented as displacements, subtraction goes smoothly but care is necessary to make sure that instructions are correctly interpreted. Consider $5 - {}^{-}3$. To take away a ${}^{-}3$ displacement from a ${}^{+}5$ displacement does not make much sense. Missing addend is better. "What displacement joined to a ${}^{-}3$ displacement equals a ${}^{+}5$ displacement" is clear. The arrow diagram shows that ${}^{+}8$ is the answer.

$^{-}3$ $^{+}5$ $^{+}8$

Take away can be used with certain preliminaries. Let ${}^{+}5 = {}^{+}5 + {}^{+}3 + {}^{-}3$. If ${}^{-}3$ is taken away, this leaves ${}^{+}5 + {}^{+}3 = {}^{+}8$. The arrow diagram illustrates.

$^{+}5$ $^{+}3$ $^{-}3$

21.7 SUMMARY Integers, or positive and negative whole numbers and zero, describe directed situations. Some lead easily to addition and subtraction, while others do not. It is sad that such common examples as points on a number line, elevations above and below sea level, and temperatures above and below zero do not lead to addition and subtraction. Perhaps the best examples are values of beans or other objects, and displacements. These lead to rules that are easy to phrase for adding and subtracting signed numbers.

EXERCISES

1. Plot the following numbers on a number line: $^{+}3$, $^{-}2$, $^{+}\frac{7}{3}$, $^{-}1$.

2. Form 2 sets of beans that show $^{+}4$ with one set having 7 or more beans in it.

3. Form 2 sets of beans that show $^{-}3$ with one set having 7 or more beans in it.

4. Describe translations (commands to a chorus line) that show $^{+}5$, $^{-}3$, $^{-}6$, 0.

5. Where do the translations in exercise 4 move a gnome at point $^{+}17$?

6. Where do the translations in exercise 4 move a beauty queen at $^{-}8$?

7. Where do the translations in exercise 4 move a scholar at $^{-}2$?

8. Form and arrange sets of beans to show $^{+}5 + {}^{+}6$.

9. As in exercise 8, form and arrange sets of beans to show $^{+}4 + {}^{-}3$. Why does your demonstration show addition and not subtraction?

10. As in exercise 8, use beans to show $^{+}3 + {}^{-}7$.

11. Form and arrange sets of beans to show $^{+}5 - {}^{+}8$. Why does your demonstration show subtraction? What subtraction attitude do you use?

12. As in exercise 11, use beans to show $^{-}5 - {}^{-}8$, using a take-away attitude.

13. Show the number relation in exercise 12 using a missing-addend attitude with beans.

14. What would you do with two translations, P and Q, which you would call "adding P and Q"?

15. Interpret the following as "adding translations":
(*a*) $^{+}5 + {}^{+}7$ (*c*) $^{-}3 + {}^{-}6$
(*b*) $^{+}5 + {}^{-}6$ (*d*) $^{-}3 + {}^{+}7$

16. What would you do with two translations, S and T, which you would call "subtracting S from T"?

17. Interpret the following as "subtracting translations" from the missing-addend attitude:
(*a*) $^{+}7 - {}^{+}5$ (*b*) $^{-}6 - {}^{-}3$

18. Interpret as "subtracting translations" from the take-away attitude:
(*a*) $^{+}4 - {}^{+}9$ (*b*) $^{-}4 - {}^{-}6$

19. Make up several rules for adding and subtracting signed numbers.

20. Suppose Winifred lives on floor $^{+}5$ and Xavia lives on floor $^{-}3$. Can you do anything with W's floor and X's floor to show $5 + {}^{-}3$?

21. On October 1 Amy received checks from people who owed her money in amounts of \$5, \$8, \$3, and \$10 and bills she will have to pay in amounts of \$4, \$6, \$2, \$12.

(*a*) What is her net worth?
(*b*) Michelle gives Amy a check for \$7. What change does this make in Amy's net worth?
(*c*) Lisa brings her a bill amounting to \$9. What change does this make in her net worth?
(*d*) John takes the bill for \$6 from her. What change does this make in her net worth?

22. Make a mathematical sentence like $^{+}5 - {}^{-}3 = 8$ for each of the four short stories in exercise 21.

23. In explaining the rules you made up in exercise 19, is it easier to illustrate with beans, translations, or checks and bills?

24. Compute:
(*a*) $^{+}378 + {}^{-}492$; (*b*) $^{-}251 - {}^{-}348$

Answers to selected questions: **3.** 1*W*4*R*, 5*W*8*R*; **4.** move 5 steps to right; 3 steps to left; 6 steps to left; stand; **8.** 5*W*0*R*, 7*W*1*R* union 12*W*1*R*; **10.** 4*W*1*R* unite with 2*W*9*R*; **12.** from 6*W*11*R* take 2*W*10*R*; **15.** (*a*) 5 steps to right followed by 7 steps to right; (*d*) 3 steps to left followed by 7 steps to right; **17.** (*a*) follow 5 to right with 2 to right to make 7 to right; (*b*) follow 3 to left with 3 to left to make 6 to left; **19.** *Example:* To add numbers with unlike signs, consider the numerals without signs. Subtract the smaller number from the larger and give the answer the sign of the larger. To subtract, change the sign of the number subtracted and add; **21.** (*a*) +\$2; (*b*) adds \$7 to it; (*c*) subtracts \$9 from it; (*d*) adds \$6 to it; **24.** (*a*) $^{-}114$; (*b*) $^{+}97$.

22

Multiplication of Signed Numbers

22.1 INTRODUCTION New as sunshine yet so old that the folk memory of man runneth not to the contrary is the problem of why a minus times a minus is a plus. The problem is especially vital for students just learning about signed numbers, and for their teachers. Other scholars have learned that this question, like a woman's age, is never asked in polite society.

Every book on arithmetic or beginning algebra has some discussion with one or two approaches. It appears that some people have quite strong feelings about the appropriateness, or "correctness," of one approach as contrasted with another. This chapter will give many approaches, with the hope that some of them will illumine the problem for the student, and give fresh ammunition to the teacher. We start with a statement of attitude, then give answers based on these approaches. These approaches are through the CAD properties, or the algebraic structure so dear to the hearts of all mathematicians, through concrete examples to soothe the physically minded, and through construction, beloved of the stratospherists. Each approach is a process of generalization.

22.2 THE ATTITUDE Our attitude is that a negative number times a negative number is a positive number because we define it that way. We define it that way because we want it that way. We could just as well define the product of two negative numbers to be zero. We don't, because nobody wants that definition.

This attitude—that we defined negative numbers and that we therefore have every right to define their products as we please—doesn't really get us anywhere. It only begs the question and transfers it to a deeper level. The deeper question is "Why do people *want* to define multiplication the way we do?" This shows that the entire question is one of taste, and not of logic founded on negative numbers. (We will, of course, use logic in our explorations.) Taste shows itself as a generalization process. We want multiplication of signed numbers to be an appropriate generalization of previous experience.

22.3 STRUCTURE OR PATTERN APPROACH This is the approach of the lawyer's *stare decisus,* or "What was good enough for grandpa is good enough for me." Grandpa, in this case, is the counting numbers, 1, 2, 3, 4, ... We say that we are so in love with the multiplication patterns of the counting numbers, that we want these same patterns to apply to the negatives.

That is fine, we are told, but what patterns are we talking about? Is there an 1890 *Vogue* around to tell us? How do we describe the pattern for the positives so that it will also apply to the negatives? How can multiplication of positives be extended to the negatives?

Our first description comes from an array. Think of a rectangular grid with an x,y coordinate system. We'll restrict ourselves at first to integers. At each grid point in the first quadrant we write the product xy. The resulting layout lies below, where we've included the products on the axes as well. We've not written in the individual values of the factors x,y, leaving that for the imagination.

$$\begin{array}{cccccccc}
 & & & \uparrow & & & \\
0 & 5 & 10 & 15 & 20 & 25 & \\
0 & 4 & 8 & 12 & 16 & 20 & \\
0 & 3 & 6 & 9 & 12 & 15 & \rightarrow \\
0 & 2 & 4 & 6 & 8 & 10 & \\
0 & 1 & 2 & 3 & 4 & 5 & \\
0 & 0 & 0 & 0 & 0 & 0 &
\end{array}$$

As we look at this array of products, we see arithmetic progressions all over. For example, the products xy for $y = 2$ form the progression 0 2 4 6 8 10 . . . . These progressions can be extended backward as well. What better pattern to follow! We extend these progressions to the left and down:

			↑			0			↑			
	−20	−16	−12	−8	−4	0	4	8	12	16	20	
←	−15	−12	−9	−6	−3	0	3	6	9	12	15	
	−10	−8	−6	−4	−2	0	2	4	6	8	10	→
	−5	−4	−3	−2	−1	0	1	2	3	4	5	
	0	0	0	0	0	0	0	0	0	0	0	
						0	−1	−2	−3	−4	−5	
						0	−2	−4	−6	−8	−10	
						0	−3	−6	−9	−12	−15	→
						0	−4	−8	−12	−16	−20	
						0			↓			

Many of our pattern-minded people feel quite happy in defining $(-a)(b)$ and $(a)(-b)$ so that they fit these arithmetic progressions.

The lower left quadrant is still empty. Do we get any suggestions for the product of two negatives? Again we see interesting arithmetic progressions, such as the row for $y = -3: \{0, {}^-3, {}^-6, {}^-9, {}^-12, {}^-15, \ldots\}$. Clearly this row can be extended to the left, adding 3 each time.

	0	0	0	0	0	0	0	0	0	0	0	
	5	4	3	2	1	0	−1	−2	−3	−4	−5	
	10	8	6	4	2	0	−2	−4	−6	−8	−10	
←	15	12	9	6	3	0	−3	−6	−9	−12	−15	→
	20	16	12	8	4	0	−4	−8	−12	−16	−20	
			↓			0			↓			

It is now possible to define $(-a)(-b)$ to fit these extended arithmetic progressions. But our Civil Liberties and Fair Play for Numbers friends object. They claim we should not give special parental rights to the numbers in the lower right quadrant, but should also define the products $(-a)(-b)$ by extending the progression patterns of the upper left quadrant downward.

We are a little disgruntled and reluctant to do this. After all, we want to keep things on the level. Further, we are just a little afraid that if the progressions in the upper left quadrant are extended downward we'll get another set of numbers below. Who wants *two* products for $(-2)\,(-4)$? Let's just extend one way, and be happy with what we have.

Old Judge Symmetry raises his head and says we must look at both extensions and put up with all results.

What is this! Mirabile dictu! We see that extending the upper left quadrant down gives exactly the same numbers in the lower left quadrant as extending the lower right quadrant sideways.

This delightful agreement convinces our array thinkers that there must be something sacred about the whole thing. Hence they plump for defining multiplication by this array chart. Sobeit.

We leave it as an exercise for the reader to draw the array of products in all four quadrants and to verify the various arithmetic progressions.

22.4 THE CAD PATTERN "That array business is just a geometric superstition. A hocus pocus fit only for dullards," sneer some of our algebraic-minded friends. "Arithmetic shouldn't depend on geometry for its inspiration," they say. "Furthermore, how do you know those progressions will extend in just that pattern? When they extend into negatives, maybe the progressions will be different." Such knot-headed nit-picking exasperates our array thinkers, but they politely listen when the algebraist continues, "The only appropriate way to describe a pattern in arithmetic is by an algebraic pattern, or formula. You really ought to base your definitions on the CAD, or commutative, associative, and distributive, properties." We follow his advice next.

We recall that the commutative properties are expressed $A + B = B + A$ and $C \cdot D = D \cdot C$. The associative properties are $(E + F) + G = E + (F + G)$ and $(H \cdot I) \cdot J = H \cdot (I \cdot J)$. The distributive property is $K(L + M) = KL + KM$. Whole numbers with ordinary addition and multiplication have these properties, or satisfy these laws. Do the negative integers?

When negative numbers are defined, they are usually defined only in terms of opposites. $^-5$ is opposite $^+5$, and represents a point on the number line opposite the point $^+5$ on the other side of 0. In terms of addition $^-5 + 5 = 0$. In words, opposite 5, or negative 5, or $^-5$ is defined to be a number which added to 5 gives 0 as the sum. In our role as algebraists, we have waved our magic wand, and presto, there is a new number. $^-5$ is its name. $^-5 + 5 = 0$ is its property. Anything which added to 5 gives 0 we will call $^-5$, or negative 5, or minus 5.

This definition says nothing about addition to numbers other than their opposites, and nothing about multiplication. However, we decide we love the CAD properties so much and we think they describe the patterns in the positive integers so well that we wish to define addition and multiplication so that the negatives have the CAD properties. Maybe we'll have further freedom of definition even after this requirement.

The following chain of steps shows how we use CAD to define sums. It helps to use a bit of judicious guessing and then to verify that the guess was correct.

Is $(^-3) + (^-4) = (^-7)$?

Since anything that gives zero when added to 7 is called $^-7$, we see what happens.

$(^-3) + (^-4) + 7 = ?$	
$= (^-3) + \{(^-4) + 7\}$	associative property
$= (^-3) + \{(^-4) + (4 + 3)\}$	renaming
$= (^-3) + \{[(^-4) + 4] + 3\}$	associative property
$= (^-3) + \{0 + 3\}$	definition of $^-4$
$= (^-3) + 3$	property of 0
$= 0$	defining property of $^-3$

Hence we see that $(^-3) + (^-4)$ and $^-7$ are two names for the same number, and are equal. Actually the only structure property we used was the associative property and not the CD properties.

Could we have defined $(^-3) + (^-4)$ as anything other than $^-7$? Our argument shows that if we want the associative property to hold, then we could not. Can you show that $^-2 + {^-5}$ must be $^-7$? How about $^-3 + 8 = ?$ $^-8 + 2 = ?$ In each case, use only the associative property, the defining property of a negative number, renaming the sum of two positive integers or zero, as in the example.

As we look to multiplication, we will discuss only specific numbers, but in a manner that extends to any numbers. The reader should repeat the discussion with other pairs of integers.

To get a suggestion for defining $0(^-3)$ we consider

$$\begin{array}{rl} 0 \cdot 0 = 0 & \text{accepted property of zero} \\ 0(^-3 + ^+3) = 0 & \text{substitution} \\ 0(^-3) + 0(^+3) = 0 & \text{distributive property} \\ 0(^-3) + 0 = 0 & \text{0 times a positive number} \\ 0(^-3) = 0 & \text{0 is the identity for addition} \end{array}$$

Hence we define $0 \cdot a = 0 = a \cdot 0$ for all integers a so that the CAD properties will hold.

For a lead to defining $(^-4)5$ so that the CAD properties will hold, consider

$$\begin{array}{rl} 0 \cdot 5 = 0 & \text{accepted property of zero} \\ (^+4 + ^-4) \cdot 5 = 0 & \text{substitution} \\ ^+4 \cdot 5 + (^-4)5 = 0 & \text{distributive property} \\ 20 + (^-4)5 = 0 & \text{renaming} \\ (^-4)5 = ^-20 & \text{definition of opposite 20} \end{array}$$

Hence our CAD properties insist we define $(^-4)5 = {}^-20 = 5(^-4)$.

With these warmups, we tackle old Fearsome himself, $(^-3)(^-7)$:

$$\begin{array}{rl} ^-3 \cdot (7 + ^-7) = 0 & \text{properties of opposite and zero} \\ ^-3 \cdot 7 + ^-3 \cdot ^-7 = 0 & \text{distributive} \\ ^-21 + ^-3 \cdot ^-7 = 0 & \text{rename} \\ ^-3 \cdot ^-7 = 21 & \text{definition of opposite} \end{array}$$

Here is another pattern using CAD properties:

$$\begin{array}{rl} (^-3)(^-7) + (^-3)7 + 3 \cdot 7 = (^-3)(^-7) + (^-3)7 + 3 \cdot 7 & \text{identity} \\ (^-3)(^-7) + (^-3 + 3)7 = {}^-3(^-7 + 7) + 3 \cdot 7 & \text{associative–distributive} \\ (^-3)(^-7) = 3 \cdot 7 & \text{properties of opposites and zero} \end{array}$$

So we define $(^-3)(^-7)$ to be 21.

Since these proofs all require using some preplanned pattern, some people like to suggest a product $(^-1)(^-1) = {}^+1$ by one of these methods, then reduce other products to this one. For example:

$$\begin{aligned}(^-3)(^-7) &= (^-1)3(^-1)7 \\ &= (^-1)(^-1)3 \cdot 7 \\ &= 1 \cdot 3 \cdot 7 \\ &= 3 \cdot 7\end{aligned}$$

Can you see where the commutative property was used here?

22.5 CONCRETE OR PHYSICAL APPROACHES Usually the teacher or engineer who has come this far is horrified. "You are destroying my faith," he says. "I want to build radios, drive cars, save money. I've used numbers to describe concrete events because I thought numbers came from concrete events. Now you tell me that numbers and their operations are but shadows on the human mind, to be defined by the mathematician's whim.

"This is completely unsatisfactory. I need concepts I can rely on to actually apply to concrete events. No airplane flies on Monday, Wednesday, and Friday because a mathematician happens to be thinking the right thoughts, and falls on Tuesday and Thursday because he has changed his mind.

"I don't want to keep you algebraists from playing your little parlor games, but I want an honest, a genuine mathematics."

When we press him more closely for what he means, he says, "I believe that numbers are used to describe concrete events. The properties of numbers are descriptions of the properties of the concrete objects or events the numbers describe. They are not just figments of the imagination.

"Indeed, this is what all this so-called modern mathematics is all about. Counting numbers are defined in terms of sets. Addition of counting numbers is defined in terms of unions of disjoint sets. Multiplication is defined in terms of arrays, Cartesian products, trees, and segments. We accept the laws because they describe physical situations. The diagram

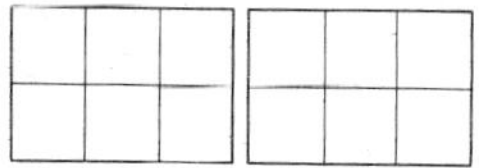

leads us to the distributive law $2(3 + 4) = 2 \cdot 3 + 2 \cdot 4$. And so on. It is wrong to shift from a good, reliable, concrete approach to pure mental gymnastics. No sane person will want to risk his life on that."

Before our engineering friend blows his cool completely and goes back to the stone age, we consider his approach. We review some of the common physical objects or situations that are named or described by numbers. We'll include both positive and negative integers to get the full picture in front of us.

1. *Sets.* The number of elements in a set is described by a positive integer. We do not have sets with a negative number of members. Many of the properties of counting numbers are introduced and discussed in terms of properties of sets, as the engineer said. The failure of sets to extend to negatives is quite a blow.

2. *Points on a number line.* Negatives appear naturally here.

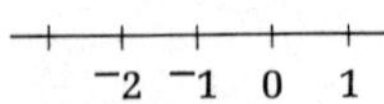

3. *Temperature.* Degrees above and below zero.

4. *Elevations above and below sea level.*

5. *Instants in time.* If dinner is served at 0, the salad is made at $^-180$, the roast put on at $^-90$, the spinach put on at $^-4$, grace said at $^+2$, and dessert served at $^+55$.

The last four examples, 2 through 5, are the ones usually used to introduce negative numbers. They all involve a linearly ordered set. Something is arbitrarily chosen as zero. One direction is chosen as positive, a unit is chosen, and the numbers are assigned.

Unfortunately none of these examples serve our engineer's needs as the basis for multiplication. There is nothing that happens to points, or weather, or instants of time that leads to anything like multiplication.

Of course, we can move about on the number line according to number patterns we have in mind. But that is not what the engineer wants. He wants the physical, concrete things to be done first, and then he wants their description to lead to multiplication. In the multiplication of counting numbers, for example, he sees an array:

× × × ×
× × × ×
× × × ×

He counts the rows, 3; the columns, 4; and the number in the array, 12. He then names the relation between the three as multiplication: $3 \times 4 = 12$. We see no such relation in examples 2 through 5 above.

We look deeper.

Instead of looking for things named with negative numbers, to see if they support multiplication, let us look for places where multiplication is used with counting numbers, and see if the operation makes sense with negatives. First we will describe several models of $p \cdot q = r$ for positive factors. In each case, we'll do 4×6, $^-3 \times {}^-5$ and $^-2 \times 7$, leaving other examples for exercises.

22.6 RECEIVING BLUE CHIPS Suppose blue chips are worth $^+1$ and red chips are worth $^-1$ each. We regard receiving as positive and paying as negative. Suppose in each case we start with 100 blue and 60 red chips. Our net worth is 40.

Let p be the number of stacks of chips we receive.

Let q be the number of blue chips in each stack.

Let r be the increase in our net worth.

If p is positive (negative) we receive (pay out) the stacks.

If q is positive (negative) the chips are blue (red).

If r is positive (negative) our net worth has gone up (down).

We receive 4 stacks of 6 blue chips. We now have 124 blue chips, 60 red. Our net worth is 64, an increase of 24. The number triple (4, 6, 24) is written $4 \times 6 = 24$.

We pay out 3 stacks of 5 red chips each. We now have 100 blue and 45 red chips. Our net worth is 55, an increase of 15. The number triple ($^-3$, $^-5$, 15) is written $(^-3) \times (^-5) = {}^+15$.

We pay out 2 stacks of 7 blue chips each. We now have 86 blue and

60 red chips. Our net worth is 26, a decrease of 14. The number triple $(^-2, 7, ^-14)$ is written $(^-2) \times {}^+7 = {}^-14$.

22.7 RECEIVING CHECKS AND DUNS We assume our friend, X, has a money box. In it he has a number of checks of various sizes that he can deposit or give out like money. He also has some duns that have to be paid. If he gives out a dun he doesn't have to pay it. He starts with a net value of $100 in the box. Numbers are assigned as follows:

(p, q, r) are (the number of checks received, the value of each check, the increase in X's worth), respectively.

p is positive (negative) $\Leftrightarrow$ X receives (gives out) the checks.

q is positive (negative) $\Leftrightarrow$ the check is actually a check (dun).

r is positive (negative) $\Leftrightarrow$ X's net worth goes up (down).

X receives 4 checks of $6 each. His worth goes up $24. The number triple is (4, 6, 24), written $4 \times 6 = 24$.

X gives out 3 duns of $5 each. His worth goes up $15. The number triple is $(^-3, ^-5, 15)$, written $^-3 = {}^-5 = 15$.

X gives out 2 checks of $7 each. His worth goes down $14. The number triple is $(^-2, 7, ^-14)$, written $^-2 \times 7 = {}^-14$.

228 RATE × TIME = QUANTITY Suppose we have a water tank with water "running in" for a period of time. How much more is in the tank at the end of the period than at the start? We suppose the tank contains 100 gallons at, say, 1:20 P.M., the start. We suppose that p is the number of gallons per minute running into the tank, q is the directed length of the period of time, and r is the number of gallons more in the tank at the end of the period than at the start. We recall that the time interval may be negative with the end of the period before the start.

p is positive (negative) $\Leftrightarrow$ flow is into (out of) the tank.

q is positive (negative) $\Leftrightarrow$ the end of the period is after (before) the start.

r is positive (negative) $\Leftrightarrow$ there is more (less) in the tank at the end than at the start.

4 gallons per minute run into the tank. How much more is in the tank at 1:26 P.M., 6 minutes after the start at 1:20 P.M.? The number triple is (4, 6, 24), written $4 \times 6 = 24$.

3 gallons per minute run out of the tank. How much more is in the tank at 1:15 P.M., 5 minutes before the start at 1:20 P.M.? In the 5 minutes, 15 gallons have run out. Since there were 100 gallons at 1:20 P.M., there must have been 115 gallons in the tank at 1:15 P.M. Hence there were 15 more gallons in the tank at 1:15 P.M., the end of the period, than at 1:20 P.M., the start. The triple of numbers is $(^-3, ^-5, 15)$, written $^-3 \times {}^-5 = 15$.

Some people like a box layout to keep the numbers straight. Here is a possible form:

	Rate	**Time**	**Tank**
End	╳	1.26	124
Start	╳	1.20	100
Interval	$^+4$	$^+6$	$^+24$

The form has three rows and three columns, labeled as shown. In the first, or rate, column nothing goes in the first two rows, and the rate of flow into the tank goes in the third row. In the second, or time, column the first row has the time of the end of the period, the second row has the time of the start of the period, and the third row has the directed time interval. This figure is row 1 minus row 2. In column 3, or tank column, the quantities of water in the tank at the end and at the start of the period go in rows 1 and 2, respectively. The amount by which row 1 exceeds row 2 goes in row 3.

This is the form for the second illustration:

	Rank	**Time**	**Tank**
End	╳	1:15	115
Start	╳	1:20	100
Interval	$^-3$	$^-5$	15

We see that the number triple appears in the third row.

If water flows out at 2 gallons per minute, how much more is in the tank at 1:27 P.M. than at 1:20 P.M.? 14 gallons have flowed out. The number triple is ($^-2$, 7, $^-14$), written $^-2 \times 7 = {}^-14$. The appropriate box form is as follows:

	Rate	Time	Tank
End		1:27	86
Start		1:20	100
Interval	$^-2$	7	$^-14$

Some people like to present this rate × time = quantity example with a film. A film can be run backward to show a negative time interval. Suppose the film shows water running out for 7 minutes at 2 gallons per minute. At the end there are 14 gallons less in the tank. Now if the film is run backward, there will be 14 gallons more in the tank at the end of the showing than at the beginning, leading to $^-2 \times {}^-7 = 14$.

22.9 VELOCITY × TIME = DISTANCE Suppose we have a car moving back and forth along a road. If it moves with a velocity p for a directed time interval q, what will be its displacement r? We recall that a car moving with constant speed of 40 miles per hour for 3 hours goes 120 miles. Velocity is directed speed. We choose one direction (to the right) as positive. Velocity and displacement in this direction are positive. We suppose there are mile posts along the road that will tell us where the car is at any instant. We assume the car starts at 1 P.M. at mile post 200.

Numbers are assigned with the following meaning:

p is positive (negative) ⇔ the car travels in the positive (negative) direction.

q is positive (negative) ⇔ the end of the period is after (before) the start.

r is positive (negative) ⇔ the car at the end of the period is on the positive (negative) side of its position at the start of the period.

Suppose the car goes 4 miles per hour in a positive direction. Where is it at 7:00 P.M.? What is the displacement in this period?

mileposts 200 224
1 P.M. start → velocity 7 P.M. end

We can make a box form:

	Velocity	Time	Place
End		7 P.M.	224
Start		1 P.M.	200
Interval	$^{+}4$	$^{+}6$	$^{+}24$

Suppose the car goes 3 miles per hour in the negative direction. Where is it 5 hours earlier, at 8 A.M.? Since it traveled a distance of 15 miles, it must have been at milepost 215. We make a box form:

	Velocity	Time	Place
End		8 A.M.	215
Start		1 P.M.	200
Interval	$^{-}3$	$^{-}5$	15

The number triple is ($^{-}3$, $^{-}5$, 15), which we write $(^{-}3) \times (^{-}5) = 15$.

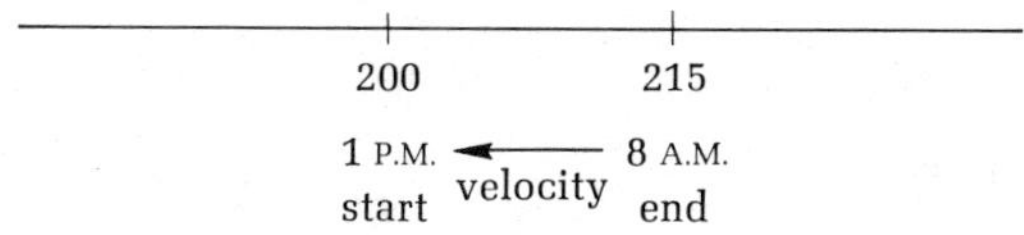

Suppose the car goes 2 miles per hour in a negative direction. What is the displacement at 8 P.M., 7 hours later? We make a box form:

	Velocity	Time	Place
End		8 P.M.	186
Start		1 P.M.	200
Interval	$^{-}2$	7	$^{-}14$

place 186 200

time 8 P.M. ← velocity 1 P.M.

end start

The number triple is $(^{-}2, 7, ^{-}14)$, written $^{-}2 \times 7 = {}^{-}14$.

22.10 SIMILAR TRIANGLES The Greek geometers thought of numbers as line segments. They developed multiplication using similar triangles. This awkward way of finding a product was one of the reasons Greek and later sciences did not advance very rapidly for over two thousand years. We show by example how they could find the product $c = a \times b$, where a and b are positive.

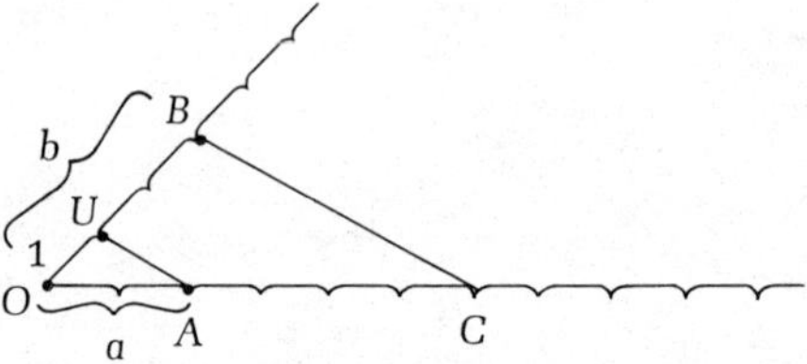

1. Locate points O,U,A,B on two intersecting lines as shown, with $OU = 1$, $OA = a$, $OB = b$.
2. Draw UA, and $BC \parallel UA$, with C on line OA.
3. Then $OC = c = a \times b$.

The figure shows the example $2 \times 3 = 6$. Can you draw the diagram for 2×4?

The method generalizes to negative numbers by extending the lines.

1. Draw two axes intersecting at O, meeting at any angle. For convenience, we call them x and y axes.

2. Locate points U and Q on the x axis and P on the y axis so that $OU = 1$, $OP = p$, $OQ = q$.

3. Draw UP.

4. Parallel to UP draw QR, with R on the y axis. Call $OR = r$.

5. The triple of numbers (p, q, r) we write as $p \times q = r$. The figure shows $^{-}2 \times {}^{-}3 = 6$.

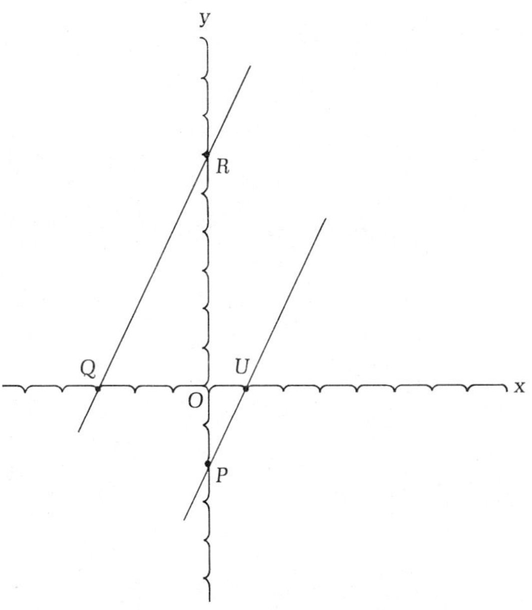

Similarity of $\triangle$s OUP and OQR shows that $\frac{OR}{OQ} = \frac{OP}{OU}$, or $\frac{c}{b} = \frac{a}{1}$, or $c = ab$, where a,b,c are the (positive) lengths of the sides. We determine the sign relations as follows. If O lies between U and Q, then it also lies between P and R. That means that, if q is negative, r will have a sign different from p. If O is not between U and Q, then P and R are on the same side of O, on the y axis. Hence if q is positive, p and r have the same sign.

22.11 TORQUES Physical observation tells us that if a weight (force) is placed on a rod balanced on a knife edge, the rod will turn about the knife edge. This tendency to turn is called torque. The size of the torque is given by weight $\times$ distance.

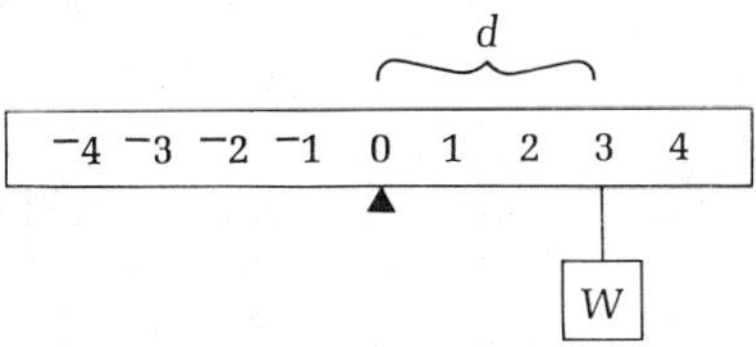

The torque generated by a weight (force) of 8 ounces at point 3 has the same effect as the torque generated by a force of 6 ounces at point 4.

This multiplication, force $\times$ distance $=$ torque, generalizes to negative numbers. Negative forces, or up forces, can be produced by helium-filled balloons. Negative distances are shown by placing the weights at the negative numbers on the scale. While it might appear that the actual apparatus is necessary, we may explore the action mentally. We use the fact that force × distance = torque, and keep the signs correct from the geometry.

A weight of 4 grams placed at 6 on the rod will have the same effect as a weight of 24 at 1 on the scale. The number triple (4, 6, 24) we write as 4 × 6 = 24.

A balloon lifting 3 grams is placed at ⁻5 on the rod. The effect is the same as placing a 15 gram weight at 1 on the rod. The number triple (⁻3, ⁻5, 15) is written as ⁻3 × ⁻5 = 15.

A balloon lifting 2 grams is placed at 7 on the rod. The effect is the same as placing a balloon lifting 14 grams at 1. The number triple (⁻2, 7, ⁻14) is written ⁻2 × 7 = ⁻14.

COMMENT We assume that the rod is level and that the forces act perpendicularly to the rod.

Our engineer is happier. These relations between physical objects that showed multiplication of positive numbers also made sense when applied to objects described by negative numbers. Since the numbers corresponding to the products satisfy the rules by the algebraist, he is more confident that the multiplication table can be relied upon.

The engineer realizes, though, that the fitting of the multiplication rules may be just luck. Why should forces, distances, velocities, and times follow a multiplication table written by a person in love with the CAD properties?

There is no supernatural reason they should, for if these relations between physical objects are examined closely, it will be seen that they satisfy the CAD properties. Hence their measures and the number operations also satisfy the CAD properties. It follows that any property logically proved by CAD also holds for the physical objects. We shall not explore this further, but will merely be grateful that the several reasonable extensions of multiplication to negative numbers all yield the same multiplication table.

22.12 EXISTENCE QUESTIONS Some readers may be surprised by such an extensive use of individual whim as appears in these discussions. They may say that they have confidence in the counting numbers, but are not so sure there are things such as negative numbers.

As with fractions, negative numbers can be constructed from positive numbers by the use of sets of ordered pairs. Suggested by the relations such as $^{-}5 = (4 - 9) = (3 - 8)$, we define an integer as a set of ordered pairs of counting numbers (a,b). Two pairs $(a,b),(c,d)$ are in the same set if and only if $a + d = b + c$. We define not only the integers but also operations called addition and multiplication in terms of these operations in counting numbers.

$$(a,b) + (c,d) = (a + c, b + d)$$
$$(a,b) \cdot (c,d) = (ac + bd, ad + bc)$$

The counting numbers are identified in the integers by identifications such as $5 = \{(10,5),(9,4),(23,18),(6,1),(7,2),(8,3),\ldots\}$.

Integers defined in this way have all the properties we have so far asked them to have. Hence we can feel confident that negative numbers and integers have just as firm a foundation as the counting numbers.

The examples in this text have been restricted to integers. We do, of course, wish to operate with signed real numbers. Some of the examples extend to nonintegral real numbers; others do not. In the cases that do extend, the multiplication table that develops is the standard one, as follows:

Suppose a,b are positive real numbers with product c. Then

$$(^{-}a)b = a(^{-}b) = {}^{-}c$$
$$(^{-}a)(^{-}b) = c$$

and

$$0 \cdot a = 0 \cdot (^{-}a) = 0$$

22.13 SUMMARY We have seen that multiplication may be generalized to both positive and negative numbers from many points of view. Fortunately, all the interesting attitudes yield the same generalization. Hence we may confidently follow patterns, use CAD rules, physical events, or social structures to illustrate why we define multiplication as we do. The definitions are

$$a \times b = ab, \qquad (-a)b = a(-b) = -ab, \qquad \text{and } (-a)(-b) = ab$$

EXERCISES

1. Draw all four quadrants of a coordinate grid out to ($^{\pm}4$, $^{\pm}4$) showing the product xy at each grid point. Identify the arithmetic progressions going through (2, 3); through ($^-4$, $^-1$); through ($^-2$, 3).

2. Which of the following best describes why a negative number times a negative number is a positive number?

(*a*) It has been proved to be so.
(*b*) Teachers and mathematicians have said so.
(*c*) The laws of nature require it.
(*d*) It is more convenient that way.
(*e*) People define it that way.
(*f*) It follows from the laws of counting numbers.
(*g*) It is one of the properties of numbers.
(*h*) The laws of physics say it is so.

3. Asuming you want negative numbers to satisfy the CAD rules as well as their defining properties, such as $^-2 + {}^+2 = 0$, find the following: (HINT: What would you add to make 0?)

(*a*) $^-2 + {}^-6 =$ (*b*) $^-7 + {}^+2 =$ (*c*) $^+6 + {}^-5 =$

How did you use CAD properties?

4. While $0 \cdot x = 0$ for $x \geq 0$, show why CAD properties $0 \cdot (^-4) = 0$. (HINT: Recall $0 + z = z$. Consider $0(^+4 + {}^-4) = 0$)

5. Use CAD properties to find (HINT: $(^+6 + {}^-6)3 = ?$)

(*a*) $(^-6)3 = {}^-18$ (*b*) $7(^-4) =$

6. Do interesting things to the obviously true statement

$$(^-6)(^-3) + (^-6)(^+3) + (^+6)(^+3) = (^-6)(^-3) + (^-6)(^+3) + (^+6)(^+3)$$

to show $(^-6)(^-3) = (^+6)(^+3)$, assuming you love CAD rules.

7. Use steps as in exercise 6 to show $(^-5)(^-7) = (^+5)(^+7)$.

8. Let white chips have value $^{+}1$ and red chips have value $^{-}1$. Interpret the following in terms of receiving or giving stacks of valued chips:

(*a*) $^{+}4 \times {}^{+}6$ (*b*) $^{+}5 \times {}^{-}3$ (*c*) $^{-}4 \times {}^{+}7$ (*d*) $^{-}3 \times {}^{-}5$

9. Interpret the following in terms of receiving or giving checks and duns:

(*a*) $^{+}3 \times {}^{+}8$ (*b*) $^{-}7 \times {}^{+}5$ (*c*) $^{+}4 \times {}^{-}3$ (*d*) $^{-}5 \times {}^{-}4$

10. If a time interval of $^{-}3$ hours starts at 4 P.M., when does it end?

11. Water flows into a tank at a rate of 8 gallons per minute for 7 minutes. How much more water is in the tank at the end of the period? Write a multiplication sentence which models this flow.

12. Water flows into a tank at 400 gallons per hour for $^{-}3$ hours, beginning at 4 P.M. How much more water is in the tank at the end of the period than at the beginning? Write a multiplication sentence modeling this flow.

13. Gasoline flows into a tank at $^{-}4$ gallons per minute for $^{-}5$ minutes, starting at 6 A.M. (*a*) When did the period end? (*b*) How much more gasoline was in the tank at the end of the period than at the beginning? (*c*) Write a mathematical sentence modeling this flow.

14. Make up a story about water flowing into a tank for each of the following multiplication sentences.

(*a*) $^{+}6 \times {}^{-}3 = {}^{-}18$ (*b*) $^{-}7 \times {}^{+}2 = {}^{-}14$ (*c*) $^{-}8 \times {}^{-}3 = {}^{+}24$

15. At 4 P.M. Christopher has been riding his bike south for 5 hours at a rate of 6 miles per hour. He is now at milepost 40. Chris just recently passed post 41, so he knows the milepost numbers increase to the north.

(*a*) Draw a diagram of the highway.
(*b*) What is Chris's velocity, or directed speed?
(*c*) What is the directed time interval, beginning now, 4 P.M., to the time he started?
(*d*) What is the change in the milepost readings?
(*e*) Write a multiplication sentence modeling Chris's ride.

16. Make up a story about a bicycle rider riding along a mileposted bikeway for each of the following:

(*a*) $^{+}4 \times {}^{+}10 = {}^{+}40$ (*c*) $^{-}6 \times {}^{+}8 = {}^{-}48$
(*b*) $^{+}5 \times {}^{-}7 = {}^{-}35$ (*d*) $^{-}3 \times {}^{-}7 = {}^{+}21$

17. Discuss why temperatures are poor illustrations for signed numbers if you wish to discuss multiplication.

18. Signed numbers are often used to label the following. Choose a signed number that may be nicely used to illustrate multiplication and point out one that would be poor for this. Give examples and reasons.
(*a*) floors in a building (*c*) values of objects
(*b*) elevation above sea level (*d*) displacements along a line

19. Draw similar triangles to show
(*a*) $^{+}\frac{3}{2} \times {}^{+}2 = {}^{+}3$ (*b*) $^{+}2 \times {}^{-}3 = {}^{-}6$ (*c*) $^{-}\frac{4}{5} \times {}^{-}\frac{1}{2} = {}^{+}\frac{2}{5}$

20. Draw pictures of a teeter-totter to show
(*a*) $^{+}3 \times {}^{+}4 = {}^{+}12$ (*c*) $^{-}3 \times {}^{+}2 = 6$
(*b*) $^{+}4 \times {}^{-}2 = {}^{-}8$ (*d*) $^{-}4 \times {}^{-}5 = {}^{+}20$

21. Which of the following attitudes, if any, apply readily to $5 \times {}^{-}2$?
(*a*) array, (*b*) repeated addition, (*c*) tree

22. Which of the attitudes in exercise 21 apply readily to $^{-}3 \times 6$?

23. A military ruler, in order to make life easy for young people, decreed that the product of any two negative numbers shall be 0. Do one of the following:

(*a*) Describe three situations in which you would be inconvenienced by this decree. State what you would have to do to be able to abide by the decree.

(*b*) If you would not be inconvenienced, state three situations in which people important to you would be affected. What would you advise them to do in these situations?

(*c*) Send $1 to the Society for the Resurrection of the Stone Age.

24. Name two reasons why people want $(-3)(-4)$ to be $^{+}12$.

25. Can you make up an illustration to show multiplication of signed numbers based on whether it is a good thing or a bad thing when good guys or bad guys enter or leave a classroom?

26. Can you make up an illustration based on facing east or west, walking forward or backward, and ending up east or west of your starting point?

27. Score the following multiplication attitudes from chapter 10 on a 0–10 scale according to the ease with which they generalize to the four signed-number multiplication types, say A $(^{+}3)(^{+}5)$, B $(^{+}4)(^{-}7)$, C $(^{-}2)(^{+}4)$, D $(^{-}4)(^{-}5)$.

(*a*)	array	(*f*)	segment-segment-segment
(*b*)	repeated addition	(*g*)	operator-segment-segment
(*c*)	Cartesian product	(*h*)	operator-set-set
(*d*)	tree	(*i*)	operator-operator-operator
(*e*)	segment-segment-region	(*j*)	formal manipulation

Answers to selected questions:

1.

−16	−12	−8	−4	0	4	8	12	16
−12	−9	−6	−3	0	3	6	9	12
−8	−6	−4	−2	0	2	4	6	8
−4	−3	−2	−1	0	1	2	3	4
0	0	0	0	0	0	0	0	0
4	3	2	1	0	−1	−2	−3	−4
8	6	4	2	0	−2	−4	−6	−8
12	9	6	3	0	−3	−6	−9	−12
16	12	8	4	0	−4	−8	−12	−16

through (2,3) . . . −3, 0, 3, 6, 9
. . . −2, 0, 2, 4, 6
through (−4,1) . . . 4, 3, 2, 1, 0, −1, −2
. . . 8, 4, 0, −4, −8, −12
through (−2,3) . . . −9, −6, −3, 0, 3
. . . 4, 2, 0, −2, −4, −6

2. *e* best; *b*, *d* next; **5.** (*a*) $0 = (-6 + {}^{+}6)\,3 = -6 \cdot 3 + 6 \cdot 3 = -6 \cdot 3 + 18$ therefore $-6 \cdot 3 = -18$.

7.
$$(-5)(-7) + (-5)(+7) + (+5)(+7) = (-5)(-7) + (-5)(+7) + (+5)(+7)$$
$$(-5)(-7) + (-5 + {}^{+}5)(+7) = (-5)(-7 + {}^{+}7) + (+5)(+7)$$
$$(-5)(-7) = (+5)(+7)$$

9. (*b*) give 7 checks for 5 each; (*d*) receive 5 duns for 4 each; **12.** −1200 gallons; $400 \times {}^{-}3 = {}^{-}1200$.

15. (*a*)

(*d*) from 4 P.M. to 11 A.M., 30 mi.

17. There is little physical meaning to $3 \times {}^{-}10°$, since the temperature is measured from an arbitrary 0.

19. (*b*)

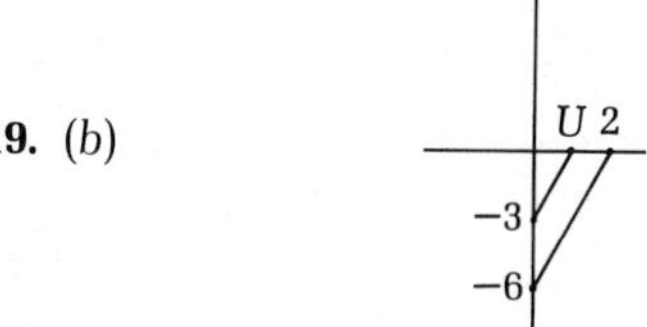

22. none; **24.** Want CAD to hold; want the progressions in the product pattern to extend. See text for others.

27. Different people will have different scores. Here is a possible result:

	A	B	C	D
(*a*)	10	0	0	0
(*b*)	10	8	0	0
(*c*)	10	0	0	0
(*d*)	10	0	0	0
(*e*)	10	4	4	3
(*f*)	10	8	8	8
(*g*)	10	4	4	2
(*h*)	10	7	2	2
(*i*)	10	1	1	1
(*j*)	10	9	9	9

23

Ratio and Proportion

23.1 INTRODUCTION Ratio was introduced in chapter 17 as a rational number describing the comparative sizes of objects. A proportion is a statement that two fractions are equal—that is, name the same ratio. Mathematically the fact that $\frac{2}{3} = \frac{6}{9}$ has already been discussed and is nothing to write home about. It gets exciting when the $\frac{2}{3}$ and the $\frac{6}{9}$ describe different things and their equality shows an equivalence between the things described. What is this equivalence and how is it used?

More than just description, proportions are a guide to design—making things similar to others, being fair and just, changing from feet to inches, and so on. Can you see the ratios and proportions in the following ten sentence stories?

1. Billy looks like his picture.

2. How many eggs are needed for a Boy Scout pancake feast?

3. A cottage has more windows than a castle.

4. What is a fair amount to pay for 5 pounds of oranges?

5. Do quarter horses eat more than trotters at Santa Anita?

6. Please enlarge this picture of my grandfather.

7. Mary and Michele drove equally fast, but Michele drove twice as far.

8. The Dodgers are better hitters than the Mets.

9. Perry Mason wanted a diagram of the room where the body was found.

10. Sadie is for the metric system because she will weigh less in kilograms than she does in pounds.

In what follows we review ratio, and then introduce proportions, show how they are used, and give their properties.

23.2 REVIEW OF RATIO A ratio arises when we compare two numbers by division. We say, "The ratio of 6 to 8 is $\frac{6}{8}$." Note that order is important, for a ratio is a ratio between a first number and a second. The first number is the numerator of the fraction, the second is the denominator. The ratio of 8 to 6 is $\frac{8}{6}$ and is different from the ratio of 6 to 8.

The two ratios determined by a pair of numbers are reciprocals. Their product is one. In the example, $\frac{6}{8} \times \frac{8}{6} = 1$. In general, the ratio of x to y is $\frac{x}{y}$. The ratio of y to x is $\frac{y}{x}$. Their product is $\frac{x}{y} \cdot \frac{y}{x} = 1$.

Ratio is often used to compare the sizes of two quantities. (Subtraction is another way.) We say the ratio of the length to the width of a 3×5 filing card is $\frac{5}{3}$. If \$60 interest is paid on a principal of \$200, the ratio of interest to principal is $\frac{60}{200} = \frac{3}{10} = 30$ percent.

Architects often make scale drawings of rooms where $\frac{1}{4}$ inch represents 1 foot in the room. A table 4 feet wide would be shown by a drawing 1 inch wide. The ratio of length in the room to length on the drawing is $\frac{48}{1}$. To determine ratios of sizes, both objects are measured in the same units. The ratio of a 48-inch (4-foot) length to a 1-inch ($\frac{1}{12}$-foot) length is $\frac{48}{1}$ or $\frac{4}{\frac{1}{12}}$.

In general, if objects A and B are measured in the same units and have sizes a and b of these units, the ratio of these sizes is $\frac{a}{b}$. When it is clear which size is meant we may say the ratio of A to B is $\frac{a}{b}$. We might say "The ratio of the table to its drawing is $\frac{48}{1}$" rather than "The ratio of the length of the table to the length of the drawing is $\frac{48}{1}$."

We recall the equivalent definition of ratio based on multiples, rather than on units of measure. Suppose we have several books 30 centimeters long and 12 centimeters wide. The ratio of length to width as defined above is $\frac{30}{12} = \frac{15}{6} = \frac{10}{4} = \frac{5}{2}$. Now, suppose we put several books end to end alongside several others placed side by side. What would we expect to see? Listing multiples of 30 and 12, we have 30, 60, 90, 120, 150, . . . and 12, 24, 36, 48, 60, 72, 84, 96, 108, 120, 132, ... Clearly 60, 120, 180, . . . are common multiples. This means that 2, 4, 6, . . . books end to end match 5, 10, 15, . . . books side by side, respectively. We see the ratios $\frac{5}{2}, \frac{10}{4}, \frac{15}{6}$ appearing as the ratios of the number of books put side by side to the number of books put end to end that match exactly.

In general, if m of the A objects match exactly n of the B objects, then the ratio of A to B is $\frac{n}{m}$. We note that if a,b are the sizes of one A and one B respectively, then $ma = nb$ and $\frac{a}{b} = \frac{n}{m}$, the ratio of A to B.

Notice that the ratio of the smaller object to the larger is less than one, and the ratio of the larger to the smaller is greater than one. Sometimes you may know the ratio is, say, either $\frac{3}{4}$ or $\frac{4}{3}$. If it is the ratio of a smaller to a larger object, the ratio must be $\frac{3}{4}$.

23.3 RATIO IN MEASUREMENT While as mathematicians we are most interested in ratios as numbers and their number properties, as non-mathematicians (which we are most of the time) we may be surprised to find that measurements of distance, weight, area, volume, and time are ratios. We shall describe how some of these measurements are actually made, then see how the process can be thought of as a ratio concept. The correspondence will help people understand the useful relation between arithmetic and life.

1. *Distance*. The water in a swimming pool is $4\frac{1}{2}$ feet deep. While there may be a label "$4\frac{1}{2}$" on the side of the pool, more likely a dry stick is pushed to the bottom. A tape measure is placed along the wet part of the stick. Some numeral such as $4\frac{1}{2}$ or 4 ft. 6 in. is printed on the tape opposite the end of the wet part.

This process of reading a number at a mark does not seem a ratio process. We neither "find a common part" or "find multiples." Yet if we take 1 inch as a unit and find the ratio between the wet part of the stick and a standard foot segment, we would find 54 in the stick and 12 in the standard foot. The ratio is $\frac{54}{12} = 4\frac{1}{2}$.

2. *Weight.* Sue purchased 5 pounds of sugar. Actually she probably purchased a sack labeled "5 pounds." Or sugar was placed on a scale until a pointer pointed at a mark labeled "5 pounds." There is no ratio process in this description. However, we can again imagine 5 standard 1-pound weights balanced against the sugar. The ratio of sugar to a standard 1-pound weight is $\frac{5}{1}$.

3. *Area.* The area of a rug is 8 square yards. Probably no direct area measures were made. If the rug is rectangular, the product length × width (in yards) is 8 or close enough to 8. If the rug is circular, then $\pi r^2 = 8$, where the length r was measured directly.

On the other hand, perhaps the rug is covered (approximately) with little squares 9 inches on a side, like floor tiles. There are 128 tiles. Then a square 1 yard on a side is covered by 16 tiles. The regions, rug and square, are separated into equal parts. The ratio of the rug size to a square yard is $\frac{128}{16} = \frac{8}{1}$. We say the area of the rug is 8 square yards. You may want to consider various shapes to see how you can get a numeral for the area practically, and how you can interpret this as the ratio of the size of the region to a standard region such as a square 1 yard on a side.

4. *Volume.* Cassandra put $3\frac{1}{2}$ gallons of gasoline in her tank. Actually, a little sign on the pump read 3.5. But if the gasoline had been in quart cans and a gallon container had been filled from the quart cans, we would see that 14 quart cans went into the car and 4 quarts filled the gallon container. The ratio is $\frac{14}{4}$, or $3\frac{1}{2}$. Alternatively, if the same amount was put into two cars, 7 one-gallon cans would be just right.

5. *Time.* The water boiled for ten minutes. Probably this measurement was made by looking at a dial with a pointer. To get the ratio concept, we look more deeply into the way a clock operates. Perhaps there is a pendulum or a balance wheel or an alternating electric current. Some repeating process is performed by machinery and each repetition is regarded as taking the same length of time. The clock counts the number of these cycles or repetitions and records the ratio of this number and the number of such repetitions in the standard time interval—one minute.

23.4 GENERALIZING THE RATIO CONCEPT We started by recalling ratio as a fraction, comparing two whole numbers. Then ratios were used to compare two quantities of the same kind, such as lengths or weights. It is but a blink of the eye to extend the use of ratio to compare two fractions. We compare $\frac{2}{3}$ and 5, for example, by saying their ratio is $\frac{2}{3}/5 = \frac{2}{15}$, since this is the standard result of simplifying complex fractions. Similarly, the ratio of $\frac{4}{5}$ to $\frac{3}{7}$ is $\frac{4}{5}/\frac{3}{7} = \frac{28}{15}$.

Not quite so simple but usually accepted without question are the following comparisons. (*a*) The ratio of the hypotenuse to one side of an isosceles right triangle with 45° angles is $\sqrt{2}$. (*b*) The ratio of the circumference to the diameter of a circle is π, or $\frac{C}{D} = \pi$.

Now it happens that $\sqrt{2}$ and π are irrational numbers. That is, there are no whole numbers A, B, C, D such that $\sqrt{2} = \frac{A}{B}$ or $\pi = \frac{C}{D}$. (This is not proved here.) There is no small interval that fits evenly into both the leg and the hypotenuse of the right triangle. So, because we want to compare segments such as these in this way, we extend the concept of ratio to include irrational numbers. We find a segment exactly $\frac{1}{10}$ the leg of the right triangle will fit into the hypotenuse 14 times but not 15. Hence we say that

$$\frac{14}{10} < \frac{\text{hypotenuse}}{\text{leg}} < \frac{15}{10}, \text{ or } \sqrt{2} = 1.4^{+}.$$

So far, ratio is extended from comparing whole numbers to comparing like quantities, fractions, and irrational numbers. Now comes an extension to gag a fire-eating sword swallower. What sense does it make to speak of the ratio of 30 feet to 5 seconds? For such unlike quantities the idea of a common unit is nonsense. Nor can we find numbers S and T so that $S \cdot 30$ feet means the same thing as $T \cdot 5$ seconds. No distance is a time interval. However, ideas such as velocity and other rates are often thought of as 30 feet divided by 5 seconds to give 6 feet per second. This appears to be a lot like "dividing" 30 feet by 5 feet to get a ratio of 6. So we extend the ratio idea to statements like "The ratio of 30 feet to 5 seconds is the speed 6 feet per second."

Numbers with units, such as 8 feet, are often called *denominate numbers*. Such number-unit pairs arise throughout applied mathematics, for example, from measurement. The ratio of two like quantities—that is, the ratio of two denominate numbers with the same units—we regard as a pure number. A pure number is a number without a unit. *Pure* is used solely to contrast with *denominate*. *Denominate* means "with a name." *Pure* means "without name or units."

The ratio of two denominate numbers with different units, however, is another denominate number. The number in the number-unit pair is the ratio of the two numbers; the unit may have a name of its own, or it may be a hybrid, like feet per second.

Other examples of extended ratios are prices (\$48 for 6 dozen golf balls is a price of $\frac{48}{6}$ dollars per dozen), automobile economy figures (miles per gallon, cents per mile), and interest rates (dollars per year). We discuss these further under proportions.

23.5 DEFINING AND SOLVING PROPORTIONS Mathematically, a proportion says two ratios are equal. For example, $\frac{9}{6} = \frac{12}{8}$ is a proportion and the four numbers—9, 6, 12, 8—are said to be *proportional* or *in proportion*.

These numbers are from a recipe. It is claimed that 9 teaspoons of pineapple juice mixed with 6 teaspoons of cranberry juice will taste exactly like a drink made from 12 cups of pineapple juice and 8 cups of cranberry juice. Is this so?

The order of the numbers—9, 6, 12, 8—is important. Sometimes we say, "9 is to 6 as 12 is to 8." The quantity of pineapple juice is to the quantity of cranberry juice in one mixture as the quantity of pineapple juice is to the quantity of cranberry in the other.

In letters we would say, "the four numbers a, b, c, d and the idea $\frac{a}{b} = \frac{c}{d}$." If $a = 2$, $b = 5$, $c = 40$, $d = 100$, we observe that $\frac{a}{b} = \frac{2}{5} = \frac{40}{100} = \frac{c}{d}$; hence 2, 5, 40, 100 are proportional. If $a = 6$, $b = 7$, $c = 8$, $d = 9$, we see that $\frac{6}{7} \neq \frac{8}{9}$, so 6, 7, 8, 9 are not proportional.

The proportion idea is extremely useful when we know three of the four numbers and know that the four numbers are proportional. Then we can find the missing number.

EXAMPLE 1 Suppose we wanted to make a small sample of the drink in the above recipe, using two small glasses of cranberry juice. How much pineapple juice would we use when we don't know the size of the small glass? Even though we don't know the size, we can say we will use p glasses, and the number p is in the proportion $\frac{12}{8} = \frac{p}{2}$. Since we know from fractions that $\frac{12}{8} = \frac{3}{2}$, then p must be 3. We use three glasses of pineapple juice.

In case we don't remember that $\frac{12}{8} = \frac{3}{2}$, we can rewrite the two fractions with a common denominator: $\frac{12}{8} = \frac{p}{2} = \frac{4p}{8}$. Since equal fractions with the same denominator must have equal numerators, $12 = 4p$. The only number satisfying this equation is $p = 3$.

EXAMPLE 2 Suppose we wish to find a number x from the proportion $\frac{9}{15} = \frac{x}{35}$. There are several common ways of finding x. We list some.

1. We may just happen to know that $\frac{9}{15} = \frac{21}{35}$. Then we conclude that $x = 21$.

2. We rewrite the two fractions with the same denominator:

$$\frac{63}{105} = \frac{3x}{105}$$

Hence,

$$63 = 3x$$

Division shows that $x = 21$.

3. Not wishing to struggle finding the *least* common denominator, we use 15×35 as the denominator:

$$\frac{9}{15} = \frac{x}{35} \text{ becomes } 9 \times \frac{35}{15 \times 35} = \frac{15x}{15 \times 35}$$

Hence,

$$9 \times 35 = 315 = 15x.$$

Division again shows $x = 21$.

4. Knowing that if $\frac{a}{b} = \frac{c}{d}$, then $ad = bc$, we move directly from $\frac{9}{15} = \frac{x}{35}$ to $9 \times 35 = 15x$.

Division gives $x = 21$.

Going directly from $\frac{a}{b} = \frac{c}{d}$ to $ad = bc$ without thinking of the fractions rewritten with a common denominator is called *cross multiplying*, from the pattern

$$\frac{a}{b} \times \frac{c}{d}$$

EXAMPLE 3 Find y from the proportion $\frac{y}{7} = \frac{5}{8}$. Cross multiplying, we go directly to

$$8y = 35 = 7 \cdot 5$$

From a definition of fraction (or dividing both sides by 8):

$$y = \frac{35}{8}$$

EXAMPLE 4 The missing number may be in the denominator of a ratio:

$$\frac{4}{6} = \frac{10}{t}$$

Rewriting with the same denominator and equating numerators, or cross multiplying, we get

$$4t = 60$$

Dividing both sides by 4, or using the missing factor division attitude,

$$t = 15$$

23.6 SOME USES OF PROPORTIONS Because proportions are so useful in so many places, we can give only a few examples. It will clarify your idea of proportion if you imagine similar situations in your own life and use the corresponding proportions. Our examples come from pricing, recipes and prescriptions, scale drawings, similar triangles, photographs, constant rates and speed situations, and converting units of measure. Our next examples will emphasize the indirect measurement of quantities such as distance, weight, time, and speed by means of proportions. Many people are surprised to recognize that they usually measure time and weight indirectly by measuring a distance first. Percentage is a magnificent use of proportion and is discussed in a later chapter.

A. Pricing In buying groceries such as apples, the amount received is proportional to the amount paid. Mary purchased 8 pounds, Karen 12 pounds. Mary paid 72 cents, Karen 108 cents. The proportion $\frac{8}{12} = \frac{72}{108}$ is true. If Peter paid 54 cents for 6 pounds, we would also have $\frac{8}{6} = \frac{72}{54}$ and $\frac{12}{6} = \frac{108}{54}$.

In each of the ratios—$\frac{8}{12}, \frac{72}{108}; \frac{8}{6}, \ldots, \frac{108}{54}$—like quantities are compared, pounds to pounds or cents to cents. If we use generalized ratios, pretending to compare cents to pounds, we find for Mary, Karen, and Peter the ratios

$$\frac{72}{8}\frac{\text{cents}}{\text{pounds}};\ \frac{108}{12}\frac{\text{cents}}{\text{pounds}};\ \frac{54}{6}\frac{\text{cents}}{\text{pounds}}$$

These are denominate numbers with unit cents/pounds. The ratios are equal, since $\frac{72}{8} = \frac{108}{12} = \frac{54}{6}$.

We see three proportions

$$\frac{72}{8} = \frac{108}{12} \qquad \frac{72}{8} = \frac{54}{6} \qquad \frac{108}{12} = \frac{54}{6}$$

The continued equality

$$\frac{72}{8} = \frac{108}{12} = \frac{54}{6} = \frac{9}{1}$$

is called a *continued proportion with common ratio* 9. The common ratio is often called the *constant of proportionality*. This constant is usually quite significant. In this example, the constant is the price of the apples, 9 cents per pound.

If someone else purchases A pounds of apples for C cents, A and C are proportional to a corresponding pair. We have, say $\frac{108}{C} = \frac{12}{A}$ and $\frac{108}{12} = \frac{C}{A} = 9$.

Often we write the formula $C = 9A$. Indeed, we see that in any constant price situation each purchase gives a ratio, $\frac{C}{A}$, equal to the corresponding ratio for every other purchase. Hence many proportions are generated. For example:

1. Sharon bought 6 glasses for 38 cents. Louise bought 9 for 57 cents. We have $\frac{6}{9} = \frac{38}{57}$. How many did Grace buy for 95 cents? We have the proportion, $\frac{6}{G} = \frac{38}{95}$. Cross multiplying gives $38G = 6 \cdot 95 = 570$. Dividing

$38G$ and 570 by 38 gives $G = 15$. Some related proportions are $\frac{G}{95} = \frac{6}{38}$ and $\frac{G}{95} = \frac{9}{57}$. We have three equal ratios, $\frac{G}{95} = \frac{6}{38} = \frac{9}{57}$. The proportionality constant is $\frac{3}{19}$ and is often quoted as "3 glasses for 19 cents." $G = 95 \times \frac{3}{19} = 15$.

2. Peanuts are quoted at 73 cents per pound. This means that purchases of 4, 3, and 7 pounds will cost 292 cents, 219 cents, and 511 cents, respectively. We have the continued proportion

$$\frac{4}{292} = \frac{3}{219} = \frac{7}{511} = \frac{1}{73}$$

Some people may prefer to use proportions related to these to describe the cost of peanuts, such as $\frac{292}{4} = \frac{73}{1}$. For example, how many pounds can Sally buy for 400 cents? If M is the amount, then we have the proportion $\frac{M}{4} = \frac{400}{292}$, or $\frac{M}{400} = \frac{1}{73}$, and so forth. We get $M = \frac{400}{73} = 5\frac{35}{73}$, or slightly less than $5\frac{1}{2}$ pounds.

B. Recipes A circus made a thorough study of food consumption by horses. They concluded that white horses ate more than black horses. Asked why, the hostler said, "The only reason we could think of is there are more white horses."

In planning menus, the amount of food cooked is proportional to the number of servings planned. Twenty-five pounds of ground beef are used to make 100 hamburgers. How much beef is required to make 40 hamburgers? Cross multiplying gives $100B = 1000$. Hence, $B = 10$ pounds of ground beef are required.

The related proportion $\frac{25}{100} = \frac{B}{40}$ leads to the same result and is preferred by many who like the continued proportion $\frac{25}{100} = \frac{B}{40} = \frac{M}{H}$, where M is the number of pounds of ground beef for H number of hamburgers.

The proportionality constant $\frac{25}{100} = \frac{1}{4}$ is clear. We may use the equation $\frac{1}{4} = \frac{M}{H}$ or $H = 4M$.

In the proportion $\frac{25}{B} = \frac{100}{40}$, the fractions are ratios comparing like quantities, pounds to pounds and hamburgers to hamburgers. The ratios are pure numbers and would have the same value even if we measured in ounces or grams or counted in dozens.

In the related proportion $\frac{25}{100} = \frac{B}{40}$, the ratios are not a comparison of like quantities but are the number of pounds of beef divided by the number of hamburgers. This ratio is not a pure number; the unit is pounds/hamburger and is read "pounds (of beef) per hamburger." The number will change if we measure in ounces or grams or count in dozens. The convenience of the continued ratio and formula is a major reason for extending the idea of ratio to unlike quantities.

We might ask how many hamburgers we would get from 7 pounds of ground beef. Using the proportion $\frac{M}{H} = \frac{1}{4}$, we get $\frac{7}{H} = \frac{1}{4}$ and $H = 4 \cdot 7 = 28$.

C. Prescribing medicines Medicine, say, aspirin, goes into the blood stream where it is distributed to all parts of the body. The appropriate dosage is approximately proportional to the weight of the sufferer. If a 5-grain tablet is right for a 100-pound woman, how much should a 180-pound man take? If A is the number of grains of aspirin for him, we have the proportion $\frac{180}{100} = \frac{A}{5}$. Cross multiplying gives $900 = 100A$. He should take 9 grains—slightly less than two tablets.

How much aspirin would you give a 60-pound dog? Assuming dogs react like people, we have $\frac{60}{100} = \frac{A}{5}$. Cross multiplying gives $300 = 100A$. We give the dog 3 grains, perhaps breaking the standard 5-grain tablet to do this.

A 120-pound mother says, "I take two tablets (10 grains). I'll give my crying baby two tablets also." If the baby weighs 30 pounds, what would have been the appropriate dose? We have the proportion $\frac{120}{30} = \frac{10}{A}$.

Hence, $120A = 300$. The appropriate dose is $A = \frac{300}{120} = 2\frac{1}{2}$ grains. The baby was given an overdose four times the right size.

Years ago, in place of $\frac{a}{b} = \frac{c}{d}$, people wrote $a:b::c:d$. Instead of using the memory device *cross multiply,* they said, "The product of the means equals the product of the extremes," $bc = ad$. The means (b and c) were the middle numbers, and the extremes (a and d) were the outside numbers in the sequence a, b, c, d. Can you justify this rule?

In the examples that follow we will use the steps: from $\frac{a}{b} = \frac{c}{d}$, we get $ad = bc$. Divide by the appropriate number to find the desired value. You may wish to tell why these steps are reliable in the examples.

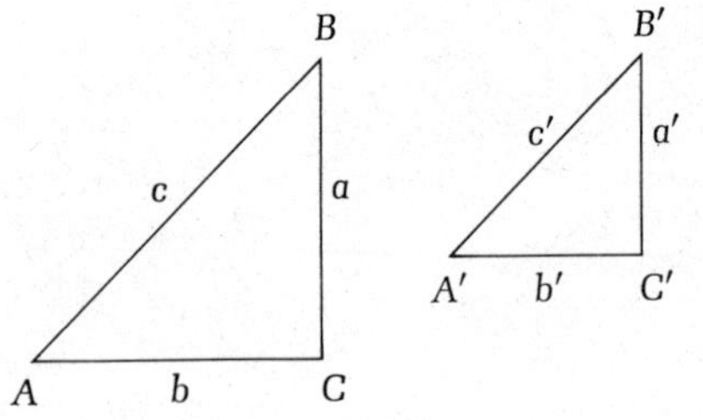

A, b, C, c, a, B; A′, b′, C′, c′, a′, B′

D. Similar Triangles In plane geometry we learned that if the corresponding angles of two triangles are equal, the triangles are called similar and the corresponding sides are proportional. In triangles ABC and $A'B'C'$, the corresponding angles are equal and the corresponding sides are labeled as shown. Then proportions like $\frac{a}{c} = \frac{a'}{c'}$ are true. So are the related proportions like $\frac{a}{a'} = \frac{c}{c'}$. If $c = 6$, $c' = 4$, $a' = 2$, what is a? Using the first proportion above, we have $\frac{a}{6} = \frac{2}{4}$. Cross multiplying we get $4a = 6 \cdot 2 = 12$. Dividing by 4 yields $a = 3$.

In the continued proportion $\frac{a}{a'} = \frac{c}{c'} = \frac{b}{b'} = \frac{s}{1} = s$, s (the proportionality factor) is also called the *scale factor* or scale ratio. The scale factor gives the relative sizes of the triangles. The ratios $\frac{a}{c} = \frac{a'}{c'}$ and $\frac{a}{b} = \frac{a'}{b'}$ relate to the shape of the triangle and do not depend on the relative sizes. To contrast with scale factor, this ratio could be called a *shape ratio* or *shape factor.* This term is not used much, since many such ratios are needed to give the total shape of a figure.

If angle C is a right angle, the ratio $\frac{a}{c}$ is determined by angle A. The ratio is so widely used, and has the same value in all right triangles with an angle equal to A, regardless of size of the triangle, that the ratio is named *sine* A (pronounced like sign A and abbreviated $\sin A$).

How big should side q be in the right triangle in the margin if m is 6 and the ratio $\frac{m}{q} = \frac{3}{4}$? From the proportion $\frac{6}{q} = \frac{3}{4}$, we get $q = 8$.

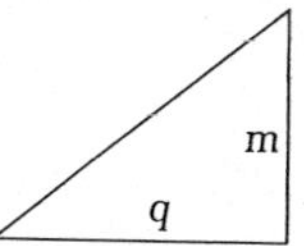

E. Scale Drawings A scale drawing of a bus looks like this:

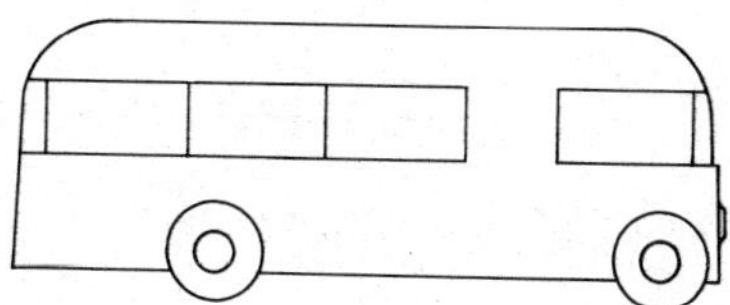

A ratio is given for each scale drawing. Suppose the scale ratio is $\frac{1}{100}$. That means that corresponding distances on the drawing and on the bus are in the ratio $\frac{1}{100}$. Suppose the tires are 25 inches in diameter. Then the pictures of the tires are $\frac{1}{4}$ inch in diameter to fit the proportion $\frac{\frac{1}{4}}{25} = \frac{1}{100}$.

If the pictures of the windows are $\frac{3}{8}$ inches wide, how wide are the windows? Let w be the width of a window. Then $\frac{\frac{3}{8}}{w} = \frac{1}{100}$. Cross multiplying gives $w = \frac{300}{8} = 37\frac{1}{2}$ inches.

Interior decorators often make scale drawings of rooms to show how they will look after the furniture is placed. The scale may be written $\frac{1}{4}$ in. to 1 ft., which equals a scale ratio of $\frac{1}{48}$. How long is the drawing of a 9 by 12 foot rug? The proportion is $\frac{1}{48} = \frac{L}{12}$. Cross multiplication gives $L = \frac{12}{48} = \frac{1}{4}$. The picture is $\frac{1}{4}$ foot long, or 3 inches.

On the drawing of the wall is a spot $\frac{3}{8}$ inches long for a mirror. How long a mirror will go there? The proportion is $\frac{1}{48} = \frac{\frac{3}{8}}{p}$. Cross multiplying, $p = \frac{48 \cdot 3}{8} = 18$ inches.

Decorators who use this ratio make the conversions automatically. They even use a ruler for their drawings that reads "one foot" for a distance of $\frac{1}{4}$ inch. Architects use a prism ruler showing six commonly used scales.

Photographs of flat objects are, in fact, scale pictures. A painting 4 feet wide and 6 feet tall of a face appears in a photograph as a picture 2 inches by 3 inches. The eye in the picture is 5 millimeters long. How long is the eye in the painting?

We set up the scale ratio in two ways, using inches in one fraction, $\frac{2}{48}$, and millimeters in the other, $\frac{5}{E}$. The ratio is 2 inches to 48 inches (4 feet). E is the length of the eye in millimeters in the painting. Since the scale ratio is the same, we have the proportion, $\frac{2}{48} = \frac{5}{E}$. Cross multiplying gives $2E = 240$, or $E = 120$ millimeters.

We may have been tempted to set up the proportion $\frac{2}{4} = \frac{5}{E}$. Such ratios would not be pure numbers. The unit on the left is inches/foot, while the unit on the right would be some muddied concept. This proportion is of no value because of confused units.

Photographs of three-dimensional objects are not quite scale pictures. The picture of a pencil placed sideways to the camera may be much longer than a picture of the same pencil placed endways.

F. Constant Speed With millions of families taking trips, frequent questions are "How far are we going?" and "How long before we get there?" If we travel at a constant speed, we find that the distance traveled is proportional to the time spent. If a person travels 8 miles in 10 minutes, how long does he take to go 12 miles? We have the proportion $\frac{8}{12} = \frac{10}{M}$. Cross multiplying gives $8M = 120$ and $M = 15$. Hence he takes 15 minutes to go 12 miles.

Similarly, if we ask how far he could go in 18 minutes, we have the proportion $\frac{d}{8} = \frac{18}{10}$. Multiplying both sides by 8 gives the distance, $d = \frac{8 \cdot 18}{10} = \frac{144}{10} = 14\frac{2}{5}$ miles.

Frequently we prefer to use related proportions, with distance on the top and time on the bottom. These ratios lead to the continued proportion $\frac{8}{10} = \frac{12}{M} = \frac{d}{18} = r$. The proportionality constant r is called *speed*, and has value $\frac{8}{10} = \frac{4}{5}$ miles per minute. The proportionality constant is a denominate number with measure $\frac{4}{5}$ and unit as miles per minute. Of course, r can also be read "4 miles in 5 minutes" or "8 miles in 10 minutes."

G. Constant rate Constant speed is a special case of constant rates in general. In fact, the idea of proportions leads to the idea of constant rates. We give further examples:

Many salesmen are paid an amount proportional to the value of goods sold. Jason Jones had sales of \$3400 and \$4000 and was paid \$306 and \$360 for two different weeks. We see that the numbers are proportional—that is, $\frac{3400}{4000} = \frac{306}{360}$. The ratio also measures the relative success in the two weeks.

If he sells \$3000 during a third week, he will expect p dollars, where $\frac{3000}{4000} = \frac{p}{360}$. Multiplying by 360 gives $p = 360 \cdot \frac{3000}{4000} = \$270$.

Most often the related proportions are used, giving the continued proportion, $\frac{306}{3400} = \frac{360}{4000} = \frac{p}{3000} = r$. The common ratio, r (the proportionality factor) is $\frac{9}{100}$. People often say the salesman is paid a commission of $\frac{9}{100}$ or 9 percent.

On an auto trip the gasoline used is proportional to the distance traveled, approximately. Sallie notes from her gas gauge that she has used 8 gallons since starting and can use 10 more gallons before refilling. If she has traveled 104 miles, how much further can she go before buying gas? Sallie uses the proportion $\frac{8}{10} = \frac{104}{S}$, where S is the distance she can go before buying more gas. Cross multiplying gives $8S = 1040$. Dividing by 8, Sallie finds $S = \frac{1040}{8} = 130$ miles.

Sallie may prefer to write the related proportion $\frac{104}{8} = \frac{S}{10} = r$. She gets S as before, but the proportionality constant, r, is also interesting in itself, since r has units of miles per gallon and is often called the "gas mileage."

While cruising, an airliner uses fuel proportional to the time in flight. If 800 pounds are used in 5 minutes, how much fuel is used in 33 minutes? The engineer uses the proportion $\frac{F}{33} = \frac{800}{5}$. Multiplying by 33 gives $F = \frac{800}{5} \cdot 33 = 160 \cdot 33$, and gives $F = 5280$ pounds. The common ratio or proportionality constant is 160 pounds per minute and is the rate of fuel consumption.

A merchant sets his selling prices proportional to his costs as a convenient way of setting prices. If he sells for \$2.50 a book that costs him \$1.60, how much can he spend for a record he will sell for \$3.00? He uses the proportion $\frac{C}{300} = \frac{160}{250}$, where C is his cost. Multiplying by 300 gives $C = \frac{160}{250} \times 300$, or $C = \$1.92$. He can pay \$1.92.

H. Converting Units We frequently wish to change the units in a denominate number. How many inches in 4 feet? What is the weight in ounces of $2\frac{1}{2}$ pounds of sugar? How many hours in 270 minutes? In each case we get conversion ratios from a table of weights and measures and then use proportions.

The length in inches is proportional to the length in feet for any segment. The table of weights and measures tells us that a segment 1 foot long is 12 inches long. We set up the proportion 12 inches/1 foot = x inches/4 feet, where x is the number of inches in 4 feet. Since the units are the same on both sides, we have $\frac{12}{1} = \frac{x}{4}$ and $x = \frac{12}{1} \times 4 = 48$. If we write the units for the last equation, we see

$$\frac{12 \text{ inches}}{1 \text{ foot}} \cdot 4 \text{ feet} = 48 \text{ inches}$$

It appears that the numbers multiply and the words *feet* and *foot* cancel. While it seems strange to speak of words canceling each other, this pattern does hold. Indeed, one of the main reasons we name the units of denominate numbers as the indicated product and quotient of fundamental units is that this cancellation is a valid guide.

Similarly, the ratios of y ounces to $2\frac{1}{2}$ pounds and 16 ounces to 1 pound form the proportion $\frac{y \text{ ounces}}{2\frac{1}{2} \text{ pounds}} = \frac{16 \text{ ounces}}{1 \text{ pound}}$. Since the units are the same on both sides of the proportion, we have $\frac{y}{2\frac{1}{2}} = \frac{16}{1}$. Cross multiplying gives $y = 2\frac{1}{2} \cdot 16 = 40$.

Alternatively, from having pounds and wanting ounces, we see that we must multiply by $\frac{\text{ounces}}{\text{pounds}}$. Hence,

$$2\tfrac{1}{2} \text{ pounds} = 2\tfrac{1}{2} \text{ pounds} \cdot \frac{16 \text{ ounces}}{1 \text{ pound}} = 2\tfrac{1}{2} \cdot 16 \text{ ounces} = 40 \text{ ounces}$$

Note that the pounds cancel, leaving ounces in the unit name.

What is the speed in miles per hour of a runner who runs 100 yards in 9 seconds? In a table we have the following: 1 yard = 3 feet; 1 hour = 60 minutes; 1 minute = 60 seconds; 1 mile = 5280 feet. We argue as follows. Two names for the speed are

$$x \frac{\text{miles}}{\text{hour}} = \frac{100 \text{ yards}}{9 \text{ seconds}} \cdot F$$

where F is a factor that changes the units and has value one. Here are some useful ways of writing one: 3 feet/1 yard; 1 mile/5280 feet; 60 seconds/1 minute; 60 minutes/1 hour. We write

$$\begin{aligned} x \frac{\text{miles}}{\text{hour}} &= \frac{100 \text{ yards}}{9 \text{ seconds}} \cdot \frac{3 \text{ feet}}{1 \text{ yard}} \cdot \frac{1 \text{ mile}}{5280 \text{ feet}} \cdot \frac{60 \text{ seconds}}{1 \text{ minute}} \cdot \frac{60 \text{ minutes}}{1 \text{ hour}} \\ &= 100/9 \cdot 3/1 \cdot 1/5280 \cdot \text{yds/sec ft/yd miles/ft sec/min min/hr} \\ &= 250/11 \text{ miles/hour} \end{aligned}$$

Clearly, the factors in F were chosen so that canceling units would leave the ones shown on the left.

23.7 PROPORTIONS IN INDIRECT MEASUREMENT

1. *Indirectly measuring distance.* How far is the moon? How tall is Mount Whitney? How high is the flagpole? Can these distances be measured directly? We recall two methods of direct measurement. First, we put the segment next to a scale and subtract the scale reading at one end of the segment from the scale reading at the other end. Second, we count the number of unit segments we can fit into the given segment.

Conceivably we could pile a stack of blocks, each one foot high, at the base of a flagpole until the pile was as tall as the pole. Or we could find a long tape measure (scale), climb the flagpole, and let one end of the tape measure down. Such direct measures are difficult, subject to large error, and practically impossible as far as the moon and Mount Whitney are concerned. Usually we measure a related segment, then find the desired distance using a proportion. This process is called *indirect measurement.*

The height of a flagpole is often found by proportions from similar triangles and shadows. In the figure, let AB be a flagpole and AC its shadow on a level region. Let XY be a 5-foot vertical stick and XZ its shadow. Suppose at the same time we measure AC as 18 feet and XZ as 3 feet. From geometry we know the sun's rays are practically parallel. The triangles ABC and XYZ are similar and their sides proportional. Hence, if the flagpole is h feet high, we have the proportion

$$\frac{h}{18} = \frac{5}{3} \text{ and } h = \frac{18 \cdot 5}{3} = 30 \text{ feet}$$

The distances to the moon and to the top of Mount Whitney can be measured as in the next figure.

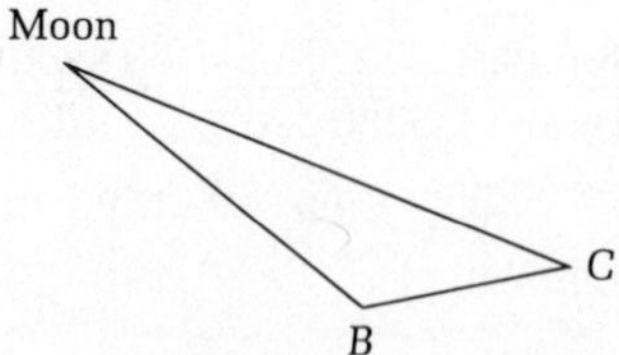

Let B, C be two places on the earth 1000 miles apart. At a given moment angles MBC and MCB are measured. A triangle $M'B'C'$ similar to MBC is drawn with $B'C' = 1$ inch. $M'B'$ is found to be 238 inches. Hence we have the proportion

$$\frac{MB}{1000} = \frac{238}{1}, \text{ and } MB = 238{,}000 \text{ miles}$$

These are simulated figures. Further, it would be extremely difficult to draw the triangle $M'B'C'$ accurately enough. Using trigonometry, we could determine MB without physically drawing triangle $M'B'C'$. The principle, however, is essentially the same.

People use radar beams to measure the distance to the moon by measuring the length of time it takes a beam to go the moon and return. Very sophisticated instruments can measure this time, as well as the speed of light, very accurately. Their product gives an excellent indirect measure of the distance.

2. *Indirect measures of weight.* When we measure weight we often use a scale with a pointer. The weight of an object is proportional to the distance the pointer moves. Suppose a 3-pound weight causes the pointer to move 2 inches. What is the weight, w, of an object that moves the pointer 5 inches? We use the proportion $\frac{w}{5} = \frac{3}{2}$. Multiplying by 5 gives $w = \frac{15}{2} = 7\frac{1}{2}$.

Usually the scale is marked to read directly in pounds, so one does not need to calculate.

If you look at the various scales you use, you may find many where the weight is measured indirectly by measuring a length which is then converted to weight by proportions. A home letter scale can be made easily with rubber bands using this principle.

3. *Indirect measure of time.* This is measured by clocks, sundials, burning candles, and various other ways. In many of these devices, the time interval is measured by a segment. The segment may be an arc on a circle, a length of candle, or a shadow. The length of time is often proportional to the segment length, but not always.

The above examples can be extended to show that many physical qualities are measured by lengths. Often the lengths form a scale with the markings made to give the physical values directly, with no calculating required.

23.8 PROPORTIONS AND LINEAR EQUATIONS A continued proportion leads to linear equations such as $y = mx$, and vice versa. Often the equation is the easiest way to describe the situation. Consider these examples.

1. Candy costs 7 cents a bar. The money spent is proportional to the number of bars. If 4, 9, 6, 3 bars are purchased, they would cost 28, 63, 42, 21 cents. We have the continued proportion $\frac{28}{4} = \frac{63}{9} = \frac{42}{6} = \frac{21}{3}$. The common ratio is 7. We have four equations,

$$28 = 7 \times 4, \quad 63 = 7 \times 9, \quad 42 = 7 \times 6, \quad 21 = 7 \times 3$$

all special cases of $y = 7x$.

2. A sculptor is making a statue 6 inches high of a man. The man is 72 inches tall. Distances on the man and on the statue are proportional. Let the length of the man's forearm be 16 inches, the circumference of his head 24 inches, the length of his ear 3 inches, and the length of his foot F. The sculptor makes the statue with forearm length a, head circumference h, ear length e, foot length f, where these numbers are in the continuing proportion,

$$\frac{6}{72} = \frac{a}{16} = \frac{h}{24} = \frac{e}{3} = \frac{f}{F} = m$$

The constant of proportionality, m, is $\frac{1}{12}$. Instead of the proportions, we write $y = mx = \frac{1}{12}(x)$, where x is a distance on the man and y is the corresponding distance on the statue. We get

$$a = \frac{1}{12} \cdot 16 = \frac{4}{3};\ h = \frac{1}{12} \cdot 24 = 2;\ e = \frac{1}{12} \cdot 3 = \frac{1}{4};\ f = \frac{1}{12} \cdot F$$

Conversely, if we have an equation $y = mx$, where $m = \frac{3}{5}$, the number pairs satisfying it are proportional:

x	2	5	6	10	20	25
y	$\frac{6}{5}$	3	$\frac{18}{5}$	6	12	15

We have the continuing proportion $\frac{2}{\frac{6}{5}} = \frac{5}{3} = \frac{10}{6} = \frac{25}{15}$, and several related proportions, such as $\frac{5}{6} = \frac{3}{\frac{18}{5}}, \frac{10}{25} = \frac{6}{15}$. We see that the coefficient m (or its reciprocal $\frac{1}{m}$) in the equation $y = mx$ is the constant of proportionality.

23.9 PROPORTION AND VARIATION Sometimes we say things like "The costs vary as the amounts bought," as well as "The cost is proportional to the amount bought." Suppose four people bought 5, 7, 4, and L pounds of nuts paying 375 cents, 525 cents, 300 cents, and C cents, respectively. We would have proportions such as

$$\frac{5}{7} = \frac{375}{525}; \frac{7}{L} = \frac{525}{C}$$

and the continuing proportion

$$\frac{375}{5} = \frac{525}{7} = \frac{300}{4} = \frac{C}{L} = r$$

In the continuing proportion, the common ratio $r = 75$ cents per pound is a denominate number. In the other proportions the ratios are pure numbers.

The continuing proportion leads to several linear equations:

$$\begin{aligned} 375 &= 5r = 5(75) \\ 525 &= 7r = 7(75) \\ 300 &= 4r = 4(75) \\ C &= Lr = L(75) \end{aligned}$$

Usually when people say "varies as," they go directly to the linear equation. "Cost C varies as the amount L purchased" is then expressed as $C = rl$, with r called the *constant of variation* or the *coefficient of variation*. We see the close tie to proportions, where r is called the constant of proportionality. In either case, the value of r is found from any known values of C and L by substitution. Since $C = 300$ cents and $L = 4$ pounds, we get $300 = r \cdot 4$, or $r = 75$, with the unit as cents/pound.

It is also true that areas of similar figures vary as the squares of corresponding lengths. Suppose a, b, c are three similar triangles with areas A, B, C. Then we have the proportions

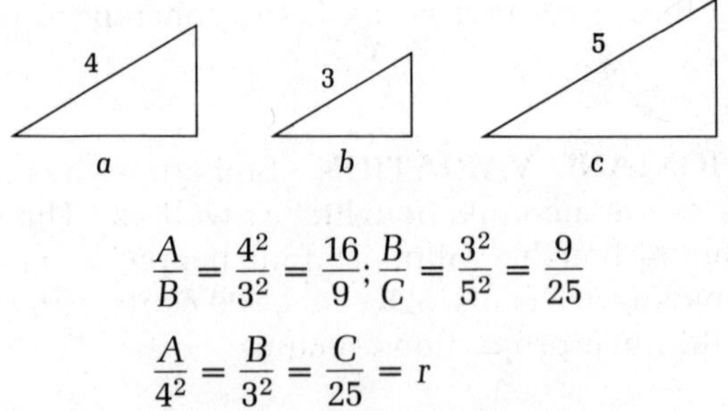

$$\frac{A}{B} = \frac{4^2}{3^2} = \frac{16}{9}; \frac{B}{C} = \frac{3^2}{5^2} = \frac{9}{25}$$

$$\frac{A}{4^2} = \frac{B}{3^2} = \frac{C}{25} = r$$

If we know r, the constant of proportionality (or variation), we can determine A, B, C. Suppose $A = 3$. Then $r = \frac{3}{16}$. We get

$$\frac{B}{9} = \frac{3}{16} = \frac{C}{25}$$

and

$$B = 9 \cdot \frac{3}{16} = \frac{27}{16}$$

$$C = 25 \cdot \frac{3}{16} = \frac{75}{16}$$

We know the speed to travel 100 yards is given by the formula $s = \frac{100}{t}$ or $s = 100\frac{1}{t}$. When we have a relation such as $y = k\left(\frac{1}{x}\right)$, we say y varies *inversely* as x. Speed varies inversely as the time it takes to go 100 yards.

23.10 SUMMARY Because a ratio that arises one way may be equal to a ratio that arises in another, proportions are a powerful tool in understanding events. Indeed many people feel they are the capstone of arithmetic, a benchmark to be reached by all citizens.

We have seen that ratios and proportions are involved in practically all direct and indirect measurement. Their central role is shown in many everyday activities such as pricing, recipes, medicine, similar triangles, scale drawings, constant speeds and other rates, and converting units. Proportions are closely related to linear equations.

EXERCISES

1. Find the ratio of the width to the length of $8\frac{1}{2}$-by-11-inch paper

(*a*) by division of fractions.

(*b*) by subunits. Find two subunits and give the two corresponding ratios.

(*c*) by multiples. Give two pairs of multiples of the lengths, giving equal segments and the corresponding ratios.

2. Interpret the following measurements as ratios, identifying subunits, or multiples.

(*a*) Melissa is $3\frac{1}{3}$ feet tall.

(*b*) David weighs 84 pounds.

(*c*) Brian drank $4\frac{1}{2}$ cups of juice.

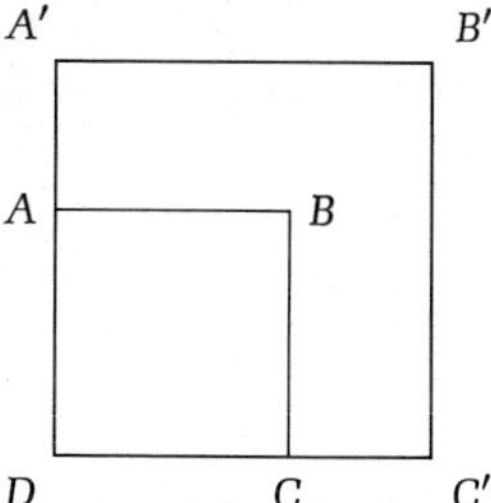

3. The smaller square, *ABCD*, has area 1. The larger square, *A′B′C′D* has area 3. What is the ratio *A′B′*/*AB*?

4. Find the value of the letter in the following proportions:

(*a*) $\frac{15}{6} = \frac{x}{8}$ (*c*) $\frac{t}{15} = \frac{24}{5}$

(*b*) $\frac{18}{25} = \frac{27}{y}$ (*d*) $\frac{25}{u} = \frac{15}{6}$

5. Find the value of the letter in these proportions:

(*a*) $\frac{2}{3} = \frac{w}{12}$ (*c*) $\frac{x}{48} = \frac{7}{16}$

(*b*) $\frac{14}{9} = \frac{21}{z}$ (*d*) $\frac{12}{s} = \frac{3}{4}$

6. If 4, 12, t, 18 are proportional, what is t?

7. Heather says the numbers 20, 30, 40, 50 are proportional. If so, show why. If wrong, (*a*) change the last number, 50, so that the numbers are proportional. (*b*) Instead of 50, change the third number, 40, so that the four are proportional. (*c*) Change the second number, 30, so that the numbers are proportional. (*d*) Change the first number, 20, so that they are proportional.

8. Set up proportions for the following purchases if walnuts sell for 63 cents a pound.

(*a*) How much do 5 pounds of walnuts cost?
(*b*) How many pounds of walnuts can be purchased for $2.52?
(*c*) How much do $3\frac{1}{4}$ pounds cost?

9. Set up a continued proportion for the facts in exercise 8.

10. Apples are 3 pounds for 97 cents. Use proportions to find the following.

(*a*) How much do 12 pounds of apples cost?
(*b*) How much do 8 pounds 4 ounces cost?
(*c*) How many pounds would the clerk say I had, if he charged me $2.46?

11. A cloth print costs $2.37 per meter. Use proportions to find the following.

(*a*) How many inches can Kelly buy for $3?
(*b*) What is the cost per yard?
(*c*) What is the cost for 250 centimeters?

12. A picture $2\frac{1}{2}$ inches by $3\frac{1}{2}$ inches is to be enlarged to make a poster 4 feet tall.

(*a*) How wide will the poster be?
(*b*) Another poster 20 centimeters wide is to be made from the same picture. How tall will it be?

13. On an architect's drawing, $\frac{1}{4}$ inch represents one foot in reality. What would be the dimensions of a rug if its drawing were 2 inches by 3 inches?

14. A map shows a scale ratio of 1:100,000.

(*a*) How many miles apart are two places that are 3 inches apart on the map?

(*b*) How many kilometers apart are two towns that are 5 centimeters apart on the map?
(*c*) How big is the map if the region it covers is a rectangle 60 miles by 100 miles? Give dimensions of the map in inches.

15. An airplane uses about 8 pounds of fuel in 3 seconds. How much fuel does it use in 40 minutes of flight?

16. Five pounds of dried soybeans are mixed with water and then added to 20 pounds of ground beef to make 120 hamburgers.

(*a*) How much beef is there in 3 hamburgers?
(*b*) How much soybean is in 3 hamburgers?
(*c*) How much beef will Nicole need for 200 hamburgers for a picnic?
(*d*) How much soybean will Nicole need for those 200 hamburgers?

17. Kimberley liked a drink made of 2 cups of pineapple juice mixed with $\frac{1}{2}$ cup of orange juice. How many cups of pineapple juice will she need to mix with 4 cups of orange juice?

18. A manufacturer finds the ratio between the selling price and the manufacturing cost of an item is 15/2.

(*a*) How much can he spend to manufacture an item that sells for $4.50?
(*b*) If it costs 30 cents to make an item, what will it have to sell for?

19. 480 square feet of wall can be painted with $\frac{3}{4}$ gallon of paint. (*a*) How many gallons will be needed to cover 2300 square feet? (*b*) How many square feet of wall can be covered with 4 gallons?

20. $\frac{7}{8}$ cup of dried milk is used with $3\frac{3}{4}$ cup of water to make one quart of reconstituted nonfat milk of standard grade. (One quart equals four cups.)

(*a*) How much water should be mixed with one cup of dried milk to make reconstituted nonfat milk of standard grade?
(*b*) How much dried milk is needed to make 3 pints of liquid milk?
(*c*) How much dried milk should be mixed with one gallon of water?
(*d*) How much liquid milk is obtained in *c*?

21. If a 5 grain aspirin tablet is about the right strength for a 100-pound woman, how much aspirin should a 40-pound child take?

22. 10 feet equals about 305 centimeters. Make a continuing proportion involving lengths of 3 feet, 200 centimeters, and 47 inches.

23. Convert the following measurements as directed:
(*a*) 20 miles per hour to kilometers per hour
(*b*) 80 meters per minute to miles per hour
(*c*) 100 feet per second to miles per hour

24. How fast do you walk? Make an estimate or rough measurement: (*a*) in feet/second, (*b*) in meters/minute, (*c*) in miles per hour.

25. If 7000 grains = 1 pound, and 22 pounds = 10 kilograms, what is the weight of one grain in grams?

26. Interest on a revolving loan account of $100 is $2.50 for one month.
(*a*) Set up a proportion to determine the interest for 100 days.
(*b*) What is the monthly interest on $150?
(*c*) What is the annual interest on $150?

27. The areas of similar figures vary as the squares of corresponding sides. Their volumes vary as the cubes of corresponding sides. Two similar triangles have corresponding sides of 3 centimeters and 5 centimeters, respectively. The area of the smaller triangle is 6 square inches.
(*a*) What is the area of the larger triangle?
(*b*) Another side of the small triangle is 4 centimeters. What is the corresponding side of the larger triangle?

28. Two statues of Buddha are similar, but one is 4 times as tall as the other.
(*a*) One part of the small statue is 3 inches long. How long is the corresponding part of the large statue?
(*b*) It costs $200 for enough gold leaf to gild the small Buddha. How much will gold leaf cost for the large statue?

29. There are two similar drinking glasses. The top of one is 3 inches in diameter and the top of the other is 4 inches. If the smaller glass holds 8 fluid ounces, how much does the larger glass hold?

30. A mother and daughter are look-alikes. The mother is 64 inches tall and weighs 120 pounds. The daughter is 48 inches tall. How much would you expect the daughter to weigh? What assumptions are you making that are probably not true?

Answers to selected questions:
1. (*a*) $\frac{8\ 1/2}{11} = \frac{17/2}{11} = \frac{17}{22}$
(*b*) 1/2 inch goes into width 17 times, length 22 times; ratio = 17/22; 1/4 inch unit gives 34/44; (*c*) 22 widths = 17 lengths = 187 inches, 17/22; 44 widths = 34 lengths = 374 inches, 34/44; **3.** $\sqrt{3}$; **5.** (*a*) 8; (*b*) 27/2; (*c*) 21; (*d*) 16; **8.** (*a*) $3.15; (*b*) 4 lbs.; (*c*) $2.05; **10.** (*a*) $3.88; (*b*) $2.67; (*c*) 7.6 lbs.; **12.** (*a*) 2.86 ft.; (*b*) 28 cm; **13.** 8 ft. × 12 ft.; **16.** (*b*) 1/8 lb.; (*d*) 8 1/3 lb.; **18.** (*a*) 60¢; (*b*) $2.25; **21.** 2 grains; **23.** (*b*) 2.98; **24.** varies with person; (*a*) 5 feet/second; (*b*) 91.4 meters/minute; (*c*) 3.4 miles/hour; **27.** (*a*) 16 2/3 in.^{2}; (*b*) 6 2/3 cm; **29.** 19 oz.

24

Decimal and Basimal Fractions

24.1 INTRODUCTION Everyone knows that, like death and taxes, fractions are inevitable. They are useful, necessary ideas, but nuisances to be avoided if possible. We shall discuss the choices people make to do their jobs and yet to dodge their friendly $\frac{a}{b}$'s.

Many are led to decimal fractions. We shall define them and show how common fractions and decimal fractions can be converted to each other. We discuss operations in decimal fractions with emphasis on the structure and easily locating the decimal point. We close by generalizing to *basimal* fractions.

24.2 FRACTION TRADE-OFFS Annoying though they are, there appears to be no numeral system as easy as common fractions, introduced in chapter 17, which will do all the things common fractions will do. Hence common fractions will remain important no matter what further numeral system may also be used.

Yet man the magnificent is often willing to trade power for convenience. To guide our choices in this trade-off, let us list some of the good things and bad things about common fractions.

We like	**We don't like**
1. the ability to represent any fractional part of a whole neatly	**1.** the peculiar three-part symbol using two integers and a bar (it does not fit easily with symbols for integers)
2. the ability to compare sizes	**2.** the difficulty in comparing (which is larger, $\frac{5}{11}$ or $\frac{6}{13}$?)

We like	**We don't like**
3. the ability to compute (add, multiply, . . .)	**3.** the many names for the same number ($\frac{2}{4} = \frac{1}{2} = \frac{12}{24}$)
	4. the awkward algorithms for addition and subtraction

To reduce these difficulties, most people are willing to give up something, say, the power to represent $\frac{121}{137}$ or $\frac{32}{193}$ exactly. However, even here we would want to have a reasonable substitute for $\frac{121}{137}$, since sometime we might want it—say at 13 P.M. on February 30.

Here are some proposed sacrifices used by everyone to some degree:

1. Limit yourself to integers. Rule fractions out of your life. Different units such as inch, foot, mile help us to avoid fractions. Many people adopt this solution and just put up with the losses that arise.

2. Limit yourself to a few strategic fractions such as $\frac{1}{2}, \frac{1}{3}, \frac{2}{3}, \frac{1}{4}, \frac{3}{4}$, and percentages. Many people ardently recommend this, especially for early primary grades. Can you list some advantages to this? Disadvantages? Can you add? Multiply?

3. Egyptians tried to face the problem by using only *unit* fractions, fractions with numerator one. This was one of history's many blind alleys. While it simplified some difficulties, it made others worse.

4. Limit the numbers that may be used as denominators. But what numbers should be used? Some people think that powers of 10 (10, 100, 1000, . . .) would be good. Other people think 12 would be a good denominator, since this allows $\frac{1}{2} = \frac{6}{12}, \frac{1}{3} = \frac{4}{12}, \frac{1}{4} = \frac{3}{12}$. The Babylonians used 60 and they could express $\frac{1}{2}, \frac{1}{3}, \frac{1}{4}$, and $\frac{1}{5}$ exactly. This led to minutes and seconds in time measure, where *minute* means "first small time interval" and *second* means "second small time interval."

In modern times we restrict ourselves to denominators with the powers of ten, 10, 10^2, . . . , 10^n, ... Then we use a special notation to avoid writing the denominators; for example, $\frac{1}{2} = .5$, $\frac{1}{4} = .25$ and others. These fractions are called *decimal fractions* or just *decimals*.

NOTE You recall that $10^2 = 10 \times 10$, $10^3 = 10 \times 10 \times 10$, and $10^n = 10 \times 10 \times 10 \times \cdots \times 10$ (n factors); n is called the exponent with 10^n defined for n, any counting number. Clearly $10^1 \times 10^2 = 10^3$, $10^2 \times 10^5 = 10^7$ and, in general, $10^s \times 10^t = 10^{s+t}$. Division and counting factors show that $10^p \div 10^q = 10^{p-q}$ if $p > q$. We can extend the meaning of 10^y for $y = 0$ and $y < 0$, yet an integer, by defining $10^{p-q} = 10^p \div 10^q$ when $q \geq p$. Then $10^0 = 1$, $10^{-1} = \frac{1}{10}$, $10^{-2} = \frac{1}{100}$, and $10^{-m} = \frac{1}{10^m}$. Similarly, $b^n = b \times b \times \ldots \times b$ (n factors), $b^0 = 1$, $b^{-m} = \frac{1}{b^m}$ for any number $b \neq 0$ and m = a counting number. Such terms are called *powers* by saying, "10^3 is the third power of ten," "b^7 is the seventh power of b," and so on.

24.3 DECIMAL FRACTION NOTATION Decimal comes from *decem*, which means ten in Latin. Strictly speaking, any number in base ten is a decimal. In common use, however, decimals are usually fractions. A typical decimal, .372 for example, has a decimal point and means $\frac{372}{1000}$. This notation can be discussed by means of a dish abacus with ratio ten. You will recall that a bean in any dish has the value of 10 beans in the dish on its right. For example, 798 is represented by

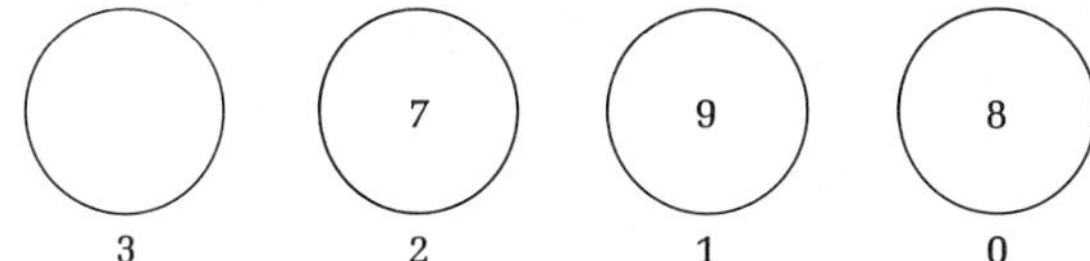

with 8 beans in dish 0 (the units or 10^0 dish), 9 beans in dish 1 (the tens or 10^1 dish), and 7 beans in dish 2 (the hundreds or 10^2 dish). We see that the dish number is the power of ten that is the value of each bean in that dish. We know that more dishes, labeled 4, 5, . . . , can be placed to the left as desired. No more dishes were placed to the right, because whole numbers come in steps of one, shown by single beans in dish 0.

This is a pattern just begging for extension. People react as though they were offered dollar bills for 50 cents. Suppose we just placed a dish to the right of dish 0. What label would we put on the dish and what value on the beans in the dish?

Can you see the sequence that suggests labeling dish $^-1$? Why do we say a bean in dish 0 is worth 10 beans in dish $^-1$? We recall that for the other dishes, a bean in any dish is worth 10 beans in the dish on the right.

We insist that this pattern continue when an extra dish is placed on the right. Then each bean in dish $^{-}1$ is worth $\frac{1}{10}$ of 1—the value of a bean in dish 0. We note that $^{-}1$ is the exponent in $10^{-1} = \frac{1}{10}$, so the values are consistent. Some abacus examples are shown in the margin.

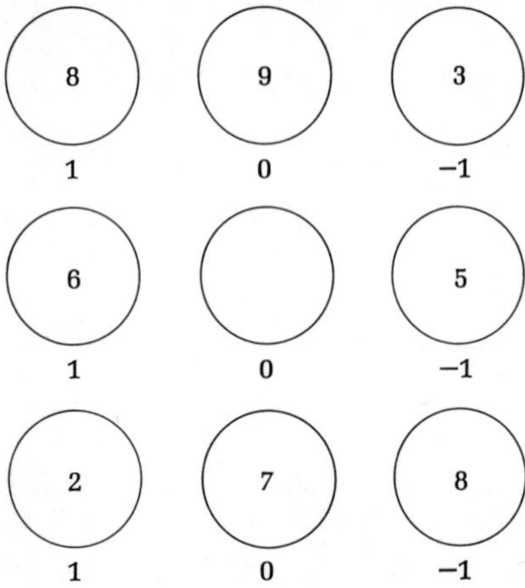

In recording the digits for the beans in the dishes without the circles, we place a dot (called the decimal point) to the right of the digit for the beans in dish 0, or units dish; for example, 89.3 is the decimal for $89\frac{3}{10}$. 89.3 represents 8 beans in the tens dish, 9 in the units dish, and 3 in the tenths dish.

More dishes may be placed to the right.

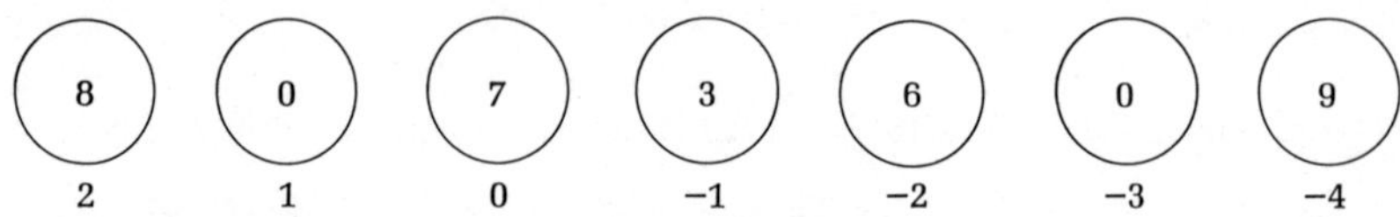

represents $807\,\frac{3}{10} + \frac{6}{100} + \frac{9}{10000} = 807\,\frac{3609}{10000} = 807.3609$. The decimal numeral shows 3 in the tenths dish, 6 in the hundredths dish, 0 in the thousandths dish, and 9 in the ten-thousandths dish.

Any fraction having a power of ten as denominator can easily be shown by an abacus, and hence by a decimal. We illustrate:

	dish 0	dish −1	dish −2	dish −3	
3/10 =		3			= .3
4/10 =		4			= .4
12/1000 =				12	
			1	2	= .012

	0	−1	−2	−3	
40/1000 =				40	
			4		= .04
36/100 =			36		
		3	6		= .36
5/1000 =				5	= .005

We did do some trading of beans from dish to dish, replacing ten beans in one dish by one bean in the dish to the left. We did not reduce any fraction like $\frac{4}{10}$ to lowest terms. Here it is important to have the denominator a power of ten.

A few uses of the dish abacus will show you the structure so that you can write the decimal fraction without using the abacus. You should make up some rules for yourself for this conversion. These rules might be like the following.

1. Write the numerator with as many digits as there are zeros in the denominator. $\frac{3}{10}$ already satisfies this requirement, $\frac{12}{1000}$ and $\frac{5}{1000}$ do not. We put 0s on the left, with the idea that 12 = 012 = 0012 . . . and 5 = 05 = 005 . . . . That is, we may write as many zeros as we please on the left of a whole number without changing its value. Now we have, $\frac{3}{10}$, $\frac{012}{1000}$, $\frac{040}{1000}$, and $\frac{005}{1000}$.

2. Place a small dot to the left of the numerator and omit the denominator and bar. Here we have, $\frac{3}{10} = .3$; $\frac{12}{1000} = \frac{012}{1000} = .012$; $\frac{40}{1000} = \frac{040}{1000} = .040$; and $\frac{5}{100} = \frac{005}{1000} = .005$, where the dot is the decimal point.

A mixed number is written with the integer part to the left of the decimal point and the fraction part to the right, as shown by these examples:

$$4\tfrac{2}{10} = 4.2 \qquad 34 + \tfrac{13}{100} = 34 + .13 = 34.13$$
$$5\tfrac{3}{100} = 5.03 \qquad 23 + \tfrac{579}{1000} = 23.579$$
$$475\tfrac{46}{1000} = 475.046$$

Just as the dishes in the abacus are called names such as *hundreds dish, units dish, tenths dish,* and *hundredths dish,* so the places in a decimal have similar names. For example, in 23.579, 2 is in the hundreds place, 3 is in the units place, 5 in the tenths place, 7 in the hundredths place, and 9 in the thousandths place.

In general, the name for a decimal is read the same as for a common fraction. We read .2 as "two-tenths," the same as for $\frac{2}{10}$. We read .0035 as "thirty-five ten-thousandths, just like $\frac{35}{10000}$. To a person who is copying the number, we may read "point zero zero three five," since we write from left to right.

In reading a mixed decimal, we indicate the decimal point by saying *and*. We read 200.013 as, "two hundred and thirteen thousandths." Sometimes people put in several *ands*. This leads to misunderstanding. For example, does "seven thousand and nine hundred and eighty three thousandths" mean 7900.083 or 7.983 or 7000.983? Some people, knowing that $5467 = 5000 + 400 + 67$, have fallen into the habit of reading it as 5 thousand and 4 hundred and 67. Including these extra *ands* is a habit hard to break. However, to avoid the confusion, we agree to use no more than one *and* in the number name, and to use that only to show the decimal point.

Extra zeros may be placed to the right in a decimal fraction. We know that $\frac{4}{10} = \frac{40}{100} = \frac{400}{1000}$, which means $.4 = .40 = .400$. This cannot be done with whole numbers, since $5 \neq 50 \neq 500$. On the other hand, we may put zeros to the left of whole numbers, for $6 = 06 = 006$. We can also write zeros to the left of the decimal point for decimal fractions, since $.32 = 0.32 = 00.32$. Care is necessary, however, for $.73 \neq .073 \neq .0073$, where the zeros are to the right of the decimal point and to the left of the digit 7.

Extra zeros are of no value when writing a single number. However, we may wish, say with certain calculating machines, to write all numbers with ten digits. Then the extra zeros allow us to use the number of digits required. Sometimes in a problem with numbers of varying lengths of decimals we find it easier to write all decimals with the same number of places. We can write .047, .3, .05, .79 all with three places as .047, .300, .050, .790.

24.4 MULTIPLYING AND DIVIDING BY TEN, THE NUMERAL BASE

Perhaps the easiest algorithms are for multiplication and division by 10, the base of the numeral system. We recall that to multiply a whole number by 10, we join a zero on the right: $23 \times 10 = 230$. If we join a zero on the right of a decimal no change occurs, $.53 = .530$. Can you see what to do?

If we multiply $47 = 47.00$ by 10, we get $470 = 470.0$, and $36.80 \times 10 = 368.0$. It appears that multiplying by 10 moves the decimal point one place to the right:

$$.038 \times 10 = \frac{38}{1000} \times 10 = \frac{38}{100} = .38$$

Using expanded notation and CAD properties, $.706 \times 10$ is $(\frac{7}{10} + \frac{6}{1000}) \times 10 = \frac{7}{10} \times 10 + \frac{6}{1000} \times 10 = 7 + \frac{6}{100} = 7.06$. Multiplying by 10 increases by one the power of 10 associated with each digit. Since moving the decimal point one place to the right has the same effect, the two acts are equivalent.

Let us divide 10. Dividing 5 or 5.00 by 10 yields $\frac{5}{10}$ or .50. Dividing $23 = 23.0$ by 10 gives $2\frac{3}{10} = 2.30$. Comparing (5.00, .50) and (23.0, 2.30), can you see that dividing by 10 has the effect of moving the decimal point one place to the left? Can you use expanded notation and the CAD properties to show this? Multiplying by 10 and dividing by 10 are inverse operations. Are moving the decimal points in these ways inverse operations?

24.5 CHANGING DECIMALS TO COMMON FRACTIONS Cinderella's Fairy Godmother changed rats into handsome footmen, and Pinnochio's friends were changed into donkeys for not studying. While we can't do this, we can change decimals into common fractions.

We show by examples:

1. $.5 = \frac{5}{10} = \frac{1}{2}$

2. $.25 = \frac{2}{10} + \frac{5}{100} = \frac{25}{100} = \frac{1}{4}$

3. $.375 = \frac{375}{1000} = \frac{75}{200} = \frac{15}{40} = \frac{3}{8}$

4. $.016 = \frac{16}{1000} = \frac{4}{250} = \frac{2}{125}$

5. $.0125 = \frac{125}{10000} = \frac{1}{80}$

6. $3.45 = \frac{345}{100} = \frac{69}{20} = 3\frac{9}{20}$

In each of the above, the decimal is first rewritten as a common fraction with a power of ten in the denominator. Then when possible the fraction is reduced to lower terms. You may like to think of the first step as follows:

1. $.5 = .5 \times \frac{10}{10} = \frac{.5 \times 10}{10} = \frac{5}{10}$

2. $.25 = .25 \times \frac{100}{100} = \frac{.25 \times 100}{100} = \frac{25}{100}$

3. $.0125 = .0125 \times \frac{10000}{10000} = \frac{125}{10000}$

Usually we reduce fractions by dividing top and bottom by common prime factors. Here is another way. Can you see what is done?

1. $\frac{5}{10} = \frac{10}{20} = \frac{1}{2}$

2. $\frac{25}{100} = \frac{100}{400} = \frac{1}{4}$

3. $\frac{375}{1000} = \frac{750}{2000} = \frac{1500}{4000} = \frac{3000}{8000} = \frac{3}{8}$

4. $\frac{16}{1000} = \frac{80}{5000} = \frac{8}{500} = \frac{40}{2500} = \frac{4}{250} = \frac{20}{1250} = \frac{2}{125}$

24.6 CHANGING COMMON FRACTIONS TO DECIMALS The following conversions are made in a similar way. Can you see what is done?

1. $\frac{1}{4} = \frac{5}{20} = \frac{25}{100} = .25$
2. $\frac{3}{4} = \frac{15}{20} = \frac{75}{100} = .75$
3. $\frac{5}{8} = \frac{25}{40} = \frac{125}{200} = \frac{625}{1000} = .625$
4. $\frac{17}{25} = \frac{68}{100} = .68$
5. $\frac{3}{16} = \frac{15}{80} = \frac{75}{400} = \frac{375}{2000} = \frac{1875}{10000} = .1875$
6. $\frac{37}{40} = \frac{185}{200} = \frac{925}{1000} = .925$

This method of multiplying both top and bottom repeatedly by 2 or 5 works if the denominator has prime factors of 2 and 5 only. Fractions such as $\frac{1}{3}$, $\frac{5}{6}$, $\frac{3}{7}$, $\frac{1}{99}$ cannot be written as a fraction with denominator that is a power of ten. Hence, there is no exact decimal form. (Later we will show how a decimal with an infinite number of digits does the job.) But we can find decimals that are close to these fractions. We explore several ways.

What number n gives the fraction $\frac{n}{100}$ which is closest to $\frac{1}{3}$? We may also ask, what whole number n comes closest to solving $\frac{n}{100} = \frac{1}{3}$ or $3n = 100$? Clearly, $n = 33$ is too small and $n = 34$ is too large. 99 is closer to 100 than 102, so we say,

$$\frac{1}{3} \approx \frac{33}{100} = .33$$

We use $\approx$ to mean, "is approximately equal to."

Similarly for $\frac{5}{6}$, we ask for the whole number p which makes $\frac{p}{1000}$

closest to $\frac{5}{6}$. We consider $6p = 5000$. Clearly, $p = 833$ and $p = 834$ are the candidates since

$$6 \times 833 = 4998 < 5000 < 5004 = 6 \times 834$$

Choosing 833 gives $\frac{5}{6} \approx .833$.

A two-place decimal approximation to $\frac{3}{7}$ comes from $\frac{3}{7} = \frac{m}{100}$ or $7m = 300$. Candidates are $m = 42$ and 43 with $294 < 300 < 301$. Hence we choose $\frac{3}{7} \approx \frac{43}{100} = .43$.

If we had wanted a three- or four-place decimal approximation to $\frac{3}{7}$, we would consider equations $7m = 3000$ or $7m = 30000$. These are division problems, and we may consider them all at once in the following layout.

$$\begin{array}{r} 42 \\ 7\overline{)300} \\ \underline{294} \\ 6 \end{array} \qquad \begin{array}{r} 428 \\ 7\overline{)3000} \\ \underline{2996} \\ 4 \end{array} \qquad \begin{array}{r} 4285 \\ 7\overline{)30000} \\ \underline{29995} \\ 5 \end{array}$$

Can you see that $\frac{3}{7} = \frac{4^+}{10} = .4^+$, $\frac{3}{7} = \frac{42}{100^+} = .42^+$, $\frac{3}{7} = \frac{428^+}{1000} = .428^+$, and $\frac{3}{7} = .4285^+$? The + sign at the right means "and a bit more, but not enough to increase the last digit." If we wish the best approximations, we would choose from the following:

$$\frac{3}{7} = .4^+ = .43^- = .429^- = .4286^-$$

where the − sign at the right means "and a bit less."

A thinking chain is: "I wish a two-place approximation for $\frac{3}{7}$. Dividing $\frac{300}{7}$ gives $42\frac{6}{7} = 42^+ = 43^-$. Hence, $\frac{3}{7} = \frac{43^-}{100} = .43^-$. I now wish a three-place decimal for $\frac{3}{7}$. Division gives $\frac{3000}{7} = 428\frac{4}{7}$. Hence $\frac{3}{7} = \frac{428^+}{1000} = .428^+$. For a four-place decimal, I find $\frac{30000}{7} = 4285\frac{5}{7}$. Hence, $\frac{3}{7} = \frac{4285^+}{10000} = .4285^+$."

Instead of making separate problems, we are tempted to say, "Instead of dividing 7 into 300, then 3000, then 30,000, let me write 3 with many zeros, and go through the division algorithm as many times as I choose." This layout is convenient. Can you read the quotients, including the fractional parts for the four division problems, $\frac{30}{7}, \frac{300}{7}, \frac{3000}{7}$, and $\frac{30000}{7}$? Each problem arose from approximating $\frac{3}{7}$ by $\frac{n}{d}$, where d was one of the numbers 10, 100, 1000, 10000. Can you see which value of d goes with $n = 428$?

```
    4285
7)300000000
  28
   20
   14
    60
    56
     40
     35
      5
```

An easy memory device is to place a little mark ($_\wedge$) after the 3. Then the upper lines in the layout look like

```
     428|57
7)3,000|00
```

(with the mark $_\wedge$ after the 3)

If the process is stopped, say as indicated by the vertical line, the three zeros between the mark and the line show that the appropriate denominator is $1000 = 10^3$. Then $\frac{3}{7} = \frac{428}{1000} = .428$.

Notice that the decimal point is just to the left of the 4. If we anticipate this result, and place the decimal point in the division layout, we see that it falls just above the mark shown.

```
    .428
7)3^00000
```

Since the mark in the dividend comes in exactly the place (to the right of the units digit) the decimal point would go, we may write the layout

```
    .428
7)3.00000
```

We use the same method to get decimals close to $\frac{1}{3}$ and $\frac{5}{6}$:

$$3\overline{)1.0000}^{\;.3333} \qquad 6\overline{)5.0000}^{\;.8333}$$

We see that $\frac{1}{3} \approx .3$ or $.33$ or $.333$ and $\frac{5}{6} \approx .8$ or $.83$ or $.833$.

If we apply the method to get a decimal close to $\frac{1}{99}$, a new problem arises.

$$\begin{array}{r} 10101 \\ 99\overline{)1_{\wedge}000000} \\ \underline{99} \\ 100 \\ \underline{99} \\ 100 \\ \underline{99} \\ 1 \end{array}$$

Since $\frac{1}{99} \times \frac{10101}{1000000} = .010101$, a zero before the first 1 is necessary. Careful placement of decimal points bears this out.

$$\begin{array}{r} .01 \\ 99\overline{)1.00000} \\ \underline{99} \end{array}$$

You may want to make up rules like the following for approximating common fractions by decimals.

1. Write the fraction in division form, say $13\overline{)10}$ for $\frac{10}{13}$.

2. Place a decimal point after the units digit in the dividend and another directly above it in the quotient. $13\overline{)10.}^{\;\;.}$

3. Put as many zeros as you please and divide. Keep the digits properly lined up. The decimal point is correctly placed in the quotient. Hence

```
      .76
13)10.000
    9 1
      90
      78
      120
```
and $\frac{10}{13} = .76^{+}$

EXAMPLE 1 $\frac{3}{4}$

```
    .75
4)3.00
  2 8
    20
    20
```
and $\frac{3}{4} = .75$

EXAMPLE 2 $\frac{17}{25}$

```
     .68
25)17.00
   15 0
    2 00
    2 00
```
and $\frac{17}{25} = .68$

EXAMPLE 3 $\frac{5}{16}$

```
    .31250
16)5.00000
   4 8
     20
     16
      40
      32
       80
       80
```
and $\frac{5}{16} = .31250 = .3125$

As a check, reduce $\frac{3125}{10000}$ to lowest terms.

EXAMPLE 4 Write $\frac{3}{41}$ as a decimal.

$$\begin{array}{r} .073 \\ 41\overline{)3.000} \\ \underline{2\ 87} \\ 130 \\ \underline{123} \\ 7 \end{array} \qquad \text{and } \frac{3}{41} = .073^{+}$$

Note the zero in the quotient because $30 < 41$.

EXAMPLE 5 Write $\frac{1}{9}, \frac{1}{99}, \frac{1}{999}$ as decimals.

$$\begin{array}{r} .1111 \\ 9\overline{)1.0000} \end{array} \qquad \text{and } \frac{1}{9} = .1111^{+}$$

$$\begin{array}{r} .010101 \\ 99\overline{)1.000000} \end{array} \qquad \text{and } \frac{1}{99} = .010101^{+}$$

$$\begin{array}{r} .001001001 \\ 999\overline{)1.000000000} \end{array} \qquad \text{and } \frac{1}{999} = .001001001^{+}$$

You may have noted that only fractions whose denominators are multiples of 2 and 5 have decimal forms with a finite number of digits.

The method of *undetermined digits* is used many times in mathematics and also works here. The method writes the answer with letters for digits —the letters are the undetermined digits. Then steps are taken that determine the digits one at a time. We illustrate.

Let $\frac{1}{4} = .abcd$, where a, b, c, d are the digits in the tenths, hundredths, thousandths, . . . place, respectively. We may put in letters for as many digits as we please. Next, multiply both sides by 10, $\frac{10}{4} = 2\frac{1}{2} = a.bcd$. The mixed number on the left has a whole-number part (2) and a fractional part ($\frac{1}{2}$). The mixed number on the right has a whole-number part (a) and a fractional part ($.bcd$). The whole-number parts are equal and so are the fractional parts. Hence

$$2 = a$$

and

$$\frac{1}{2} = .bcd$$

Multiplying the last equation by 10 yields

$$5 = b.cd$$

Equating whole-number and fractional parts on both sides gives

$$5 = b$$

and $$0 = .cd$$

We have determined $b = 5$. From the last equation $c = d = 0$. Hence

$$\tfrac{1}{4} = .abcd = .2500 = .25.$$

Similarly $\tfrac{3}{7} = .pqrst$

$\tfrac{30}{7} = 4\tfrac{2}{7} = p.qrst.$

Hence $p = 4$ (whole-number parts)

and $\tfrac{2}{7} = .qrst$ (fractional parts).

Multiply by 10, $\tfrac{20}{7} = 2\tfrac{6}{7} = q.rst.$

Hence $q = 2$ and $\tfrac{6}{7} = .rst.$

Multiply by 10, $\tfrac{60}{7} = 8\tfrac{4}{7} = r.st.$

Hence, $r = 8$ and $\tfrac{4}{7} = .st.$

At this state, we have $\tfrac{3}{7} = .pqrst = .428^{+}$.

24.7 REPEATING DECIMALS AND COMMON FRACTIONS You probably noticed that fractions with denominators whose prime factors are 2 and 5 have an exact decimal form, while others like $\tfrac{1}{3}$ and $\tfrac{5}{6}$ do not. As you looked at several decimal forms for $\tfrac{1}{3}$ and $\tfrac{5}{6}$ you noticed that the digits repeated. For example, $\tfrac{1}{3} = .333^{+}$ and $\tfrac{5}{6} = .8333^{+}$. Similarly, $\tfrac{4}{11} = .363636^{+}$. Looking at the division process, we see

```
     .36
 11)4.00
    3 3
     70
     66
      4
```

The new dividend digit is the same as the original. Hence, we expect the digits in the quotient to repeat.

Alternatively, using the method of undetermined digits, $\frac{4}{11} = .abcdef$. Multiplying both members by 100, we get

$$\frac{400}{11} = 36\tfrac{4}{11} = ab.cdef$$

The whole-number parts are equal and the fractional parts are equal. Hence, $a = 3$, $b = 6$, and $\frac{4}{11} = .cdef$. But $\frac{4}{11} = .abcdef$. So $.abcdef = .cdef$ and $c = a$ and $d = b$. Hence the cycle repeats and we know $\frac{4}{11} = .363636 \ldots$.

To avoid confusion as to how the decimal is continued, we often write $\frac{4}{11} = .\overline{36}$ or $.36\overline{36}$, where the digits under the bar are repeated indefinitely. The set of digits that repeats, in this example the 36, is called the repetend. Of course, we could also write $\frac{4}{11} = .\overline{3636} = .3\overline{63}$, and so on. The number of digits in the shortest repetend is called the *period*. Such decimals are called *repeating decimals* and $\frac{4}{11}$ is represented by a repeating decimal of period 2. Repeating decimals are called infinite decimals, since we can always find another nonzero digit for any decimal approximation.

While no finite decimal exactly equals $\frac{4}{11}$, the more digits we take, the closer approximation we get. For example,

$$\frac{4}{11} - .36 = \frac{400 - 396}{1100} = \frac{4}{1100}$$

$$\frac{4}{11} - .3636 = \frac{40000 - 39996}{110000} = \frac{4}{110000}$$

showing that .3636 is much closer to $\frac{4}{11}$ than is .36.

The appearance of infinite repeating decimals leads us to two useful theorems. (1) Every fraction has an infinite repeating decimal representation. (The repeating digits will be zero for fractions such as $\frac{2}{5}$.) (2) Every infinite repeating decimal is another name for a fraction. Finding the repeating decimals for the following fractions will show what these theorems mean. Divide until you are sure of the repetend.

$$\frac{5}{9}, \frac{7}{11}, \frac{3}{8}, \frac{59}{99}, \frac{247}{1111}, \frac{22729}{30123}, \frac{1}{7}$$

What convinced you that you had the repetend, and that you knew what the period was? Can you make some rule about the period for a fraction? For example, will the period for $\frac{3}{17}$ be greater than 3? Greater than 17? Smaller than 17?

The steps are easy in theorem 1: Every fraction $\frac{a}{b}$ can be represented by an infinite repeating decimal. Look at the division algorithm for $\frac{39}{11}$.

$$\begin{array}{r} 3.54 \\ 11\overline{)39.00} \\ \underline{33} \\ 60 \\ \underline{55} \\ 50 \text{ etc.} \end{array}$$

Can you see in this a series of division problems with single-digit quotients and remainders less than 11? For example, using the multiplication form related to division, we have

$$39 = 3 \times 11 + 6$$
$$60 = 5 \times 11 + 5$$
$$50 = 4 \times 11 + 6$$

The successive quotients are 3, 5, 4, ... The remainders are 6, 5, 6, ... The dividends are 39, 60, 50, . . . and the divisor is always 11. We note that each dividend is 10 times the previous remainder. Let us take a few more steps:

$$39 = 3 \times 11 + 6$$
$$60 = 5 \times 11 + 5$$
$$50 = 4 \times 11 + 6$$
$$60 = 5 \times 11 + 5$$
$$50 = 4 \times 11 + 6$$
$$60 = 5 \times 11 + 5$$

We have $\frac{39}{11} = 3.54545 \ldots$. Can you predict what the next digits will be? The next 3 digits? We notice that not only the quotient digits repeat, but also the remainders.

We look at $1\frac{1}{8}$. Divide and identify the following steps:

$$11 = 1 \times 8 + 3$$
$$30 = 3 \times 8 + 6$$
$$60 = 7 \times 8 + 4$$
$$40 = 5 \times 8 + 0$$
$$00 = 0 \times 8 + 0$$

The fourth remainder and all further quotients are zero. Hence,

$$\tfrac{11}{8} = 1.375000\ .\ .\ . = 1.375\overline{0}$$

Consider $\frac{5}{9}$. Do you get the following?

$$5 = 0 \times 9 + 5$$
$$50 = 5 \times 9 + 5$$
$$50 = 5 \times 9 + 5 \text{ and } \tfrac{5}{9} = .555\ .\ .\ .$$

Each division is the same. The remainders are the same. Hence, the next dividends are the same, and the next quotients.

For $\frac{5}{27}$, do you get the following?

$$5 = 0 \times 27 + 5$$
$$50 = 1 \times 27 + 23$$
$$230 = 8 \times 27 + 14$$
$$140 = 5 \times 27 + 5$$
$$50 = 1 \times 27 + 23$$
$$230 = 8 \times 27 + 14$$
$$140 = 5 \times 27 + 5$$
$$50 = 1 \times 27 + 23$$

We might guess that $\frac{5}{27} = .185185185\overline{185}$.

When do we first see what the decimal will be? We see that if the same dividend is divided by 27, the quotients will be the same and the remainders the same. Since the next dividends are 10 times the remainders, these next dividends are the same. The process continues and we have the repeating decimal as soon as we have two remainders (and hence next dividends) the same.

Will this always happen? We appeal to a famous principle of counting numbers known as the Pigeon Hole Principle. It is said that a famous pigeon keeper watched his 28 pigeons fly home to their 27 nests every night. He noticed a nest with at least two pigeons in it. Wanting his pigeons to sleep alone, he tried various devices. He fed them onions and garlic, painted the nests different colors, but to no avail. Always, there was at least one nest with two pigeons. So he announced the principle, "If $n + 1$ (or more) pigeons fly to n nests, at least one nest will have at least two pigeons."

Using this principle we could have predicted that $\frac{5}{27}$ would have a repeating decimal representation. Each of the remainders of the sequence of divisions must be less than 27—that is, must be one of the 27 numbers 0, 1, 2, . . . , 26. But if we keep dividing we'll eventually have 28 remainders (and more). Hence, at least two of these 28 remainders must be the same by the PHP. (Can you identify pigeons and nests?)

For $\frac{5}{27}$ there were only 3 remainders—5, 23, and 14, rather than all possible 27. The pigeons used only three nests.

The same process can be applied to any fraction. Hence every fraction is represented by a repeating decimal. This theorem tells us that a nonrepeating infinite decimal is not a fraction. For example, .121121112111121111112 . . . is not a repeating decimal and does not represent a fraction. This number is irrational.

The other major theorem is: Every repeating decimal represents a fraction. We use front-end multiplication and subtraction to great advantage. We show the process by examples.

EXAMPLE 1

$$\text{Let } n = .666\overline{6}$$
$$\text{Then } 10n = 6.66\overline{6}$$

Subtracting the first from the second gives $9n = 6$. Hence $n = \frac{6}{9} = \frac{2}{3}$.

EXAMPLE 2

$$\text{Let } m = .31515\overline{15}$$
$$\text{Then } 100m = 31.515\overline{15}$$
$$\text{Subtracting, } 99m = 31.2 \text{ or } 990m = 312$$
$$\text{Hence } m = \frac{312}{990} = \frac{52}{165}$$

NOTE $100m - 1m = (100 - 1)m = 99m$.

You may wish to divide these fractions to show they yield the given repeating decimal.

Here is another way of finding the common fraction for a repeating decimal. We show by examples.

$$.333\ldots = 3 \times .111\ldots = 3(\tfrac{1}{9}) = \tfrac{1}{3}, \text{ since } \tfrac{1}{9} = .\overline{1}$$

Similarly, $.777\ldots = 7 \times .111\ldots = 7(\tfrac{1}{9}) = \tfrac{7}{9}$

Also, $.342342\overline{342} = 342 \times .001001001001\ldots.$

Can you guess what $.001001001\ldots$ is? Recall

$$\tfrac{1}{9} = .111\ldots$$
$$\tfrac{1}{99} = .01010101\ldots$$
$$\tfrac{1}{9999} = .0001000100010001\ldots$$

Hence $.342\overline{342} = 342 \times .001001001\ldots = 342(\tfrac{1}{999}) = \tfrac{38}{111}$

In $\tfrac{1}{3} = .333\overline{3}$ and $\tfrac{5}{11} = .454545\ldots$, every digit is part of a repetend. This is not true in general as shown by $\tfrac{5}{6} = .833\overline{3}$ and $\tfrac{211}{900} = .234444\ldots.$

Some fractions have two repeating decimals. For example,

$$\tfrac{1}{2} = .50\overline{0} = .49\overline{9} \text{ and } 1 = 1.00\overline{0} = .99\overline{9}$$

Here are two arguments that $.9\overline{9} = 1.0\overline{0}$.:

1. Let $x = .99\overline{9}$ then

$10x = 9.99\overline{9}$ subtracting

$9x = 9$

$x = 1$

2. The sequence of numbers, 9, .99, .999, . . . gets closer and closer to 1.00. It seems reasonable to assert that the infinite decimal is equal to 1. Hence, we see that any finite fraction (which is a repeating fraction with repetend $\overline{0}$) also has a repeating fraction with repetend $\overline{9}$. With the exception of these cases, all fractions have only one repeating decimal form.

Some people like to point out that while a finite decimal such as .345

is a definite set of digits, a repeating decimal is more like a machine which, on demand, produces another decimal. The notation $.1\overline{6}$ is looked on as telling you how to get the fifth digit rather than itself telling you what the fifth digit is. In .2364891 we are specifically told the fifth digit is 8. Does the notation $.3\overline{7694}$ tell you what the seventeenth digit is or how to find the seventeenth digit?

24.8 ADDING DECIMAL FRACTIONS Adding decimal fractions is much easier than adding common fractions. Recall that in adding $\frac{3}{7} + \frac{2}{9}$ we first rename with a common denominator and then write the sum of the numerators over that common denominator. $\frac{3}{7} + \frac{2}{9} = \frac{27}{63} + \frac{14}{63} = \frac{41}{63}$. Similarly, in adding $\frac{3}{10} + \frac{13}{100}$ we write $\frac{30}{100} + \frac{13}{100} = \frac{43}{100}$. In decimal form, $.3 + .13 = .30 + .13 = .43$. We can easily use a vertical layout.

$$\begin{array}{r} .30 \\ \underline{.13} \\ .43 \end{array}$$

We keep the decimal points in a vertical line or column. In this way the digits with the same place value are in a vertical column. Then to add digits with the same place value, we merely add those in a column. Some people compare this with writing common fractions with common denominators.

Suppose we wish to add several fractions. For example, $\frac{2}{10} + \frac{23}{100} + \frac{18}{1000} + \frac{1}{10}$. Writing in decimal form, we have $.2 + .23 + .018 + .1$. The most convenient common denominator is 1000. To write all fractions with this denominator is easy—all we need to do is join zeros on the ends of those decimals with fewer than three places. Hence we get $.200 + .230 + .018 + .100$. The vertical layout is in the margin. The sum is $.548 = \frac{548}{1000}$.

$$\begin{array}{r} .200 \\ .230 \\ .018 \\ \underline{.100} \\ .548 \end{array}$$

Carrying (or converting or regrouping) and the adding of mixed numbers goes smoothly and automatically. Suppose we wished to add $23\frac{1}{4} + 13\frac{4}{5} + 4\frac{7}{1000} + 21\frac{19}{1000}$. In common fraction notation, this involves (1) finding a common denominator for the fractions, (2) renaming all the fractions, (3) adding the fractions, (4) finding the integral part of the sum, (5) adding all the whole numbers, (6) writing the sum of the whole numbers next to the fractional part of the sum of the fractions. We can't avoid these acts but they become almost automatic with the great ease brought by decimals. We rename the numbers as mixed decimals, 23.25 + 13.8 + 4.007 + 21.019. In vertical format we keep the decimal points in a vertical line. It helps to join some zeros.

$$\begin{array}{r} 23.250 \\ 13.800 \\ 4.007 \\ \underline{21.019} \\ 62.076 \end{array}$$

Compare the steps in the addition in common fractions and with these in addition in decimal fractions.

24.9 SUBTRACTING AND COMPARING DECIMAL FRACTIONS Subtraction is similarly easy. We write the two numbers with the decimal points lined up, subtract as though they were whole numbers, putting the decimal point in line in the difference. For example,

$$\begin{array}{r} 23.78 \\ -\ 18.24 \\ \hline 5.54 \end{array} \quad \text{and} \quad \begin{array}{r} 12.003 \\ -\ 8.742 \\ \hline 3.261 \end{array}$$

Decimal fractions are easy to compare. The larger of .07619 and .07621 is .07621. The larger of .0183 and .0097314 is .0183. We merely start at the decimal point and compare the digits place by place. The larger number is the one that first has a larger digit. In comparing .07619 with .07621, we think, "Both tenths-place digits are the same, 0. Go to the next place, the digits are the same, 7. Go to the next place and the digits are the same, 6. Go to the next place and $2 > 1$. Therefore $.07621 > .07619$." Note that we don't worry about either the size or the number of digits further to the right.

24.10 MULTIPLYING DECIMAL FRACTIONS We recall that

$$\frac{a}{b} \times \frac{c}{d} = \frac{ac}{bd} \quad \frac{3}{7} \times \frac{8}{5} = \frac{24}{35} \quad \frac{3}{10} \times \frac{15}{100} = \frac{45}{1000}$$

In decimals we have

$$.3 \times .15 = .045$$

$$.4 \times .02 = \frac{4}{10} \times \frac{2}{100} = \frac{8}{1000} = .008$$

$$.4 \times .25 = \frac{4}{10} \times \frac{25}{100} = \frac{100}{1000} = \frac{1}{10} = .1$$

The new denominator, the product of the old denominators, is 10 raised to the sum of the old powers, $10^1 \times 10^2 = 10^3$. Since the power of 10 is shown by the number of places in the decimal, the number of places to the right of the decimal point in the product is the sum of the numbers of places in the two factors. The numerator is the product of the numerators, the numbers shown by the digits if there were no decimal point. Further examples illustrate this. Consider 3.6×4.3. Now $36 \times 43 = 1548$. Since there is one decimal place in each factor, there are $1 + 1 = 2$ in the product. Hence $3.6 \times 4.3 = 15.48$. Multiply 10.5×3.12; we note that $105 \times 312 = 32760$. Counting decimal places, we get $1 + 2 = 3$. Hence the product of $10.5 \times 3.12 = 32.760$.

Sometimes counting decimal places is a nuisance. People locate the decimal point by approximation. Since the digits in 10.5×3.12 are 32760, the product must be one of the following values: 3.2760, 32.760, 327.60, 3276.0, ... We can see that 10.5×3.12 is about $10 \times 3 = 30$, so the exact product must be 32.760—the only one of the possible values that is close to 30.

PROBLEM $1.708 \times .3125$. *Think* $1708 \times 3125 = 5337500$. The product is about $2 \times .3 = .6$, which locates the decimal point, so $1.798 \times .3125 = .53375$. Counting decimal places, there are $3 + 4 = 7$ in the product, which is then .5337500. Note that in counting decimal places one must be careful with zeros at the right end, even though they do not affect the value otherwise.

Multiplication of infinite decimals raises questions. How do you count the number of places in a factor? How do you mark off the number of places in the product? It can't be done. Many people are satisfied to replace the infinite decimals by finite approximations and use these as factors. A common rule of thumb is to take one or two more significant figures in each factor than you wish in the product. For example, consider $.\overline{33} \times .\overline{27}$.

We try

$$.33 \times .27 = .0891$$
$$.333 \times .272 = .090576$$
$$.3333 \times .2727 = .09089091$$

It seems the decimal product begins .09 . . . , with the next place uncertain. We can get a guarantee by using upper and lower bounds. Since $.333 < .\overline{33} < .334$ and $.272 < .\overline{27} < .273$, we can be sure that

$$.333 \times .272 < .\overline{33} \times .\overline{27} < .334 \times .273$$

or

$$.090576 < .\overline{33} \times .\overline{27} < .091182$$

This guarantees that the product begins either .090 or .091. More accurate results will come from bounds that are closer; that is, contain more than three digits.

```
.27272727 . . .
.33333333 . . .
---------------
.06
 21
  6
----
.087
   6
  21
   6
-----
.0903
   21
    6
   21
    6
------
.09084
     6
    21
     6
    21
     6
-------
.090900
```

We may wish to use front-end multiplication. Any layout is messy, but we can go as far as we wish and have a good idea of the correctness of any digit as we go along. We need not start over when we want more accuracy. At the first step we have $.3 \times .2 = .06$. We can be confident of the decimal point, but not of the 6. At the next stage our product is .087. We note that at each stage we add one more partial product than at the previous stage. Each partial product is either 6 or 21. At the next stage we add three partial products, which will add about 3 to 7—we don't feel confident of 7 in .087 and the 8 might become 9. The zero will not become 1.

At the next stage we have .0903. The next four partial products will change the 3, perhaps the zero—but certainly the 9 is secure. At the next stage we have .09084. The next five partial products will change 4, may change 8 to 9, but the .090 is secure.

We can always convert two repeating decimals to common fractions, multiply these fractions, and revert to a repeating decimal. In this example, $.\overline{33} \times .\overline{27} = \frac{1}{3} \times \frac{3}{11} = \frac{1}{11} = .09\overline{09}$.

24.11 DIVISION WITH DECIMAL FRACTIONS Division is much like division with counting numbers, with the added problem of locating the decimal point in the quotient. First we consider division by a counting number that comes out even.

EXAMPLE .36 ÷ 4. We may think of this as $\frac{1}{4} \times \frac{36}{100} = \frac{36}{4} \times \frac{1}{100} = 9 \times \frac{1}{100} = .09$. That is, we change the problem to dividing counting numbers and then dividing by 100. The process is simplified by the layout,

$$4\overline{).36} \rightarrow 4\overline{)36}^{\,9} \text{ and divide the quotient by 100} \rightarrow 4\overline{)\ 36}^{\,.09}$$

We can keep track of these steps by the mechanical trick of (*a*) writing the division layout, (*b*) dividing as though the dividend were an integer, (*c*) placing a decimal point in the quotient directly above the decimal point in the dividend and inserting zeros as needed for placeholders.

To reduce oversights and careless error, many people place the decimal point in the numerator before dividing. This is recommended.

EXAMPLE .056 ÷ 7

$$7\overline{).056} \rightarrow \overset{.}{7\overline{).056}} \rightarrow \overset{.008}{7\overline{).056}}$$

We check by multiplying: 7 × .008 = .056, so .056 ÷ 7 = .008.

In the next example, the division does not come out even.

EXAMPLE .49 ÷ 6. It is convenient to rewrite with several extra zeros as .490000 ÷ 6. Following the steps *a*, *b*, *c*, above, we have

$$\overset{.}{6\overline{).490000}} \rightarrow \overset{.08}{6\overline{).490000}} \rightarrow \overset{.081666}{6\overline{).490000}}$$

We can imagine continually putting in extra zeros. The repetition in the division process shows that the quotient will be a repeating decimal, $.0816\overline{6}$.

This pattern always happens. Fortunately, the position of the decimal point is decided at the start, and is not affected by putting extra zeros on the dividend.

Dividing a mixed decimal by a divisor ending in one or more zeros is straightforward.

EXAMPLE 4.2 ÷ 290.

$$4.2 \div 290 \rightarrow \overset{.}{290\overline{)4.20000}} \rightarrow \overset{.014}{290\overline{)4.20000}}$$

```
     .014
290)4.20000
    2 90
    1 300
    1 160
      140
```

While we know the answer is a repeating decimal, the period may be large, since the divisor is 290. We will not determine the repetend, but satisfy ourselves with $4.2 \div 290 = .014^+$.

We can check the result roughly by rounding the terms to $4 \div 300$, which is about $\frac{1.3}{100}$, or about .013. Since the digits 14 are about the same as 13, and the decimal point shows each a little bigger than .01, we have a good check.

Division of a finite decimal by a finite decimal is very similar to what has gone before. It is usually convenient to change to the division of a decimal by a counting number by multiplying both terms by convenient powers of 10. We show by example.

EXAMPLE 1 $14.4 \div 2.4$. Multiplying $\frac{14.4}{2.4} \times \frac{10}{10}$, we get $144 \div 24$, or 6.

EXAMPLE 2 $1.5\overline{)2.85}$. Multiplying both terms by 10 or 100, we get either $15\overline{)28.5}$ or $150\overline{)285}$. The first is considered easier, although some may prefer the second. Everyone is free to choose the one he prefers, although to fix ideas for a discussion a teacher may ask members of a class to use just one. Following the earlier examples,

$$15\overline{)28.5} \rightarrow 15\overline{)\overset{.}{28.5}} \rightarrow \begin{array}{r} 1.9 \\ 15\overline{)28.5} \\ \underline{15} \\ 13\;5 \\ \underline{13\;5} \end{array}$$

Hence, $2.85 \div 1.5 = 1.9$.

$$.072\overline{)3.46} \rightarrow 072.\overline{)3460.} \rightarrow 72\overline{)\overset{.}{3460.}} \rightarrow \begin{array}{r} 48.05 \\ 72\overline{)3460.000} \\ \underline{288} \\ 580 \\ \underline{576} \\ 4\;00 \\ \underline{3\;60} \\ 40 \end{array}$$

EXAMPLE 3 An example that does not come out even is $3.46 \div .072$. The sequence of shed layouts showing the steps appears in the margin. In the first step we multiplied both terms, divisor and dividend, by 1000. This changed the divisor to a counting number. In step 2 we dropped the now useless zero in 072 and placed a decimal point above the shed directly above the decimal point in 3460. In step 3 we placed some extra zeros in the dividend as placeholders and divided as though the numbers were counting numbers. We get $3.46 \div .072 = 48.05^+$.

As a check, we round the digits 346 to 35 and 072 to 7 and $\frac{35}{7} = 5$. The leading digits in the product are 48, very close to 5 as a leading digit. Increasing both terms slightly, we have $3.46 \div .072 \approx 4 \div .1 = 40$. This checks the decimal point in the result. With the first four digits being 4805 and the answer being about 40, the answer should be 48.05^+ and not 4.805^+ or 480.5^+ or . . . .

A close look will show that the answer is truly a repeating decimal. Can you find the repetend?

After multiplying the terms by a power of 10, instead of rewriting the layout we often insert a new decimal point in the divisor and dividend as follows:

```
            .
1.5^)3.4^5
```

We use a different symbol for the new decimal point. With this notation it is easy to check for copying errors. (See the examples below.)

```
                   630.
EXAMPLE 4   .006^)3.780^   and 3.78 ÷ .006 = 630.
```

```
                4.7444 . . .
EXAMPLE 5   .09^).42^7000 . . .
```

```
                  .028+
EXAMPLE 6   3.02^).08^650
                   6 04
                   2 610
                   2 416
                     194
```

We leave discussing the division of a repeating decimal by a counting number or a finite decimal as an exercise for you. Review multiplication of repeating decimals to figure out what should be done and said.

Division of a repeating decimal by a repeating decimal is a dull task, so wretchedly awkward that if we really had to do it often we'd probably junk decimals altogether and use common fractions. People often do one of the following:

1. Round off the repeating decimals to a convenient size, and then divide these finite decimals. The relative error in the quotient is approximately the difference in the relative errors of rounding off in the given numbers. Using upper and lower bounds will give more certain information.

EXAMPLE 7 $7.2\overline{2} \div .5\overline{5}$. Using rounded figures

$$\frac{7}{.6} = 11.666^{+} \qquad \frac{7.2}{.56} = 12.85 \qquad \frac{7.22}{.556} = 12.985$$

2. Convert the repeating decimals to common fractions and divide.

EXAMPLE 8 $7.\overline{2} \div .5\overline{5} = 7\frac{2}{9} \div \frac{5}{9} = \frac{65}{5} = 13$

EXAMPLE 9 $7.4\overline{76} \div 13.35\overline{123}$. It is easier if all the numbers after the decimal point repeat, so we multiply both by 100. Further, we write so that both numbers have the same period.

$$\begin{aligned} 7.4\overline{76} \div 13.35\overline{123} &= 747.\overline{67} \div 1335.\overline{123} \\ &= 747.\overline{676767} \div 1335.\overline{123123} \\ &= 747\,\frac{676767}{999999} \div 1335\,\frac{123123}{999999} \\ &= \frac{74767020}{1335121788} = .5600058561\ \ldots \end{aligned}$$

See chapter 26 for more on computation with decimals.

24.12 LOCATING DECIMAL POINTS BY APPROXIMATION The decimal point in a product or quotient is ordinarily located by counting places, keeping decimal points in line, and other careful steps in the layout. For many it is quicker and more accurate to approximate the answer. This is especially true when using a slide rule or desk calculator. Ignoring the decimal point, the calculator operates with the given numbers, obtaining the leading sequence of digits for the answer. Then the decimal point is placed to get the proper size.

EXAMPLE $.04378 \times 2.6035$. The sequence of digits in the product is 113981230. The answer will be one of: 11.398123, 1.1398123, .11398123, .011398123, .0011398123. Rounding the factors to $.04 \times 3$, we see that the product is about .12. The number close to .12 is .11398123. Hence .04378 × 2.6035 = .11398123.

Similarly, $\frac{36.3}{.104}$ produces the sequence of digits 3490+. Rounding the terms to $\frac{40}{.1} = 400$, we place the decimal point to get an answer close to 400, $\frac{36.3}{.104} = 349.0^{+}$.

24.13 BASIMAL FRACTIONS This is the name for what corresponds to decimal fractions in other bases. Strictly speaking, we should have a separate name for each base, such as "binary fractions" in base two, "quinary fractions" for base five, and so on. Since such fractions are used so seldom, we will not give them separate names. Where clarity is needed, many people would say "basimal fractions in base five," specifically naming the base. Some say "decimal fractions in base five," extending the word *decimal* to mean any fraction of the form .*abc* . . . , but this is frowned on.

Perhaps the main value of basimal fractions is pedagogical. The general structure and theory is exactly the same as for decimal fractions. Do you understand the structure of decimal fractions? From long experience in calculating with decimal fractions, many people feel fairly easy with them. Still they occasionally make the work harder than necessary because they don't fully see the pattern. When similar problems are worked in basimal fractions the pattern and structure become clearer. Warning: a worker must be looking for the pattern. It is possible to learn to manipulate basimal numbers by rote just as to manipulate decimal numbers by rote and never see the relation between them.

We limit our discussion to transforming fractions to base-five form. We restrict fractions to denominators of 10, 10^2, . . . as before, but now these denominators are $b, b^2, \ldots$, where b is the number base.

EXAMPLE 1 $\frac{2}{5} = \frac{2_{\text{five}}}{10} = .2_{\text{five}}$ (read "point 2 base five")

EXAMPLE 2 $\frac{10_{\text{ten}}}{25} = \frac{20_{\text{five}}}{100} = .20_{\text{five}} = .2_{\text{five}}$

EXAMPLE 3 $1\frac{23_{ten}}{125} = 1\frac{43_{five}}{1000} = 1.043_{five}$

The last example is read, "one point zero four three, base five." In particular we do not read $.2_{five}$ as "two-tenths base five." We could say, "two-fifths."

To convert a base-five decimal, we convert first to a common fraction, then to a decimal as follows:

EXAMPLE 4 $.32_{five} = \frac{32_{five}}{100} = \frac{17_{ten}}{25} = .68_{ten}$

EXAMPLE 5
$$\begin{aligned} .234_{five} &= (\frac{2}{5} + \frac{3}{25} + \frac{4}{125})_{ten} \\ &= (.4 + .12 + .032)_{ten} \\ &= .552_{ten} \end{aligned}$$

Alternatively,

$$.234_{five} = \frac{234_{five}}{1000} = \frac{69_{ten}}{125} = \frac{552_{ten}}{1000} = .552_{ten}$$

To convert a base-ten decimal to a base-five decimal, the easiest way is by the method of undetermined digits. To convert $.76_{ten}$ to base five, we suppose as follows:

EXAMPLE 6 $.76_{ten} = .abcd_{five}$ where $a, b, \ldots$ are digits to be determined. Multiplying both sides by 5, on the left we follow common practice and on the right move the basimal point one place to the right.

$3.80_{ten} = a.bcd_{five}$ Then $3 = a$ and $.8_{ten} = .bcd_{five}$. Multiplying the fractional parts by 5,

$4.0_{ten} = b.cd_{five}$. Hence $4 = b$ and $.0 = .cd$.

Hence, $.76_{ten} = .abcd_{five} = .34_{five}$

EXAMPLE 7 Convert $.325_{\text{ten}}$ to base five.

Let $.325_{\text{ten}} = abcd_{\text{five}}$	Multiply fractional part by 5.
$1.625_{\text{ten}} = a.bcd_{\text{five}}$	Hence, $1 = a$ and $.625_{\text{ten}} = bcd_{\text{five}}$
$.625 \times 5 = 3.125_{\text{ten}} = b.cd_{\text{five}}$	Multiply fractional part by 5. So, $3 = b$ and $.125_{\text{ten}} = .cd_{\text{five}}$
$.125 \times 5 = .625_{\text{ten}} = c.def_{\text{five}}$	Multiply fractional part by 5 and make room for more digits.

Hence, $0 = c$ and $.625_{\text{ten}} = .def_{\text{five}}$. Since $.625 = .bcd$ above, we see $d = b$, $e = c$, $f = d$, etc. Hence,

$$.325_{\text{ten}} = .abcdef_{\text{five}} = .1303030\ldots_{\text{five}}$$

EXAMPLE 8 Convert 284.67_{ten} to a base-five numeral. We treat the integral and fractional parts separately. Dividing successive quotients by 5, we get

```
5|284
 5|56    4
 5|11    1
  5|2    1
    0    2
```

and $284_{\text{ten}} = 2114_{\text{five}}$. Multiplying successive fractional parts by 5, we get

```
 .67
   5
3.35
   5
1.75
   5
3.75
   5
3.75
   5
3.75
```

Hence $.67_{\text{ten}} = .3133\overline{3}_{\text{five}}$ and $284.67_{\text{ten}} = 2114.3133\overline{3}_{\text{five}}$.

EXAMPLE 9 Convert $\frac{2}{3}$ to base-five basimal. To convert a common fraction to a base-five basimal fraction, we may change to base-five integers and divide in base five or use the method of undetermined digits as before. We illustrate the latter:

$$\tfrac{2}{3} \times 5 = \tfrac{10}{3} = 3\tfrac{1}{3} = .abcd \times 5 = a.bcd$$
$$\tfrac{1}{3} \times 5 = \tfrac{5}{3} = 1\tfrac{2}{3} \text{ and } \tfrac{2}{3} \times 5 = \tfrac{10}{3} = 3\tfrac{1}{3}$$

Do you see a pattern? Do you see that $\frac{2}{3} = .31\overline{31}_{\text{five}}$?

24.14 PERCENTAGE Price reductions at sales, interest rates, increases in the cost of living, and other such changes are often best measured in percents. While people are interested in the actual change, the relative change often is a more reliable guide number. *Percent* means "per hundred," and is a useful common language of relative change.

EXAMPLE 1 Matilda lost 50 pounds. If Matilda is a girl who originally weighed 125 pounds, her relative weight loss is $\frac{50}{125} = \frac{2}{5}$. Our study of proportions would give $\frac{50}{125} = \frac{m}{100}$, where m is the number of parts per hundred, or percent, that Matilda lost. Clearly,

$$\frac{50}{100} = \frac{2}{5} = \frac{40}{100} = \frac{m}{100}$$

So, $m = 40$, or Matilda lost 40 percent of her weight.

In the language of percent, Matilda's original weight, 125 pounds, is called the *base*. The 50 pounds lost is called the *percentage*, and the relative loss in parts per hundred, 40, is called the *percent*.

EXAMPLE 2 If Matilda is a horse whose original weight was 1000 pounds, her relative loss is $\frac{50}{1000} = \frac{5}{100} = 5\%$. For Matilda the horse, the base is 1000 pounds, the percentage is 50 pounds, and the percent loss is 5 percent.

We see the relation between percent, percentage, and base is given by the proportion

$$\frac{\text{percent}}{100} = \frac{\text{percentage}}{\text{base}}.$$

EXAMPLE 3 The price of a coat was reduced 15 percent. Originally it was \$60. How much was it reduced? The base is \$60, the percent is 15%, and we wish to know the percentage. Substituting, we have

$$\frac{15}{100} = \frac{\text{percentage}}{60}$$

Multiplying both members by 60, we get

$$\text{percentage (reduction)} = 60 \times \frac{15}{100} = 9$$

Hence we see the coat was reduced \$9.

EXAMPLE 4 Cerise received \$40 interest from a bank that paid 5 percent interest on savings. How much did she have on deposit? The proportion is $\frac{5}{100} = \frac{40}{d}$, where d is the deposit—the base in this problem. Cross multiplying and dividing by 5 gives $d = 800$. So Cerise had \$800 deposited in her account.

The proportion $\frac{p}{100} = \frac{c}{b}$ where p = percent, c = percentage, and b = base, can be solved for any one variable in terms of the others.

1. $p = 100 \times \frac{c}{b}$

2. $c = \frac{p}{100} \times b$

3. $b = \frac{100 \times c}{p}$

Many authors regard these as three different kinds of problems to be treated separately. We recommend thinking of them as three ways of looking at the basic proportion.

Percents are a form of fractions. For example, 17 percent $= \frac{17}{100} = .17$. The proportion is often written in multiplication form as in (2) above. Examples such as 60 is 20% of 300 show the form $60 = \frac{20}{100} \times 300$.

We often see questions such as "24 is what percent of 60?" "15 is 16 percent of how much?" "60 percent of 80 people is how many?" In each case we have $c = \frac{p}{100} \times b$, read "$c$ is p percent of b." Substituting for the three questions we have

$$24 = \frac{p}{100} \times 60 \qquad 15 = \frac{16}{100} \times b \qquad c = \frac{60}{100} \times 80$$

Solving each equation in turn we get

$$p = 40 \qquad b = \frac{1500}{16} = 93.75 \qquad c = 48$$

24.15 SUMMARY Decimals are the most widely used form of fractions. They are easier to use than common fractions because the denominators are omitted or rather are indicated by the decimal point.

This ease comes at a cost. Many rational numbers require infinite repeating decimals. Sometimes the algorithms for addition, subtraction, multiplication, and division are more difficult than in common fractions. Often we are satisfied with an approximation where the algorithms are easier.

We saw how to convert fractions from common to decimal form and back again. We saw how to use decimals in arithmetic operations with special hints on locating the decimal point in answers.

Decimal fractions are used with base ten. There are corresponding basimal fractions for numbers in any base. While there is little commercial need for basimal fractions in base three, say, their investigation is a good check on our understanding of the structure of base-ten decimal fractions.

Percents are introduced as ratios, parts per hundred that are used widely. Percent problems are worked from a proportion,

$$\frac{p}{100} = \frac{c}{b}, \quad \text{or an equation,} \quad c = \frac{p}{100} \times b.$$

With the ever growing use of the metric system, skill in understanding and using decimals will be more valuable for more people at an earlier age than ever before.

EXERCISES

1. List two reasons why people prefer decimal fractions to common fractions.

2. List two reasons why people prefer common fractions to decimal fractions.

3. Write the following as common fractions in their lowest terms

(*a*) .10 (*c*) .25
(*b*) .40 (*d*) .375

4. Write the following as decimal fractions

(*a*) $\frac{3}{4}$ (*c*) $\frac{5}{8}$
(*b*) $\frac{1}{2}$ (*d*) $\frac{3}{16}$

5. Set up a dish abacus to show the following

(*a*) 3.45 (*c*) .00472
(*b*) 47.03 (*d*) 2.034_{five}

6. Of all common fractions equal to $\frac{2}{3}$, (*a*) which has the largest denominator less than 100? less than 1000? (*b*) Which has the smallest denominator greater than 100? than 1000?

7. Of all common fractions equal to $\frac{3}{8}$, which has the largest denominator less than 100? Which has the smallest denominator greater than 100?

8. Of all fractions with denominator 100, which fraction is closest to (*a*) $\frac{1}{4}$, (*b*) $\frac{1}{6}$, (*c*) $\frac{3}{8}$?

9. What fraction with denominator 1000 is closest to (*a*) $\frac{1}{2}$, (*b*) $\frac{1}{3}$, (*c*) $\frac{5}{8}$, (*d*) $\frac{1}{16}$, (*e*) $\frac{3}{11}$?

10. Use the method of undetermined digits to find three-place decimal approximations of (*a*) $\frac{3}{4}$, (*b*) $\frac{1}{3}$, (*c*) $\frac{3}{11}$, (*d*) $\frac{5}{8}$.

11. Write the following as repeating decimals:

(*a*) $\frac{1}{3}$ (*c*) $\frac{1}{11}$ (*e*) $\frac{17}{33}$
(*b*) $\frac{5}{6}$ (*d*) $\frac{2}{3}$ (*f*) $\frac{14}{37}$

12. Add:
(*a*) 3.47 + 14.5 (HINT: use dish abacuses.)
(*b*) 2.63 + .263 + .0263 (*d*) $.3\overline{3} + .0\overline{17}$
(*c*) .14 + 2.14 + 3.014

13. Subtract:
(*a*) 23.47 − 9.29
(*b*) 2.347 − 1.29
(*c*) 6.4 − 8.0

14. Estimate to one digit, using zeros elsewhere, and locating the decimal point:

(*a*) 437 × 894 (*c*) .006378 × .07074
(*b*) 5.283 × .3492

15. Multiply:
(*a*) 4.6 × .007 (*c*) 1.7 × 3.2
(*b*) .063 × 4.5 (*d*) .35 × .44

For exercises 16–18

A. Estimate the product.
B. Using two-digit upper and lower bounds for the factors, find upper and lower bounds for the product.
C. Using front-end multiplication, continue until you feel sure you have a periodic product. Overbar the repetend.

16. $.\overline{25} \times .40$

17. $.\overline{3} \times .\overline{1}$

18. $.\overline{1} \times .\overline{1}$

19. Estimate the quotients to one figure, placing the decimal point correctly:
(*a*) .323 ÷ 1.7 (*d*) .799 ÷ .47
(*b*) .371 ÷ .7 (*e*) .01703 ÷ .0013
(*c*) .0418 ÷ 19

20. Divide to 3 significant digits:
(*a*) .323 ÷ .17 (*c*) .799 ÷ 47.0
(*b*) .799 ÷ .047 (*d*) 18.17 ÷ 2.3

21. Divide to 3 significant digits:
(*a*) 3.4 ÷ .7
(*b*) .423 ÷ .018

22. Divide, drawing a bar over the repetend in the quotient:
(*a*) $.4\overline{6} \div .2$ (*c*) $.\overline{43} \div .22$
(*b*) $.\overline{3} \div 5$

23. Divide: $.\overline{45} \div .\overline{6}$.

24. Find the decimal (base-ten) form of:
(*a*) $.3_{\text{five}}$ (*c*) $.003_{\text{five}}$
(*b*) $.24_{\text{five}}$ (*d*) $.2\overline{2}_{\text{five}}$

25. Find the base-five basimal for the base-ten decimals:
(*a*) .4 (*c*) .16
(*b*) .7 (*d*) $.3\overline{3}$

26. Find the base-ten decimal for (*a*) $.12_{\text{three}}$, (*b*) $.\overline{2}_{\text{three}}$.

27. Find the base-three basimal for (*a*) $.5_{\text{ten}}$ (*b*) $.\overline{6}_{\text{ten}}$.

Solve percentage problems by using ratios in a proportion.

28. A book, originally priced at $15, is reduced 20 percent. How much is cut from the price? What is the reduced selling price?

29. Salesman Nick received a commission of 23 percent on what he sold. How much did he earn for selling $87 worth of brushes?

30. Nick (see exercise 29) earned $66.70 one day. How many dollars' worth of brushes did he sell?

31. Joseph received $41.80 for selling $220 worth of fencing. What commission rate (percent of sales) did he earn?

32. The Cleanwater and Rusty creeks join, forming the Little Murky. If the Cleanwater flows at the rate of 2 cubic feet per second and the Rusty flows at 5 cubic feet per second, what is the rate of flow in the Little Murky? What percent of the flow in the Little Murky comes from the Cleanwater? What percent from the Rusty?

Answers to selected questions: **3.** (*a*) 1/10; (*b*) 2/5; (*c*) 1/4; (*d*) 3/8; **4.** (*b*) .5; (*d*) .1875; **7.** 36/96, 39/104; **9.** (*b*) 333/1000; (*e*) 273/1000; **11.** (*b*) $.8\overline{3}$; (*f*) $.\overline{378}$; **14.** (*a*) 400,000; **17.** (*B*) $.0363 < p < .0408$; (*C*) $.\overline{037}$; **19.** (*b*) .5; (*e*) 10; **21.** (*a*) 4.86; **23.** $.6\overline{81}$; **25.** (*a*) .2; (*b*) $.3\overline{2}$; (*c*) .04; (*d*) $.13\overline{13}$; **27.** (*a*) $.1\overline{1}$; (*b*) .2; **29.** $20.01; **31.** 19%.

25

Roots and Radicals

25.1 INTRODUCTION Radicals are symbols like $\sqrt{2}$ and $\sqrt[3]{17}$, which name the positive roots of equations such as $x^2 = 2$ and $x^3 = 17$. These numbers are named "square root of 2" and "third, or cube, root of 17." In general a radical $\sqrt[n]{k}$, where n and k are counting numbers is called the nth root of k. It stands for a positive number such that the product of n of them is k.

Do these peculiar-looking symbols stir up questions in you? Perhaps—

1. What good are they? Why did anyone ever think of them?

2. How do we find the numerical value of a radical? (This is better stated as "How do we find a rational number close to the radical?" "What do we mean by 'close'?")

3. Are radicals new numbers, or are they rational numbers at a masquerade?

While we can square negative numbers, it is customary in elementary discussions of radicals to restrict the symbol $\sqrt{\ }$ to mean the positive square root.

25.2 WHAT GOOD ARE RADICALS? Square roots come up in many important formulas. For example, $A = s^2$ gives the area of a square in terms of a side s. Conversely, $s = \sqrt{A}$ gives the side of a square of area A. The radius of a circle of area Q is $r = \sqrt{\frac{Q}{\pi}}$. The hypotenuse of a right triangle with unit legs is $\sqrt{2}$ long. Even more important, many formulas in engineering require solving quadratic equations, such as $2x^2 + 7x \quad 10 = 0$. Such solutions are most easily discussed with square roots. In chemistry rather unusual radicals come up. Since practically all uses of radicals are easier after the student has studied algebra, they are only introduced in arithmetic.

25.3 APPROXIMATING RADICALS BY FRACTIONS Some people used to speak of "computing the value of $\sqrt{5}$." Without stopping to tar and feather such bastardizing, we think more careful speakers would say "finding a fraction (perhaps decimal) close to $\sqrt{5}$." More sophisticated people would say, "finding two close fractions with $\sqrt{5}$ between them." If the two fractions are so close together their difference is negligible, then either fraction or any between them, we will call a good approximation to $\sqrt{5}$.

Some people prefer to say "Find the 3-digit decimal closest to $\sqrt{5}$" or "Find the largest 3-digit decimal smaller than $\sqrt{5}$."

An observation will tell us the first digit and the number of digits in the integer part of the root. Recall that in 46.37, 46 is the integer part and .37 is the fractional part. Imagine a huge table of squares for all numbers having one nonzero digit. Some of the entries in this table are listed in tha margin. We imagine that all the other entries are there also. We can, of course, read the table both ways. This is indicated by a double heading.

n	n^2
$\sqrt{z}$	z
8000	64,000,000
3000	9,000,000
80	6400
30	900
8	64
3	9
.03	.0009
.08	.0064

EXAMPLE Suppose we wish $\sqrt{23}$. Can you see that in the z column, 23 is between 16 and 25? Hence, $4 < \sqrt{23} < 5$, the corresponding entries in the $\sqrt{z}$ column.

EXAMPLE We wish $\overline{)4376}$. Can you see that in the z column 4376 is between 3600 and 4900? Hence, $60 < \sqrt{4376} < 70$. It is easy to find the place in the z column as follows. There is always an even number of integer zeros in the n^2 column. So we pair off the digits in the radicand, starting from the decimal point. The radicand is the number under the radical sign whose root we seek. The left "pair" may have only one number in it. Now we need compare only the left pair with the nonzero digits in the z column, replacing the other digits with zeros.

EXAMPLE 3273 → 32′73 → 3200, which falls between 2500 and 3600.

EXAMPLE 83764 → 8′37′64 → 8′00′00, which falls between 40,000 and 90,000.

EXAMPLE .000426 → .00′04′26 → .00′04+, which falls between .0004 and .0009.
Looking in the z column, $50 < \sqrt{3273} < 60$, $200 < \sqrt{83764} < 300$, and $.02 < \sqrt{.000426} < .03$.

After a few trials you will see that there is one digit in the root for each pair marked off in the radicand. The left digit in the root is the largest number whose square is less than the number in the left "pair" of the radicand.

Hence
$$\sqrt{8'47.30} = 2a.b = 29.bc$$
$$\sqrt{63'90.05} = 7a.b = 79.bc$$
$$\sqrt{2'73'49.26} = 1ab.c = 16b.c$$

where a, b, c are digits to be determined. Here some of the more important ways of determining a, b, c, . . . are illustrated for several roots.

1. *Look up $\sqrt{5}$ in a table of square roots.* Here are some results. $\sqrt{5} = 2.24$. We probably aren't sure whether this is the value closest to $\sqrt{5}$, or the largest 3-digit number less than $\sqrt{5}$. To check we square $2.24^2 = 5.0176$. Since $5.0176 > 5$, we know that $2.24 > \sqrt{5}$, so we know that $\sqrt{5}$ is in the interval (2.23, 2.24).

2. *Look up $\sqrt{5}$ in a bigger table.* We find $\sqrt{5} = 2.23606^+$. From this we can see that $\sqrt{5}$ is in the interval (2.23, 2.24) and is closer to 2.24.

3. *Use a slide rule or calculator.* (These first three methods, of course, require special equipment one just may not have.)

4. *Guess.* This is the most important method of all. In fact, every method is essentially guessing combined with testing.

5. *Guess again.* Of course, in your second (and later) guesses you have the advantage of having tested your first guesses. Here's how the process might work to approximate $\sqrt{13}$:

Since $3^2 < 13 < 4^2$ we know $3 < \sqrt{13} < 4$. We might

guess $g_1 = 3.1$ test $3.1^2 = 9.61 < 13$ g_1 is too small
guess $g_2 = 3.2$ test $3.2^2 = 10.24 < 13$ g_2 is too small
guess $g_3 = 3.3$ test $3.3^2 = 10.89 < 13$ g_3 is too small
guess $g_4 = 3.4$ test $3.4^2 = 11.56 < 13$ g_4 is too small
guess $g_5 = 3.5$ test $3.5^2 = 12.25 < 13$ g_5 is too small
guess $g_6 = 3.6$ test $3.6^2 = 12.96 < 13$ g_6 is too small
guess $g_7 = 3.7$ test $3.7^2 = 13.69 > 13$ g_7 is too large

At this point we know that $\sqrt{13}$ is in the interval (3.6, 3.7). Do you see how to follow this method? Do you think you could have found this interval in fewer than seven guesses? What do you guess will be the next digit?

6. *Halving the interval is often a good routine way of improving a guess.* Since $\sqrt{6'92}$ is in the interval (20, 30), we guess $\sqrt{692} = 25$, because 25 is is the midpoint of the interval. Test $25^2 = 625 < 692$. 25 is too small. Now we know $\sqrt{692}$ is in the interval (25, 30). We make a table:

Interval	**Midpoint**	**Test**
(20, 30)	25	$25^2 = 625 < 692$
(25, 30)	27.5	$27.5^2 = 756.25 > 692$
(25, 27.5)	26.25	$26.25^2 = 689.^{+} < 692$
(26.25, 27.5)	26.875	

At this point we know $\sqrt{692}$ is in (26.25, 27.5), and is approximately 26.875, with error no greater than .625.

The halving process reduces the interval by half at each step, but does introduce some awkward calculations in decimal numerals. Clearly, we may continue until the interval is as short as we please.

7. *This method is called the* divide and average *method.*

A moment's reflection will convince you that if g is less than $\sqrt{5}$, $\frac{5}{g}$ will be greater. If $g < a$, then $a < \frac{a^2}{g}$. If g is less than 3, then $\frac{9}{g} > 3$ is an easy example to test. Similarly, if $g > \sqrt{5}$, then $\frac{5}{g} < \sqrt{5}$.

$$g_1 = 2, \frac{5}{2} = 2.5, \sqrt{5} \text{ is in } (2, 2.5), \text{ midpoint is } g_2$$

$$g_2 = 2.25, \frac{5}{2.25} = 2.222, \sqrt{5} \text{ in } (2.222, 2.25), \text{ midpoint is } g_3$$

$$g_3 = 2.2361111, \frac{5}{2.2361111} = 2.236024 \ (2.236024, 2.2361111)$$

Clearly, at this point we know $\sqrt{5} = 2.236 \ldots$, whose first four digits are the first four digits of both ends of the interval. This method of approximating square roots is easy to understand and to remember, and it gives good values (narrow intervals) quite quickly. Of course, the better the starting guess, the shorter the resulting intervals.

Suppose you had a table that gave you $\sqrt{5} = 2.23$ but you wished a better value. Dividing $\frac{5}{2.23} = 2.242152$. The average of 2.23 and 2.242152 is 2.236076. The first seven-digit approximation is 2.236068, which differs by only .000008. This illustrates the more general fact that the divide-and-average method almost doubles the number of correct digits at each step. Not having a table, we may divide $\frac{5}{2.236076} = 2.236059^+$. This clearly guarantees that $\sqrt{5} = 2.2360^+$, since these are the first five digits in both divisor and quotient. If we continued the division to several places we would find the average of 2.236076 and 2.236059^+ would have about ten correct digits.

Radicals other than square roots can be approximated by ways similar to these. We limit our discussion to shrewd guessing, which should always be done as a check for more sophisticated ways.

We illustrate by $\sqrt[3]{37964}$. We want a number whose cube is about 37964. Now the cubes of numbers between 1 and 10 are between 1 and 1000. Any number between 10 and 100 will have a cube between 1000 and 1,000,000. 37964 lies in this interval. Considering $10^3 = 1000$, $20^3 = 8000$, $30^3 = 27{,}000, \ldots, 90^3 = 9^3 \times 1000 = 729{,}000$, we see that we can estimate the first digit in the root by considering $\sqrt[3]{37}$ (marking off the three places, 964, in 37′964). Since $3^3 = 27$ and $4^3 = 64$, $\sqrt[3]{37}$ is between 3 and 4 and $\sqrt[3]{37964} = 3x$, a number between 30 and 40.

Can you make a similar first-digit-correct estimate of $\sqrt[3]{78621}$? Of $\sqrt[4]{78621}$? More accurate estimates require extra equipment, such as tables or calculators, some algebra beyond the scope of this book, or tedious testing of guesses.

25.4 THE TRADITIONAL SQUARE-ROOT METHOD Traditionally a method was taught for approximating square roots by decimals using a layout that looked a lot like division. It was taught by rote in the seventh and eighth grades, even though the reasons behind the method are often obscure even to those with a good background in algebra. It was taught (1) because some people felt it was a useful skill and (2) because it was a genuine prestige item. Anyone who could "do square root" was a genius. The method is rapidly being dropped from modern curricula. We include it here for its historical value, and because some readers may make a great impression on people who were exposed as children.

A geometrical diagram makes the process clear. We seek, say, $\sqrt{15}$. Suppose we have a square of area 15. Clearly, we may carve a square of side 3 and of area 9, from the upper right corner. What remains looks like an L with area $15 - 9 = 6$. How much can we increase the side of the smaller square, and still have it inside the square of area 15? Suppose we increase the small side by x to $3 + x$. We think of x as being the largest possible of 0, .1, .2, .3, . . . , .9. Another diagram (not to scale), shows that the extra region carved out of the L is two rectangles and a square of total area $x(3) + x(3) + x^2$ or $[2(3) + x]x$. We want the largest x such that $[2(3) + x]x \leq 6$. It appears that x is merely the quotient when 6 is divided by $[6 + x]$. Annoyingly, x is part of the divisor as well as the quotient. Perhaps a bit of shrewd guessing and trial will do.

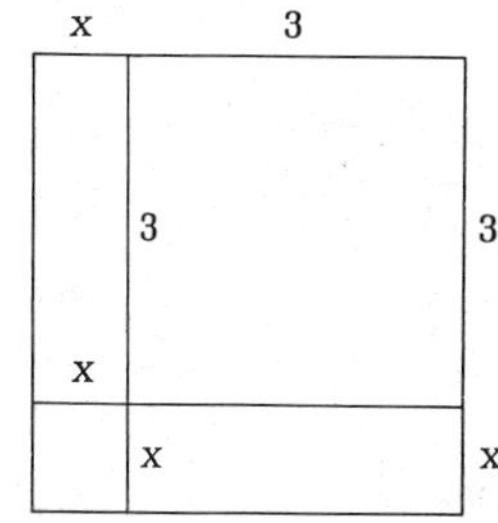

Since $(6 + x)$ is a bit more than 6, then $6 \div (6 + x)$ is a bit less than 1, say .9 or .8. We test trial divisors 6.9 and 6.8.

$(6 + .9) \times .9 = 6.21 > 6$ .9 is too big for x.

$(6 + .8) \times .8 = 5.44$ .8 is just right.

So the square with the side 3.8 has the area $9 + 5.44 = 14.44 < 15$. If we draw the square (not to scale) inside the square of area 15, the leftover L contains $15 - 3.8^2$, or .56. We repeat. We increase the side by y, where y is the largest possible of .00, .01, .02, . . . , .09. The amount carved from the L is $y(3.8) + y(3.8) + y^2 = [2(3.8) + y]y$. We want the largest y so that $[2(3.8) + y]y = [7.6 + y]y \leq .56$. It appears that y is the quotient when .56 is divided by $(7.6 + y)$. The divisor is slightly more than 7.6. Since $\frac{.56}{8} = .07$, we guess that y is about .07. Using trial divisors of 7.67 and 7.68, we get $.56 \div [7.67] = .07^{+}$, $.56 \div 7.68 = .08^{-}$. Hence we take

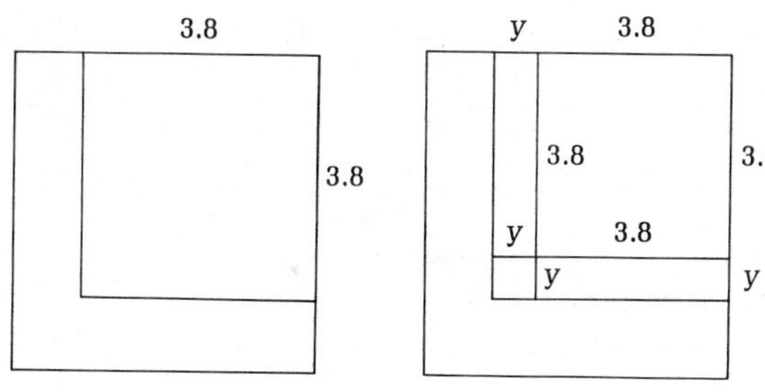

$y = .07^{+}$. We may carve out a square of side 3.87, and leave an L of area $.56 - .5369 = .0231$, and so on.

Traditionally, the squares were not drawn, but the following layout was used, with steps as noted.

1. Draw a shed. Write 15. under the shed. Directly above the decimal point put a $_{\wedge}$ for the decimal point in the answer.

$$\overset{\wedge}{\sqrt{15.\quad\;}}$$

2. What is the largest number whose square is less than 15? Put that number, 3, above the shed, $3^2 = 9$ below 15, and subtract, getting 6.

$$\begin{array}{r} 3_{\wedge} \\ \sqrt{15.\;\;} \\ \underline{9} \\ 6 \end{array}$$

3. To avoid decimals, write two 0s beside 15. Bring them down beside 6.

$$\begin{array}{l} \phantom{\sqrt{1}}3_{\wedge} \\ \sqrt{15.00} \\ \phantom{\sqrt{1}}\underline{9} \\ \phantom{\sqrt{1}}6\;00 \end{array}$$

4. Double the "quotient" (3). Multiply the product by 10. This number (60) is a "trial divisor." Find x, a little less than $\frac{600}{60}$, say 9, as your next quotient. Then $60 + x = 69$ is another trial divisor, and $9 \times 69 = 621$. $621 > 600$, so 9 was too big.

$$\begin{array}{rl} & \phantom{\sqrt{1}}3_{\wedge}9 \\ & \sqrt{15.00} \\ & \phantom{\sqrt{1}}\underline{9} \\ & \phantom{\sqrt{1}}6.00 \\ 69 & \phantom{\sqrt{1}}6.21 \end{array}$$

IT'S SOME KIND OF MODERN ARCHITECTURE
IT SHOULD GET TO THE ROOT OF THE PROBLEM

5. Try $x = 8$. Then $60 + x = 60 + 8 = 68$ is the divisor, 8 is the quotient. $8 \times 68 = 544 < 600$. O.K.

```
        3‸8
      √15.00
        9
        ─────
        6 00
  68    5 44
        ─────
           56
```

6. Write two more zeros after the 15. Bring them down. Double the quotient, 38, and multiply by 10. The trial divisor 760 divides 5600 less than 8. The quotient $y = 7$ works with divisor $760 + y = 767$.

```
        3‸8   7
      √15.00 00
        9
        ─────────
        6 00
  68    5 44
        ─────────
           56‸00
 767       53‸69
           ──────
            2 31
```

7. The largest 3-digit number less than $\sqrt{15}$ is 3.87.

We suggest you match the steps in the layout with the diagrams.

This method has the advantage of being straightforward, the accuracy is clear, the work is in one place. However, hardly anyone in the United States uses it any more, since tables, slide rules, and calculators are readily available.

25.5 ARE RADICALS RATIONAL? Since radicals do not arise by dividing counting numbers, it is not clear whether they are rational or not. Certainly there is no whole number whose square is 2, but couldn't there be a rational number? If the processes of earlier sections were carried out far enough, perhaps with a bit of luck, couldn't they possibly produce a

fraction whose square is 2? Perhaps one with a large denominator? From experience with squares and right triangles we know there are lengths and ratios we wish to label $\sqrt{2}$.

The ancient Greeks knew about radicals and very much wanted them to be rational numbers. Indeed, we are told that at least one man was murdered because he publicly claimed that some radicals were irrational.

He was correct. Indeed we can show that $\sqrt{N}$ is irrational for all counting numbers N except perfect squares such as 4, 9, 16, ... We limit ourselves to showing that $\sqrt{2}$ is irrational.

The method is to make guesses $\frac{a}{b}$ for $\sqrt{2}$, and show that any such guess must fail.

For example, is $\frac{7}{5} = \sqrt{2}$? If so, squaring both sides and multiplying would give us $49 = 2 \times 25$. This is an equation to test which is free of radicals. But $49 \neq 2 \times 25$. Hence, $\frac{7}{5} \neq \sqrt{2}$. This argument is convincing for $\frac{7}{5}$, but tells us nothing about $\frac{71}{50}$, or $\frac{141}{100}$.

Now we knew that $49 \neq 2 \times 25$, because we recognized those names. Here are some other ways:

1. Factor 49 and 2×25 into prime factors. We test $7 \times 7 = 2 \times 5 \times 5$. Without looking at the others we count the factors of 2. There is one, an odd number, on the right and zero, an even number, on the left. Two equal numbers must have the same number of factors of 2. Hence, $49 \neq 2 \times 25$.

We can use this method to show two numbers are not equal even when we don't know what either is. Suppose we wonder if $\frac{a}{b} = \sqrt{2}$. Then $a^2 = 2b^2$ is an equation without radicals we can test. Any factor of a appears twice as a factor of a^2, and similarly for factors of b. Simple counting shows the number of factors of 2 on the left is even and on the right is odd. Hence we know that $a^2 \neq 2b^2$ even though we know neither counting number a nor b.

2. Write 49 and 2×25 in base three, getting $49 = 1211_{\text{three}}$ and $2 \times 25 = 2 \times 221_{\text{three}} = 1212_{\text{three}}$. The units digit is 1 in 1211 and 2 in 1212. Since the units digits are different, $49 \neq 2 \times 25$.

To test whether $a^2 = 2b^2$ we write a^2 and b^2 in base three. Of course we can't know the digits for these because we don't know a or b. But if you convert $1^2, 2^2, 3^2, 4^2, 5^2, 6^2, \ldots$ to base three you quickly see that in each case the first nonzero digit is 1. For example, $15^2 = 22100_{\text{three}}$. If you try, you will see that multiplying a square in base three by 2 gives a number whose first nonzero digit is 2. Testing $a^2 = 2b^2$ we see that even if we don't know what a and b are, the first nonzero digit on the left is 1 and on the right is 2. Hence, $\frac{a}{b} \neq \sqrt{2}$. We say $\sqrt{2}$ is irrational.

Method 1 extends readily to tests of other radicals, but method 2 does not. We are using the fact that any counting number can be written as a product of primes in only one way.

Are you convinced no fraction exists so that $\left(\frac{a}{b}\right)^2 = 2$? Most people are. This type of argument is called *indirect proof.* But because people still want $\sqrt{2}$, they are led to extend their number system beyond rational numbers so that it will include radicals.

25.6 GENERALIZING RADICALS Physical events and solutions to equations such as $x^3 = 17$ cause people to generalize, or enlarge, the number system to include all radicals $\sqrt[n]{k}$, where n and k are positive integers. In advanced mathematics, we assign meaning to $\sqrt[n]{k}$, where n and k are any rational numbers; we look for solutions of equations with more than two terms, say $x^4 - 3x^3 + x = 42$. We also expect that compound symbols such as $\sqrt{4 + \sqrt{2}}$ would be worth investigating. We shall not explore them here.

EXERCISES

1. (*a*) What symbol represents the length of the side of a square of area 5?
(*b*) What symbol represents the volume of a cube of side 2?
(*c*) What symbol represents the side of a cube of volume 9?

2. Is $\sqrt{16} + \sqrt[3]{27} - (\sqrt{8})^2$ positive or negative?

3. Recall that Pythagoras taught that in any right triangle $a^2 + b^2 = c^2$ where a, b, and c, are the lengths of the sides.

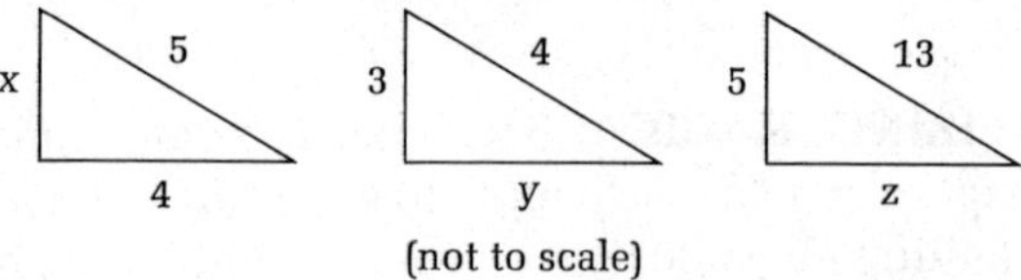

(not to scale)

(*a*) How long is x?
(*b*) How long is y?
(*c*) How long is z?

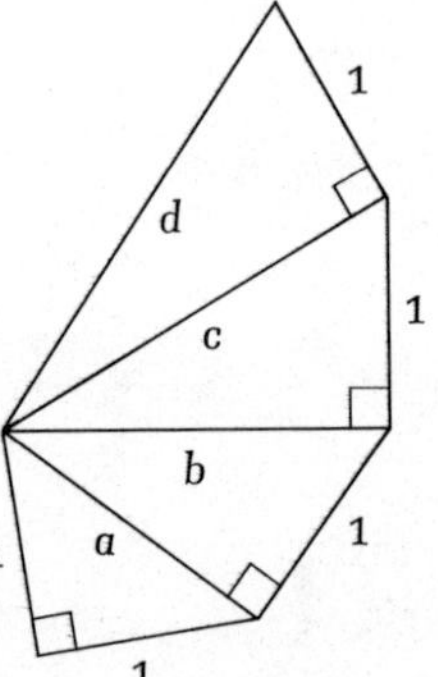

4. The sides in the figure marked 1 are 1 long. The angles marked with ∟ are right angles. How long are sides a, b, c, d?

5. Locate each of the following radicals between two consecutive numbers with one nonzero digit, as in $.5 < \sqrt{.31} < .6$. If the radical is equal to such a number, so state.

(*a*) $\sqrt{2}$
(*b*) $\sqrt{7}$
(*c*) $\sqrt{596}$
(*d*) $\sqrt{.18}$
(*e*) $\sqrt{.000245}$
(*f*) $\sqrt{625}$
(*g*) $\sqrt{613089}$
(*h*) $\sqrt{.7396}$
(*i*) $\sqrt{.0961}$
(*j*) $\sqrt{.0841}$
(*k*) $\sqrt[3]{8000}$
(*l*) $\sqrt[3]{24389}$
(*m*) $\sqrt[3]{.046656}$

How would you check your answers?

6. By repeated guessing, find the largest counting number less than the radicals below.

(*a*) $\sqrt{257}$ (*b*) $\sqrt{3456}$

7. State whether the following inequalities are true or false. Can you tell at a glance?

(*a*) $\sqrt{9341} < \frac{9341}{93}$

(*b*) $\sqrt{.0249} < \frac{.0249}{.16}$

8. Use the divide-and-average method to approximate the following radicals to two significant digits.

(*a*) $\sqrt{3.456}$ (*b*) $\sqrt{.0678}$

9. Use the traditional square-root process to compute $\sqrt{3.456}$ to two significant digits. Are three steps necessary?

10. Knowing $\sqrt{3} = 1.73$, find (*a*) $\sqrt{12}$, (*b*) $\sqrt{75}$.

11. What would you look up in a square-root table to get (*a*) $3\sqrt{7}$ and (*b*) $2\sqrt{13}$ with the easiest multiplication?

12. (*a*) Using $\sqrt{2} = 1.414214$ find $\frac{1}{\sqrt{2}}$ to 5 decimal places.
(*b*) Using $\sqrt{3} = 1.732051$, find $\frac{1}{\sqrt{3}}$ to 5 decimal places.

13. What symbol shows the solution to $x^7 = 27$?

14. Prove that $\sqrt{3}$ is irrational.

15. Explain why in the traditional square-root process the value of the root at any stage is doubled to form a trial divisor for finding the next digit.

16. Using the method described in the text to prove that $\sqrt{2}$ is irrational, try to prove that $\sqrt{4}$ is irrational. Where does the proof break down?

Answers to selected questions: **1.** (*a*) $\sqrt{5}$; (*c*) $\sqrt[3]{9}$; **3.** (*b*) $\sqrt{7}$; **5.** (*c*) $20 < \sqrt{596} < 30$; (*h*) $.8 < \sqrt{.7396} < .9$; (*l*) $20 < \sqrt[3]{24389} < 30$; square the numbers and see if the inequality holds; **7.** (*b*) false; yes $.0249/.16 < .16$; **10.** (*a*) $\sqrt{12} = 2\sqrt{3} = 3.46$; (*b*) $\sqrt{75} = 5\sqrt{3} = 8.65$; **12.** (*a*) $\sqrt{2}/2 = .707107 = .70711$; **14.** If $\sqrt{3} = a/b$, then $3b^2 = a^2$ has odd number of factors of 3 on left, even number of factors on right: impossible; **16.** From $\sqrt{4} = a/b$, $4b^2 = a^2$. We have an even number of factors: 2 on both sides. If a has a factor 2, we cannot conclude that b^2 also has this factor.

26

Real Numbers

26.1 INTRODUCTION On the number line we have located all the integers and rationals. These are dense, which means that between any two rationals we can always find a third. Even though dense, they do not completely fill the line, because we found radicals such as $\sqrt{2}$, which are on the line but are not rational. It may be there are other kinds of numbers on the line besides rationals and radicals.

Hence we have annoyingly different kinds of numbers, different names, different algorithms, and different ways of locating them on the number line. In working with these numbers it is difficult to keep a clear idea of how big they are. For example, which is bigger, $\sqrt[3]{13} - \sqrt[2]{15} + \pi$ or $\sqrt[2]{13} + \sqrt[3]{15} - \pi$? We can work nicely with decimals and other fractions, so we will seek some way of transferring our convenience with rationals to all the numbers on the number line.

This grand collection of numbers is called the set of real numbers. We will define them—$\sqrt[2]{3}$, $\sqrt[3]{117}$, and all—in terms of rationals in three ways. First as an infinite decimal, second as the common point in a set of nested intervals, and third as the least upper bound of a set of rationals. These ways are equivalent but they look different, so sometimes one way is preferred, sometimes another. We will discuss computing with infinite decimals, but not with the other approaches.

26.2 INFINITE DECIMALS We saw that rational numbers may be represented by infinite repeating decimals. For example, $\frac{3}{22} = .1\overline{36}$. Roots are approximated by finite decimals as $\sqrt{2} \approx 1.414$. People sometimes say $\sqrt{2} = 1.414\ldots$ continued forever, with 1.414 a good approximation.

In neither of these cases do we actually have an infinity of digits. Instead we have a rule for finding each of as many digits as we please. In the examples we are given the first five and first four, respectively. The thousandth digit for $.1\overline{36}$ is easily seen to be 3 on observing that the 2d, 4th, 6th, 8th, . . . digits are 3. The thousandth digit for $\sqrt{2} = 1.414\ldots$ can be found with several applications of the divide-and-average algorithm, lots of time, paper, pencils, or computing machines.

We now assert: Every infinite decimal names a real number, and every real number is named by an infinite decimal. Each number has one and only one infinite decimal, with the exception of rationals one of whose decimals ends in 0s ($.50\overline{0} = .49\overline{9}$ for example).

There are many problems in arithmetic of the form "A number is characterized by a certain property. Find the first few digits of its infinite decimal." For example, "3 times a number is 8" and "the square of a number is 5" characterize two numbers. Can you find the numbers to three decimal places?

The method of finding the digits in an infinite decimal may be as simple as following a repeating decimal. Or it may be quite complicated, depending on how the number is characterized. There is no single simple way that works for all numbers.

We have been using decimals. A similar theory exists for numbers to bases other than ten. Naturally a number will be represented by different infinite basimals if the numerals are in different bases.

26.3 NESTED INTERVALS How does an irrational number, $\sqrt{17}$ say, fit into the set of rationals? It is not equal to any of them. But it does separate the rationals into those bigger and those smaller than it is. It seems we ought to be able to locate $\sqrt{17}$ by giving one rational which is too big, 4.13, and another which is too small, 4.12, and saying that $\sqrt{17}$ is the irrational number between them.

This idea is fine and gives us an excellent idea about the size of $\sqrt{17}$. Unfortunately there are many irrational numbers in the interval (4.12, 4.13). Thus we have not located 17 exactly. We might say "take a shorter interval." But any interval will contain an infinite number of points, and we want exactly one point. Here is the scheme.

Think of not one interval, but an infinite sequence of intervals, each inside the preceding one, each containing $\sqrt{17}$, and such that we can find one with length as small as we please by looking far enough in the sequence.

While each interval has an infinite number of numbers, there is only one number that is in all the intervals. For any two numbers there is always an interval in the sequence too short to contain both. The infinite sequence can be used to locate the single number contained in all intervals, $\sqrt{17}$.

The halving method for approximating square roots gave a sequence of intervals. Each interval was half as long as, and was contained (or nested) in, the preceding interval. The desired square root was contained in each interval. By going far enough we could get an interval as short as we pleased.

We assert that such a sequence of nested intervals defines a real number. That is, a real number is a process that produces a sequence of intervals, each inside the previous interval and the lengths becoming as short as we please if we go far enough.

Clearly, a nested interval machine could be used to locate any number on the number line, rational numbers as well as irrational. Ordinarily, of course, we would use common fractions or decimals to name a rational number, because they are easier.

An infinite-decimal machine is easily changed to a nested-interval machine. We think of the successive finite decimals as the left ends of the intervals. The right end of each interval is found by adding 1 to the right digit of the left end. For example, $\frac{3}{22} = .1\overline{36}$. A sequence of nested intervals for $\frac{3}{22}$ is (.1, .2), (.13, .14), (.136, .137), (.1363, .1364), ... Do you see the pattern?

While each real number has only one infinite decimal (with minor exceptions), there are many sequences of nested intervals that determine the same number. For example, $(\frac{1}{2}, \frac{3}{2})$, $(\frac{3}{4}, \frac{5}{4})$, $(\frac{7}{8}, \frac{9}{8})$, . . . and $(\frac{6}{7}, \frac{8}{7})$, $(\frac{8}{9}, \frac{10}{9})$,

$(\frac{10}{11}, \frac{12}{11}), (\frac{12}{13}, \frac{14}{13}), \ldots$ are two sequences of nested intervals determining the number 1.

A sequence of nested intervals can become an infinite-decimal machine as follows. For each interval, write the endpoints as repeating decimals. Some of the left digits for both endpoints will agree. The finite decimal matching that interval is the decimal formed with the digits in which both endpoints agree.

The finite decimals in the sequence so formed will get longer, unless the endings of the left decimals tend to all 9s and the endings of the right tend to all 0s. The appropriate action is clear. Example: A nested interval sequence for $\sqrt{10}$ is (3.0, 3.4), (3.125, 3.2), (3.160, 3.165), (3.162, 3.163), (3.1621, 3.1625). The associated decimals are 3., 3., 3.16, 3.16, 3.162.

26.4 LEAST UPPER BOUND Nested intervals are fine for locating a point on the number line, but it appears that we are overkilling our quarry. Both the lower end and the upper end of the intervals get arbitrarily close to the number of interest, say $\sqrt{2}$. Can't we define the number using the lower endpoint only? Anyone who ever tried to catch a fly from one side only can see the difficulty.

This leads us to the idea of *least upper bound*. We locate $\sqrt{2}$, for example, by describing a set of rationals for which $\sqrt{2}$ is the least upper bound.

Let L be the set of all rational numbers $\frac{a}{b}$ such that $\frac{a^2}{b^2} < 2$. Each fraction in L is $< \sqrt{2}$, so $\sqrt{2}$ is an upper bound to L. There is no number smaller than $\sqrt{2}$ that is an upper bound to L, because we can find a fraction just as close to $\sqrt{2}$ as we please. Hence $\sqrt{2}$ is the least upper bound to L.

Note that $\sqrt{2}$ is not used in describing L. L is completely described with rational numbers. Thus $\sqrt{2}$ described as the *LUB* of L is defined in terms of rationals.

There are other sets of rationals that have $\sqrt{2}$ as a least upper bound and could just as well have been used to locate $\sqrt{2}$. These would be subsets of L.

We assert: Every infinite set of rationals bounded above has a unique least upper bound, a number that is greater than or equal to every number in the set, and smaller than any other upper bound. This LUB is a real number.

It is easy to define some numbers as a least upper bound. For example, $\sqrt{5}$ is the least upper bound of all rationals whose squares are less than 5. This is much easier than describing the infinite decimal or an appropriate set of nested intervals.

Why not say "$\sqrt{5}$ is that number whose square is 5"?

This definition would not guarantee that there was such a real number. The phrase "$\sqrt{-1}$ is that number whose square is -1" defines $\sqrt{-1}$, but we know there is no such real number.

The LUB idea is easily seen to be equivalent to the infinite decimal. The set of finite decimals is an infinite set of rational numbers, for which the infinite decimal is the least upper bound. If we have a bounded set, we can find the appropriate infinite decimal as in the next example. Suppose 1 is in the set and 8 is an upper bound. Test the numbers 2, 3, 4, 5, 6, 7, 8. At least one, maybe all, are upper bounds. Suppose 5, 6, 7, 8 are upper bounds but 4 is not. There is some member in the set greater than 4.

By test, we find the largest of 4.0, 4.1, 4.2, . . . , 4.9 that is not an upper bound. Suppose this is 4.2. Then 4.2 are the first two digits of the infinite decimal. Next find the largest non-upper bound of 4.20, 4.21, 4.22, . . . , 4.29. This number shows the first three digits of the infinite decimal. And so on.

26.5 CALCULATING WITH REAL NUMBERS

If our real numbers are finite decimals (the repetend is $\overline{0}$) we compute as in an earlier chapter. Sums, differences, and products are finite decimals, while quotients may be infinite repeating decimals.

The fun comes in computing with infinite decimals. We create a nested interval sequence for the outcome, by operating with the right and left ends of the intervals given us. For example, let

$$A = .121121112111121111112\ \ldots \text{ and } B = .344344434444344444\ \ldots$$

Do you see the pattern? Can you write the next ten digits?

We can get $A + B = .465465546555^{+}$ and $B - A = .223223322333223333$ . . . easily using front-end addition and subtraction.

Using front-end multiplication for the product (see margin), we see that no matter how far we go, the product would be $.041^{+}$, with the next digit being either 6 or 7.

```
   .3443444344443444443 . . .
 × .1211211121111211111 . . .
 ----------------------------
.0‸3443
   4688
    034
    153
     68
```

We may also use nested intervals:

Since $.1 < A < .2$,	$.3 < B < .4$,	then $.03 < AB < .08$
Since $.12 < A < .13$,	$.34 < B < .35$,	then $.0408 < AB < .0455$
Since $.121 < A < .122$,	$.344 < B < .345$,	then $.041624 < AB < .04209$
Since $.1211 < A < .1212$,	$.3443 < B < .3444$,	then $.04169473 < AB < .04174128$

At this stage we can be sure the product is $.041^+$.

Many people are surprised that the product does not have as many significant figures as the factors. This shows what we saw in chapter 13, that the relative error in products was the sum of the relative errors in the factors.

As we take more digits in the factors, we make the interval for the product as small as we please. From the endpoints, we get a decimal having as many digits as we please. This is the infinite decimal for the product.

While front-end multiplication gives a way of directly computing the digits in the product one by one, there appears to be no similar division algorithm. To find $\frac{A}{B}$, we might try $\frac{.12}{.34} = .35294 \ldots$. Because of our approximation of A and B, we do not know how many of the digits in .35294 are significant—that is, are those of the correct infinite decimal for $\frac{A}{B}$. Further, if we want a more accurate approximation we must start over. These computations are of no use.

A nested set for A/B is found by the intervals

$$\frac{.1}{.4} < \frac{A}{B} < \frac{.2}{.3},\ \left(\frac{.12}{.35}, \frac{.13}{.34}\right),\ \left(\frac{.121}{.345}, \frac{.122}{.344}\right), \ldots .$$

To get a satisfactory approximation for $\frac{A}{B}$, we pick one of these intervals and see if the endpoints are close enough together. If they are not, we choose a shorter interval, using more places in A and B.

As in multiplication, we appear to have about one fewer significant figures in the quotient than we had in the divisor and dividend. For example, $\left(\frac{.1211}{.3444}, \frac{.1212}{.3443}\right)$ is (.3516, .3520). We have 3^+ significant figures in .1211, .3444, . . . , yet only 2^+ significant figures in .3516, .3520.

In actual practice, we only ask a finite decimal approximation to a product or quotient of two infinite decimals. Hence if we want k significant digits in the product we multiply the $k + 1$ significant digit approximations to the factors. Similarly for the quotient.

EXERCISES

1. (*a*) What is the third digit in $.3\overline{45}$?
(*b*) the fifth digit?
(*c*) the thousandth?

2. (*a*) What is the third digit in $.6\overline{789}$?
(*b*) the fifth?
(*c*) the thousandth?

3. Describe how, using only paper and pencil, you could find the 14th digit in the decimal expansion of $.\overline{9864}$.

4. Form an infinite decimal by writing the counting numbers in order:

1, 2, 3 . . . 998, 999, 1000, 1001, . . .

and then placing a decimal point at the left and moving the digits together:

.123 . . . 99899910001001 . . .

(*a*) Name the 5th digit; (*b*) the 15th digit; (*c*) the 25th digit without writing the number and counting.

5. Is the number in exercise 4 another name for a rational number $\frac{A}{B}$, where A and B are counting numbers? Why?

6. You know n lies in the interval (.34264, .34138). Give your best estimate of the value of n and an upper bound to the error in your estimate.

7. Each of the following is a way to describe a real number. For each, state whether it seems to you primarily to

A. produce a series of digits forming an infinite decimal,

B. produce a sequence of shrinking nested intervals all containing the number, or

C. produce an increasing sequence whose least upper bound is the number.

(*a*) Expanding an infinite repeating decimal.

(*b*) The divide-and-average method of calculating square roots.

(*c*) Finding the area of a region, such as a circle, by completely tiling the inside of the region with squares. (You shift to smaller squares when no more of those you are using will fit.)

(*d*) The halving process for finding a cube root.

(*e*) The continued division which does not come out evenly, such as 2 divided by 7.

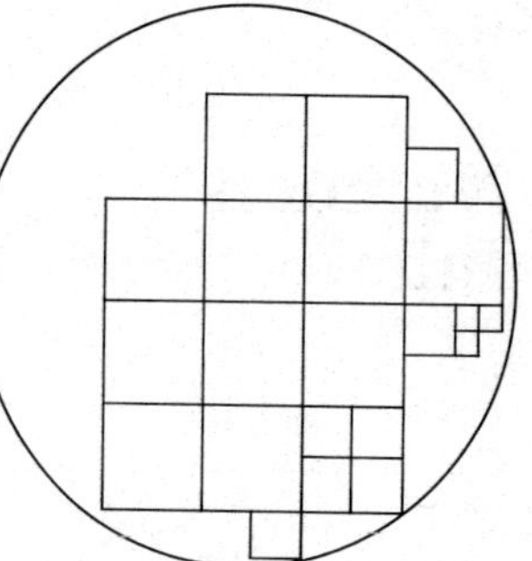

8. From a table, you find $\pi = 3.1415927$ and $\sqrt{10} = 3.16227766$. Calculate the first three digits in

(*a*) $\sqrt{10} - \pi$ and (*b*) $\sqrt{10} \times \pi$

Prove that if these numbers were calculated by a perfect calculator who knew all the digits, the calculator would get the same first three digits.

Answers to selected questions: **2.** (*b*) 7; **4.** (*a*) 5; (*b*) 2; (*c*) 7; **6.** .34201, error $\leq$ .00063; **7.** (*b*) *B*; (*c*) *C*; (*e*) *A*.

27

Large Numbers and Infinity

27.1 INTRODUCTION "Who can name the largest number?" asks the kindergarten teacher, hoping that Mamie will be distracted from putting sand in Peter's peanut butter. Rod claims he can, because he will add 1 to whatever number anyone else says. Jerry claims that his number, infinity, is bigger than Rod's, because no number is bigger than infinity. The teacher decides that sand in the peanut butter may not be so bad after all, and changes the subject. In years to come, will Rod still wonder if infinity plus 1 is larger than infinity?

To help Rod and his friends, we shall consider examples of large sets, define what we mean by infinity, and see if there is more than one size of infinity. We give some examples showing that infinite sets have such startling properties we will not want to include infinity among the counting or real numbers.

27.2 LARGE SETS AND NUMBERS Here are some examples of large sets that have been suggested:

1. The people in the United States
2. The people in California
3. The hairs on a man's head
4. The dollars owed by the federal government
5. The counting numbers: 1, 2, 3, . . .
6. The fractions
7. The grains of sand in a liter bucket
8. The grains of sand on all the beaches
9. The points on a line
10. The seconds in a man's lifetime

Can you order these sets in order of increasing size?

Some of these sets are not well defined. For example, if we counted the people in the United States today, then recounted tomorrow, we would get different numbers. No matter; we can imagine the set and make a close estimate.

Do any of these sets have Rod's property, that if we start counting them 1, 2, 3, . . . , there will always be another one not yet counted no matter how far we go? Some do and some don't. The counting numbers, fractions, and points on a line have this property. The other sets do not. Sets with this property are called *infinite sets.*

Perhaps we can estimate the numbers in the other sets. There are about 200,000,000 people in the United States and about 20,000,000 in California. More accurate figures can be obtained from census reports or almanacs in any public or school library.

Look closely at someone's head. How far apart are the hairs? About one millimeter? Suppose one hair is growing at each intersection of lines one millimeter apart. That is one hair per square millimeter. The top of a man's head is something like a hemisphere of radius 100 millimeters. Then the area is $4\pi r^2 \times \frac{1}{2} = 4 \times \frac{22}{7} \times 100 \times 100 \times \frac{1}{2} = 63{,}000$ square millimeters. The man would have about 63,000 hairs on his head. You may wish to measure more accurately. However, we'd be greatly surprised if the normal person didn't have between 1,000 and 1,000,000 hairs on his head.

The federal government owes about $500,000,000,000. A fantastic amount. See your librarian for more accurate figures.

Even to count the grains of sand in a teaspoon is a tough job, let alone a bucket. What we suggest is that you spread out a teaspoon of sand on a sheet of paper. You can separate the sand into two approximately equal parts quite well. Take one of the parts and separate it into two equal parts in the same way. Continuing, you soon will have a part with about 30 grains of sand. Count these and work backward to get the number of grains in a teaspoon, and hence in a bucket. Suppose you divided the spoonful of sand into halves 5 times, and in the part left you had 30 grains. In the spoonful there were about $2 \times 2 \times 2 \times 2 \times 2 \times 30 = 960$ grains. Suppose 100 of your spoonfuls make a half liter. Then your bucket would hold about $960 \times 100 \times 2$ or about 190,000 grains.

How much sand on all the beaches? Let us be overly generous and suppose the whole world is covered with sand 2 meters deep. The earth is approximately a sphere of radius 6300 kilometers and hence has area

$4\pi r^2 = 4 \times \frac{22}{7} \times 6300 \times 6300 \times 1000 \times 1000$ square meters. (π is about $\frac{22}{7}$.) The area is about 5×10^{14} square meters. The layer of sand is 2 meters deep, 1000 liters in a cubic meter; then there are about $5 \times 10^{14} \times 2 \times 1000 \times 190{,}000$ = about 19×10^{22} grains of sand in a layer over the entire earth. We can be very sure, then, that there are fewer than 190,000,000,000,000,000,000,000 grains of sand on all the beaches.

You can make a better estimate. But we can see that as many as this number shows, it is still a finite number. If we counted rapidly, and had lots of time, we would run out of sand in our counting.

A nine-year-old nephew of a very delightful mathematician, Edward Kasner, made up the name *googol* for 10^{100}—1 followed by 100 zeros. You might want to compare it with the number of grains of sand there would be if the whole universe as far as the most distant star detected were solidly packed with sand.

Kasner also suggested *googolplex* for 1 followed by a googol of zeros.

These are very large numbers—but they are still finite.

We saw that the set of counting numbers, the set of fractions, and the set of points on a line have an infinite number of members. There are more in each of these than in any of the finite sets. We know how to add finite numbers. How does infinity act? Do we want to call infinity a real number?

27.3 A TALE OF INFINITY There is a story told that a rich Arab built a hotel with an infinite number of rooms. Each room had a counting numeral on the door. No room had two numbers. Every counting number labeled some room.

One day, an infinite number of salesman came to the hotel for a convention. There was Mr. One, Mr. Two, ... Could they have rooms for a week?

"No trouble," said the desk clerk, and he put Mr. One in room 1, Mr. Two in room 2, Mr. Thousand in room 1000, and so on.

Every man had a room, and every room was full.

Then Mr. Selby, the sales manager, arrived. "Sorry to be late," he apologized. "May I have my room." "I'm so sorry," said the clerk, "but every room is full. You will have to go to the hotel down the street." Mr. Selby called the owner and suggested that if he couldn't have a room, he'd take all infinity of his salesmen to the hotel down the street. "No

problem," said the owner. "You may have room 1. Mr. One will go to room 2, Mr. Two will go to room 3, and so on." So every salesman moved to the room with the next-highest number.

Now every man had a room, and every room was full.

Who was put out to make room for Mr. Selby? No one. Does this suggest that infinity plus one is infinity? That adding one to infinity does not make infinity any larger?

On Wednesday it was seen that the salesmen would need some letters typed. So each man sent for his secretary and told the clerk he'd need a room for her. "Wow," said the clerk. "We need not one, but an infinity of extra rooms!" and off he ran to the owner. "No problem," said the owner. "Easily settled." (Can you suggest where he might have put the secretaries?) The man in room 1 moved to room 2. The man in room 2 moved to room 4. The man in room 3 moved to room 6 . . . . The man in room k moved to room $2k$.

Every man had a room, and rooms 1, 3, 5, 7 . . . were ready for the secretaries.

At the end of the week, the owner said to the sales manager, "The charge is one gold piece per man. I'd like to give you a 10 percent commission for bringing me all this business." "Fine, said Mr. Selby, "I'll have my men pay their bills in order. Each man will label his gold piece. When the first ten men have put their gold pieces in the till, I will take out the gold piece Mr. One put in. When the next ten men have put their money in the till, I will take Mr. Two's gold piece."

After all the men had paid their bills, how many gold pieces were in the till? Would you believe none? The sales manager took Mr. Two's gold piece when Mr. Twenty paid. He took Mr. Three's gold piece when Mr. Thirty paid. Did he get Mr. Four's gold piece? Mr. Forty-Eight's gold piece? Mr Thousand One's? Is there anyone whose gold piece he didn't get? If so, name him . . . . Mr. Selby got away with all the money.

This story shows us that infinity doesn't behave like the counting numbers, so we say that infinity is not a counting number, not a real number. Arithmetic involving infinity will be quite different from the arithmetic of real numbers.

27.4 ARE ALL INFINITE SETS THE SAME SIZE? We turn to the infinite sets *e, f, i*—the counting numbers, the fractions, and the points on a line. Do we wish to say that all infinite sets have the same number of members? Is there only one "infinity"? Are there different infinities, with some being larger than others? What will we accept as evidence that two infinite sets have the same number of members, or that one is larger?

It seems reasonable to say that two sets which can be matched one to one are the same size, have the same number of members, and are equivalent.

Strangely, we may be able to match a set with a proper subset; hence a set can have the same number of members as a proper subset. For example, since the even counting numbers are a subset of the counting numbers 1 2 3 4 5 6 7 8 9 10 . . . , one might expect there to be more counting numbers. Yet we can match the members in the two sets:

$$\begin{matrix} 1 & 2 & 3 & 4 & 5 & 6 & \cdots & n & \cdots \\ 2 & 4 & 6 & 8 & 10 & 12 & \cdots & 2n & \cdots \end{matrix}$$

Hence the sets are the same size.

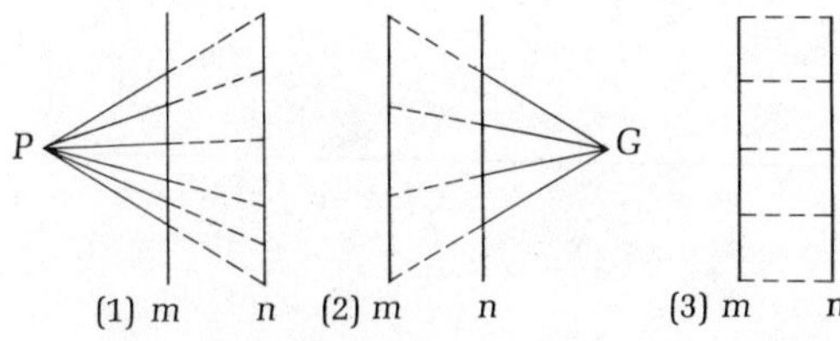

If a finite set A matches a subset of B with some left over, we can conclude that no matter how we try to match A with B there will always be some in B unmatched. If the capital letters A, B, C, D are matched with the small letters x, y, z, there will always be a capital letter unmatched. This is not so with infinite sets. Consider m,n. In diagram 1, every point of n is matched with a point of m as the intersections with a line through P. But there are some points of m unmatched with a point of n. In diagram

2, there are points left over in n. In diagram 3, there are no points left over in either m or n.

If we are to assert that a set S is greater than a set T, we need to show that *every* matching will always leave some out in S. It is not enough to show that for one or two matchings there are elements of S unmatched with an element from T. Someone else might find another matching where S and T paired exactly.

There are two fundamental theorems that follow from these ideas: (1) the set of fractions and the set of counting numbers are the same size; and (2) the set of real numbers is larger than either of the other two.

We sometimes say (1) the fractions are countable, meaning they can be matched with the counting numbers. Similarly we say (2) the real numbers (points on a line) are uncountable. An easy corollary to (1) is that rational numbers are countable. We restrict our argument to nonnegative numbers.

We use the ordering of fractions by height and numerator. We recall the height of $\frac{a}{b}$ is $a + b$. We list possible heights and fractions with the height. The pattern is soon clear.

TABLE 1

Height	Fractions
1	$\frac{0}{1}$
2	$\frac{0}{2}\ \frac{1}{1}$
3	$\frac{0}{3}\ \frac{1}{2}\ \frac{2}{1}$
4	$\frac{0}{4}\ \frac{1}{3}\ \frac{2}{2}\ \frac{3}{1}$
5	$\frac{0}{5}\ \frac{1}{4}\ \frac{2}{3}\ \frac{3}{2}\ \frac{4}{1}$
6	$\frac{0}{6}\ \frac{1}{5}\ \frac{2}{4}\ \frac{3}{3}\ \frac{4}{2}\ \frac{5}{1}$
.	.
.	.
.	.

Matching the fractions with counting numbers yields

TABLE 2

1	2	3	4	5	6	7	8	$\cdots$
$\frac{0}{1}$	$\frac{0}{2}$	$\frac{1}{1}$	$\frac{0}{3}$	$\frac{1}{2}$	$\frac{2}{1}$	$\frac{0}{4}$	$\frac{1}{3}$	$\cdots$

We are ordering the fractions by height and numerator. Table 2 can easily be extended as far as we please. You could find, for example, that fraction #601 is $\frac{5}{30}$ and that $\frac{2}{29}$ is fraction #468. It is tedious, but possible, to find the matching partner for any counting number or any fraction.

This order of fractions is also called *corner clipping* order. We can get the order by cutting the corner off a rectangular array. Form all the fractions into a rectangular array going indefinitely in two directions as shown:

$$\begin{array}{cccccccccc}
\frac{0}{1} & \frac{1}{1} & \frac{2}{1} & \frac{3}{1} & \frac{4}{1} & \frac{5}{1} & \frac{6}{1} & \cdot & \cdot & \cdot \\
\frac{0}{2} & \frac{1}{2} & \frac{2}{2} & \frac{3}{2} & \frac{4}{2} & \frac{5}{2} & \frac{6}{2} & \cdot & \cdot & \cdot \\
\frac{0}{3} & \frac{1}{3} & \frac{2}{3} & \frac{3}{3} & \frac{4}{3} & \frac{5}{3} & \frac{6}{3} & \cdot & \cdot & \cdot \\
\frac{0}{4} & \frac{1}{4} & \frac{2}{4} & \frac{3}{4} & \frac{4}{4} & \frac{5}{4} & \frac{6}{4} & \cdot & \cdot & \cdot \\
\cdot & \cdot & \cdot & \cdot & \cdot & \cdot & \cdot & \cdot & \cdot & \cdot \\
\cdot & \cdot & \cdot & \cdot & \cdot & \cdot & \cdot & \cdot & \cdot & \cdot \\
\cdot & \cdot & \cdot & \cdot & \cdot & \cdot & \cdot & \cdot & \cdot & \cdot
\end{array}$$

You can see the pattern of the numerators and denominators.

Imagine cutting off the upper left corner, one slice of numbers at a time. The slices are as follows:

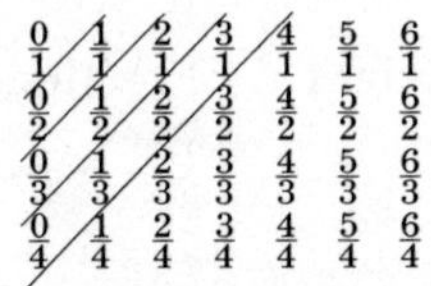

These slices are exactly the rows in table 1.

A bit of fall-out knowledge is that any set that can be arranged in a rectangle going to infinity in two ways can be corner clipped to give a single sequence. Hence the set is countable.

In algebra you may find shortcuts for finding the matching partners for a given fraction or counting number. This is beyond the purpose of this book.

Many fractions have equal values. By eliminating fractions of duplicate value, we get the theorem that the rational numbers are countable. To eliminate duplicate values, we drop all fractions that can be reduced to lower terms. The reduced fraction will have already appeared in the list.

The real numbers are uncountable. We prove by showing that anyone who thinks he has made a list of the infinite decimals is sadly mistaken. We show him one he has left out. This is an indirect proof.

To simplify the work, we show that the real numbers between zero and one are uncountable. If these are uncountable, then a set even larger is uncountable. Recall that a set is countable if it can be matched with the counting numbers—that is, arranged in a list, first, second, . . . and so on. Suppose Horace tells us he has listed all the infinite decimals and we look at his list. Let us say it begins as follows:

1	.247635 . . .
2	.337045 . . .
3	.450612 . . .
4	.001232 . . .
5	.800394 . . .
6	.230296 . . .
.	.
.	.
.	.

The dots mean that while we don't know what the digits are, Horace has a machine for each infinite decimal that gives him as many digits as he pleases.

We ask, "Horace, did you not leave $M = .3223 \ldots$ out of your list?" It would take a rare genius, like George Cantor who developed this proof, to see what the next digits in $M = .3223 \ldots$ would be. Here is how it is done. Form the number D with digits from the "diagonal." $D = .23029 \ldots$. We see the 50th digit in D is the 50th digit from the 50th number in Horace's list. D might be the 6th number in Horace's list, or it might not.

Compare D and M.

$$D = .23029 \ldots$$
$$M = .3223 \ldots$$

The rule is: "For every place where D has 2, put 3 for M. For all other places, where D does not have 2, put 2 for M. Can you show that the 5th digit in M is 2?

Horace is quite annoyed—for M is an infinite decimal, and M is not in his list. For suppose he thought M was the 4th number. The 4th digit in the 4th number is 2, the 4th digit in D. The 4th digit in M is 3. Since M and the 4th number differ in the 4th digit, at least, M is not the 4th number.

Similarly Horace might think *M* is the 500th number in his list. But clearly the 500th digit in *M* is different from the 500th digit in the 500th number. Hence *M* is not the 500th number. *M* appears nowhere in the list.

Since every effort to list the infinite decimals is bound to leave at least one out, we have shown the real numbers are not countable. The infinity of the real numbers is greater than the infinity of the integers.

27.5 SUMMARY Bigness has charm. We have seen that very large sets may still be finite, even though they stretch our imagination. Other sets with which we feel familiar are infinite. We have seen that infinite sets have startling properties quite different from finite sets. Hence we do not want to include infinity as an integer, or even a real number. We have seen that the set of counting numbers, the set of even integers, and the set of fractions are all the same size. However, the real numbers, the points on a line, form a definitely larger set. We cannot list all the numbers between 0 and 1!

EXERCISES

1. Determine or estimate these large numbers:
- (*a*) the number of seconds in a year
- (*b*) the number of heartbeats in a lifetime
- (*c*) the number of hairs on Mildred Plantagenet's head
- (*d*) the number of particles there would be in the room if the room were full of beans, sand, dust
- (*e*) the number of millimeters you can run in one hour
- (*f*) the largest number you can remember using, or seeing used, in a count or a measurement

2. (*a*) Estimate the number of grams of food you have eaten in your life.
   (*b*) If the moon were made of green cheese weighing one gram per cubic centimeter, how long would it take you to eat it at the rate implied in *a*?

3. Suppose that each of the infinity of salesmen in the hotel with an infinite number of rooms wanted two extra rooms. Show how the manager could arrange this.

4. Suppose the 7th number on Horace's list (section 27.4) were .322332332 . . . What is the 7th digit of *D*? the 7th digit of *M*? Could it be that this number is *M*? Why?

5. Write the first 10 fractions listed by increasing height and numerator.

6. What position does $\frac{3}{4}$ have in the listing by height and numerator?

7. Set up a matching between points on segments *a* and *b*.

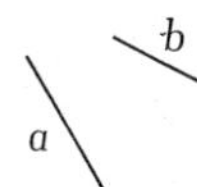

8. After Mr. Selby's salesmen leave, a group of valentine manufacturers ask for rooms. There is one valentine manufacturer for every point on a line segment one centimeter long. Show that no matter how the manager assigns them to rooms there is always a valentine manufacturer without a room.

Answers to selected questions: **1.** answers vary, (a) 32,000,000; (e) 10,000,000; **3.** Move the person in room *n* to room 3*n*; **5.** 0/1, 0/2, 1/1, 0/3, 1/2, 2/1, 0/4, 1/3, 2/2, 3/1.

28

Congruences and Modular Numbers

28.1 INTRODUCTION Yang and yin, male and female, even the counting numbers are divided into two sets, odd and even. Children learn early that odd and even have properties, such as odd times odd is odd, and even plus even is even. These properties are interesting, useful, and may even be said to form a number system.

Other systems similar to odd-even exist. We shall point them out showing that–

1. They fit some human situations nicely, especially repetitive or cyclic situations.
2. They help us understand the structure of the rational numbers better.
3. They give tests for divisibility.

You may have heard of the systems called *modular number systems, clock arithmetic, congruences modulo n,* and *remainder arithmetic.*

28.2 WHAT MODULAR NUMBERS ARE When we count, or measure time or distance, the numbers keep getting bigger and bigger. Yet we live in a world with many cyclic, or repeating, events and are interested in position in the cycle rather than distance from the starting point. For example, we use Sunday, Monday, . . . , Saturday to label the weekly cycle of days. We say, "We go to church on Sunday," rather than, "We go to church on days 3, 10, 17, . . . , 6324, 6331, . . ." Similarly, we label clock times 1, 2, . . . , 12, 1, 2, ... We label months in a cycle. The Chinese labeled years in cycles, but in Christendom we count from approximately the birth of Christ. On odometers and adding machines the digits go in cyclic order: 3, 4, 5, . . . , 9, 0, 1, 2, 3, 4, ...

Here are some typical problems:

1. It takes 23 days for a ship to cross the ocean. If it leaves on Friday, on what week day will it arrive? Do you say Saturday? or Sunday?

2. A Chinese delicacy is slowly baked for 33 hours. At what time does it go in the oven to be ready at 7 P.M.?

3. Can you see at a glance that

$$3{,}647{,}897 + 13{,}268{,}648 = 16{,}964{,}466$$

is wrong?

4. Can you see at a glance that

$$7934 \times 3694 = 29{,}118{,}194$$

is wrong?

In (1) we might say that Friday + 23 = Sunday. In (2) 7 − 33 = 10. In (3) odd + even ≠ odd. In (4) the units digit must be 6 since $4 \times 4 = 16$.

These are all examples of remainder arithmetic, or modular arithmetic. Since $23 = 3 \cdot 7 + 2$ we merely go two days from Friday to get Sunday, the day of arrival. We can ignore the three full weeks. In (2), since $33 = 2 \cdot 12 + 9$, we need only count back 9 hours from 7 P.M. In $(793 \times 10 + 4) \times (369 \times 10 + 4)$, we see that the units digit in the product must be the units digit in 4×4.

The common idea in these examples is seen by thinking about remainders after division, or repeated subtraction.

Two was the remainder on dividing 23 by 7. Nine was the remainder on dividing 33 by 12. Odd and even refer to remainders 1 and 0 on division by 2, and the units digit in a number is the remainder on division by 10.

To fix our ideas, we talk about division by 3, which leads to numbers modulo 3. All the integers that leave remainder 2 when divided by 3 we place in a set that we call "2," or "2, modulo 3," or "2 mod 3."

$$2, \bmod 3 = \{\ldots,-7,-4,-1,2,5,8,11,\ldots\}$$

Similarly,

$$0, \bmod 3 = \{\ldots,-6,-3,0,3,6,\ldots\}$$
$$1, \bmod 3 = \{\ldots,-5,-2,1,4,7,\ldots\}$$

Any two numbers that differ by 3 or a multiple of 3 are in the same set. This is similar to using week names, where any two days that are a multiple of 7 days apart have the same name (Tuesday, for instance).

Two integers in the same modulo number set, such as 4 and 7, are said to be "congruent mod 3," written "$4 \equiv 7 \bmod 3$". All Wednesdays are congruent mod 7.

Similarly, there are six mod 6 numbers:

$$0, \bmod 6 = \{\ldots,-12,-6,0,6,12,18,\ldots\}$$
$$1, \bmod 6 = \{\ldots,-11,-5,1,7,13,19,\ldots\}$$
$$2, \bmod 6 = \{\ldots,-10,-4,2,8,14,\ldots\}$$
$$3, \bmod 6 = \{\ldots,-15,-9,-3,3,9,15,\ldots\}$$
$$4, \bmod 6 = \{\ldots,-14,-8,-2,4,10,\ldots\}$$
$$5, \bmod 6 = \{\ldots,-7,-1,5,11,\ldots\}$$

We have $7 \equiv 19 \bmod 6$ (read "7 is congruent to 19 modulo 6"), $14 \not\equiv 4 \bmod 6$ (read "14 is not congruent to 4 mod 6"), and $13 - (-5) \equiv 0 \bmod 6$. Ordinarily the name we give to the modular number is the smallest nonnegative integer in the set. It is the smallest nonnegative remainder when any integer in the set is divided by 6. It may seem strange to call a set of integers a number. It is convenient to do so because we can define numberlike operations of addition and multiplication. These are usefully related to operations in integers. Also we've called a set of numerals a number before—for example, when we said the set $\{\frac{1}{2},\frac{2}{4},\frac{3}{6},\ldots\}$ is a rational number.

When using modular numbers, we must know the modulus, since 2 mod 3 is different from 2 mod 6. Usually we write the modulus, but if it is otherwise clear we may write "2" rather than "2 mod 3," and "$7 \equiv 25$" rather than "$7 \equiv 25 \bmod 3$."

Any integer may be taken as the modulus of a modular number system. If 0 is the modulus, the system is the same as the integers. Hence we see that modular numbers are a generalization of integers.

28.3 ADDITION AND MULTIPLICATION It is easy to define the sum of two mod 3 numbers a, b as the remainder when $a + b$ is divided by 3. This leads to the following addition table:

+	0	1	2
0	0	1	2
1	1	2	0
2	2	0	1

Similarly we define the product of two numbers as the remainder when $a \cdot b$ is divided by 3. This leads to the multiplication table:

×	0	1	2
0	0	0	0
1	0	1	2
2	0	2	1

Some people prefer to define these operations from the following observations:

1. The sum of any number in the set 1 and a number in set 2 is a number in the set 0.

Indeed, the sums of any number in one set, a, and any number in another set, b, will always be numbers in a single set, c. Hence we say that c is the name for $a + b$. Consider $5 + 4 = 9$, $-4 + 7 = 3$, $11 + -5 = 6$, $-1 + -5 = -6$. In each case the first addend is from set 2. The second addend is from set 1. The sum is from set 0. Hence we say in mod 3 numbers $2 + 1 = 0$.

2. The products of any number from set a with any number from set b will all be in a single set, d.

Hence we say $a \cdot b = d$. For example, $5 \cdot 4 = 20$, $-4 \cdot 7 = -28$, $11(-5) = -55$, $-1 \times -5 = 5$. In each case the first factor is from set 2, the second factor is from set 1, and the products are all in set 2. Hence we say in mod 3 numbers $2 \times 1 = 2$.

The addition and multiplication tables for modulo numbers look much like the tables for counting numbers. Indeed, for addition, only 3 of the 9 addition facts appear different: $1 + 2 = 0$, $2 + 1 = 0$, $2 + 2 = 1$. Only one of the nine multiplication facts looks different, $2 \times 2 = 1$. Hence many of our intuitive ideas about these operations in counting numbers will carry over to modular numbers. We are probably not surprised when told that these operations satisfy the CAD laws as well.

Life might be a bit more startling if we took the next larger number in each set as the name. Then we would get addition and multiplication tables:

+	3	4	5
3	3	4	5
4	4	5	3
5	5	3	4

×	3	4	5
3	3	3	3
4	3	4	5
5	3	5	4

None of the eighteen addition and multiplication facts look like the corresponding fact in counting numbers.

You will recall that the additive identity is a number p such that $x + p = x$ for all numbers x in the set. Can you point out the additive identity from the tables? Similarly can you point out the multiplicative identity? Is there one? Is it easier using the names (0, 1, 2) or the names (3, 4, 5)? These show the convenience of naming modulo numbers as we do.

Both the remainder definition and the set definition for modular numbers and the operations give the same addition and multiplication tables. Some people prefer the set name definitions, because it is a bit more flexible, allowing startling results, such as $3 + 3 = 3$, which may help people see things a bit more deeply. We can no longer say, "$2 + 0 = 2$, because 0 means nothing, and if you don't add anything to 2, you don't change it, so naturally $2 + 0 = 2$." Now 0, mod 3 (or 3, if you prefer) does not stand for "nothing." Addition is not simply a statement about the union of sets. It just happens that 0, mod 3, or 3, mod 3 turns out to be the additive identity.

Some people like to call operations with modular numbers "clock arithmetic." What time will it be when Mylene wakes up if she goes to bed at 9 and sleeps 8 hours? Turning the hands on a watch yields 5. We also get this number as the remainder on dividing $9 + 8$ by 12. We can also think of this as subtracting 12 from $9 + 8$, since division can be thought of as successive subtraction.

In many elementary books a 5-hour clock is introduced, since 12 is thought to be too large and too fixed to our timepieces. A 5-hour clock

relates nicely to modular 5 numbers. Where does one end if, starting at 0, one adds 3 + 4 + 1? Counting steps around the circle one lands on 3, then 2, then 3. So 3 + 4 + 1 = 3. Subtracting 5 from 3 + 4 + 1 gives 3 as a remainder. Similarly we find 1 + 3 + 2 + 4 + 2 + 4 = 1.

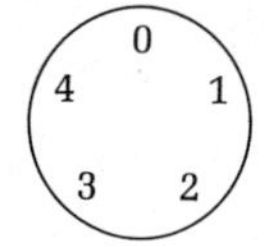

If multiplication is treated like repeated addition, we find 4 × 3 = 3 + 3 + 3 + 3. Counting around the circle we land on 3, 1, 4, 2. Hence 4 × 3 = 2. Also the remainder on dividing 4 × 3 by 5 is 2.

Strictly speaking, the clock analogy breaks down when we speak of multiplication. Modular numbers are not counting numbers, although we may use the same names. The modular number 4 in 4 × 3 = 2 is not the operator–counting number in 4 × 3 = 3 + 3 + 3 + 3. This difference probably causes little difficulty, but it is good for teachers to know it exists in case a pupil asks.

Some people like to call modular arithmetic *remainder arithmetic* rather than *clock arithmetic*." They feel this term more actually describes what happens and is just as vivid in its suggestiveness.

28.4 CHECKING WITH REMAINDER ARITHMETIC Two properties make remainder arithmetic useful in checking numerical computations: (1) the remainder on dividing the sum of two numbers is the sum of the remainders, and (2) the corresponding statement for products. The same divisor is used throughout.

For example, the remainders on dividing 201 and 11 by 9 are 3 and 2, respectively. The remainder on dividing 201 + 11 = 212 by 9 is 5 = 3 + 2 and on dividing 201 × 11 = 2211 is 6 = 3 × 2.

Is it obvious that $367 \times 489 \neq 179{,}436$? The answer is about the right size, since $400 \times 500 = 200{,}000$. But an odd number times odd number cannot be even. This mod 2 argument convinces us that an error exists somewhere.

Similarly the sum of a column of figures can be checked by counting the odd addends. Is $28 + 17 + 35 + 46 + 19 = 144$? Since the sum of three odd numbers is odd, an error has been made.

One can check with a modulus other than 2. The checks may catch more errors, but they may not be quite so easy. Here's how the check with mod 9 works.

You will recall that the sum of any integer in the 2, mod 9 set with an integer from 4, mod 9 will be in the 6, mod 9 set. For example $11 + 13 = 24$. The check says to replace the addends with more convenient numbers from the sets. Since $11 \equiv 2$, $13 \equiv 4$, we check $11 + 13$ by computing $2 + 4$. Since $2 + 4 = 6 \equiv 24 \bmod 9$, we have satisfied the check.

Similarly, we check $182 + 301 = 483$. We find $182 \equiv 2$, $301 \equiv 4$ and $483 \equiv 6$. Since $2 + 4 = 6$, the sum checks.

The corresponding property holds for products. $13 \times 11 = 133$. Since $133 \equiv 7$ and $4 \times 2 = 8 \not\equiv 7$, an error has been made. Can you correct it?

You may complain, "The check works, but it is so hard to find out what mod 9 number includes 133 (or almost any number) that the check takes too long." True, in general, but did you notice that each of these numbers was congruent mod 9 to the sum of its digits? Note $13 \equiv 4$, $133 \equiv 1 + 3 + 3 = 7$.

That this is always true we see from expanded notation.

We note that $10 \equiv 1 \bmod 9$, and hence $100 = 10 \times 10 \equiv 1 \times 1 = 1$. Indeed $10^3 \equiv 10^4 \equiv \cdot\cdot\cdot\ 10^n \equiv 1$. Substituting, we get

$$\begin{aligned} 133 &= 1 \cdot 100 + 3 \cdot 10 + 3 \\ &\equiv 1 \cdot 1 + 3 \cdot 1 + 3 = 1 + 3 + 3 = 7 \end{aligned}$$

We might also note that

$$\begin{aligned} 133 &= 100 + 30 + 3 \\ &= 99 + 1 + 3 \cdot 9 + 3 + 3 \\ &= (99 + 3 \cdot 9) + (1 + 3 + 3) \end{aligned}$$

Since the number in the first parenthesis is clearly divisible by 9, the remainder on dividing 133 by 9 is the remainder on dividing $(1 + 3 + 3)$ by 9.

These methods can show us that any base-ten number is congruent to the sum of its digits, mod 9.

EXAMPLE

$3764 = 3 \times 10^3 + 7 \times 10^2 + 6 \times 10 + 4$	expanded notation
$\equiv 3 \times 1 + 7 \times 1 + 6 \times 1 + 4$	replacing 10 by 1, since $10 \equiv 1 \bmod 9$
$= 3 + 7 + 6 + 4 = 20$	facts
$= 2 \times 10 + 0$	expanded notation
$\equiv 2 \times 1 + 0$	replacing 10 by 1, since $10 \equiv 1 \bmod 9$
$= 2$	

Hence $3764 \equiv 2$.

We summarize: In any sum (or product) of base-ten numerals, replace each addend (or factor) by the sum of the digits. The sum of the digits in the sum (or product) equals the sum (product) of the replaced addends (factors), provided that whenever any sum of digits has more than one digit, we replace this number by the sum of its digits.

Finding the remainder is often called *casting out*. If a number is represented by a set of beans, division by 9, say, can be shown by taking away sets of 9 as many times as possible. The number of beans left is the remainder. Taking away vigorously can be called casting out. Hence, checking by using numbers modulo 9 is often called casting out nines.

EXAMPLE Is $365 \times 24 = 8860$? The product is even as it should be. Rounding off, $400 \times 20 = 8000$, so the product is about the right size. Adding the digits to check by congruence mod 9 (casting out nines):

$$365 \equiv 14 \equiv 5 \qquad 8860 \equiv 22 \equiv 4$$
$$24 \equiv 6$$

$5 \times 6 = 30 \equiv 3$. Alas $3 \not\equiv 4$. There is an error. Can you find it?

EXAMPLE Similarly $93 \times 192 = 17856$ is checked from

$$93 \equiv 12 \equiv 3,\ 192 \equiv 12 \equiv 3,\ 17856 \equiv 27 \equiv 9 \equiv 0, \text{ and } 3 \times 3 \equiv 0$$

EXAMPLE We can check addition by casting out nines.

$$732 + 489 + 251 + 1700 = 3172.$$

$$\begin{aligned} 732 &\equiv 3 \\ 489 &\equiv 3 \\ 251 &\equiv 8 \\ \underline{1700} &\equiv 8 \\ 3172 &\equiv 4 \end{aligned}$$

Since $3 + 3 + 8 + 8 = 22 \equiv 4$, we have a check.

Just as odd and even checks do not catch all errors, so casting out nines will not catch all errors. In particular, if in the example above one transposed two numbers and got 3712 as the sum, this error would not be caught. Sometimes two or more errors are made whose effects cancel the check. For example, 3262 would be accepted as correct. Such exquisite multiple errors are uncommon, however; so except for transpositions, the check is quite effective.

Checking by casting out nines has long been popular for hand calculations. In recent years the test has not received as much attention as formerly, because those who do lots of calculating use mechanical equipment with more sophisticated checks, while those who do little repeat the operation several times as a check. They are willing to use this relatively inefficient method because the total time is not much and the surer check of several repetitions is worthwhile.

28.5 DIVISIBILITY TESTS Everyone knows that an even integer is one that is divisible by 2, and that every base-ten number is even if, and only if, the units digit is even. Such rules for telling if a number is divisible by another number by looking at the digits in its numeral are called divisibility tests. Here are some other tests:

If the units digit is 0, the number is divisible by 10.

If the units digit is 0 or 5, the number is divisible by 5.

If the two-digit number formed by the tens and units digit is divisible by 4, the number is divisible by 4.

If the sum of the digits is divisible by 3, the number is divisible by 3.

If the sum of the digits in even places equals the sum of the digits in the odd places, the number is divisible by 11. This may also be expressed as follows: If the sums of alternate digits are equal, or if the digits alternately added and subtracted sum to zero, the number is divisible by 11.

If the number formed by the hundreds, tens, and units digits is divisible by 8, the number is divisible by 8.

These rules can be verified and proved separately. However, they all derive from one idea and can be found by using modular numbers or remainder arithmetic.

Consider 12,345. In expanded notation:

$$12{,}345 = 1 \times 10^4 + 2 \times 10^3 + 3 \times 10^2 + 4 \times 10 + 5$$

A number divisible by 3 is congruent to 0, mod 3. So we need merely reduce this number mod 3. Since $10 \equiv 1 \bmod 3$ and $1 \times 1 \times 1 = 1$, all the powers of 10 can be replaced by 1. Hence

$$12{,}345 \equiv 1 + 2 + 3 + 4 + 5 = 15 \equiv 6 \equiv 0 \bmod 3$$

So 12,345 is divisible by 3. This argument does not say what the quotient is.

Is 13,452 divisible by 5? We look to see if

$$1 \times 10^4 + 3 \times 10^3 + 4 \times 10^2 + 5 \times 10 + 2 \equiv 0 \bmod 5$$

Now $10 \equiv 0 \bmod 5$, so $10^3 = 10 \times 10 \times 10 \equiv 0 \times 0 \times 0 = 0$. All the terms involving powers of 10 can be replaced by zero. Hence, $13{,}452 \equiv 2 \bmod 5$. This argument shows that 13,452 is not divisible by 5.

For divisibility by 4, we note that $10 \equiv 2 \bmod 4$. Hence, $10^2 \equiv 2 \times 2 \equiv 0 \bmod 4$, and every higher power of 10 is congruent to 0. Hence

$$34{,}125 = 3 \times 10^4 + 4 \times 10^3 + 1 \times 10^2 + 25 \equiv 0 + 0 + 0 + 25 \bmod 4.$$

Hence we need look only at 25. Since 25 is not divisible by 4, neither is 34,125.

We conclude these examples by asking if 361,042 is divisible by 11. Now $10 \equiv -1 \bmod 11$.

Hence, $10^2 \equiv 1$, $10^3 \equiv -1$, $10^4 \equiv 1$, ... Then

$$\begin{aligned} 361{,}042 &= 3 \times 10^5 + 6 \times 10^4 + 1 \times 10^3 + 0 \times 10^2 + 4 \times 10 + 2 \\ &\equiv 3(-1) + 6(1) + 1(-1) + 0(1) + 4(-1) + 2 \\ &= {}^-3 + 6 - 1 + 0 - 4 + 2 = 0. \end{aligned}$$

Hence 361,042 is divisible by 11.

We have used large numbers in the examples so that the structure would be clear. You may wish to make up statements of your own for the various divisibility tests.

You may have noticed these divisibility tests give remainders on dividing by the modulus. If the remainder is zero the modulus is a factor.

28.6 SOME DIFFERENT NUMBER STRUCTURES Some modulo number systems are similar to the rational number system; others are not, but they differ in ways that throw light on the structure of numbers. The principle of learning from contrasts with counterexamples by seeing what the ideas mean in different and unusual situations gets a boost here. We use modular numbers to explore subtraction, integers, division, fractions, cancellation, roots, and solutions of quadratic equations.

Consider the modulo-5 numbers, named 0, 1, 2, 3, 4. They act a lot like counting numbers. By successively adding 1 we get 0, then 1, 2, 3 . . . just as with counting numbers. Only here the ellipses (. . .) means something different. While we continue to add 1, the sequence repeats 0, 1, 2, 3, 4, 0, 1, 2, 3, 4, 0, 1, 2, ... Still we have the "there's always one more" structure of the counting numbers.

Are there any negatives? At first glance we'd say no. No names have dashes in front of them. But is it the name that makes a number negative? What did we mean by $^-2$? We recall that in discussing integers the negatives arose in two ways:

1. If we had a sequence extending in two directions from the starting point 0, then naming with numbers in both directions gave us the pattern . . . , −3, −2, −1, 0, 1, 2, 3, . . . , where $^-2$ was "opposite 2."

2. We wished additive inverses. $^-2$ is the number that added to 2 gives zero. $^-2$ is the solution to $2 + x = 0$, and $y + 2 = 0$.

+	0	1	2	3	4
0	0	1	2	3	4
1	1	2	3	4	0
2	2	3	4	0	1
3	3	4	0	1	2
4	4	0	1	2	3

Are there numbers with one or both of these properties in mod 5 numbers? If we consider the sequence of mod 5 numbers . . . , 3, 4, 0, 1, 2, . . . we see that 3 is opposite 2. Looking at the addition table, we see that 3 is also the solution to $2 + x = 0$ and $y + 2 = 0$.

Is it reasonable to say that 3 and $^-2$ are merely different names for the same mod 5 number, and that there is in fact a $^-2$ in mod 5 numbers? Many authors say yes.

Similarly we ask if there is a number "$\frac{2}{3}$" in mod 5 numbers. Is there division? We search in vain for the label $\frac{2}{3}$ among 0, 1, 2, 3, 4. Let us look behind the labels at the attitudes that brought fractions to the fore.

1. *The name of an object with a history.* A whole is separated into three parts and two are taken. This appears like nonsense as far as modulo-5 numbers are concerned.

2. *An instruction.* Separate (something) into three parts and take two. Again nonsense.

3. *Ratio.* Apparently meaningless.

4. *Missing factor.* $\frac{2}{3}$ is a number which multiplied by 3 gives the product 2. It is the solution to the equations $3z = 2$ and $w3 = 2$. A look at the modulo-5 multiplication table shows that 4 satisfies $3 \cdot 4 = 2$ and $4 \cdot 3 = 2$. Are we going to say that 4 and $\frac{2}{3}$ are both names of the same number?

×	0	1	2	3	4
0	0	0	0	0	0
1	0	1	2	3	4
2	0	2	4	1	3
3	0	3	1	4	2
4	0	4	3	2	1

We see that from some fraction attitudes $\frac{2}{3}$ is nonsense; but from others, $\frac{2}{3}$ is another name for 4. Is object-with-a-history or missing factor the more important attitude? You must choose. Probably no one will slit your throat if you disagree with him. The author thinks the topic is more interesting if we say that $\frac{2}{3}$ and 4 name the same number. Is $\frac{2}{3} = \frac{4}{6}$?

If $2x = 2y$, can you conclude that $x = y$?

Certainly this is true for integers. But this "cancellation law" is not always true. Consider the multiplication table for mod 8 numbers. We see that

$$2 \cdot 1 = 2 \cdot 5 \qquad 2 \cdot 2 = 2 \cdot 6$$
$$2 \cdot 3 = 2 \cdot 7 \qquad 2 \cdot 2 = 2 \cdot 0$$

×	0	1	2	3	4	5	6	7
0	0	0	0	0	0	0	0	0
1	0	1	2	3	4	5	6	7
2	0	2	4	6	0	2	4	6
3	0	3	6	1	4	7	2	5
4	0	4	0	4	0	4	0	4
5	0	5	2	7	4	1	6	3
6	0	6	4	2	0	6	4	2
7	0	7	6	5	4	3	2	1

Hence if $2x = 2y$, x and y might well be different. Is the cancellation law true in mod 5 numbers?

Are there square roots? Is $\sqrt{2}$ another name of a number in a mod system? Is there a number n such that $n^2 = 2$ in mod 5? We see that there is not, since $0^2 = 0$, $1^2 = 1$, $2^2 = 4$, $3^2 = 4$, $4^2 = 1$. However, in mod 7 there is. For $0^2 = 0$, $1^2 = 1$, $2^2 = 4$, $3^2 = 2$, $4^2 = 2$, $5^2 = 4$, $6^2 = 1$. We see that there are two numbers, 3 and 4, whose squares are 2, and are applicants for the name $\sqrt{2}$.

Many of you learned in high school algebra that "every quadratic equation has two roots." While a thorough discussion of algebra is beyond the scope of this book, we point out that $x^2 = 1$ has at least three solutions in mod 8 numbers: 1, 3, and 7. You may also see that $x^2 + x = 0$ has more than two solutions in mod 6 numbers. Can you find them?

A quadratic equation having exactly two roots is a problem of advanced algebra. In modular numbers, it is closely related to whether the modulus of the system is composite, such as 8 and 6, or is prime, such as 5.

You may find other paradoxes in modulo numbers, such as the ability to divide by zero, and the existence of some fractions but not all. Much depends on the particular modulo system you are in.

28.7 SUPPORT FROM ALGEBRA We've used the properties that the remainder of a sum (product) is the sum (product) of the remainders. This is easy to show. Suppose we are casting out nines—that is, working with remainders on dividing by 9. Let $a = 9q + r$ and $b = 9s + t$. The remainders on dividing a and b by 9 are r and t, respectively. Then $a + b = 9(q + s) + (r + t)$. So the remainder on dividing $a + b$ by 9 is the remainder on dividing $r + t$ by 9. Similarly

$$\begin{aligned} ab &= (9q + r)(9s + t) \\ &= 81qs + 9qt + 9rs + rt \\ &= 9(9qs + qt + rs) + rt \end{aligned}$$

Hence the remainder on dividing ab by 9 is the remainder on dividing rt by 9.

A similar argument holds when we divide by a number other than 9.

Because congruence of integers is so closely related to equality of modular numbers, many people use $\equiv$ as well as $=$ for equality in modular numbers. Strictly speaking, we use = for the same integer, or for the same modulo number (a set of integers). We say two integers are congruent, such as $4 \equiv 13 \bmod 3$.

EXERCISES

1. List four areas in your life where affairs go in cycles, like days in the week. What is the modulus, or period of the cycle, such as 7 for days in the week?

2. When should a dish be set to simmer if it is to be ready at 6:15 P.M. and is to simmer for 25 minutes?

3. List the members of the set $\{1, 2, \ldots, 10\}$ that have the remainder 1 when divided by 2. What is the ordinary name for the quality these numbers share?

4. Describe even numbers in terms of division and remainders.

5. List the numbers in $\{1, 2, \ldots, 10\}$ that are congruent to 1, mod 3.

6. Find the remainder when $58 = 31 + 27$ is divided by 3. What equation in numbers mod 3 corresponds to this problem?

7. Make an addition table for odd and even numbers. Use 1 for the odd numbers and 0 for even.

8. Make a multiplication table for odd and even numbers.

9. Make an addition table for numbers mod 2. Use 2 and 3 for the names of even and odd.

10. Make a multiplication table for numbers mod 2.

11. Make addition and multiplication tables for numbers mod 6.

12. Using the tables from exercise 11, find the names for the mod 6 numbers -3, -17, $\frac{1}{5}$, $\frac{1}{2}$, if they exist.

13. Find numbers mod 5 that are also named -3, -17, $\frac{1}{4}$, $\frac{2}{3}$, $\frac{1}{2}$.

14. Is $\frac{4}{6} = \frac{2}{3}$ in mod 7 numbers? In mod 8 numbers?

15. What is the remainder when 4683 is divided by 9? Find without dividing by 9.

16. Find the remainders when the following numbers are divided by 9:
(*a*) 13 (*c*) 6923
(*b*) 525 (*d*) 123456789

17. Find the remainders when the numbers in exercise 16 are divided by 3, 5, 2, 11.

18. Using the fact that, when 10 is divided by 13, the remainder is −3, that is, 10 = 1 · (13 − 3), show that the remainder when the following are divided by 13 is as shown:
(*a*) 100, 9
(*b*) 1000, −1
(*c*) 10000, 3
Find the remainder when 403 is divided by 13. When 5005 is divided by 13.

19. Find all the solutions to $x^2 + x = 0$ in mod 6 numbers. In mod 5 numbers.

Answers to selected questions: **2.** 5:50 P.M.; **4.** They have remainder 0 when divided by 2; **6.** rem = 1; 1 = 1 + 0.

9.

+	2	3
2	2	3
3	3	2

10.

×	2	3
2	2	2
3	2	3

12. 3; 1; 5; none; **14.** yes (3 = 3); maybe (2 and 6 both = 4/6; only 6 = 2/3); **16.** (*a*) 4; (*b*) 3; (*c*) 2; (*d*) 0.

17.

	a	b	c	d
3	1	0	2	0
5	3	0	3	4
2	1	1	1	1
11	2	8	4	5

19. 0, 2, 3, 5 mod 6; 0, 4 mod 5.

29

Complex Numbers

29.1 INTRODUCTION Complex numbers, which involve $\sqrt{-1}$, are well beyond the usual topics considered in an arithmetic book. However, they play such an important role in both theoretical and applied mathematics that many people want to know something about them. Further, they are a prime example of how ideas are generalized.

Describing segments was one of the most powerful motivations for discovering counting numbers, fractions, and real numbers. With real numbers we are able to locate any point on the number line suggested to us by experience with tight strings, folds in paper, tall trees, and other straight objects. They are excellent for describing motion on a line. For example, "Go $2 + \sqrt{2}$ to the right" is quite clear.

Everything has its limitations. There are two limitations for real numbers that are particularly challenging. Negative numbers have no square roots, and rotations are hard to describe with real numbers. Efforts to enlarge real numbers to evade these restrictions led to complex numbers.

29.2 WHAT ARE COMPLEX NUMBERS? Suppose we imagine we have "numbers" that are square roots of negative reals. Let's name them $i = \sqrt{-1}$, $j = \sqrt{-2}$, $k = \sqrt{-3}$, $a = \sqrt{-4}$, $b = \sqrt{-5}$, and so on. About all we can say about them is (1) $i^2 = -1$, $j^2 = -2$, $k^2 = -3$, $a^2 = -4$, $b^2 = -5$, and so on; and (2) they don't represent points on the number line (some people say they represent imaginary points on the line).

These new "numbers" won't be much use if we can't combine them with real numbers. So we not only assume (1) and (2) above, but also assume that we can add, subtract, multiply, and divide them along with the real numbers and that they satisfy the CAD laws. We see we need only one new "number," $i = \sqrt{-1}$. For $(2i)^2 = 2i \cdot 2i = 4i^2 = -4$. Hence, $a = 2i$. Similarly, $b = i\sqrt{5}$, $j = i\sqrt{2}$, $k = i\sqrt{3}$, and so on.

Using multiplication and addition, we conclude that symbols like $2 + 3i$, $\sqrt{2} - 4i$, $7 + \pi i$ must represent numbers in our system. Using division, $\frac{3}{i}$, $\frac{4+i}{i}$, $\frac{2-i}{1+i}$ must be in the system. The division symbols may be simplified, since

$$\frac{3}{i} \cdot \frac{-i}{-i} = \frac{-3i}{-i^2} = -3i$$

Similarly

$$\frac{4+i}{i} \cdot \frac{-i}{-i} = \frac{-4i - i^2}{-i^2} = \frac{1-4i}{1}$$

$$\frac{2-i}{1+i} \cdot \frac{1-i}{1-i} = \frac{2 - i - 2i + i^2}{1 + i - i - i^2}$$

$$= \frac{1-3i}{2} = \frac{1}{2} - \frac{3}{2}i.$$

In each division symbol we multiplied the numerator and the denominator with what is called the conjugate of the denominator. The conjugate of a number is what one gets on replacing i with $-i$.

The numbers represented by the symbols $a + bi$ are called *complex numbers;* a and b are real numbers and i satisfies $i^2 = -1$. In $3 + 4i$, 3 is called the real part and 4, the coefficient of i, is called the imaginary part of the complex number.

Using the CAD properties, we can add, subtract, multiply, and divide complex numbers by rules shown by the examples shown in the margin.

These are special cases of the formulas:

$$(a + bi) + (c + di) = (a + c) + (b + d)i$$

$$(p + qi) - (r + si) = (p - r) + (q - s)i$$

$$(f + gi)(x + yi) = (fx - gy) + (gz + fy)i$$

$$\frac{l + mi}{u + vi} = \frac{(ul + mv) + (mu - lv)i}{u^2 + v^2}$$

$$(2 + 3i) + (5 + 7i) = (2 + 5) + (3 + 7)i = 7 + 10i$$

$$(4 + 6i) - (7 + 5i) = (4 - 7) + (6 - 5)i = -3 + i$$

$$\begin{aligned}(5 + 3i) \cdot (2 - 8i) &= 5 \cdot 2 + 5(-8i) + 3i(2) + 3i(-8i)\\ &= 10 - 40i + 6i - 24i^2\\ &= 10 - 34i - 24(-1)\\ &= 34 - 34i\end{aligned}$$

$$\begin{aligned}\frac{5+2i}{3-4i} &= \frac{5+2i}{3-4i} \cdot \frac{3+4i}{3+4i}\\ &= \frac{(15 - 8) + (20 + 6)i}{(9 + 16) + (3 \cdot 4 - 4 \cdot 3)i}\\ &= \frac{7 + 26i}{25}\\ &= \frac{7}{25} + \frac{26}{25}i\end{aligned}$$

Some people do not like introducing $i = \sqrt{-1}$. They say, "The whole value of numbers is that they can be used to talk about real, physical events. The symbol $i = \sqrt{-1}$ doesn't represent anything, and so should not be introduced. Only real numbers can be accepted as numbers."

We can accept this attitude. We then say that a complex number is an ordered pair of two real numbers. These ordered pairs, such as (2,3), (5, −1), (a,b), (c,d), which correspond to $2 + 3i$, $5 - i$, $a + bi$, $c + di$, have addition and multiplication defined by

$$(a,b) + (c,d) = (a + c, b + d)$$
$$(a,b) \times (c,d) = (ac - bd, ad + bc)$$

Thus we have complex numbers without *i*!

It is beyond our purpose to point out how multiplication of complex numbers is an excellent tool to discuss rotations, just as addition of real numbers is used to discuss translations.

Far from being unrelated to events in the physical world, it turns out that complex numbers simplify the discussion of many physical problems. You may study these in courses in algebra and engineering.

EXERCISES

1. Write symbols for the numbers satisfying the following equations (using i is helpful):

(a) $x^2 = 36$
(b) $x^2 = 10$
(c) $x^2 = -1$
(d) $x^2 = -16$
(e) $y^2 + 4 = 0$
(f) $t^2 + 3 = 0$

2. Simplify:

(a) $2(3 - i)$
(b) $(3i)^2$
(c) $(2 + 3i) + (1 - 4i)$
(d) $(5 + 3i) - (6 + i)$
(e) $(1 + i)(1 - i)$
(f) $(2 + i)(1 - 2i)$
(g) $\dfrac{2 + i}{i}$
(h) $\dfrac{3 - i}{2 - i}$

3. Where on the number line (0 1 2) would you plot $(\frac{1}{2} + i)$? Why?

4. In representing complex numbers without i, if $A = (2, 3)$ and $B = (-4, 5)$, find
   (a) $A + B$
   (b) $A \cdot B$

5. In writing $\sqrt{-36} = 6i$, did we use the following properties?
   (a) i can be multiplied by a real number.
   (b) i can be added to a real number.

Answers to selected questions: **1.** (b) $\pm\sqrt{10}$; (e) $\pm 2i$; **2.** (a) $6 - 2i$; (c) $3 - i$; (e) 2; (h) $\frac{7 + i}{5}$; **4.** (a) $(-2, 8)$; (b) $(-23, -2)$.

30

Statistics and Probability

30.1 INTRODUCTION *Statistics* are the numbers used in describing a situation. *Probability* is a number between 0 and 1 used to measure our knowledge of the likelihood of the occurrence of some future event. Statistics are also used in describing probabilities of related events and in making decisions.

The main problem in descriptive statistics is to satisfy the person who wishes to understand the information in a large set of numbers but is confused by looking at the whole set. For example, how might he look at the examination grades of 100 students? Often typical values, measures of variation, tables of the data, and graphs are satisfying. We discuss typical values and measures of variation.

The probability of some simple events can be seen at a glance. The probability of an event related to such events can often be calculated from the probabilities of the simple events. For example, we agree that the probability of getting a head on flipping a true coin is .5. Getting two heads on flipping two coins involves flipping single coins, so we expect we can calculate the probability of that event also. In other cases, such as insurance, we use the relative frequency of past events to calculate the probability of a future event.

Sophisticated people often base decisions on *expected value,* a term for the average return per trial over the long run. More accurately, expected value can be computed at once, while no one can say what the average of future returns will be.

30.2 TYPICAL VALUE A *typical value* is usually one number used to answer questions like "How well did the students do? Did they do better than 84 on the whole? How much does a 13-year-old girl weigh?" Suppose we had grades on an examination or weights of 13 year-old girls, of 88, 92, 75, 75, 85 as data for our answer. (We probably would have many such numbers, but we will restrict ourselves to five for discussion purposes.)

Several candidates present themselves as typical-value answers, including 83, 85, 75, 83.5, 92, 80. What number would you choose?

Did you recognize these candidates as the arithmetic mean or average, the median, mode (and minimum), mid-range, maximum, and a convenient rounded number? We discuss these possible typical values in turn.

The arithmetic *mean,* or *average,* is the sum of the values divided by their number. Using $\overline{x}$ to stand for the mean in the example, we have

$$\overline{x} = (88 + 92 + 75 + 75 + 85)/5$$

The mean of several numbers is shown by drawing a bar above the letter. The mean of four y values is

$$\overline{y} = (y_1 + y_2 + y_3 + y_4)/4$$

Do you agree with your neighbor on $\overline{x}$ and its value in the sample?

The average may be a fraction, even if the original values are all integers. Some quibblers do not like to call the average a typical value, because they say a typical value has to be a possible value. They prefer to call the average a measure of location or of central tendency.

Knowing the average and the number of values, we can easily compute the total. For example, if we know the average cost of 4 boxes of soap is \$1.25, we know the total spent on soap is 4 × \$1.25 = \$5.00.

EXAMPLE 1 A girl gets 23, 35, and 17 tomatoes from 3 plants she planted in 3 pots. What was the average number of tomatoes per plant?

EXAMPLE 2 A boy ran 8 laps on a track, averaging 53.75 seconds per lap. How long did it take him to run the 8 laps?

Did you get 25 tomatoes and 7 minutes 10 seconds?

The *median* is the middle value when the set is ordered by size. The values $\{7, 11, 23, 14, 5\}$ ordered by size are $\{5, 7, 11, 14, 23\}$. The median is 11. There is no middle value in a set with an even number of values, such as $\{9, 13, 15, 29, 21, 46\}$. The closest to middle values are 15 and 29. We take 22 as the median. Can you see that 22 is halfway between 15 and 29, and is their average?

The most frequent value in the set is called the *mode*. The name has the same origin as mode in fashion—that which is most popular. With the set $\{6, 9, 4, 9, 4, 7, 9\}$, the mode is 9. In some sets no single number occurs more frequently than any other, so we have more than one mode. Indeed, it is possible for every value to be a mode.

The mid-range is the value halfway between the largest and the smallest value in a set. You simply pick out the largest and the smallest values and average them. The mid-range of $\{6, 9, 4, 9, 4, 7, 9\}$ is $(9 + 4)/2 = 6.5$.

EXAMPLE 3 Find the average, median, and mid-range of $\{43, 19, 76, 0, 38, 100, 27, 93, 59, 82\}$. Which was the easiest to compute?

It is peculiar to call either the smallest or the largest value a "typical value." For example, the largest value is perhaps the desired answer for the builder of a garage who asks "How big is an automobile?"

Many people want their typical values given as rounded numbers, which are easy to remember. For the set in the first example, $\{88, 92, 75, 75, 85\}$, the rounded number that falls closest to the mean and median is 80.

Many sets of data tend to have more values near the middle. For such sets the average, median, mode, and mid-range are close together and indeed may be equal. Our examples have shown that they need not be equal. Some people call these numbers measures of location, or central tendency, as well as typical values, because they show the location or center of the set. Some people use the name typical value for *random value,* any number chosen by chance from the set.

Each of these values, average, median, and so on, is a statistic of the set of numbers, obtained in a well-defined way.

In choosing which of the typical values to use you are influenced by the purpose you have for the statistic and the ease of using it. You may also be interested in how your typical value changes as members of the set are changed.

EXAMPLE 4 Suppose the number 8 is changed in the set $\{3, 8, 11\}$. Will this change the average? The median? The mid-range? The maximum? The minimum? Try replacing 8 first with 7, then with 2, then with 20 to explore what happens.

EXAMPLE 5 Continue your explorations above by replacing 11 successively by 10, 2, and 20.

You may find there is no simple way to describe the effect of changes in the values on various typical values unless you consider specific sets and values.

EXAMPLE 6 Suppose you can increase one number of the set $A = \{50, 37, 32, 58, 43\}$ by any counting number you choose less than 11. For instance, you can replace 50 by any number in the set 51–60, or 37 by any number 38–47. If you wish the maximum of the new set to be as large as possible

1. What number would you replace?
2. What number would you replace it with?
3. What is the new maximum?
4. How much did you change the maximum?

Can you see there is just one answer in each case, and this is (*a*) 58, (*b*) 68, (*c*) 68, (*d*) 10?

Can you answer similar questions if it is the median you wish to increase? If it is the average? Suppose it is the mid-range you want to make as small as possible. Is there any case in which different changes give the same results?

30.3 VARIATION Besides a typical value, some measure of the variation of the values in a set is usually considered. The range, average deviation, standard deviation, and variance are the most common. Sometimes the maximum and minimum together are given, indicating the location of the numbers and the range. We discuss these terms below.

The range is the difference between the maximum and the minimum value. For the set $C = \{5, 7, 9, 11, 13\}$, the range is $13 - 5 = 8$. For the set $D = \{5, 3, 14, 9, 14\}$, the range is 11. Does D seem more variable to you?

The range is easy to compute, understand, and discuss, but it has some disadvantages. For example, the sets $E = \{10, 10, 10, 30, 30, 30\}$ and $G = \{10, 20, 20, 20, 20, 30\}$ have the same mean and range, yet G is much more concentrated than E. Perhaps we want a measure of variation that depends on all the values, not just the extreme ones.

The *average deviation* is the average of the deviations of the values from their mean. For the set $H = \{4, 7, 14, 15, 15\}$, the mean is 11. The deviations from the mean are $11 - 4$, $11 - 7$, $14 - 11$, $15 - 11$, $15 - 11$ or 7, 4, 3, 4, 4. The average of these deviations, the average deviation, is 4.4.

Some people prefer to use the median instead of the mean. The median of H is 14, and the deviations from 14 are 10, 7, 0, 1, 1. The average of these deviations is 3.8. It can be shown that the average of the deviations from the median is always less than or at most equal to the average deviation from the mean.

Clearly the average deviation is a typical deviation. The median, mode, or mid-range of the deviations could also be used. Can you see that 50 percent of the values in a set will fall within one median deviation of the average?

The *variance* is the average of the squares of the deviations from the mean. For set H the variance is

$$v = (7^2 + 4^2 + 3^2 + 4^2 + 4^2)/5 = 21.2$$

The variance is not a typical deviation. The usefulness of the variance is rather uncertain. In slightly more advanced statistical decision making, the variance plays a very important part. Here we find it most useful as a start towards the standard deviation.

The *standard deviation* is the square root of the variance. It can be called a typical deviation. The value for set H is $\sqrt{21.2} = 4.6$. The standard deviation is more widely used than the average deviation because (1) it plays a key role in more advanced statistics, and (2) large deviations influence the standard deviation more than they do the average deviation.

30.4 PROBABILITY The probability of a future event happening is a measure of our knowledge that it will. If we can separate the future into n equally likely outcomes, one and only one of which is sure to happen, and if our event takes place with m of these outcomes, then we say the probability of the event happening is $\frac{m}{n}$. We also say the probability of a favorable outcome is $\frac{m}{n}$. Probability is a number between 0 and 1: 0 when no outcomes are favorable and 1 when all n outcomes are favorable.

EXAMPLE 1 The probability of drawing a spade from a 52-card deck of bridge cards is $\frac{13}{52} = \frac{1}{4}$.

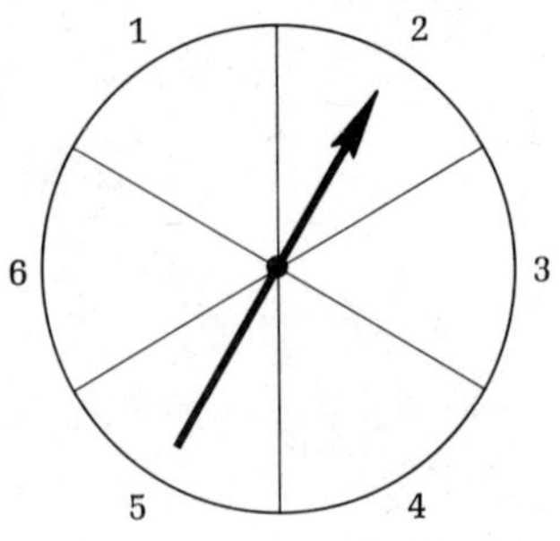

EXAMPLE 2 What is the probability the spinner will stop on 5? On an even number? Assume that all sections are of equal size.

Did you get $\frac{1}{6}$ and $\frac{1}{2}$? What is the probability the spinner will stop on an even number? We assume that stopping in each of the 6 sections is equally likely. There are 3 even numbers. Hence, $n = 6$ and $m = 3$, and the probability of an even number is $\frac{3}{6} = \frac{1}{2}$.

EXAMPLE 3 There are 5 chairs in a row, and you pick your seat at random from a box with numbered slips of paper. What is the probability you will get the middle seat? An end seat? Did you get $\frac{1}{5}$ and $\frac{2}{5}$? What are m and n? What is the probability your seat will not be an end seat?

EXAMPLE 4 You purchased 7 tickets in a drawing in which 200 were sold. What is the probability you will win the prize, a clock radio? Disguised in base seven, 10/404.

EXAMPLE 5 The probability that a flipped coin will fall tails is .5.

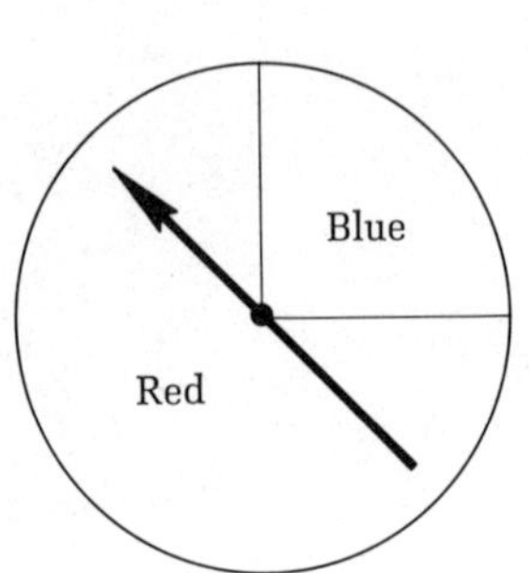

EXAMPLE 6 What is the probability the spinner will stop on red? On blue? What is the sum of these probabilities? ($\frac{3}{4}$, $\frac{1}{4}$, 1)

Odds are integers proportional to the probabilities of events. The odds on getting a spade in drawing a card from a standard deck are 1 to 3. Typically, odds compare success with nonsuccess rather than with all possibilities.

EXAMPLE 7 The odds on getting a head on flipping a coin are 1 to 1.

EXAMPLE 8 The probability of getting an end seat in example 3 is $\frac{2}{5}$. The odds are 2 to 3.

While odds are useful in comparing success with nonsuccess, probability is easier in most applications.

30.5 RELATIVE FREQUENCY The relative frequency of success in past trials is often taken as the probability of success in a future trial when we are unable to identify n equally likely outcomes of which m are favorable.

EXAMPLE 1 A thumbtack is tossed 100 times and falls point-up 34 times. We say the probability of it falling point-up on the next toss is .34. Here it is hard to even imagine n equally likely outcomes, one and only one of which is sure to happen.

EXAMPLE 2 A barrel contains red and white beans of the same size and equally likely to be drawn. What is the probability a red bean will be drawn? Theoretically, the answer is clear: we can count all the red beans and all the beans in the barrel and form the fraction with these two numbers. Instead we draw a few beans, say 100, and count the red ones. We take the ratio of the red beans to the beans drawn in this sample as the probability for the barrel as a whole.

EXAMPLE 3 Of 800 auto accidents, 20 had damage greater than \$500. We say the probability the next accident will cause more than \$500 damage is $\frac{20}{800} = .025$.

The $\frac{m}{n}$ and relative-frequency definitions are related by the theorem that if a large number of trials are to be made where success has probability $\frac{m}{n}$, the probability that the relative frequency of success is close to $\frac{m}{n}$ is very high. In other words, the relative frequency will usually be close to the probability.

EXAMPLE 4 Modern coins may not be perfect owing to their sandwich construction. If they are balanced on edge and then spun, rather than flipped in the air, they may have a relative frequency of heads that is far from $\frac{1}{2}$. Do you think the probability of heads in such a case is $\frac{1}{2}$? Are heads and tails equally likely?

30.6 EXPECTED VALUE If you received \$2 each time you drew a spade from a well-shuffled deck, how much would you have after drawing 100 cards? After each drawing the drawn card is replaced, so each draw is from a complete deck. The question cannot be answered, since you might have any even number of dollars between \$0 and \$200. But surely there is some question like this that makes sense and can be answered. Can you phrase it?

Since we want to answer the question before we actually draw the cards, the question cannot be about what is going to happen. Most people ask, "What is your expected value?" where expected value is $N \times P \times R$. N is the number of trials, P is the probability of success on one trial, and R is the reward. For our example, the expected value is $100 \times \frac{1}{4} \times 2 = 50$.

We can see that this is plausible, because the relative frequency of getting a spade should be close to $\frac{1}{4}$. You should get about $100 \times \frac{1}{4}$ spades for a total return of 25×2 dollars. The average return per draw, if this is what happened, would be $\frac{1}{4} \times 2$. We define the expected value from one trial as $P \times R$. $P \times R$ is a convenient figure for estimating what the average return on a large number of trials would be.

EXAMPLE 1 A roulette wheel has eighteen red, eighteen black, and two green holes. The ball is equally likely to fall in any of them. Ralph gets \$20 if it falls in a black hole, nothing otherwise. What is the expected value of one try? Since the probability of success is $\frac{18}{38}$, the expected value is $\frac{18}{38} \times \$20$, or \$9.48. Note that the expected value is not a possible value, since Ralph gets either \$20 or \$0.

EXAMPLE 2 If Ralph pays \$10 to play, what is his expected loss on each play? Do you see 52 cents?

EXAMPLE 3 If Ralph played 1000 times, how would you expect him to fare? You would expect him to lose about \$520, or the expected loss on each try times the number of tries. You can check this by calculating the expected relative frequency of winning and the number of wins in 1000 tries to give this relative frequency, and then subtracting the total winnings from his total payments of \$10,000.

EXAMPLE 4 The probability of a person 80 years old dying in the next year is about 10 percent. What is the expected loss to the insurance company that has insured his life for \$10,000? Do you see \$1000? Over a large number of policies the amount actually paid out will be close to the sum of the expected values of the individual policies. The company charges more in premiums than it expects to pay out in death benefits in order to cover operating expenses and to build up a cushion in case the death rate rises.

EXAMPLE 5 For a certain class of drivers, the probability is .12 that a particular driver will have an accident costing $100, .08 for an accident costing $500, and .02 for an accident costing $1000. The probability of not having an accident is .78. What is the expected loss for a driver? The expected loss on $100 accidents is $12, on $500 accidents is $40, and on $1000 accidents is $20. The total expected loss is $72.

In actual practice the insurance company will consider a larger variety of accidents and the possibility of more than one accident during the period involved. The company may charge premiums of about twice the expected loss in order to cover the expenses of selling policies, collecting premiums, and investigating and paying claims.

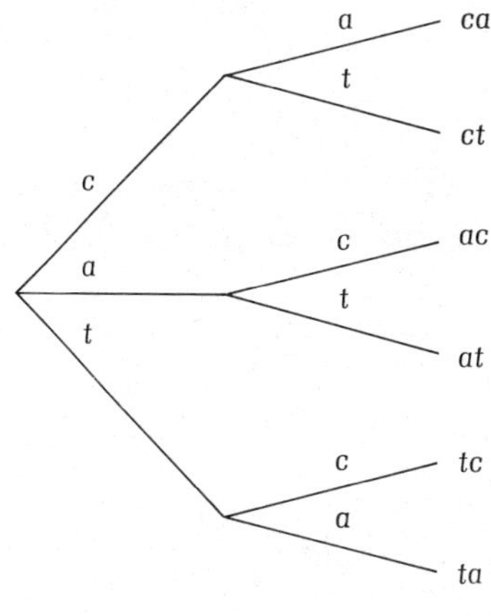

30.7 LISTING OUTCOMES In many cases eventual outcomes are compounded from simple outcomes. It is important to list the compound outcomes in terms of the simple ones. Tree diagrams are often helpful.

EXAMPLE 1 List the two-letter code words made from the letters of *cat*, using no letter more than once. The six code words, reading from top to bottom, are *ca, ct, ac, at, tc, ta*.

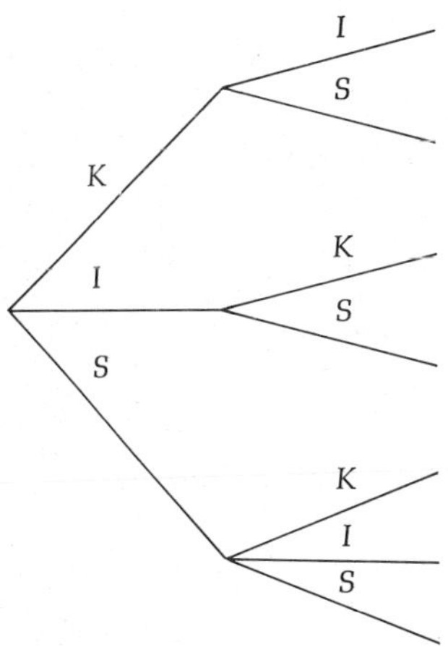

EXAMPLE 2 In Scrabble, the word KISS is made by arranging 4 small pieces showing the letters K, I , S, S, respectively. List the two-letter code words that can be made with these 4 pieces. Note that IS is only one word, although it can be made in two ways with the pieces.

EXAMPLE 3 Three coins are prepared as follows. Instead of heads and tails, a quarter has *B* on one face and *F* on the other. A nickel has *A* and *I* on its two faces. A penny has *T* and *M*. A word is made whose first letter is found by flipping the quarter, the second by flipping the nickel, and the third by flipping the penny. List the resulting words. Can you see how the tree applies? What is the probability that *BAN* will be the resulting word?

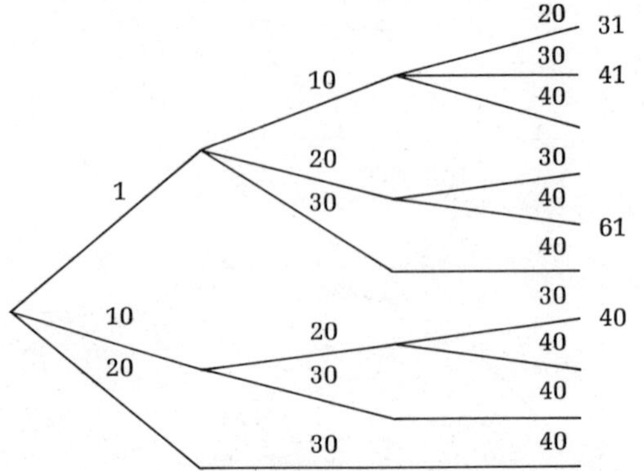

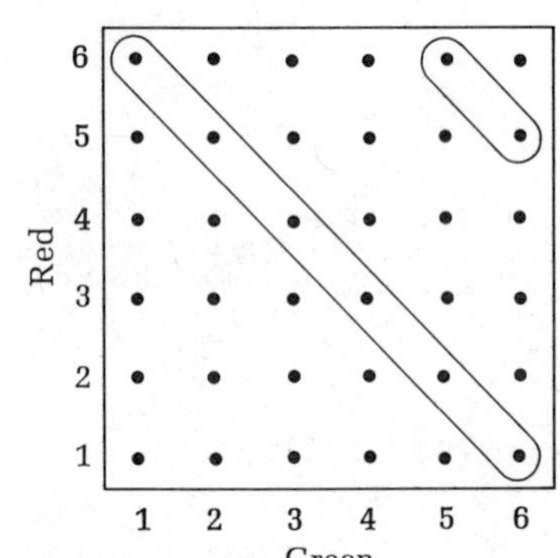

EXAMPLE 4 For a hot fudge sundae at the dairy bar, Sallie pays the sum of the prices on three slips of paper drawn at random from a set of five marked 1, 10, 20, 30, and 40 cents, respectively. What is the highest price she could pay? The lowest price? How many different prices are there?

We make a tree.

Did you get 90 cents, 31 cents, and 9 for answers to the questions? What is the probability Sallie will be charged 31 cents?

30.8 PROBABILITIES OF COMPOUND EVENTS The probabilities of compound events that are not equally likely can often be computed if the events are compounded of equally likely events.

EXAMPLE 1 What are the probabilities of 3, 2, 1, and 0 heads on flipping 3 coins? We may think of the coins as being a quarter, a nickel, and a penny, with faces marked *H* and *T*. Making a tree, we see that we get eight code words, *HHH, HHT, HTH, HTT, THH, THT, TTH, TTT*. Each of these words is equally likely. Counting tells us that one has 3 *H*'s, three have 2 *H*'s, three have 1 *H*, and one has 0 *H*'s. Hence the probabilities of 3, 2, 1, and 0 heads are $\frac{1}{8}$, $\frac{3}{8}$, $\frac{3}{8}$, and $\frac{1}{8}$, respectively. We leave it as an exercise for you to draw the tree.

EXAMPLE 2 What is the probability of getting a total of 7 on rolling two dice? The probability of getting 11? Of getting either a 7 or an 11?

We can think of one die as being red and the other green to simplify our ideas. Making a tree shows that there are 36 equally likely outcomes. Similarly, the diagram shows 36 spots, each representing one equally likely outcome. The outcomes (6, 1), (5, 2), (4, 3), (3, 4), (2, 5), (1, 6) show 7. Hence the probability of 7 is $\frac{1}{6}$. The probability of 11 is $\frac{2}{36}$. The probability of either 7 or 11 is $\frac{2}{9}$. Can you see why?

30.9 SUMMARY It has been pointed out that descriptive statistics tell us what a large set of numbers is trying to say. We discussed several ways of getting a typical value, such as finding the mean, median, or mid-range. Which typical value one wishes to use depends on his purpose. The mean is the value most widely used. It is the value from which the sum of the squares of the deviations is a minimum, and is closely related to the sum of the numbers. The median can be found by arranging and counting, and requires no operations such as addition and division. It is the value from which the sum of the deviations is a minimum, and has 50 percent of the range of values above it and 50 percent below it. The mid-range is easy to compute with rapid scanning of the data, and is usually the value thought of when we give the largest and smallest member of the set. It is the value described by go-nogo gauges in industry—cases in which a product is satisfactory if it is neither too large nor too small. Oranges and apples are sized this way.

Measures of variation—range, average deviation, and standard deviation—were also described. Students are encouraged to explore the effects on typical values and variation measures by changes in some of the values.

Probability was introduced both from the $\frac{m}{n}$ and the relative-frequency attitudes as the measure of our knowledge of the occurrence of a future event. Expected value was introduced as a well-defined concept about the future—a value often used in making choices and decisions.

The probability of compound events can be determined if the equally likely simple events can be identified along with the structure of the compound events. We examined ways of listing several compound events using trees.

Probability and statistics become more powerful and useful as one learns more about them. We hope readers will plan on future explorations into these areas.

EXERCISES

1. Find the (*a*) mean, (*b*) median, and (*c*) mid-range for the set $\{7, 3, 8, 8, 6\}$.

2. Find the (*a*) mean, (*b*) median, and (*c*) mid-range for the set $\{1, 2, 3, 4, 5, 6, 1000\}$. This shows the effect of an extreme value.

3. A banker divides the total amount on deposit in his bank by the number of depositors to get a typical deposit.

(*a*) Which kind of typical value is he using?

(*b*) Do you think most depositors would have more than, less than, or about the same deposit as the typical deposit?

4. A mixture of balls of various sizes is sorted by sieves. The first sieve has holes 6 cm across, the second has holes 5 cm across.

(*a*) What is the typical size of a ball that will go through the first sieve but not the second?

(*b*) What kind of typical value are you using?

5. Clarence caught fish of lengths 14, 13, 10, 18, 14, 16, 15 inches. What did he report as the typical length of these fish?

6. Uncle Joe's automobile agency advertises the typical price of its cars as $2950. What kind of typical value may they be using if they are inclined to exaggerate?

7. Vince's four boxes weigh 24 pounds total. What is their average weight?

8. Five counting numbers total 30. If possible, state their (*a*) mean, (*b*) median, (*c*) maximum. If not possible, show why.

9. Make up a set of 3 weights, whose average weight is 15 pounds and whose median weight is 14 pounds.

10. Three weights are 2, 4, and 7 kilograms. What two weights can you put with them so the set of 5 has a median of 6 and a mean of 5 kilograms?

11. Make up a set whose mean is 11 and whose mid-range is 10.

12. Make up a set whose median is 20 and whose mid-range is 22.

13. Make up a set whose median is 12, mid-range is 13, and mean is 14.

14. In the set $\{4, 7, 13\}$, 7 is replaced by 8. What, if any, change was made in the (*a*) mean, (*b*) median, (*c*) mid-range, (*d*) maximum, and (*e*) minimum?

15. In the set $\{6, 12, 14, 8, 10\}$, 14 is replaced by 13. What, if any, change is made in the (*a*) mean, (*b*) median, (*c*) mid-range, (*d*) maximum, and (*e*) minimum?

16. In the set $\{6, 12, 14, 8, 10\}$, 12 is replaced by 9. What, if any, change is made in the (*a*) mean, (*b*) median, (*c*) mid-range, (*d*) maximum, and (*e*) minimum?

17. For the set $\{3, 7, 4, 9, 7\}$, find the (*a*) range, (*b*) average deviation, and (*c*) standard deviation.

18. The sets $A = \{10, 10, 10, 20, 20, 20\}$, $B = \{10, 15, 15, 15, 15, 20\}$ and $C = \{10, 12, 14, 16, 18, 20\}$ have the same mean, median and mid-range. Compute the (*a*) range, (*b*) average deviation, and (*c*) standard deviation for each set.

19. Guess the answers to exercise 20. Think before guessing, but do not roll any dice.

20. Role a die 11 times, recording the number of spots shown each time. For these 11 numbers, compute the (*a*) mean, (*b*) median, (*c*) mid-range, (*d*) average deviation, (*e*) range and (*f*) standard deviation. Compare with your answers to exercise 19.

21. Compare your results for exercise 20 with those of several of your friends. Do you find more agreement on some of the values *a*–*f* than on others? Why did this happen?

22. What is the probability of (*a*) drawing a black card from a 52-card bridge deck? (*b*) drawing a face card (jack, queen, or king)?

23. (*a*) You have 4 tickets in a lottery of 10,000 tickets. What is your probability of winning?
(*b*) Three digits are chosen at random to form a 3-digit number between 000 and 999 inclusive. What is the probability the number will be 231?

24. Shirley is flipping coins on a TV giveaway show. She gets \$5 for flipping heads and \$2 for flipping tails. (*a*) What is her expected prize for one flip? (*b*) Is it possible for her to actually receive her expected prize?

25. Hugh believes the probability is $\frac{1}{3}$ that he will recruit a given person who inquires about joining his squad. If Hugh gets \$50 for each person who joins, what is his expected gain per prospect?

26. Two letters from the word *jump* are chosen at random, one after the other, and without replacing the first letter drawn. What is the probability they will spell *up*?

27. Two letters are chosen one after the other from *goat*. What is the probability they will spell a well-known English word, in the order drawn, if (*a*) the first letter drawn is replaced before the second is drawn, and (*b*) the first letter is not replaced?

Answers to selected questions: **1.** (*a*) 6.4; (*b*) 7; (*c*) 5.5; **3.** (*a*) mean; (*b*) less; **5.** 18 in.; **7.** 6 lbs.; **9.** many sets, example: 14, 14, 17; 1, 14, 30; **11.** many: 5, 15, 13; 4, 16, 13; **13.** many: 4, 11, 12, 21, 22; **15.** (*a*) decreases 1/5; (*b*) no change; (*c*) decreases 1/2; (*d*) decreases 1; (*e*) no change; **17.** (*a*) 6; (*b*) 2; (*c*) 2.19; **20.** various answers; (*a*), (*b*), (*c*) about 3.5; (*d*) about 1.5; (*e*) 5; (*f*) about 1.7; **22.** 1/2; 3/13; **24.** (*a*) \$3.50; (*b*) no; **27.** (*a*) 3/16; (*b*) 3/12 = 1/4.

31

Introduction to Algebra

31.1 INTRODUCTION Beginning algebra is concerned with the forms and patterns of arithmetic. Arithmetic is concerned with numbers, calculations, and problems with numbers. The algorithms follow certain forms studied in algebra. We've already looked at many of these forms, so much of this chapter is recognizing old friends.

There are three big ideas in algebra: (1) representing numbers and their operations by symbols; (2) the structure of the number system—that is, the properties of all numbers and the set of numbers as a whole; and (3) equations, functions, and transformations. We shall look at these ideas and show how algebra is a powerful tool.

31.2 USING SYMBOLS FOR NUMBERS AND VARIABLES Our old friends 3, 17, and M are symbols that stand for numbers in various numeral systems. We know them well, because given these symbols we can make a set with that many members. Frequently we wish to talk about a certain number, just like an old gossip, when we know something about the number, even if we can't make a set for it. Then we merely use another symbol for the number. For example, if Helen is 3 years older than Bob, we might say, $H = 3 + B$, where H and B are Helen's and Bob's ages, respectively. A child counting eggs might say $\square = 9 + 3$, where he uses $\square$ as the symbol for the number of eggs in the carton. The most common symbols used are letters.

Letters are also used as variables or *open symbols*. We say, "Let x be a whole number less than 100. Multiply by 2. Add 5. Do you get $2x + 5$?" Here x could be any one of the numbers, 1, . . . , 99. $2x + 5$ could be 7, 9, 11, . . . , 203. The set of numbers any one of which can be used to replace x is called the *replacement set*.

To fix our ideas, we define *variable.* A variable is a letter or symbol that can be replaced by or take on the value of any member of a given set, called the replacement set or domain of definition for the variable.

For contrast we repeat these two attitudes toward a letter, say y, as a number symbol. (1) y stands for a specific number whose common name is unknown. (2) y is a variable that can be replaced by any member of a replacement set.

As usual, people use both attitudes, often toward the same symbol in a problem. This ability to switch attitudes is part of the power of mathematics, but may be a source of confusion. One may set up a problem using one attitude, then switch to the other while solving. The letter is often called a variable even when the first meaning is used.

31.3 NUMBER PROPERTIES AND CAD LAWS We saw earlier that the counting numbers satisfy the CAD laws. In terms of variables these laws are easy to describe. The replacement set is the counting numbers.

$a + b = b + a$	commutative laws	$cd = dc$
$(e + f) + g = e + (f + g)$	associative laws	$(hk)j = h(kj)$
$t(r + s) = tr + ts$	distributive law	

We also know these rules hold when the replacement set is enlarged to integers, to rationals, to real numbers, and to complex numbers. Indeed, we find the CAD rules so useful that we use them as a way of defining some algorithms, such as multiplying negative numbers.

Because the CAD rules hold for all these common sets of numbers, we often find it easy to use them to explain steps we take in problem solving.

Sometimes sets of things that are not numbers support operations that are often called "addition" and "multiplication" and follow the CAD properties. For example, we may say we are "adding" switches in electric currents if we join them in parallel, and "multiplying" switches if we join two switches in series.

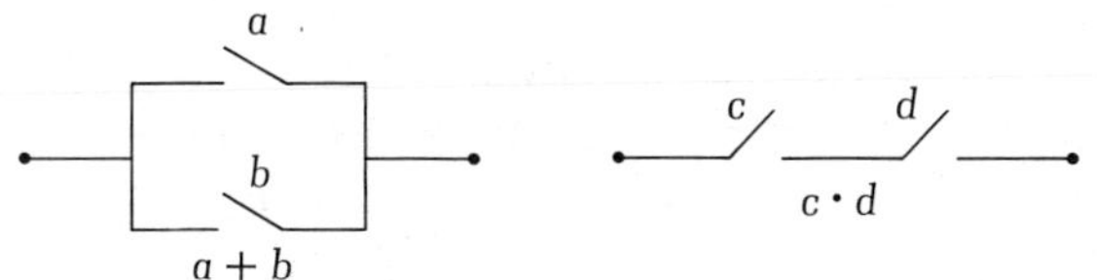

The sum and product are the switches formed by these combinations. We shall not consider this illustration further, but if you know how electric switches work, you may be able to show that the CAD properties hold with this unusual replacement set and definition of "addition" and "multiplication."

31.4 FORMULAS AND FUNCTION You know many useful formulas using letters as variables—for example, $A = s^2$ for the area of a square. Can you match the following formulas with their names in the column at right? What do the letters represent?

a. $p = 3s$	**A.** Area of a circle
b. $p = 2l + 2w$	**B.** Volume of a block
c. $A = lw$	**C.** Area of a triangle
d. $A = \pi r^2$	**D.** Perimeter of a triangle with equal sides
e. $V = lwh$	**E.** Area of a rectangle
f. $C = 18w$	**F.** Perimeter of a rectangle
g. $A = \frac{1}{2}bh$	**G.** Cost of apples @ 18¢/pound

Formulas are useful because they show the relation between quantities in a situation, as illustrated here. The formula $z = xy$ represents the binary operation multiplication. Calling it a formula doesn't add much and shows that frequently we look at old ideas in slightly different ways. Here are some other formulas:

$$y = 3x + 5 \qquad A = \frac{3B}{C} + 8 \qquad p = qr + s$$

Clearly there is no limit to the number of formulas that can be written. Can you see that these three formulas show a unary, a binary, and a ternary operation, respectively? The variables on the right side are the "holes" or "inputs" for the operator. The left sides are the outputs.

Besides thinking of formulas as showing operators, we also say they show *functions*. A function is a matching of members from one or more sets to a member in another set. Unary, binary, and ternary operators are also called functions of 1, 2, and 3 variables, respectively. A function may or may not have a formula.

Many formulas can be "worked backward." In working with rectangles, we know that $A = lw$; $l = \frac{A}{w}$ and $w = \frac{A}{l}$. That is, given the length and width, we get the area by $A = lw$. Given the area and the width, we get the length by $l = \frac{A}{w}$. Indeed, these facts are what we mean when we speak of the missing-factor attitude toward division. If we go from $A = lw$ to $l = \frac{A}{w}$, we often say we are "solving for l." We may say "$A = lw$ shows A explicitly" and "$l = \frac{A}{w}$ shows l explicitly," because A and l are placed alone on one side of the equation. *Explicit* comes from a Latin word meaning to "unfold." Solving a formula for different variables is a powerful tool. This way we can represent several facts in just one formula. Do you know how to do this? We look at some ways.

31.5 SOLVING FORMULAS It may be that you go from $A = lw$ to $l = \frac{A}{w}$ because you recall that if one formula is correct, you know that the other is correct also. Our purpose in this section is to show how you can find related formulas without having to remember them, by merely recalling some familiar rules of numbers.

EXAMPLE 1 You recall that a number has many names. For example, 5, five, V, $4 + 1$, $7 - 2$, and $\frac{20}{4}$ are all different names for the same number. We look on an equation such as $A = lw$ as saying that A and lw are two names for the same number.

We all know that, if we divide one number by another, the quotient we get doesn't depend upon the names used. There is only one number. Hence $\frac{A}{w}$ and $\frac{lw}{w}$ are two names for the same number. We write $\frac{lw}{w} = \frac{A}{w}$. From our study of fractions, and the meaning of multiplication and division, we know that for any numbers l and $w (w \neq 0)$ $\frac{lw}{w} = l$. That is, $\frac{lw}{w}$ and l are two names for the same number. Hence we say

$$l = \frac{lw}{w} = \frac{A}{w} \text{ or } l = \frac{A}{w}.$$

EXAMPLE 2 In $z = x + y$, z and $x + y$ are names of the same number. Do we wish to solve for x? Subtract y from both sides of the equation. Since the difference of two numbers does not change when we change the names, ______ Can you finish the sentence? Let us put down the steps, with a statement of what we are doing for each step:

1. $z = x + y$	**1.** Two names of same number.
2. $z - y = x + y - y$	**2.** Subtract y from that number, using both names. The resulting names are of the same number.
3. $x + y - y = x$	**3.** Subtracting y is the inverse of adding y.
4. $z - y = x$	**4.** Substituting from 3 in 2.
5. $x = z - y$	**5.** Interchanging the sides of an equality.

Now we have the formula "solved for x."

Some people may feel there has been a shifting of attitudes. In (1) they had a set or segment attitude toward z, x, and y. Then in (3) they had an operator attitude toward $+y$. Changing our attitudes this way may make it easier to follow calculations. No harm is done, since the same results will occur.

EXAMPLE 3 Solve $r = s - t$ for s. We might think r and $s - t$ are names of the same number. On the right, s is operated on by $-t$. What is the inverse of $-t$? Add t. Hence we add t to the number, using both names. We get $s = r + t$.

Listing the steps, we have

1.	$r = s - t$	**1.**	Two names of same number.
2.	$r + t = s - t + t$	**2.**	Add t to the number.
3.	$s - t + t = s$	**3.**	Using inverse operators.
4.	$r + t = s$	**4.**	Substituting from 3 in 2.
5.	$s = r + t$	**5.**	Commuting the sides of equality.

DIVERSION Many people like to think of the sides of an equation as describing weights on the pans of a balance scale. If the scale balances, the weights on the two sides are equal. If equal weights are placed in both pans, the scale continues to balance. Similarly, if the same weight is taken off both pans, the scale continues to balance. This physical fact leads to the statement "You can add (or subtract) the same number to both sides of an equation without upsetting the equality."

If this argument is used, the teacher ought to see to it the student has actually used such scales and weights so that his belief will be more than just listening to the teacher.

It is somewhat subtler, using a scale balance, to show that you can multiply and divide both sides of an equation by the same number and still maintain the balance. And it's purely a leap of faith that these facts can be extended to "You can do anything to one side if you do it to the other," which is often thought to be true.

On the other hand, the "two names of the same number" attitude toward equations clearly allows you to do anything you can do to a number to both sides of the equation without upsetting the equality. Some care is necessary in writing the terms, however.

For example, $6 = 2 + 4$. While you can multiply the number by 3 using both names, it is not permissible to write "$\times$ 3" on the right on both sides of the equation without taking precautions. If we do, we have

$$6 \times 3 \stackrel{?}{=} 2 + 4 \times 3$$

The agreements on the order of performing operations say that $2 + 4 \times 3$ means $2 + (4 \times 3) = 14 \neq 18$. The agreements say to perform multiplications and divisions first, then additions and subtractions. The way to do it is to write $6 \times 3 = (2 + 4) \times 3$.

To many it is helpful to think of operating on both sides of the equation as well as on both names of a single number. Hence we shall say "add 5 to both sides" rather than "add 5 to the number using both names."

EXAMPLE 4 Solve $c = \frac{d}{e}$ for e. Here, e is not only connected to d; it is also in the denominator. Here are three reasonable first steps:

$$c = \frac{d}{e} \Rightarrow ec = d. \qquad \text{Multiply both sides by } e.$$

We have freed e from the denominator. broken its tie to d, but put in a tie to c. It seems a net gain. We then divide both sides by c to get $e = \frac{d}{c}$.

The reciprocal of a fraction $\frac{p}{q}$ is the fraction $\frac{q}{p}$.

$$c = \frac{d}{e} \Rightarrow \frac{1}{c} = \frac{e}{d}.$$

Take the reciprocal of both sides; e is now in the numerator. If we multiply both sides by d, we get $e = \frac{d}{c}$. Recall $c = \frac{c}{1}$ so the reciprocal of c is $\frac{1}{c}$.

$$c = \frac{d}{e} \Rightarrow \frac{c}{d} = \frac{1}{e}$$

Divide both sides by d. This separates e from d. We now take the reciprocal of both sides to get $e = \frac{d}{c}$.

EXAMPLE 5 Solve $y = ax + b$ for x. We need to isolate x from a and b. We apply the inverses of the operators connecting a and b to x in the opposite order. Since in $ax + b$ we think of the multiplication done first, then the addition, we first subtract then divide.

1. $y = ax + b$
2. $y - b = ax$ 2. Subtract b from both sides.
3. $\frac{y - b}{a} = x$ 3. Divide both sides by a.
4. $x = \frac{y - b}{a}$

EXAMPLE 6 Solve $w = \dfrac{a}{bt + d}$ for t. We need to get t in the numerator and disconnect t from a, b, d. Here is one way:

1. $w = \dfrac{a}{bt + d}$
2. $w(bt + d) = a$ 2. Multiply both sides by $bt + d$.
3. $bt + d = \dfrac{a}{w}$ 3. Divide both sides by w.

Now all we need to do is peel off d and b.

4. $bt = \dfrac{a}{w} - d$ 4. Subtract d from both sides.
5. $t = \dfrac{a/w - d}{b}$ 5. Divide both sides by b.
6. $t = \dfrac{a - wd}{wb}$ 6. A slightly different form.

After step 3 we might have divided by b.

4. $t + \dfrac{d}{b} = \dfrac{a}{wb}$
5. $t = \dfrac{a}{wb} - \dfrac{d}{b}$ 5. Subtract $\dfrac{d}{b}$ from both sides.
6. $t = \dfrac{a}{wb} - \dfrac{wd}{wb}$
7. $t = \dfrac{a - wd}{wb}$

In step 5 some fractions were rewritten in different form. The expression $\dfrac{a}{w} - d$ is to be written as $\dfrac{P}{Q}$. The author thinks it easiest at this point to think of a, w, and d as being specific numbers whose standard names aren't available. We recall that $\dfrac{15}{4} - 2$ is changed by the steps

$$\frac{15}{4} - 2 = \frac{15}{4} - \frac{2 \cdot 4}{4} = \frac{15}{4} - \frac{8}{4} = \frac{15 - 8}{4} = \frac{7}{4}.$$

We rewrote the numbers as fractions having the same denominator and indicated the sum of the numerators over the denominator. Following similar steps we get

$$\frac{a}{w} - d = \frac{a}{w} - \frac{dw}{w} = \frac{a - dw}{w}.$$

Since the process holds for any members of the substitution sets for a, w and d, we think of the result as a formula good for all.

31.6 SENTENCES, TRUTH VALUE, OPEN SENTENCES, IDENTITIES, CONDITIONAL SENTENCES The statement $3 + 5 = 8$ is true. The sentence $3 + 9 = 8$ is false. *True* and *false* are called "truth values" for sentences. We can't say whether $3 + x = 8$ is true or false until we know the number substituted for x. Sentences like $w + x = 8$, whose truth value we don't know, are called *open sentences*. The set of objects that may be substituted for x is called the *replacement set*. Those objects that make the sentence true are called the *truth set* for the sentence, or *solutions of the equation*.

For example, the sentence $-2 < x - 10 < 3$ might have the integers as its replacement set. We must be told, or assume, what the replacement is. It cannot be determined from the sentence. The truth set is $\{9,10,11,12\}$, since these are the integers that make the sentence true, provided $\{9,10,11,12\}$ is a subset of the replacement set.

The truth set for $3 + x = 8$ is $\{5\}$.

The truth set for $x^2 - 2x = 3$ is $\{-1,3\}$, if the replacement set is the integers. If the replacement set is the counting numbers, the truth set is $\{3\}$.

The truth set for $ab = ba$ is the same as the replacement set—almost. If the replacement set is a set of numbers, such as the counting numbers, integers, or reals, then the statement is true. The replacement set can be of other objects for which the expressions make sense (such as matrices, not studied in this book), but for which the truth set is not the replacement set. For such objects, the commutative law of multiplication does not hold.

We see that the truth set depends not only on the open sentence, but also on the replacement set. Both must be known before it makes sense to look for the truth set.

Open sentences whose truth sets are the same as their replacement sets are called *identities*. When the truth set is smaller than the replacement set, the open sentence may be called a *conditional sentence* or a *conditional equation*. We must know the replacement set before we can say whether or not an open sentence is an identity or a conditional sentence.

1. More attention is paid these days to inequalities, the language of approximations and order. These are not equations, yet many operations are useful for both. For example, we can add the same number to both sides of an equation (inequality) without changing the truth or falseness of the equation (inequality). Hence we like to talk about both together. It is easier to see that both equations and inequalities are sentences than it is to shift our mind from thinking about equations to thinking about inequalities.

2. Finding the truth set is a major activity in mathematics. Normally we start with equations whose truth set has only one member, such as $3x = 15$. When the idea of "solution of an equation" is introduced at the same time, our Pavlovian reaction is to think that "the solution to an equation is the number that satisfies the equation." This suggests the solution is only one number. It is difficult to break out of this rut when we go on to equations such as $x^2 - 2x = 3$ whose solution set is $\{-1,3\}$, and to inequalities whose solution is an infinite set of numbers. These annoying psychological blocks are avoided if we say *open sentences* and *truth sets* rather than equations and solutions. It was hard for the author, and he suspects for some readers, to make this switch from what he was taught at his mother's knee and other such joints.

3. Most open sentences arise from trying to solve problems. The solution to the problem may or may not be the truth set. Frequently the best we can say is that the solution to the problem is contained in the truth set. We need further information to pick out which member is the solution to the problem.

4. The term *conditional sentence* arises as follows. Frequently open sentences arise with the letter regarded as representing a single value whose standard name is not known. Consider the example: How many gallons of water must be added to 10 quarts to make 30 quarts? It is usual to think "Let x be the number of gallons I must add. Do I know x? No.

Do I know anything about x? Yes. Several things. x is not a dog. x is a positive real number. Further, x must satisfy $4x + 10 = 30$." This sentence is a requirement, a condition on x. Hence the name.

31.7 SOLVING OPEN SENTENCES Finding the truth set, or solving an open sentence, can frequently be done by using the methods shown earlier in solving formulas. The method is to replace the given open sentence until we reach one whose truth set we can easily guess. Equivalent open sentences are those with the same truth set. For example, we get the following sequence of equivalent sentences:

$4x + 10 = 30$	Subtract 10 from both sides as a first step in isolating x.
$4x = 20$	Divide both sides by 4 to isolate x.
$x = 5$	

It seems fairly easy to guess that the truth set of $x = 5$ is 5.

Some people prefer a different attitude, although the equations are the same. They say

"I am searching for the standard name of a number, x. I know that

$4x + 10 = 30$	must be true. Hence
$4x = 20$	and
$x = 5$	must be true.

The standard name for x is 5, or x equals 5."

Sometimes one attitude is clearer than the other. The second is probably the attitude of those who say "Find x" as the purpose of the problem.

Here is an example using inequalities: "Find the truth set in rational numbers for $-2 < 2x - 10$." Is the truth set clear? We may proceed as follows to make it clear:

Add 10 to both sides: $8 < 2x$	The truth sets are not changed by these acts.
Divide both sides by 2: $4 < x$	

The truth set is obvious for $4 < x$. In this case, it is not helpful to think of x as a name of a specific number.

31.8 INEQUALITIES In approximations we make statements like "The weight is within 5 of 83." We may wish a different statement of possible values. Let w be the weight. Translated into inequalities, we see that w satisfies the condition

$$-5 < w - 83 < 5.$$

That is, two inequalities must be satisfied:

$$w - 83 < 5 \text{ and } -5 < w - 83.$$

Now if we start with one number smaller than another, and add the same number to both, the sums will be unequal in the same order. Let us add 83 to both sides of each of these inequalities. We get

$$w < 88 \quad \text{and} \quad 78 < w.$$

Then w is a number between 78 and 88, and we write $78 < w < 88$.

Similarly, if z is no more than 4 smaller than 48, and no more than 7 larger, then

$$-4 \leq z - 48 \leq 7.$$

Adding 48 to both inequalities gives

$$44 \leq z \leq 55.$$

The symbol $\leq$ is read "less than or equal to."

It is perhaps good to ask, "What will the equivalent inequality be if we do other things to both sides of an inequality?" We limit ourselves to just three pairs of operations.

Do	**Result**
1. Add or subtract the same number from both sides of the inequality.	**1.** The results are unequal in the same order.
EXAMPLE $8 < 12; 8 + 3 < 12 + 3$	
2. Multiply or divide both members of an inequality by the same *positive* number.	**2.** The results are unequal in the same order.

EXAMPLE $8 < 12; 8 \times 3 < 12 \times 3;\quad 8 \div 4 < 12 \div 4$
$24 < 36 \qquad 2 < 3$

3. Multiply or divide both members by the same *negative* number.

3. The results are unequal in *opposite* order.

EXAMPLE $9 < 15; 9(-2) > 15(-2);\quad 9 + -3 > 15 \div -3$
$-18 > -30 \qquad -3 > -5$
$-5 < -3$

EXAMPLE Find the truth set for $25 < 97 - 3x$. We find more obvious open sentences in two ways:

1. $25 < 97 \div 3x$	**1.** Given.
2. $-72 < -3x$	**2.** Subtract 97 from both members.
3. $72 > 3x$	**3.** Divide both members by -1.
4. $24 > x$	**4.** Divide both members by 3.

From this open sentence, we see that the members of the truth set are those members less than 24.

1. $25 < 97 - 3x$	**1.** Given.
2. $25 + 3x < 97$	**2.** Add 3x to both sides.
3. $3x < 72$	**3.** Subtract 25 from both sides.
4. $x < 24$	**4.** Divide both sides by 3.

31.9 TRANSFORMATIONS Many people like to think of formulas such as $a = b + 4$; $V = s^3$; $T = 6t - 3$ as *transformations.* In modern computer language we might say "Let $x \rightarrow x + 4$" for the first. This formula replaces 7 by 11, 13 by 17, 20 by 24, and so on. We say it transforms (changes) x to $x + 4$.

Such transformation ideas can be illustrated by jumps on the number line. Starting with 7 and applying the transformation four times is shown by

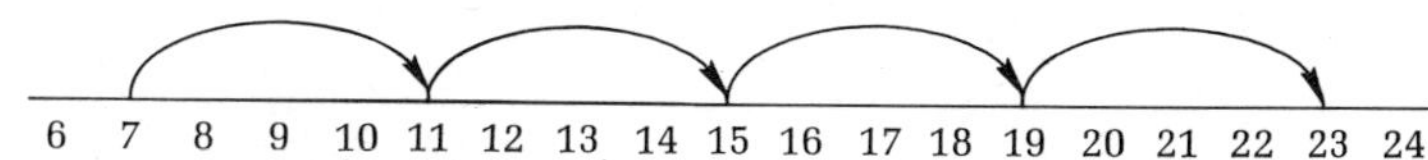

Under the transformation, each point moves 4 to the right.

Such transformations lend themselves to useful questions, such as "If we start at 9 and apply the transformation twice, where do we end?" "Where do we start so that we end at 0?" "If we start at 4 and apply the transformation a million times, where will we end?" "Is there any point that doesn't move?"

EXAMPLE 1 Suppose the jumping point follows the rule $x \to 3x - 6$. We illustrate on the number line with several jumps, starting at $x = 1$ and starting at 4.

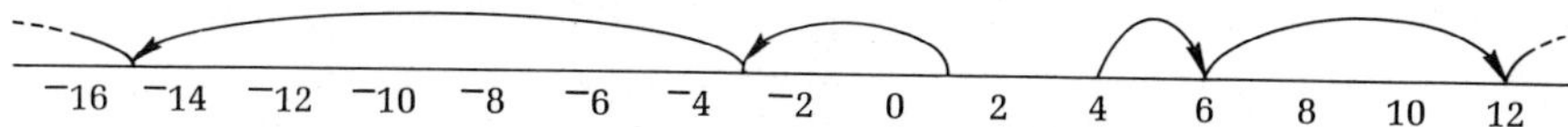

We see the jumps are not the same size; nor are they in the same direction. Is there some point that stays fixed, transforms to itself? What about $x = 2$? $x = 3$? $x = \frac{5}{2}$? If a jumper starts at $\frac{5}{2}$ will he, after a few or many jumps, land on an integer?

EXAMPLE 2 Let the transformation be $x \to \frac{16}{x}$. Starting at 1, where do we end after five jumps? Starting at 2? Starting at 3? Starting at -8? Is there any point that remains fixed?

EXAMPLE 3 Let the transformation be $x \to \frac{6}{x - 1}$. If we start at 0 and apply the transformation three times, we get $0 \to -6 \to \frac{-6}{7} \to \frac{-42}{13} = -3.2 \to$. Plotting, we get

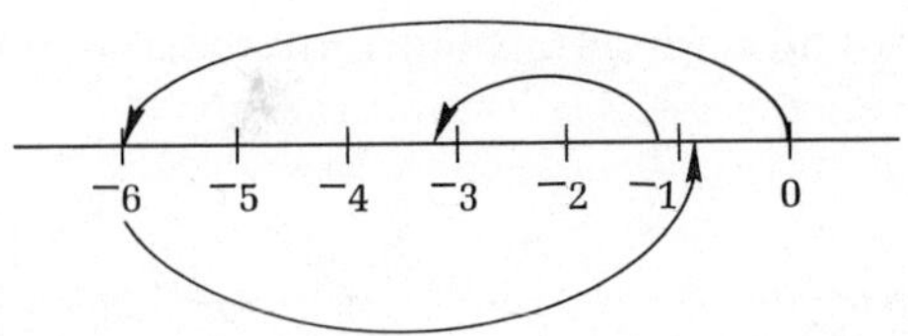

Where will the jumper go the next time? The next time?

If the jumper makes 1000 jumps where do you think he will end? If he starts at your guessed point, where will he end at the next jump?

Is there any point that remains fixed?

If there is a point that remains fixed, lets call it f. Then

$$f = \frac{6}{f - 1}$$

Multiply both sides by $f - 1$, we get

$$f(f - 1) = 6 \qquad \text{or}$$
$$f^2 - f = 6 \qquad \text{and}$$
$$f^2 - f - 6 = 0$$

Hence any point that is fixed under this transformation must be in the truth set for $f^2 - f - 6 = 0$.

EXAMPLE 4. You may recognize

$$x \rightarrow \frac{x + \frac{10}{x}}{2} = \frac{x^2 + 10}{2x}$$

as a transformation way of describing the divide-and-average process for approximating $\sqrt{10}$.

31.10 SUMMARY Elementary algebra is a closer look at the structure of a number system through letting a letter or other symbol represent a number. When used in this way, the letter is called a variable. Numbers are added and multiplied and so on, but now there is no automatic useful way of getting a single standard name for say, $A + B$, as there was for $3 + 5$, which is 8.

We are led to formulas such as $A = lw$, which are mathematical models of situations. Formulas are powerful because they help us find things we didn't know.

Formulas usually express one variable in terms of others. It is useful to "solve" formulas for another variable. We do this by repeated use of the inverses of the operators shown in the formula.

In many applications we can state something we know about a number in an equation, even if we can't give the common name of the number. Frequently we can find an equivalent equation from which we easily see the common name. This is called "solving the equation." We can treat inequalities similarly.

Formulas show special cases of functions and operators. Functions of a single variable, or unary operators, or formulas of one variable can be interpreted as transformations on a number line. This point of view helps understand solving equations.

EXERCISES

Problems 1 to 4 ask you to make up formulas by substituting letters for numbers. Different people may get different answers.

1. In a formula for Q, one substitutes 6 for x and 9 for y, then (*a*) says $6 + 9 = 15$, and (*b*) announces "$Q = 15$." Give the formula for Q.

2. If 4 and 7 are substituted for A and B respectively to get C, one takes the following steps: (*a*) $4 \times 7 = 28$, (*b*) $28 \div 2 = 14$, (*c*) $C = 14$. What is the formula?

3. After substituting 17 for x, 5 for y, and 9 for z, these steps were taken: (*a*) $9 \times 5 = 45$, (*b*) $45 \times 17 = 765$, (*c*) $765 + 2 = 767$, (*d*) $C = 767$. Write the formula.

4. When 8 is put in $\square$ and 15 in $\triangle$, the steps are: (*a*) $8 \times 8 = 64$, (*b*) $\frac{15}{3} = 5$, (*c*) $64 - 5 = 59$, (*d*) $59 - 3 = 56$, and (*e*) $\bigcirc = 56$. Write a formula for $\bigcirc$ in terms of $\square$ and $\triangle$.

5. (*a*) It is known that x is one of the numbers $\{3, 5, 7, 9\}$. What can you say about the possible values for y if $y = 3x - 2$?
(*b*) Are you thinking of x as a number whose common name is not known, or as a variable?

6. Explain to a friend how it is possible to have either attitude in exercise 5.

7. If you thought of x in exercise 5 as a variable, what is its replacement set?

8. If *a*, *b*, *c* are electric switches as in section 31.3, what does $(a \cdot b) \cdot c = a \cdot (b \cdot c)$ mean? Is the statement true?

9. List the steps you might take in solving
(*a*) $A = \frac{1}{2}BH$ for H
(*b*) $x = \dfrac{3}{y + 2}$ for y

10. List the steps you might take in solving
(*a*) $L = \dfrac{4M + 3}{5}$ for M
(*b*) $S = r^2$ for r

11. Solve $P = \frac{q}{r}$ for r in 3 different ways, using a different first step in each way.

12. Solve (*a*) $2 = \dfrac{12}{3y + 4}$ for y
(*b*) $2 = \dfrac{12}{3y + 4}$ for 4

By "solve an inequality for x" we mean "write an equivalent inequality with x alone on the left side."

13. Solve $x - 4 < 7$.

14. Solve $4 - 2x < 1$.

15. (*a*) Trace three jumps starting at 0 for the transformation

$$x \rightarrow 4 + \tfrac{1}{2}x$$

(*b*) Trace the three jumps starting at 1.
(*c*) Is there a fixed point? What is it?

16. Trace three jumps for the transformation $x \to 8 \div \frac{x}{2}$ starting from different points. Is there a fixed point? Where?

17. Illustrate the transformation $x \to 5 + 2x$ by starting several jumps at 0 and $^-7$. If there is a fixed point, find it.

18. Trace jumps for $x \to \frac{2}{x + 2}$.

19. Trace jumps for $x \to \frac{x^2 + 2}{2x}$, starting at (*a*) $x = 1$ and (*b*) $x = {}^-5$. (*c*) If you make ten jumps each for (*a*) and (*b*), would you end up at about the same place?

20. The transformation $x \to \frac{2x^3 + n}{3x^2}$ is good for estimating $\sqrt[3]{n}$. Try for (*a*) $n = 8$ and (*b*) $n = 27$.

21. If *B* is beans, *C* is corn, and *T* is tomatoes, is it true that $B(C + T) = BC + BT$? If so, why? If not, why do we say $B(C + T) = BC + BT$ in algebra?

22. The four 4s problem. The number 1 can be written using four 4s and elementary operations signs ($+$, $-$, $\times$, $\div$, $\sqrt{\ }$) as $\frac{4 + 4}{4 + 4}$. The number 2 can be written as $\frac{4 \times 4}{4 + 4}$. Can you write 3? 4? 5?

Answers to selected questions: **1.** $X + Y = Q$; **3.** $Z \times Y \times X + 2 = C$; **5.** (*a*) Y is one of {7,13,19,25} (*b*) either answer satisfactory; **9.** (*a*) multiply both sides by 2; divide both sides by *B*; (*b*) multiply both sides by $y + 2$; divide both sides by x; subtract 2 from both sides.

11.

$p = \frac{q}{r}$	$\frac{1}{p} = \frac{r}{q}$	$1 = \frac{q}{pr}$
$rp = q$	$\frac{q}{p} = r$	$pr = q$
$r = \frac{q}{p}$	$r = \frac{q}{p}$	$r = \frac{q}{p}$

13. $x < 11$

15. (*a*) 0, 4, 6, 7; (*b*) 1, 4.5, 6.25, 7.125; (*c*) yes, 8.

18. 0, 1, .667, .75, .727, .733

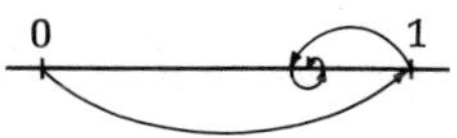

19. (*a*) 1, 1.5, 1.417, 1.414

(*b*) -5, -2.7, -1.72, -1.44, -1.414

21. not true; *B*, *C*, *T* must be numbers in algebra.

32

Introduction to Geometry

32.1 INTRODUCTION Geometry is the study of the location, shape, and size of figures. The name comes from ancient Greek and means "measurement of the earth." This suggests that even in very early times people were interested in measuring distances and areas on the earth. As each path and field differed from every other path and field, experts with much specific detailed knowledge were necessary in each community.

Then it was discovered that each unique figure could be approximated quite well by a combination of certain basic and elemental shapes such as segments, triangles, squares, and circles. These shapes also approximated other objects such as animals, trees, and manufactured things.

In this chapter we look at these basic shapes. We show how they are used in approximation and describe some of their properties, including symmetry and parallelism. Many of the ideas are generalized to three dimensions.

32.2 MAPS The author once listened to an enthusiastic young teacher from up-country Africa explain through his filed-to-points teeth how he used maps to make geometry meaningful to his pupils. Outside the palm-thatched palaver house used as a schoolroom was a smoothed patch of sand about 20 by 20 feet. Traced on the sand were the nearby trails, creeks, and other major geographic features well known to the children. Into the sand the children had stuck short twigs to mark the paths, longer twigs to mark the streams, and pebbles to mark the school, the homes, and so forth.

Here is a diagram of the child-made twig and rock land. (We have added the scales along the bottom and at the left.)

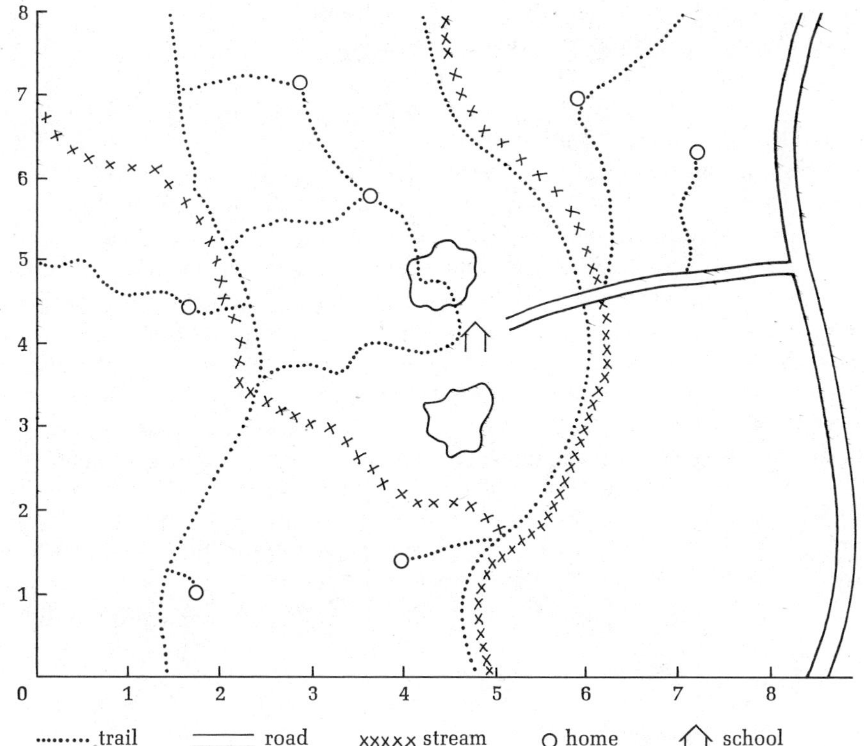

We shall use a coordinate system, or grid, to locate points on the map. Imagine a grid like the one in the margin, lightly sketched on the map or made on a transparent overlay. The lines through (0, 0) with number scales are called the x and y axes. The numbers along the bottom are x values; those at the left side are y values. We locate a point by giving a pair of numbers (x, y). For example, A is the point (2, 7), because the vertical line through A intersects the horizontal x-axis at 2, and the horizontal line through A intersects the vertical y-axis at 7. We also say that A is 2 units from the y-axis (the x-value) and 7 units from the x-axis (the y-value).

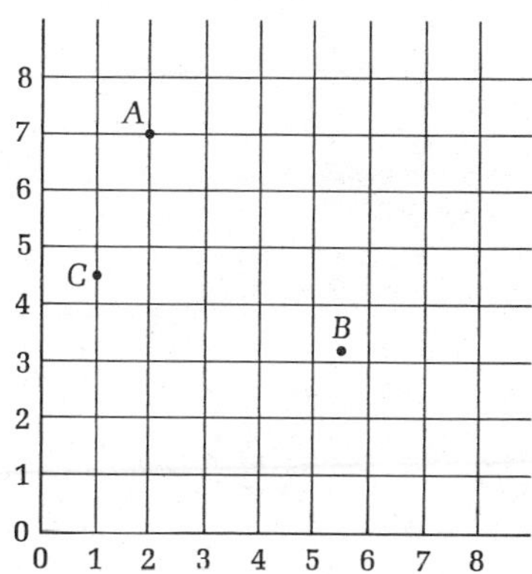

Similarly, B has coordinates (5.5, 3.2). What are the coordinates of C? Where is the point (3, 1)?

The school is located at (4.8, 4). That is, a vertical line at $x = 4.8$ intersects a horizontal line at $y = 4$ at the schoolhouse. The schoolhouse is 4.8 to the right of the vertical y-axis and 4 above the horizontal x-axis.

A stream starts at the top of the map at (4.4, 8) bends to the right to (6, 4), and leaves at the bottom at (5, 0). Another stream enters at the left at (0, 7). There are seven homes, one at (1.5, 4.5). Can you locate the other six?

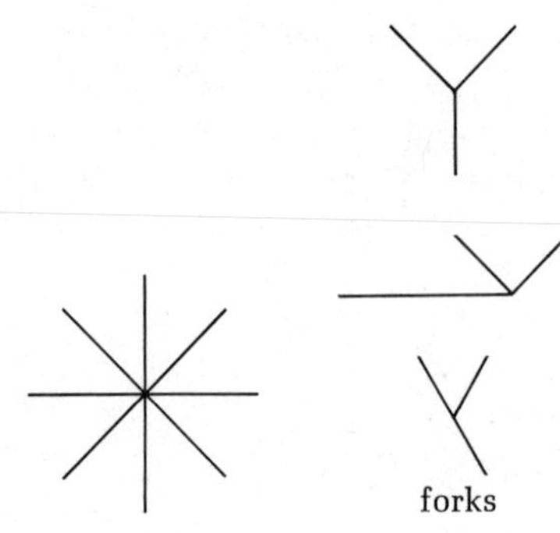

A traveler at a general point on a trail has a choice of two directions to travel, forward and back. At a simple branch point or fork he has three choices as shown by the diagram. He can go right, left, or back. At multiple branch points several trails meet and the traveler has more options. Similarly streams, roads, or other one-dimensional objects can be shown by curves which may meet or intersect. In the trails there are more than five forks. Can you find six? Can you find a fork in the stream? How many? How many forks in the road are there?

Near the school are two regions at about (4.5, 3) and (4.3, 4.8). One of these regions is a school playground and the other is a steep, rugged hill. Can you tell which is the hill and which the playground?

We see that the trails are more curved than straight. This is not only because the land is rolling but also because people like curved trails better.

Can you locate the home at (1.7, 1)? How many ways can a student who lives here get to school? A separate way from one point to another must not retrace itself or intersect itself. For example,

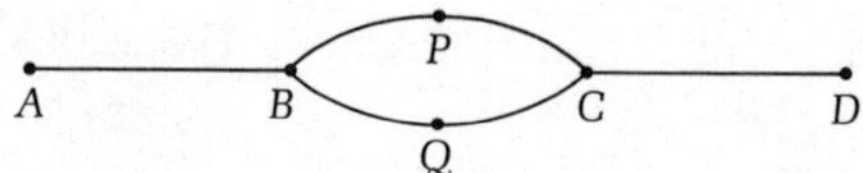

There are two ways to go from *A* to *D* as shown by the letters *ABPCD* and *ABQCD*. If a path could repeat or intersect itself, *ABPCQBPCD* would be a third path. Can you trace with your finger a person following this path?

Find the home of the child closest to school. Which child has the furthest to go? (NOTE: There may be a tie.) What are the three ways the child who lives at (2.8, 7) can take to school? How many ways can the child at (5.8, 7) take? At any place a trail or road crosses a stream there is a bridge. How many bridges are there?

Such maps have been used since the dawn of history to show the geography and geometry of regions. Travel maps and atlases show much about the shape of large areas. Real-estate maps and land-use maps show fine detail for builders and landscape architects. Maps show shape, distance, area, routes, and connections—geometric concepts useful in themselves and also important in later mathematics.

The earth is approximately a sphere. An annoying property of spheres is that it is impossible to map a region of a sphere onto a plane sheet of paper so that all distances, areas, and shapes are right. That is, measurements on the sphere will not be exactly proportional to measurements on the map. This lack of proportionality is important with large regions. For example, how does the size of Greenland on a Mercator projection compare with its size on a globe?

For small regions, however, maps show lengths, areas, and shapes well. All common maps, such as road maps and atlases, are quite satisfactory over distances up to 200 kilometers.

32.3 NETWORKS The twigtown map is drawn from memory, by children who are rather sure of some things and not quite so sure of others. For example, they are very sure of which trails, streams, and roads go to which homes, have forks, or intersect. They know the betweenness or order relations. That is, of three points on a trail, they know which point is between the other two. They know whether two points are on the same side or opposite sides of a trail or stream. The map will show all of these locations correctly.

On the other hand, the children are not so sure of measurements. The lengths of the arcs, or pieces of trail, on the map may not be proportional to the actual trail. One row of twigs may be twice as long as another, while the first trail may not actually be twice as long as the second. Similarly, the areas of regions will not be proportional to the map areas. Frequently the lengths and areas of regions of importance and interest to the drawer will, relatively speaking, be larger than those of no interest. The corners and curvatures of paths and streams may not be shown accurately.

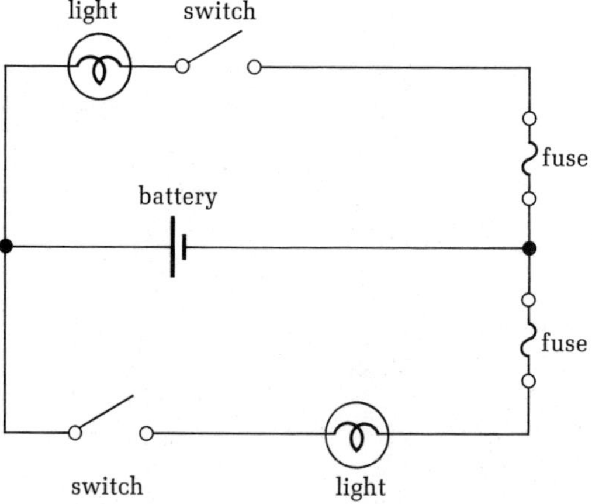

Such maps are closely related to wiring diagrams for a house, building, car, or other place with electric outlets. A wiring diagram shows wires represented by arcs. (We will see later that arcs need not be straight.) The connections, outlets, switches, and fuses are drawn large and placed on the diagram so that they can easily be seen and repairs can be made to the real circuit it represents if something goes wrong. The lengths, corners, and straightness of the wires and the sizes and locations of the parts may be quite different on the diagram from what they are in the building or car.

Such maps and diagrams, where connections and order relations are important, but not sizes and shapes, are examples of networks. Here are some other examples of networks:

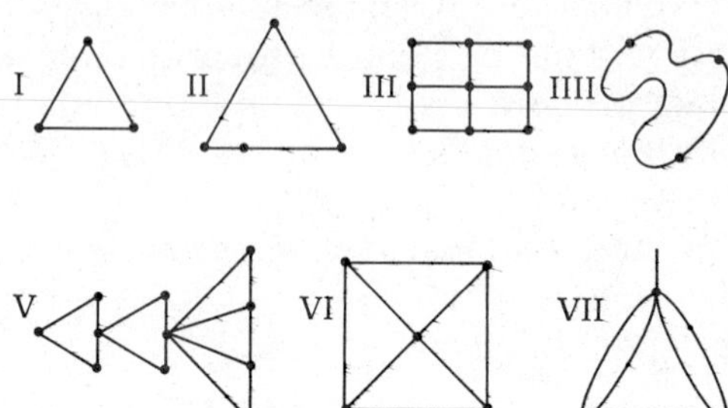

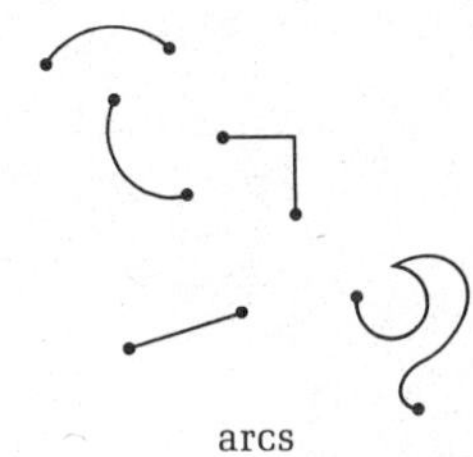
arcs

Each network is composed of *arcs,* which end at two distinct *vertices.* Some examples of arcs are shown in the margin. Note that here an arc does not have to be part of a circle and indeed may be straight. Network I above has 3 arcs and 3 vertices; network III has 12 arcs and 9 vertices.

Network II is the same as network I except that it has an extra vertex separating the bottom arc into two arcs. Are networks I and II equivalent? We are free to choose, yes or no. Let us think of a network of trails. Would it make much difference to a person using the trail if one arc of the trail was separated into two arcs by a bramble along the trail making another vertex? Clearly not. So we say that separating one arc of a network into two by inserting a vertex produces an equivalent network. Conversely, a vertex which is the common end of just two arcs can be removed and the two arcs joined into one. Corners can be straightened out without changing the network.

In the sample networks, I, II, and IIII are equivalent. Two others are equivalent also. Can you find them? Sometimes it is difficult to recognize that two networks are equivalent. One test is to see if they have the same number and same types of vertices.

A vertex can be the common endpoint of any number of arcs. Hence we can classify vertices as one-arc, two-arc, . . . , k-arc vertices. We shorten these to 1-vertex, 2-vertex, . . . , k-vertex. In the margin are some sample networks with the vertices labeled. V_3 is a 3-arc vertex, and so on.

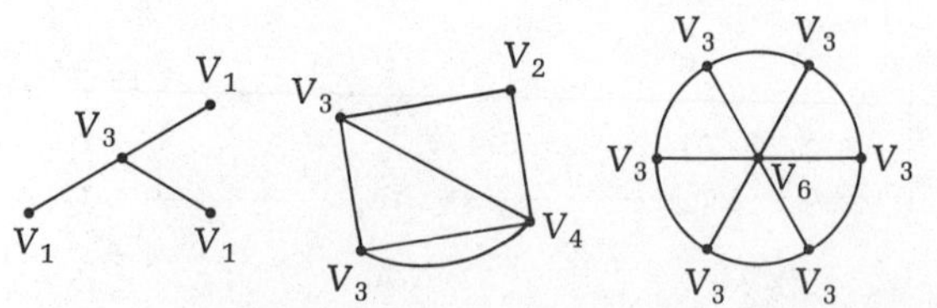

To be equivalent, two networks must have the same number of 1-vertices, 3-vertices, ... They may have different numbers of 2-vertices, since these may be inserted or removed at will without changing the network.

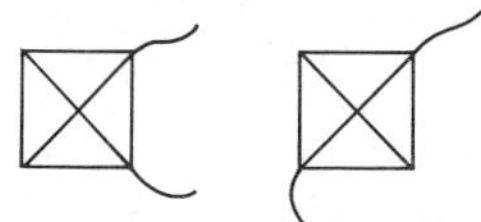

Two networks may have the same numbers of different types of vertices and yet not be equivalent. Each of the networks in the margin has two 1-vertices, two 3-vertices, and three 4-vertices, yet they are different. We invite you to make up your own definition of two equivalent networks.

A network is called *connected* if one can start at any point and move to any other point along the network. Network *A* is connected; *B* is not. It is really a matter of choice whether you wish to say *B* is one nonconnected network or two connected networks.

A network is called *one-drawable* if a copy can be drawn without lifting the pencil from the paper and without retracing any part, although intersections are allowed.

A and *B* are one-drawable starting at any point. *C* is one-drawable if you start at the upper right and end at the lower left. *D* is not one-drawable.

We argue as follows. If a curve is one-drawable starting and ending at any point, each vertex must be an even vertex—that is, must lie on an even number of arcs. For every time one traces an arc into the vertex he traces a different arc out. If a vertex is odd, then the tracing either starts or ends at that vertex. Hence we see that a one-drawable network has no more than two odd vertices and in that the tracing must start at one odd vertex and end at the other.

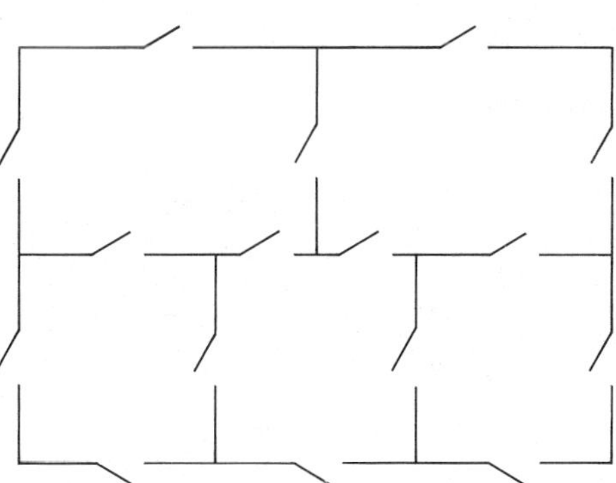

You may have seen this puzzle: The five-room house shown in the margin has a door in each wall. Find a way to start in some room (or outside) and walk through each door once and only once.

This problem can be changed into a network problem. We represent each room by a vertex. Going through a door matches going along an arc connecting two vertices. We draw the network, including one vertex for outside. We list the orders of the vertices: 9,5,5,5,4,4. There are four odd vertices, so thc puzzle cannot be solved.

We leave it as an exercise to determine a route which will go through all but one door. A closely related problem is to determine which, if any, door can be locked and then a route found that goes through all of the other doors once and only once.

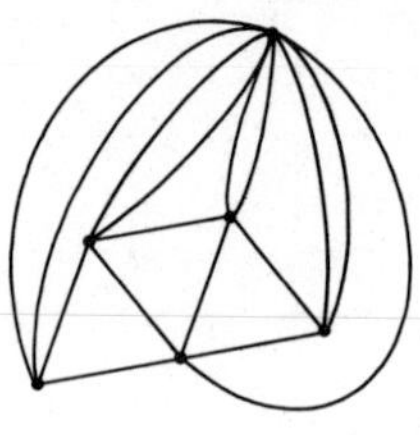

32.4 BASIC SHAPES Just about everything we see has a different shape from everything else—if we omit the tin cans and beer bottles at picnic grounds and along highways, So if we want to talk about their shapes, we must mentally smooth out and mash the outlines to some sort of standard shape which we easily understand and which will tell most of the story.

As this is written the author is seated beside a lake in the Adirondacks looking at pine trees, cabins, water skiers, and birds and thinking about the city with its buildings, streets, cars, trains, and windows. It is impressive how many of these shapes can be approximated by a combination of one or more of the following: triangle, square, rectangle, and circle.

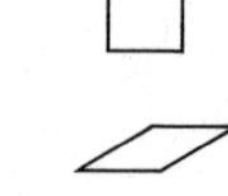

A square is a flat or plane figure with four equal sides and four equal angles. A picture of one is in the margin. Each angle is a right angle and contains 90 degrees. The square is not rigid. Four equal sticks pinned at the ends may not form a square. If each joint is free to turn, the square collapses. The square is the most common figure used to tile a region. We cover a region with squares, count the squares, and give the result as the area.

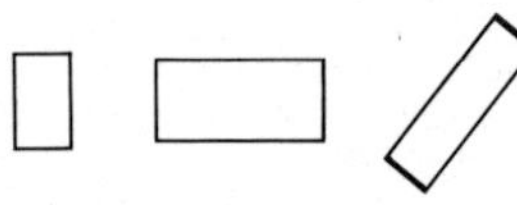

A rectangle has been flippantly called "a square with unequal sides." Some examples are in the margin. The rectangle is a flat figure with four angles, all equal, all right angles. The four sides may or may not all be equal. If they are equal, the rectangle is a square. The sides may be of any size. The opposite sides of a rectangle are equal and parallel.

Man-made things like books, sheets of paper, automobile bodies, silhouettes of cans and boxes show rectangles. Tree trunks, blades of grass, stretches of river, animal bodies show outlines closely approximated by rectangles.

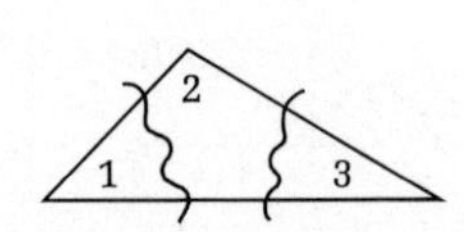

A triangle has three sides and three angles. If you tear the corners off and fit them together, you'll see that they make a straight line. In the margin is a diagram showing how to do this. We say that the sum of the angles in a triangle is 180 degrees.

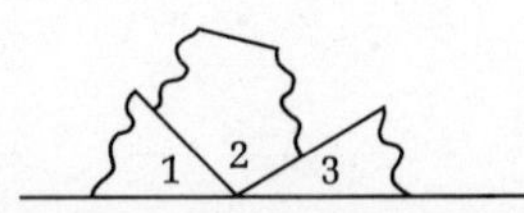

The equilateral triangle has all three sides equal and all three angles equal. Isosceles triangles (two equal sides) with included angles of 30°,

60°, and 90° are shown. A right triangle with angles 30°, 60°, and 90°, is also shown.

Since a straight line is the shortest path between two points, any of the three sides of a triangle is shorter than the sum of the other two.

The triangle is rigid. That is, any three sticks such that each is shorter than the sum of the other two can be pinned together at their ends to form only one triangle. The sticks cannot be moved. This fact is universally used in construction, from teepees to spaceships.

A triangular region is a portion of a plane bounded by a triangle. Similarly for a rectangular region and other regions bounded by shapes. Often people say "triangle" when they mean "triangular region." We urge care in using these words to make sure no confusion will follow.

A region can be nicely approximated by using triangular regions formed by connecting strategic points. See the example in the margin.

While triangles have many shapes, excellent approximations can be made by using only a few:

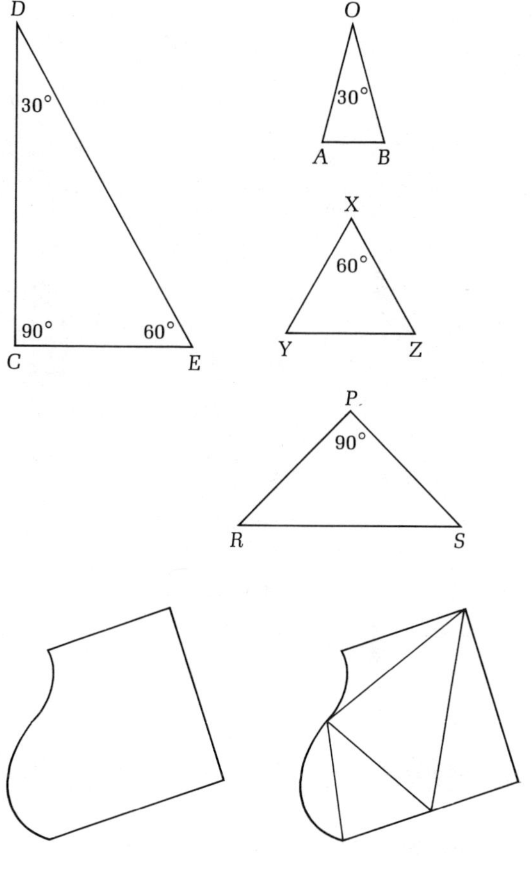

1. The equilateral, or regular, triangle with all three sides equal and each angle 60 degrees.

2. The 30-60-90 triangle, which can be formed by cutting the equilateral triangle in half.

3. The isosceles right triangle with equal shorter sides (legs) and with angles of 90, 45, and 45 degrees.

4. The 3-4-5 triangle with sides of lengths 3, 4, 5. This triangle has a right angle, and is beloved of numerologists because $3^2 + 4^2 = 5^2$.

A copy of a triangle will fit the triangle exactly. Usually the two will fit in only one way. However, if the triangles are isosceles (with two equal sides), the two fit in two ways. A copy of an equilateral triangle will fit in six ways. Can you list them? Before you do this, you first need to invent a way of describing the way one triangle can fit another.

The above figures are all made with line segments. A simple nonstraight curve is the circle. The circle is not only important in its own right, but is used to approximate other curves. Here are some of the properties of the circle:

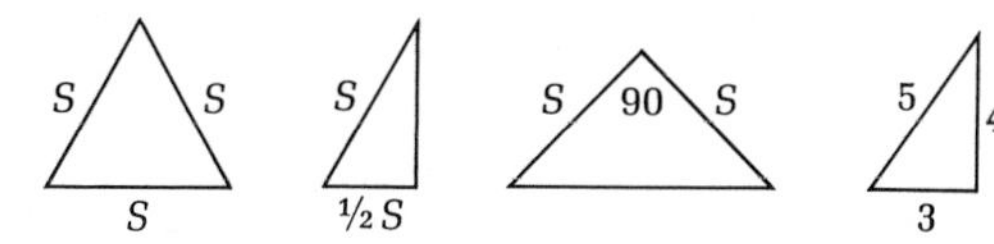

1. Every point is the same distance from the center. This distance is called

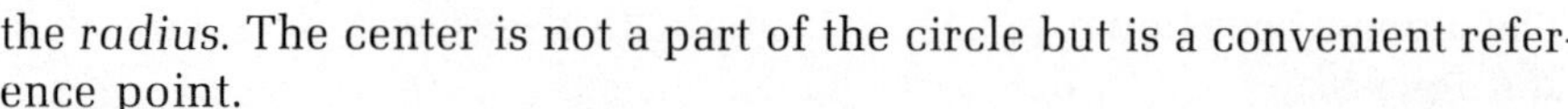

the *radius*. The center is not a part of the circle but is a convenient reference point.

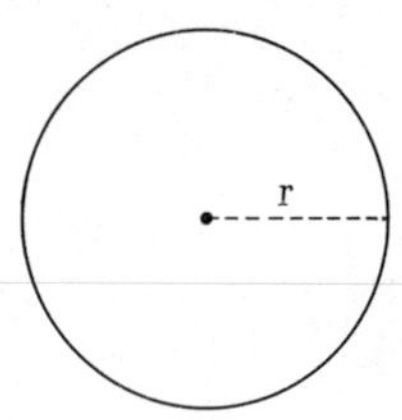

2. The circle is the shape usually chosen for wheels, bearings, and other rolling objects.

3. The copy of a circle fits the circle any way it is placed upon it. Hence the circle is a good shape for things that must be stacked, such as coins.

4. The circular region, or disk, has the largest area of all the regions having the same perimeter (distance around). If a string is tied in a loop and used to outline a region, the largest such region will be circular.

Several shapes can be shown and discussed quite nicely on the geoboard. A commonly used geoboard has 36 pegs arranged in a square grid as shown in the margin. Rubber bands are stretched about the pegs to show the shapes:

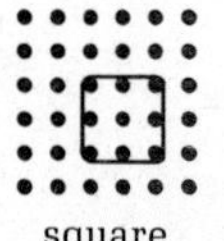
square

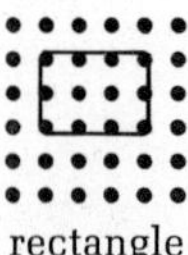
rectangle

triangle

Not all shapes can be shown on a geoboard. For example, neither a circle nor an equilateral triangle can be shown.

The shapes discussed so far have been flat. They can be drawn on a sheet of paper. Corresponding surfaces and networks in three dimensions are useful. The most basic are cubes, rectangular prisms, and spheres.

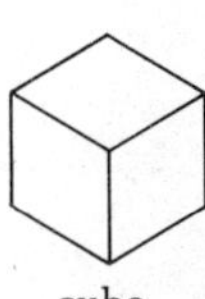
cube

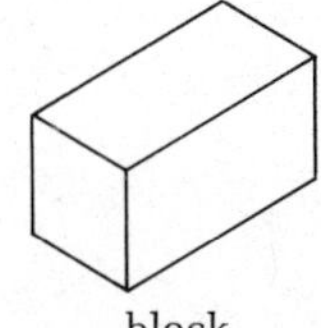
block

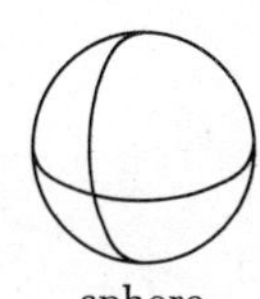
sphere

The cube is the shape of a die or sugar cube. Each face is a square and all edges are the same length. It is impossible to draw an exact diagram on paper, but a stylized diagram is shown in the margin.

A rectangular prism is a stretched rectangle. The faces are all rectangles. Bricks, building blocks, lumber, plywood, and chests are all examples of rectangular prisms.

A sphere is the surface of a ball. All points are equally distant from one point, the center. A stylized picture appears in the margin.

Just as a sphere is a three-dimensional generalization of a circle, a cube a generalization of a square, and a rectangular prism a generalization of a rectangle, so the tetrahedron generalizes the triangle. Let four points not in a plane be connected by segments. The six lines form four triangles.

The three-dimensional figure is a tetrahedron—four vertices, six edges, and four faces.

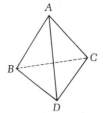

A tetrahedron is a triangular pyramid. In the figure shown in the margin, we think of *BCD* as forming a triangle and connected with *A*. *A* is the vertex of the triangular pyramid and *BCD* is the base.

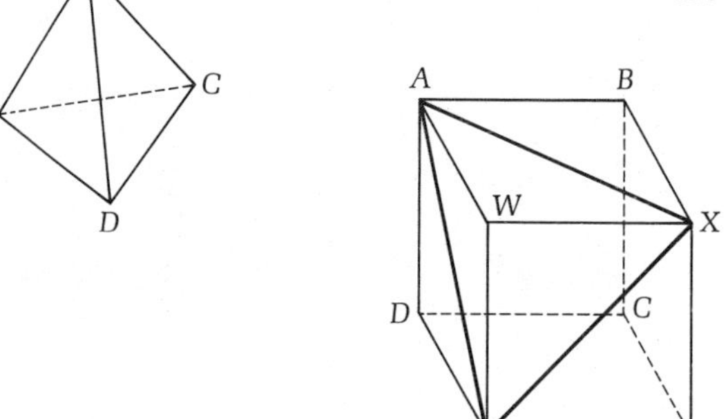

Do you think that if four corners of a cube, say *W, B, D, Y*, are cut off the cube as illustrated for corner *W*, then what is left of the cube is a regular tetrahedron whose six edges are equal and whose four faces are congruent equilateral triangles? You may want to make a model before committing yourself.

33.5 FITTING AND CUTTING The ability of one shape to fit another may be the most important relation between geometric figures. Fitting is the relation between a foot and a footprint, between a glass and the water it contains. More loosely, it is the relation between the hand and the glove, the outside of the piston and the inside of the cylinder.

Sometimes we say that two figures that fit have the same size and shape. Part of one figure may fit part of another without the entire figures fitting. Here are some examples of fitting among the basic shapes:

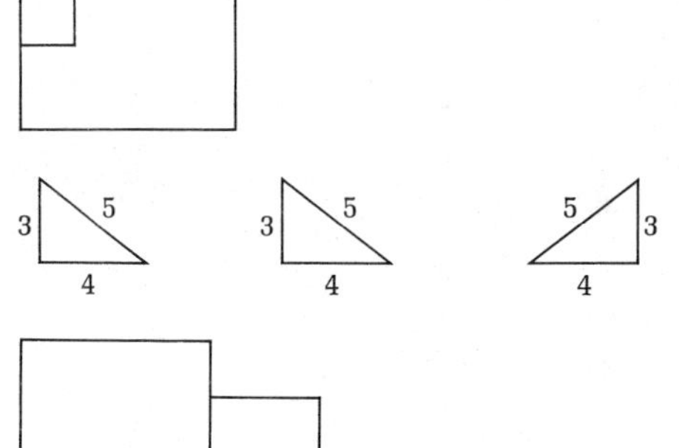

1. Any corner of a square fits any corner of a rectangle.
2. Two squares with sides of the same length fit.
3. Two triangles with sides of the same length fit. (One triangle may have to be turned over.)
4. Two circles with the same radius fit.
5. Two rectangles, side by side, fit a line.

Sometimes it is convenient to think of a figure as fitting a cutout. Imagine a 3-4-5 triangle carefully cut from a sheet of cardboard. We have a sheet with a triangular hole and a triangular piece. We think of the triangle *abc* as fitting the triangle *ABC*. There is only one way this fitting can be done, *a* on *A*, *b* on *B*, and *c* on *C*.

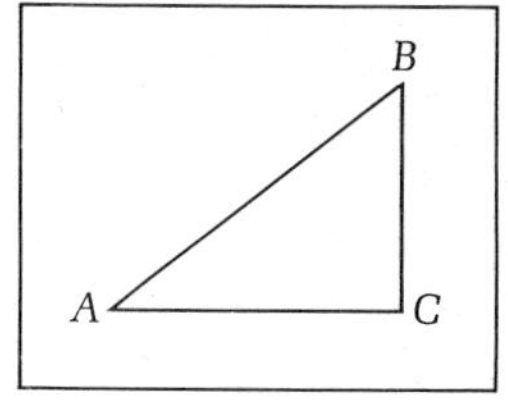

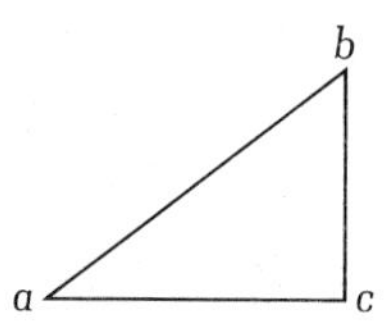

If we have two isosceles triangles *PDQ* and *pdq* with the same length of sides and included angle, then *pdq* will fit *PDQ* in two ways: *pqd* and *dqp* on PQD.

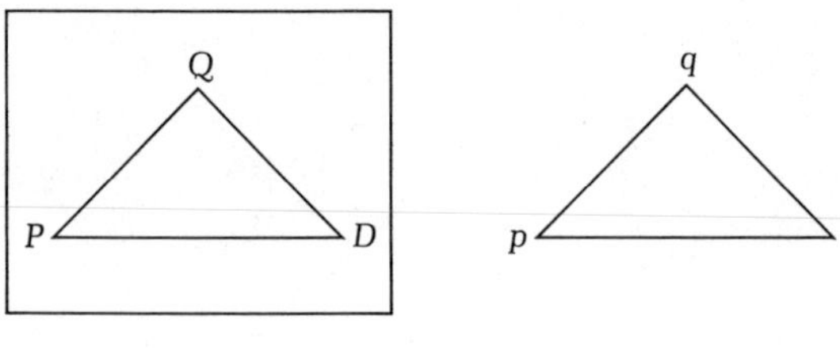

These fitting properties explain in part why the shapes are used the way they are. The fact that circles of equal radius fit in any way they are put together is one reason why things to be stacked, such as dishes, are often made circular. Building bricks are made in the form of a rectangular prism because they fit together in so many ways.

Often complicated regions can be cut up into simple basic figures. In particular, many figures can be separated into triangles just by cutting between strategic points. The triangles then fit together to fit the complicated figure. Some examples are shown in the margin.

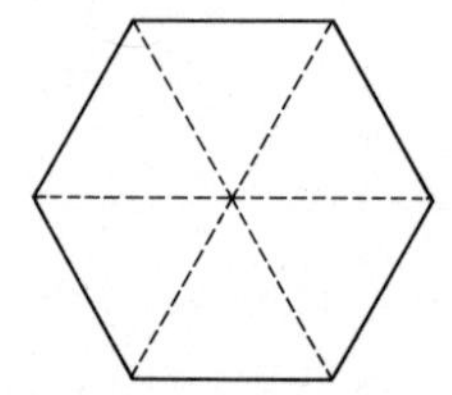

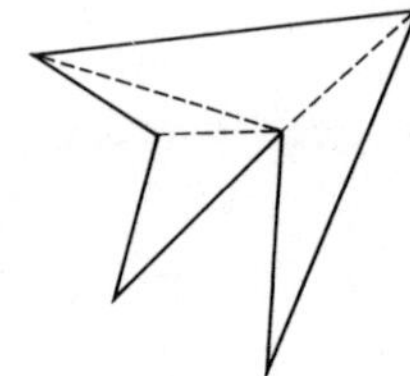

In addition to *fit,* people use the words *superimpose* and *congruent.* Traditionally, figures were said to be congruent if they could be superimposed or made to fit everywhere. In the last few years people who wished to take time and motion out of geometry have objected to fitting and superposition as valid geometric operations. They define congruent as setting a one-to-one correspondence between the points such that corresponding distances are equal. Many people prefer to be aware of the several attitudes and use the ones best for the circumstances.

32.6 ELEMENTAL FIGURES It is useful to consider figures that can be combined to make any figure. The common elemental figures are points, lines, planes, segments, rays, half lines, and half planes. Of course these are all mental constructs suggested by what we see in physical objects.

A point has position. It is suggested to us by dots, ends of needles, corners, and intersections of lines. While the concept of length does not apply to point, we sometimes extend the idea and say a point has zero length.

A line segment is suggested by needles, blades of grass, taut strings, folds in paper. The ends of a segment are points. While a segment is an entity in its own right, it is often convenient to think of it as made up of an infinite number of points. A segment has length but no width and no thickness.

There is only one line segment with two given points as end points. If two segments intersect, they intersect in either one point or a segment. The segment is the shortest path between its end points. The end points may or may not be considered part of the segment, depending on the convenience of the geometer and the way the segment is defined.

If we think of a segment extended indefinitely in one direction we get a ray or half line. If the remaining end point is part of the figure we call it a ray. If it is not, the figure is a half-line.

Extending a segment indefinitely in both directions gives us a line. Many people like to discuss lines first and then line segments as parts of a line.

A taut string suggests a line, a slack one suggests a curve. Some people think of a curve as a line with bends. It is more useful to think of a line as a special kind of curve, a straight curve. There are an infinite number of curves between two points.

A plane is a surface suggested by the tops of tables, surfaces of still ponds, floors, and ceilings. Just as a line shows our idea of straight, a plane shows the idea of flat. A rectangle, as part of a plane, is often considered comparable to a segment as part of a line. We say a plane has two dimensions because the rectangle has length and width.

A plane and a line can intersect in no points, in one point, or in all points (if the line lies in the plane). However, a curve can intersect a plane at several points, like a thread being stitched in a cloth.

These examples show why many people like to think of all figures as sets of points as well as collections of segments, arcs, portions of planes and other surfaces. Sometimes people like to think of figures as sets of lines rather than points. A triangle is a set of three lines rather than an infinite number of points. We do not explore this.

32.7 ANGLES Several examples of angles are shown in the margin. Angles appear at corners of figures such as triangles and rectangles. Each of these figures shows two line segments with an end point in common. These angles are named "angle *ACB*," "angle *PQR*," "$\angle XOY$," and so on.

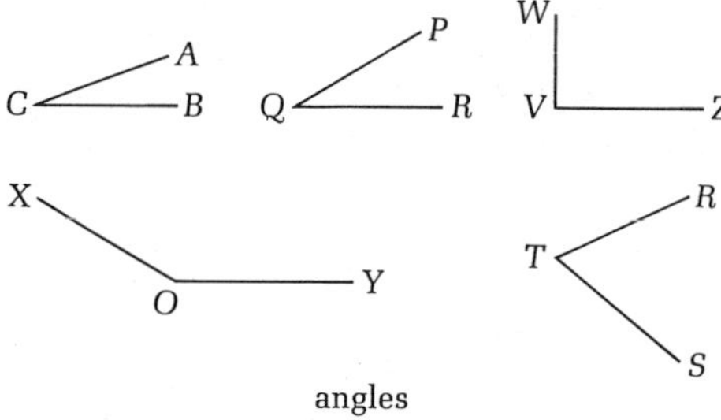

angles

It would seem reasonable to define an angle as a figure formed by two segments with a common end point. It turns out that this definition does not lend itself to several of the things people wish to do with angles. People like to compare (which is larger?), measure, add, and bisect angles, and they do not wish to worry about the lengths of the segments (sides).

Three different statements have been suggested as definitions by various people. We give them with some of their advantages and disadvantages:

1. An angle is the union of two rays with a common end point.

2. An angle is the region between two rays with a common end point.

3. An angle is the amount of turning of a ray as it rotates about its end point from an initial to a final position.

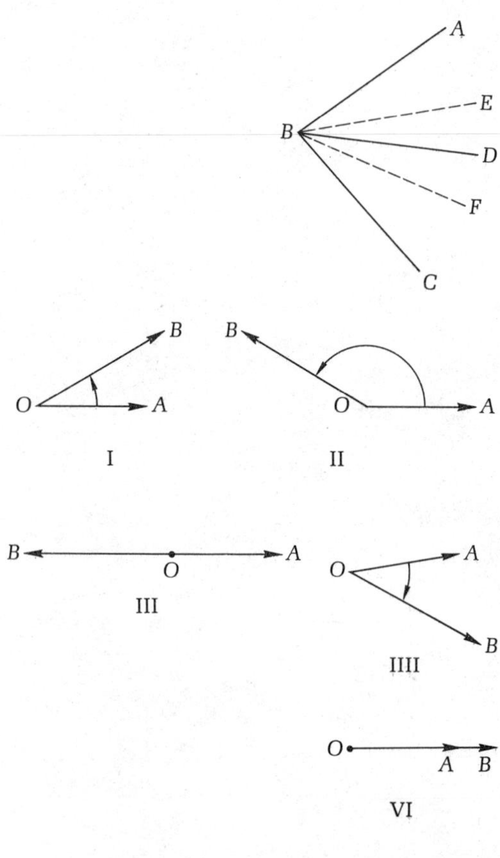

In statement 1 the line segments are replaced by rays in our description of the pictures of angles. In this way we avoid all questions about the length of the segments.

Thinking of the angle as a region, as in statement 2, makes it easy to think of bisecting an angle or joining two angles by forming a union of the regions. It is easy to think, for example, that angle *ABC* is bisected by *BD* and separated into thirds by *BE* and *BF*. This is what we wish. If we use statement 1, the set of points in angle *ABC* is not even separated by the rays *BD*, *BE*, and *BF*. So ideas like bisect and separate into three angles don't make sense.

The "amount of turning" concept of statement 3 allows for zero angles, angles greater than 180°, and negative angles. For, if *OA* is the initial position of a turning ray and *OB* is the final position, we can get angles like those in the margin.

Angle I is a typical angle and could be one of the angles in a right triangle. Angle II also satisfies all three statements and could be an angle in a triangle but not a right triangle. Angle III is called a *straight angle*. It is not clear which half plane is the region between the two rays. Angle IIII is negative, as shown by the circular arrow turning counterclockwise. Statements 1 and 2 imply only positive angles. Angle V is an angle greater than a straight angle. This is an extension of the idea of angle in both statements 1 and 2. Angle VI is a zero angle.

Statement 3 has many desirable properties. However, most people want an angle to be a geometric object, a set of points. An "amount of turning" is not a set of points.

Here, as in set theory and the empty set, we have an idea extended and generalized. We want to include all these ideas. Hence, we say an angle is the union of two rays with a common endpoint, the vertex. We think of a turning ray rotating about the vertex, starting on one of these rays and ending on the other. The region swept by this turning ray is the interior of the angle and is an angular region. If the interior is the union of two disjoint angular regions, the angle is the sum of the corresponding angles, that is $\angle AOC = \angle AOB + \angle BOC$.

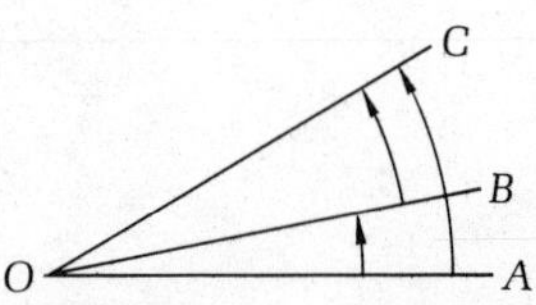

Just as there are several common units for measuring distances such as inches, centimeters, and meters, there are several common units for measuring angles. These are degrees, minutes, seconds, radians, right angles, half turns, and full turns or revolutions. There is no natural unit for length, but the full turn or revolution is a natural unit for angles. It is easy to define, requires no expensive equipment to construct, and is available anywhere. The angle shown by AOB is a full turn or one revolution.

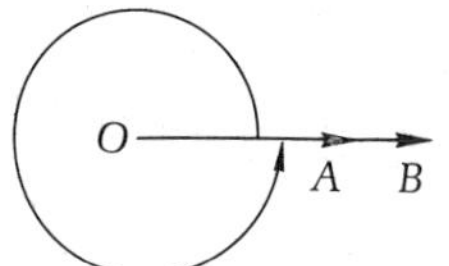

Clearly the full turn is too big a unit for most measurements. The Babylonians separated the full turn into 360 equal parts, each of which we call a degree. Why 360 rather than 100 or 1000? The Babylonians, great astronomers of antiquity, thought in terms of 60s rather than 10s. Since 360 is divisible by 2, 3, 4, 5, 6 and many other factors, the full turn could be divided into equal parts in many ways and each part measured in an integral number of degrees. Further, 360 is close to 365, so the stars would shift about one degree each night.

Angles are conveniently measured with protractors. How the protractor is used is illustrated in the margin.

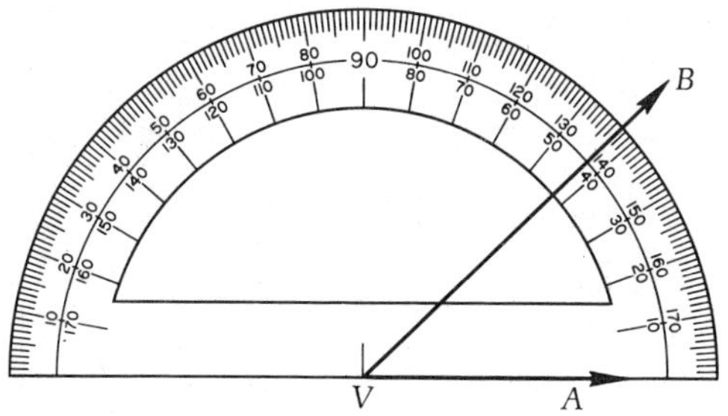

A half disk is cut out from paper or embossed in transparent plastic. The circumference has a scale going from 0 to 180. The protractor is placed on an angle to be measured with the center of the half-disk on the vertex, V, of the angle and the initial ray VA on the edge of the disk passing through the scale point 0. The scale reading where the final ray VB cuts the scale is the size of the angle. Using such a protractor an angle can be measured to within about one degree of accuracy.

With more elaborate instruments we can measure closer than one degree. Each degree is divided into 60 first minute (small) parts and each of these is divided into 60 second minute parts. *Minute* is a word for small. The names are abbreviated so the first minute parts are called *minutes* and the second minute parts are called *seconds*. Here is an example of the nominal use of numbers—an object is named second because of its position in a sequence. A city in California was named "El Segundo" because it was the place where an oil company erected its second oil refinery.

Instead of using minutes and seconds, some people use decimal parts of a degree. They say 30.4 degrees rather than 30 degrees 24 minutes, and 40.27 degrees rather than 40 degrees, 16 minutes, 12 seconds.

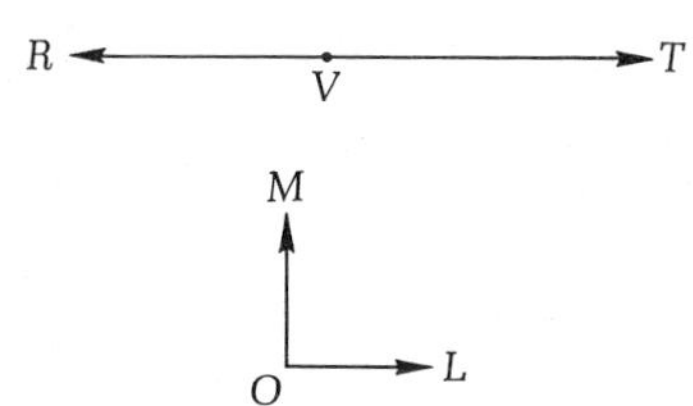

The half turn, or straight angle, is shown by RVT. The two rays of the angle lie on the same line.

The right angle or quarter turn or square corner is shown by LOM.

The right angle is a more natural unit than the full turn according to some people, because it clearly has two sides meeting at a vertex. A right angle is easily made by folding a piece of paper twice. One fold makes a line *AB*. Then fold so *A* falls on *B* making a square corner showing a right angle *BOP*. The ancient Egyptians made a loop of rope 12 units long. When the rope is stretched to form a triangle with sides 3, 4, 5 units long, the largest angle is a right angle.

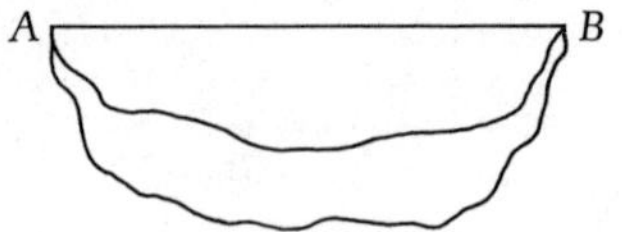

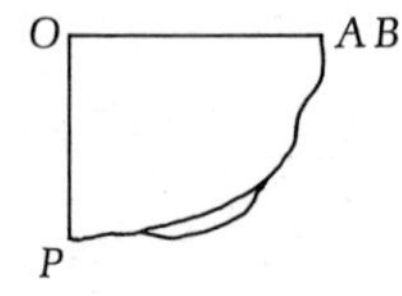

Angles are often measured in radians. Imagine a circle with center *C* and radius *PC*. If *PQ* is an arc of the circle of the same length as the radius, *PC*, then the angle *PCQ* is one radian.

Traditionally, ruler and compass were used to make angles of 60, 30, 90 degrees and to bisect and to copy angles. Today people more commonly use templates or cutouts (plastic or paper triangles with the desired shape) and trace these angles. Angles of prescribed size are made using a protractor.

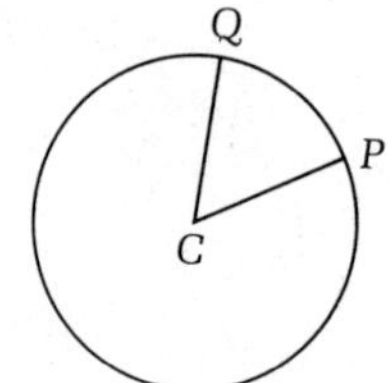

32.8 MEASUREMENT Measuring lengths, areas, and angles is a major part and purpose of geometry. Measurement of line segments may be made with a ruler or fitting unit lengths as discussed in chapter 12.

The perimeter, circumference, or distance around several segments joined together into a polygon is the sum of the lengths of the segments. There are three approaches to finding the length of a curve: (1) Straighten the curve to a segment and measure it. (2) Roll the curve along a line and measure the segment rolled upon. (3) Simulate the curve using a polygon whose sides are good approximations of the corresponding parts of the curve. For example, the circumference of a circle is close to the perimeter of either polygon *ABC* . . . *A* or *A′B′C′* . . . *A′*. As the polygon acquires more and more sides that come closer and closer to the curve of the circle, its perimeter is a better and better approximation to the length of the curve.

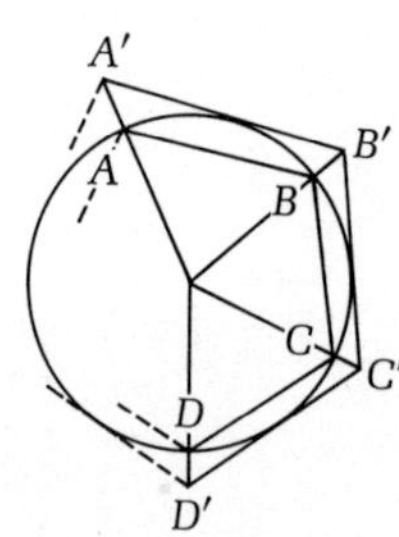

The area of a region is the number of standard squares, such as square centimeters, that will just cover it. For a rectangle with integer length sides, the diagram where the small squares are on a side clearly shows that

$$\text{area} = \text{length} \times \text{width}$$

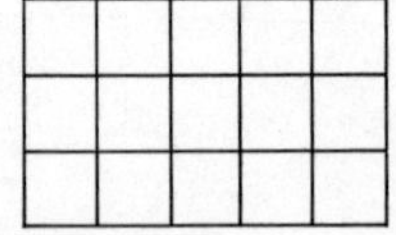

When the sides are fractions the same result follows on splitting the unit squares into the appropriate smaller squares. The diagram in the margin illustrates this with a rectangle that is $1\frac{1}{2} \times 2\frac{1}{3} = 3\frac{1}{2}$. Can you count the squares and see this?

If a right triangle is given a half turn about O, the midpoint of the hypotenuse, the side opposite the right angle, a rectangle results. This shows the area of a right triangle is $\frac{1}{2}$ base × altitude, where the base and altitude are the two perpendicular legs.

A non-right triangle such as PRQ can be divided into two right triangles by dropping a line from one vertex to the opposite side. We see that

$$\begin{aligned} \text{area } \triangle PQR &= \text{area } \triangle PLR + \text{area } \triangle LRQ \\ &= \tfrac{1}{2}PL \times RL + \tfrac{1}{2}LQ \times RL \\ &= \tfrac{1}{2}(PL + LQ) \times RL \\ &= \tfrac{1}{2}PQ \times RL \end{aligned}$$

which shows that the area for a general triangle is

$$\text{area} = \tfrac{1}{2}\text{ base} \times \text{altitude}$$

In this case, the base is one side and the altitude is the distance of the opposite vertex to that side. You may wish to explore this argument for various triangles including those with an obtuse angle (greater than a right angle).

There is no particular relation between the perimeter of a figure (distance around it) and the area. For example, different rectangles can have the same area but different perimeters. Consider the rectangles in the margin. In each case the area is 16, but the perimeters are 18, 34, $16\frac{2}{3}$. Can you find a rectangle with area 16 and perimeter greater than 34, say about 60? One with perimeter less than $16\frac{2}{3}$? What is the largest perimeter a rectangle of area 16 can have? the smallest?

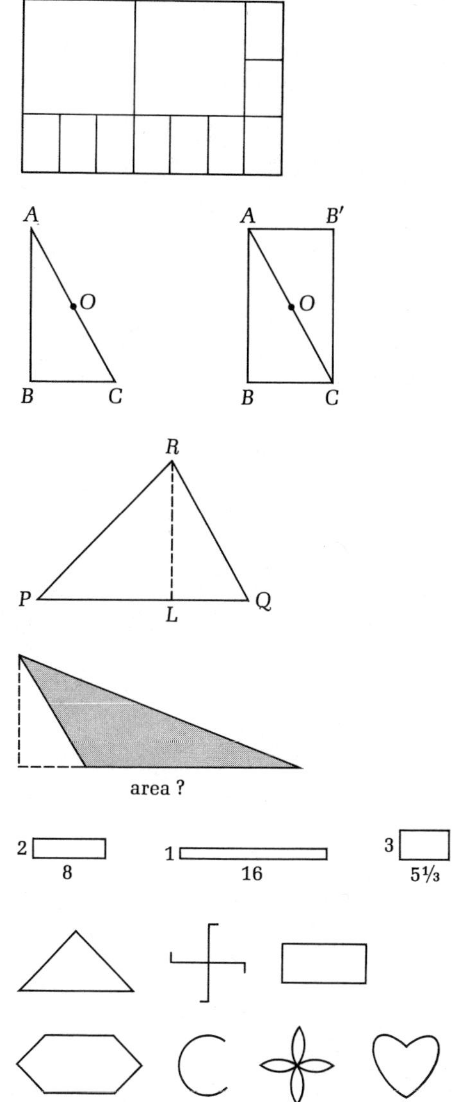

32.9 SYMMETRY Symmetry is one of the beautiful aspects of geometry. Each of the figures in the margin is symmetric, but there seems to be a difference. Is there such a thing as two different figures having different kinds of symmetry? How can we describe symmetry?

To start, we think of *two kinds* of symmetry—folding symmetry and turning symmetry. The isosceles triangle has folding symmetry, because if the triangle is folded on the line bisecting the included angle (the angle between the equal sides), the two sides fit.

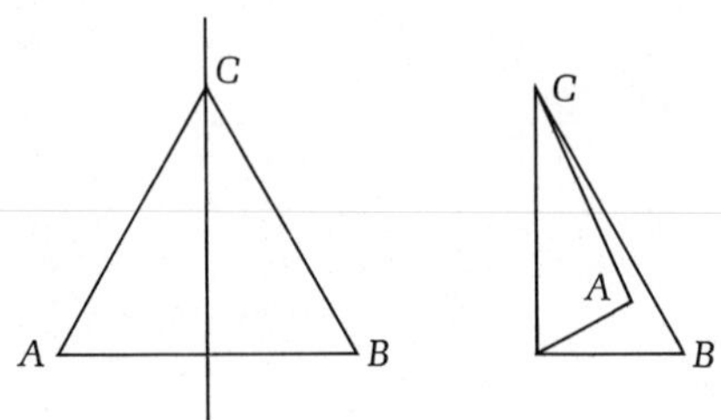

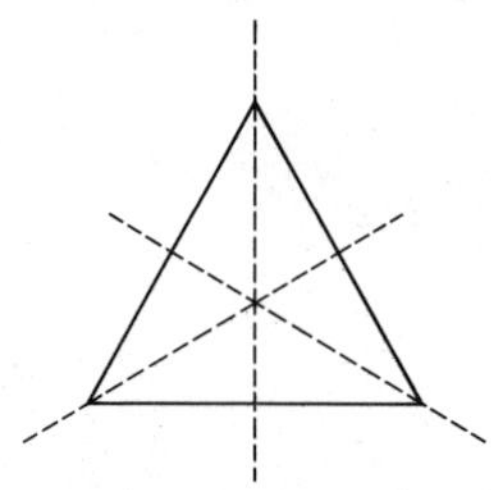

If a figure can be folded about a line so that the parts on either side fit, the figure has folding symmetry. The fold line is called the axis of symmetry. We also say the figure has line symmetry or symmetry with respect to a line. Instead of folding we may think of the plane of the figure being rotated 180° about the line of symmetry.

If a figure can be rotated about a point through an angle other than a full turn (360°) so that it coincides with itself, it has turning, or point, symmetry. The letter S has point symmetry, since it can be given a half turn (180°) about its midpoint so that it coincides with itself.

An equilateral triangle has both folding and turning symmetry. The triangle can be folded or flipped about any of the three angle bisectors. It can be rotated about the center through either 120° or 240° so that it coincides with itself.

Another way of describing a symmetric figure is to imagine it cut out of cardboard, leaving a hole to be filled. Of course, the piece can be put back the way it came out. If it can also be put back some other way, the figure has symmetry. An isosceles triangle can be put back into its template hole in two ways. An equilateral triangle can be put back in its hole six ways.

We say that two figures have the same symmetry if they will fit their holes after the same motions. An isosceles triangle and 8 have the same symmetry, since they can each be rotated 180° about a line and then will coincide with their original position. Rotating the letter S about a point shows that its symmetry is different from the symmetry of the isosceles triangle or 8

The three-cornered pinwheel Y and S both have turning symmetry. Can you see why their symmetries are different? What kinds of turns make the figures fit their original positions?

A circle has the most symmetries of all plane figures, since it coincides with itself no matter how it is placed. Every line through the center is an axis of symmetry, and every turn of any size about the center turns the circle onto itself.

32.10 PARALLELISM Two lines that do not meet are called *parallel.* Segments from two parallel lines are shown in the margin. Here are some properties of parallel lines:

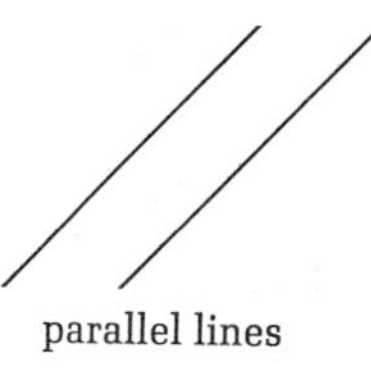
parallel lines

1. Two rays on parallel lines have the same or opposite direction.

2. Two sides of a triangle cannot be parallel.

3. If a line intersects one of several parallel lines, it intersects all the others also. In the margin are lines *a, b, c, d,* all parallel. Line *q* is known to intersect *d*. Then without looking, we can be sure *q* also intersects *a, b, c*.

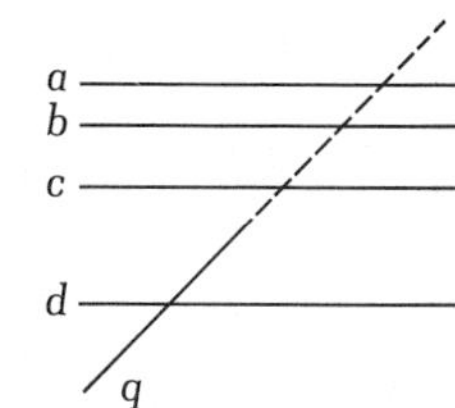

4. When parallel lines are cut by a line, the corresponding angles are equal. In the diagram, segments *p, q, r* are parallel. Angles *A, B, C* have corresponding positions and are equal. So are angles 1, 2, and 3.

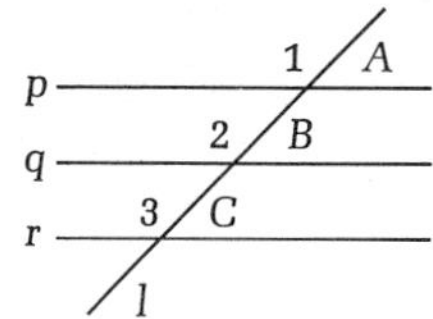

5. Suppose two segments *l* and *m* are cut by a line *r* as shown. Consider angles 1 and 2. If $\angle 1 = \angle 2$, then segments *l* and *m* are parts of parallel lines. ($\angle 1$ means angle 1.) If *l* and *m* are extended, they never meet. If $\angle 1$ is greater than $\angle 2$, then if *l* and *m* are extended, they meet on the side of $\angle 2$—to the right in this drawing. If $\angle 1$ is less than $\angle 2$, then *l* and *m* meet to the left on the side of $\angle 1$.

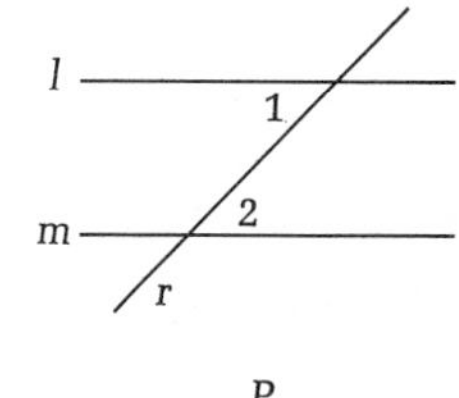

6. Through point *P*, one and only one line *m* can be drawn parallel to a given line *l*. In order for the statement to make sense when *P* is on *l*, we say $m = l$ and a line is parallel to itself. This is an extension of the definition of parallel.

Historically, the study of parallelism has been one of the great events in human thought. It appears that parallelism is a property of lines. After thousands of years of frustrating investigation, people now prefer to say it is a property of space containing the lines. The switch in thinking was traumatic, forcing people to deny the universal truthfulness of one of Euclid's postulates. This led to the study of noneuclidean geometry or the study of non-flat space. It also led to the examination of beliefs that people had accepted in fields other than mathematics, thus leading to deeper knowledge in many places.

In three dimensions we speak of planes being parallel. In a stack of sheets of paper on a table we think of the separate sheets as all showing parallel planes, all parallel to the table top.

32.11 MORE SHAPES AND FIGURES Besides the triangles, squares, rectangles, and circles there are several other figures that are fairly commonly used and named. We consider the parallelogram, rhombus, trapezoid, kite, regular *n*-gon, and ellipse. As for solid figures, in addition to the cube, rectangular prism, and sphere, we consider the cylinder, cone, and pyramid.

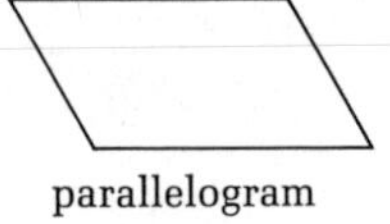
parallelogram

If two sets of parallel lines intersect, the segments cut off form a parallelogram. The opposite sides of a parallelogram are equal as well as parallel. Rectangles and squares are special cases of parallelograms.

The area of a parallelogram is the base times the height. Any of the four sides can be taken as the base. The following figures show how parallelograms may be cut into figures that recombine to form rectangles with the base and height of the corresponding parallelograms.

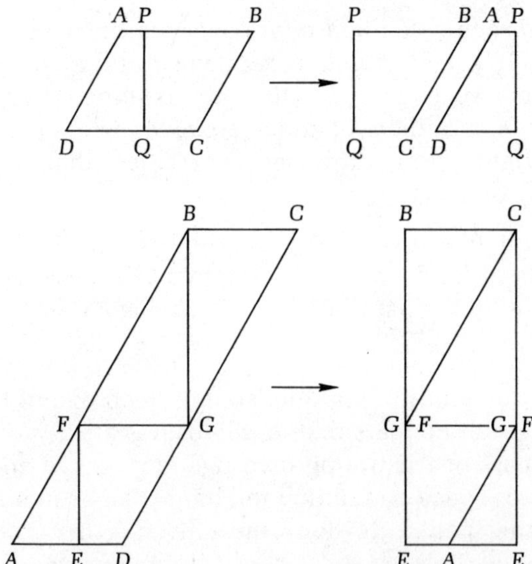

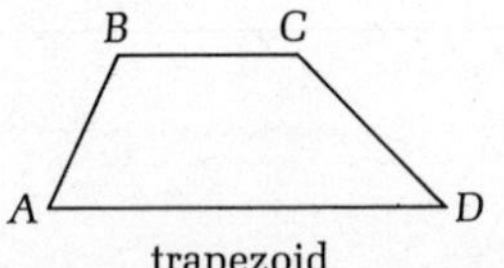

trapezoid

A trapezoid is a four-sided figure with two sides parallel (see example in margin).

If a copy of the trapezoid is given a half turn about the midpoint of one of the non-parallel sides, say *CD*, a parallelogram $ABA'B'$ is produced. Can you see how this leads to the formula for the area of a trapezoid, which is 1/2 (base 1 + base 2) × height, where base 1 + base 2, the sum of the bases, is the sum of the lengths of the two parallel sides? It is sometimes convenient to think of a parallelogram as a special trapezoid. We note that the definition is phrased so that a parallelogram fits the definition of a trapezoid.

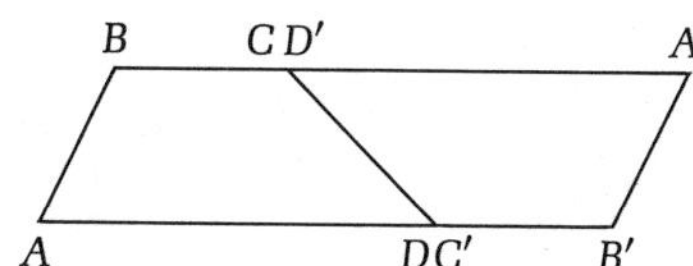

However, a parallelogram is symmetric. Given a half turn about its center, the intersection of the diagonals, it coincides with its original position. Unless it is a rectangle, there is no line of symmetry. Trapezoids have no symmetry unless the two legs are equal, as in the illustration. Can you tell whether it is turning or folding symmetry?

A picture of a kite is in the margin. One diagonal of the kite is a line of symmetry as shown. Which of the following are special cases of a kite: circle, square, rectangle, triangle, parallelogram, trapezoid?

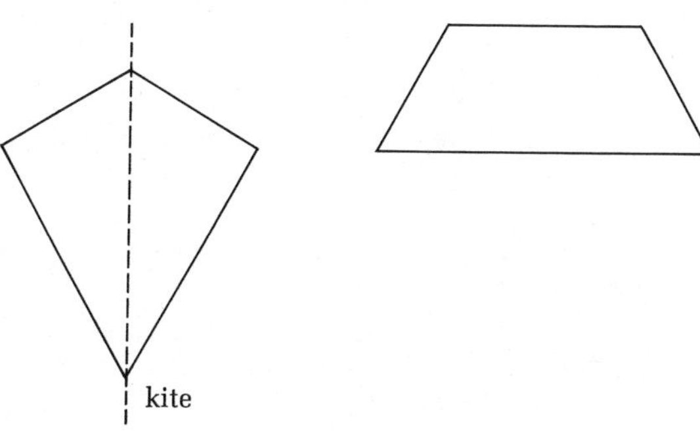

We leave it as an exercise for you to write a definition of a kite. Can you write a formula for the area of the kite in terms of the lengths of the diagonals?

If n points are equally spaced about a circle and then the adjacent points connected, the result is a regular n-gon. Here are some samples:

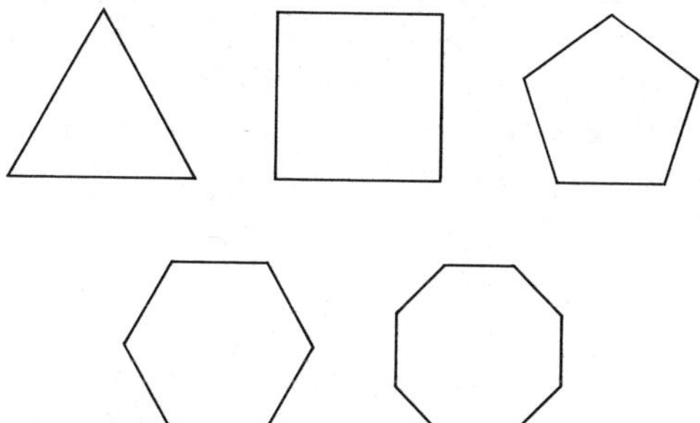

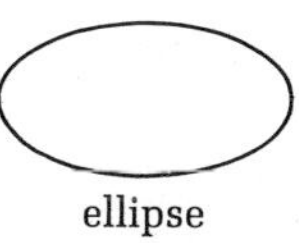
ellipse

The 3-gon is an equilateral triangle, the 4-gon is a square. Using Greek prefixes for the number, we call the others a pentagon, hexagon, and octagon. The n-gons have lots of symmetry. Indeed, there are $2n$ ways a regular n-gon can be made to fit a hole from which it is cut!

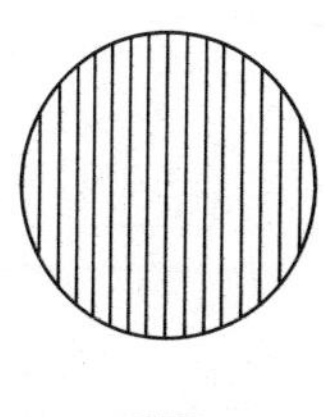

The ellipse is the most common of the oval curves. It looks like a squashed circle. Indeed, it is just that. If all the chords of a circle in a given direction are shortened proportionally, say to half their original size, the resulting curve is an ellipse. If a disk such as a coin or the top of a glass is looked at from an angle, it appears to be an ellipse.

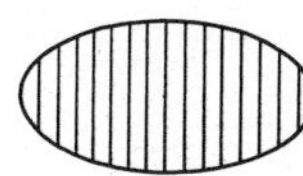

The ellipse is also a stretched circle. If a broom stick is cut perpendicularly, the slice shows a circle. If it is cut at an angle, the slice shows an ellipse.

The broom stick is an example of a right circular cylinder. So is a piece of chalk for the school room, a pipe for water, a piston or a cylinder for an engine. If a line moves parallel to itself, pointing in the same direction and always touching a curve, the line sweeps out a cylinder. If the curve is a circle, it is a circular cylinder. If the line is perpendicular to the plane of the circle it is a right circular cylinder. Take a piece of paper and roll it into a tube. Then see if you can look at the tube as swept out by a line always moving parallel to itself and always touching a circle. Can you point out the circle? Two or three positions of the line?

A cone is like a pinched cylinder. Instead of the line moving parallel to itself, it always goes through a fixed point and touches a curve. The fixed point is called the *vertex* of the cone, the curve is called the *directrix*. There appear to be two parts, called *nappes*, to a cone which touch at the vertex.

Sand falling through a hole falls in a pile that looks like one nappe of a cone. Many cones are right circular cones, where the curve guiding the moving line is a circle and the line from the vertex to the center of the circle, called the *axis*, is perpendicular to the plane of the circle.

If lines a and g intersect at a point and if the plane is revolved about line a, then g sweeps out a right circular cone with vertex at V:

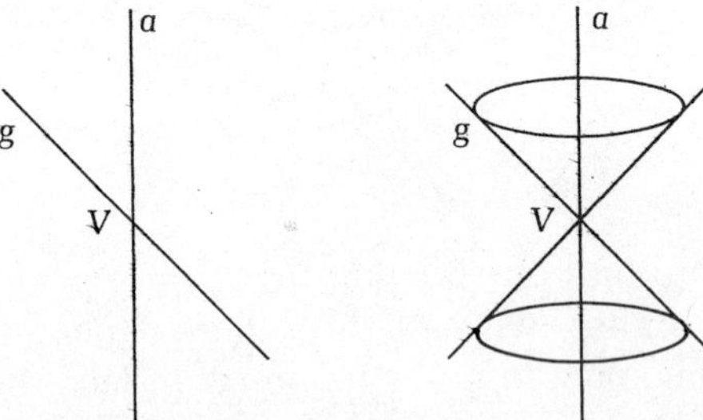

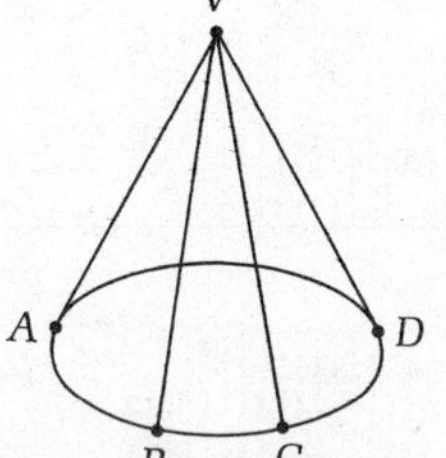

If a cone is cut off by a plane, the intersection is called a base. A cut-off nappe as shown in the figure is often called a cone with base $ABCD$.

A pyramid is like a cone with a polygon for a base. In the illustration every point on a square $PQRS$ is connected by a segment to V, a point not in the plane of the square. The surface formed by all these segments is the pyramid. It is called a square pyramid because the base is a square. Triangle VRS is a face. All faces of a pyramid are triangles. If the base is a triangle, the pyramid is a triangular pyramid; if the base is a pentagon, it is a pentagonal pyramid.

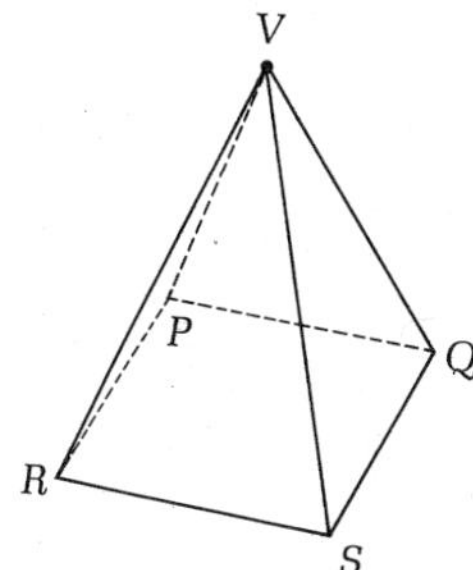

The volume of a pyramid is 1/3 area of base × height. A cone like $VABCD$ has the same volume formula. We may think of a pyramid as being a special kind of cone.

32.12 POLYGONS Squares, triangles, and regular n-gons are all examples of polygons. A polygon is a figure formed by segments intersecting at their end points. Segments can be put together in many ways, so there are many kinds of polygons. The important ones are the simple polygons, which don't intersect themselves. For example, [figure] and [figure] are simple polygons while [figure] and [figure] are not. All the vertices of simple polygons lie on either one or two segments, while there is at least one vertex of a non-simple polygon that is the end of more than two segments.

A closed polygon is one like a triangle or a square. All vertices belong to at least two segments. An open polygon has all vertices belonging to two segments except the ends, as shown by the figure in the margin. The most interesting polygons are the simple closed ones, like triangles and rectangles.

Sometimes it is convenient to approximate a curve by a polygon. For example, if a number of points are chosen on a curve, the polygon resulting by connecting them in order is an approximation of the curve. The length of a curve is often defined as a number close to the perimeter of a polygon that approximates the curve in this way.

EXERCISES

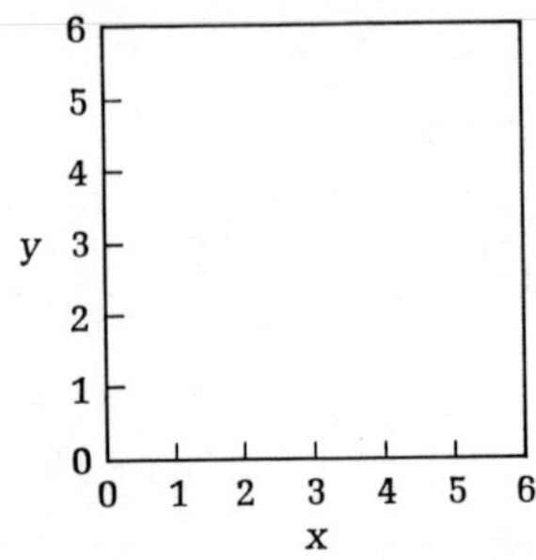

1. Draw a square with sides about 6 cm long with scales as shown in margin.
   (a) Draw a small circle at (4, 2)
   (b) Draw a small square at (1, 5).
   (c) Draw 8 small crosses at 8 points for which $x + y = 5$. (1.5, 3.5) is one such point, since $x = 1.5$ and $y = 3.5$.

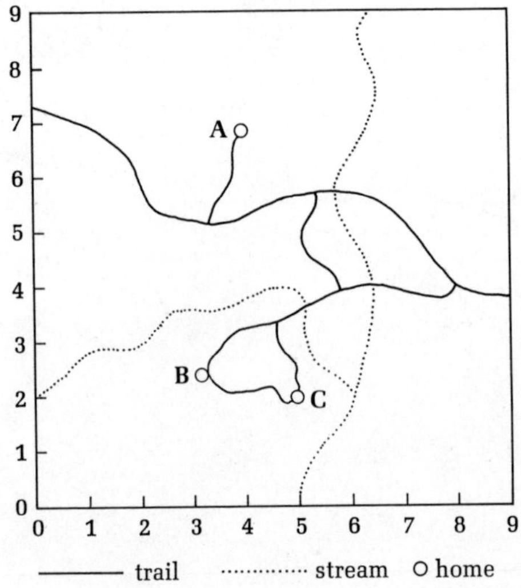

2. A twigtown is shown with scales added for reference:
   (a) Which home is at (5, 2)?
   (b) What is there at (8, 4)?
   (c) How many forks are there in the trails?
   (d) If each unit on the scale is 100 meters, how far is it from *A* to *B* "as the crow flies"?
   (e) How far is it from *A* to *B* along the trail?
   (f) (7, 4.5) is inside a region bounded by trails and streams. What is the area of this region in square meters? In hectares?

3. Consider the capital letters as networks.

   A B C D E F G H I J K L M N O P Q R S T U V W X Y Z

   (a) Which letters are equivalent to *O*?
   (b) Which letters are equivalent to *Y*?
   (c) Which of the first five letters are one-drawable?

4. Draw each of the following networks shown in the margin without lifting your pencil or retracing an arc. If not possible, state why.

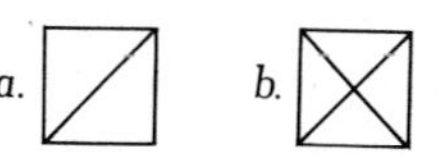

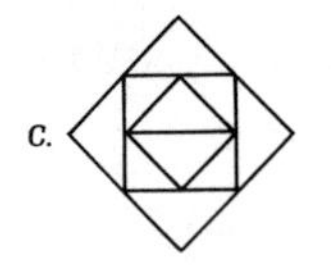

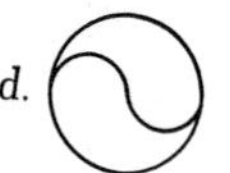

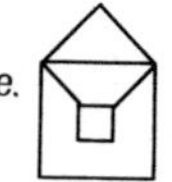

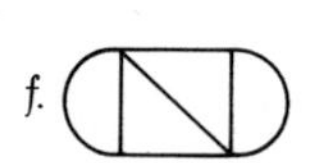

5. Name a part of the body of the figure in the margin approximated by a

(*a*) circle
(*b*) triangle
(*c*) square
(*d*) rectangle

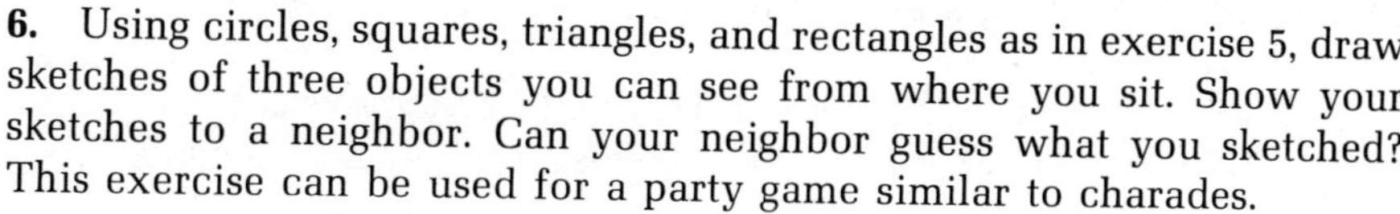

6. Using circles, squares, triangles, and rectangles as in exercise 5, draw sketches of three objects you can see from where you sit. Show your sketches to a neighbor. Can your neighbor guess what you sketched? This exercise can be used for a party game similar to charades.

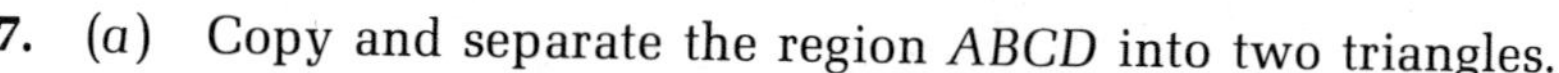

7. (*a*) Copy and separate the region *ABCD* into two triangles.
(*b*) In how many ways can this be done?

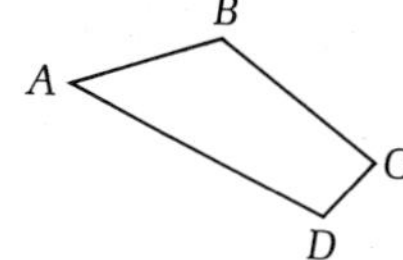

8. Copy and separate the region *PQRSTUV* into 7 triangles so that each side is a side of one and only one of the triangles.

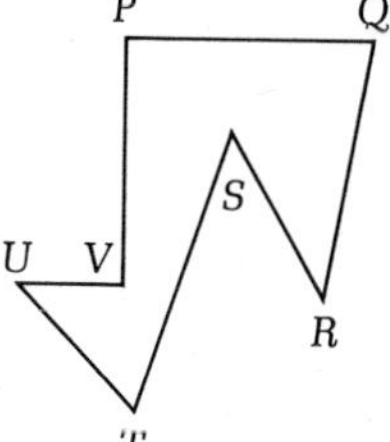

9. Trace the picture of the bird in the margin. Describe two ways the tracing will fit the picture.

10. Make the following figures on a geoboard or a 6 × 6 grid of points.

(*a*) Three isosceles triangles of different size.
(*b*) Two rectangles of different shape with perimeter 16.
(*c*) A 3-4-5 triangle.
(*d*) Six squares of different size.

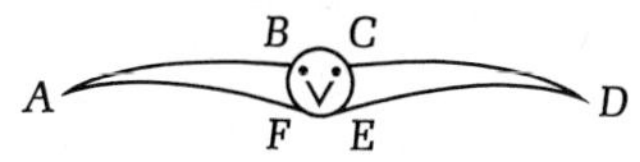

11. How many different squares with vertices at pegs can be drawn on a 6×6 geoboard? Different squares may be the same size.

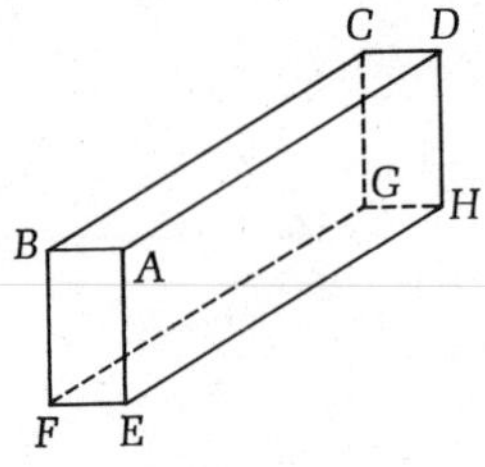

12. The rectangular prism *ABCDHEFG* has sides $AB = 1$, $AE = 2$, $AD = 5$.
   (a) Find three paths of different lengths from *A* to the opposite vertex *G* if the path follows the edges of the prism and goes through no point more than once. *ABFGC* and *ABC*, for example, are two paths from *A* to *C*.
   (b) Find the longest path from *A* to *B*.
   (c) What is the length of the path in *b*?

13. Using a protractor, or good guessing, draw angles of these sizes:

(a) 30°	(d) 45°	(g) 180°
(b) 60°	(e) 120°	(h) 270°
(c) 90°	(f) 210°	(i) −50°

14. A full turn is separated into 7 equal angles. Give the size of each angle—
   (a) in degrees to the nearest hundredth of a degree.
   (b) in degrees, minutes, and seconds to the nearest second.

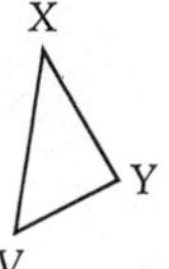

15. In XVY (not drawn to scale) angle $V = 90°$, $XV = 10$, $YV = 13$. Which of the following is closest to the size of angle XYV: 10°, 35°, 50°, 75°.

16. (a) Draw four rectangles of different shape with perimeter 16.
   (b) What are their areas?
   (c) Which has the largest area? The smallest area?
   (d) What is the largest area possible for such a rectangle? The smallest area?

17. Consider the capital letters A B C D E F G H I J K L M N O.
   (a) Which have folding symmetry about one and only one line?
   (b) Which have folding symmetry about two lines?
   (c) Which have turning symmetry?

18. Describe the symmetry of the three-leaf rose.

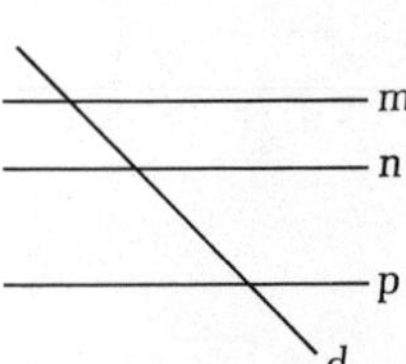

19. Copy the figure showing three parallel segments *m*, *n*, *p* cut by a fourth line, *d*. Place a 1 in one angle and all angles equal to it, a 2 in another angle and all angles equal to it, and so on until all twelve angles are marked.

20. Segments s and t are cut by h, making angles of the sizes shown. Are s and t parallel? If not, will they intersect if extended to the right? To the left?

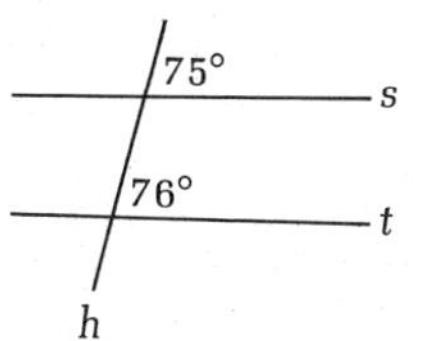

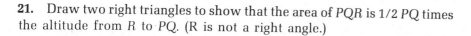

21. Draw two right triangles to show that the area of PQR is 1/2 PQ times the altitude from R to PQ. (R is not a right angle.)

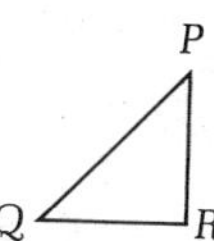

22. Cut the parallelogram $ABCD$ into parts and put the parts together in such a way as to show that the area is AB times the perpendicular distance between the parallel lines on which AB and CD lie.

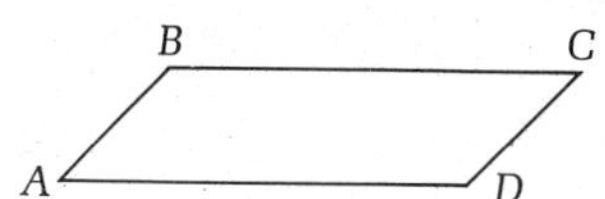

23. A right triangle ABC is folded (or flipped) about a short side, say BC. The resulting figure, disregarding BC, is $ACDB$ (not to scale). What kind of figure is $ACDB$?

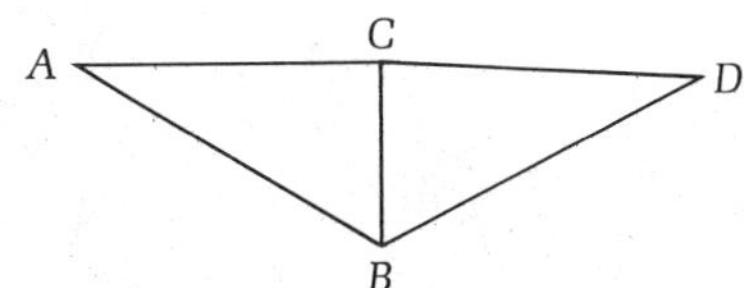

24. A triangle all of whose angles are less than 90° is folded about one side. What figure results?

25. A triangle ABC is given a half turn about the midpoint of AC, falling on $A'B'C'$, with $A'C'$ falling on CA. What figure results?

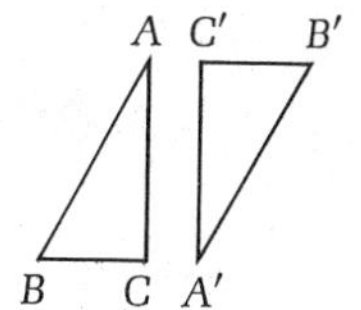

26. Ten circles of radius 2 are drawn along arc AB as follows:
- Circle 1 has center at A and intersects the arc at C_2.
- Circle 2 has center at C_2 and intersects the arc at C_3 and A.
- Circle 3 has center at C_3 and intersects the arc at C_4 and C_2 and so on.
- Circle 10 has center at B. How long is AB?

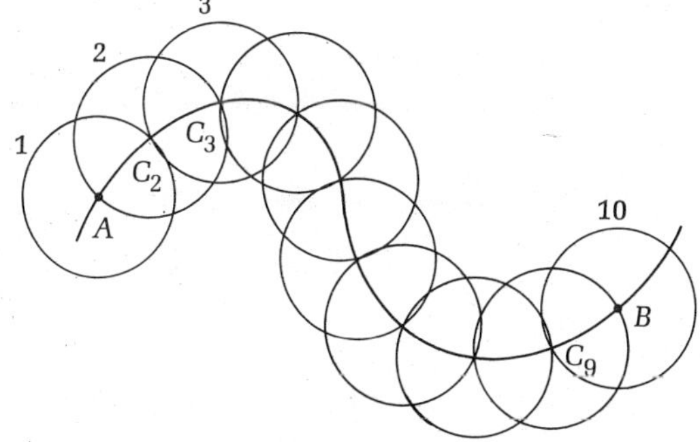

27. Make a paper triangle. Fold the corners in so they fit together showing that the sum of the angles is 180°.

28. Any triangle can be separated into 3 segments. Can any 3 segments be joined to form a triangle?
Sketch a triangle made from segments of lengths 1, 1, and 1.
Similarly for segments 2, 3, and 4.
Similarly for segments 1, 4, and 2.

29. Explain how you can tell, just by knowing the lengths of 3 segments, whether they can be put together to form a triangle.

Answers appear on page 548.

Appendix

Problem Solving

1 INTRODUCTION A famous Russian dictator was said to know only two English phrases: "What in hell is going on here?" and "Where is the men's room?" Similarly, problem solving, that fundamental technique thought by many as the most valuable fruit of mathematics, has a cognitive side involving thought and a side involving the physical condition of the solver.

Obviously there is no Aladdin's lamp we can rub that will solve every problem. If there were, the world would have used it. Haven't you seen a few problems lately? What follows will be suggestions for approaches that have been amazingly successful when applied to the kinds of exercises that appear in mathematics and science texts, and have been quite helpful elsewhere.

Many people see a difference between problems and exercises. They say an exercise is a short task carefully prepared by an author or teacher. A problem is more general. It may be vaguely stated, or not even stated at all. The exercise may be completely solved by number manipulation. The problem requires choices based on preferences and value judgments. "How shall I get to school?" "How much time shall I spend on history tonight?" "How much will the party cost?" are problems requiring greater depth.

Often the objective in a set of exercises is to get a grade. The objective in a set of problems is to relieve the tension the problem poses and to gain power and understanding for future problems. Such objectives are hard to score. Problems are more important than exercises, but are harder to organize in a course of study. Teachers assign exercises hoping students will look at them in depth and make problems of them.

Since much of what one says about solving exercises will apply to problems and vice versa, we will not make a clear distinction in what follows. In specific cases, you may ask "How does this apply to problems? How does this apply to exercises?"

We classify the thought processes into four steps: (1) understanding the problem, (2) planning the attack, (3) carrying out the plan, and (4) overview. These classifications are not hard and fast. An activity one person would put in one place another would put elsewhere. Any process as complicated as problem solving may be analyzed by different people in different ways.

A solver may wish to repeat these four steps several times before he considers the problem solved. Perhaps, on overviewing his work he has a clearer understanding that leads him to a new plan, and so on.

Under solver's conditions we list two types: (1) working conditions and (2) attitudes.

2 UNDERSTANDING THE PROBLEM Frequently people get the right answers to problems by guessing, rote, or instinct. This may lead to uneasiness, and make arithmetic a blind gamble with decreasing odds as we face more challenging problems. Hardly ever does one completely understand a problem until it is solved and checked; but we can give a collection of hints that will help you see the problem clearly enough to plan an attack. These hints are suggestions of things to do and questions to ask.

1. *State the problem.* Write it in complete sentences.

2. *Read the problem.* Read it aloud. Does it make sense? Can you define the words? Can you imagine being in a situation where the problem arises?

3. *What is wanted?* Give the unknown a name like Joe, or x, or □. What is the nature of the unknown? A number? A count? A measurement? A map? A formula? A relationship? A description? A diagram? A table? A curve? A flowchart? A point? Some geometric object? . . .

4. *What quantities are given?* What is their nature? Are lengths, weights, . . . involved? Which have numbers or shapes given? List the given information.

5. *What relations are given?* Are some things equal to, bigger than, around, between, under, connected to, derived from, multiplied by, added to others? Write the relations in declarative sentences and/or in equations (which are declarative sentences in different form). Use the name of the

unknown just as though you knew what it was. The given quantities and relations form the data.

6. *Draw a picture.* Draw a diagram, an illustration, a scale drawing, a flowchart.

7. *Restate the problem.* Use different words. Use your own pictures. Make up a story about the problem. Make up a new story having the same numbers and the same relations as those given. Tell the story aloud to a friend, to a pet, to a doll.

8. *Can the problem be solved?* Is there enough data given? Is there one solution? None? A few? Many? Is the data consistent? Perhaps it is contradictory. Do you suspect a typographical error? What is needed to solve the problem?

We comment on these eight steps.

1. Most problems in life arise in situations rather than given in a book. Or a teacher may put you in a problem situation. Stating the problem is a part of the task, and is a great help in getting your ideas clear.

2. Reading the problem is an obvious requirement, with many difficulties coming from misreading. Some typical reading errors are misreading, skipping, and short-circuiting. Misreading is a blunder such as reading "9" for "6," "23" for "32," or replacing one word by another. In skipping, one does not see certain words. If the word is a trivial connective, no harm is done. Skipped words often have vague meanings. In reading newspapers we usually get the meaning of such words from the context, or the word is not important to the story, so the damage is small. Indeed, many people recommend skipping as part of skimming in speed reading. In mathematics problems one key word skipped may block the entire problem.

Short-circuiting occurs when a solution suggests itself to the reader before he has fully read the given problem. For example, a student may have thoroughly studied the basic facts for addition, such as $5 + 8 = 13$ and so on. Asked for the basic fact $5 + 8$ in base 9 he short-circuits, writing 13 rather than 14.

Reading aloud helps the student to be aware of the whole problem, to find skipped words, and to avoid seeing things that aren't there. It does slow things, and perhaps should be done only when the student is stuck.

3. Focusing attention early on what is wanted, or the unknown, often avoids wasteful excursions. Giving the unknown a name makes it easy to make short, clarifying statements. Choose the unknown accurately. For "How many marbles has John if . . ." the unknown is the number of marbles. One says "Let x be the number of marbles John has." This is far better than "Let x be John's marbles . . . ." Then we make short statements such as $x + 3 = 5$. We indicate the answer by "$x =$ " or "x is □."

It costs so little and the value is so great in naming the unknown by a letter or other symbol that we recommend it as a routine step.

4. Listing the quantities given before asking how they are related, or how they are appropriate, helps us to avoid short-circuiting and make sure we include everything. One can do this mentally and quickly; then if solving is difficult, it helps to write the list.

5. Listing the relations between quantities is a vital step. Frequently these relations can be changed to equivalent equations as part of the solution process. "It" is frequently used with several different meanings in the same problem if one is not careful. Writing a complete sentence is very difficult, and hence annoying, especially when one is not quite clear about the meaning. Hence such writing is an excellent way of clearing up ambiguities.

6. A picture or diagram helps clarify relations. Words may give the facts, but you might not see that something at the head of a paragraph is closely related to something at the end. A picture is two-dimensional, and one is aware of relations in several directions at once. The picture may be a pseudo-photograph with stick figures or an austere diagram. Usually it pays to draw a diagram carefully somewhat to scale, using a straightedge for lines, and cutout templates, coins, or a compass for circles.

7. Restating the problem in other words helps clarify the main parts, showing that you see the relations rather than only repeating the words given. Making up a new story stimulates your imagination and extends your awareness of the ideas. Telling the problem to someone else helps you be careful and accurate in your account. Many say that the best way to understand an idea is to teach it to someone else.

8. Can the problem be solved? Looking at the broad structure tells you a lot about the problem. For example, "To meet the bread shortage, the bakery doubled its production. Did the baker succeed?" Perhaps, but we ask if the question is answerable. Does success, mean "double production," or "meet the shortage." Do we have enough data for the last? Do you see the question is unanswerable without further figures, for example, the daily shortage and the bakery production. Sometimes when you ask what information is needed, you find that you have overlooked something, or you get a hint as to where to go to get what is needed.

A typographical error or miscopying the numbers is irritating and exasperating. So, if stuck, ask, "If I made a slight change would the problem be easier, reasonable? Perhaps my slight change corrects an error."

3 PLANNING THE ATTACK

1. *Can the relations, the equations, be solved for the unknown by changing to equivalent relations?* Stating the relations differently may show a way.

2. *Have you seen the problem before?* A similar problem? A related problem? An analogous problem?

3. *What has been done with these things before?* How have these relations been used before? Do you know a theorem related to these things? Does the problem fit an attitude of some binary operator? A unary operator? Is there an algorithm for this operator?

4. *Look at the problem from the back.* What would it take to get the unknown? Could the unknown arise in ways other than these given in the problem?

5. *Change the problem.* Substitute small numbers (0, 1, 3) for the values. Drop or change some relations. Specialize the problem to a single simple case. Generalize the problem. Substitute a letter for a given value.

6. *Vary the conditions.* What are the largest values the variables or quantities could have and the problem still make sense? The smallest values? Must the numbers be integers? Rational? Now do some numbers change because you change (vary) others? Turn your figure on its side, upside down. Does it look different?

7. *Guess the answer.* Could it be 0? 100? Why?

8. *Make a mathematical model.* The model is the union of the diagrams and relations put together to make it clear how they are interconnected. Solving the problem is often done by manipulating the model.

9. *What tools do you have?* Paper, pencil, calculating machine, programmed calculator, drafting tools, references?

A plan for attack may be as vague as a hunch to try an approach you think you saw somewhere. Or it may be an elaborate flowchart with each step explicitly stated, junctions with instructions for choices, and feedback loops. For any but the shortest problems it pays to spend some time thinking through the types of steps necessary. Most often you see what to do from the statement of the problem. Since most problems in books are illustrations of the theory, your plan of attack may be similar to that used in the book.

Most problems are very similar to problems you have met before, and will yield to approaches used before. The question is to locate that other problem or approach.

One philosopher suggests that our memory files our experiences by subject, by process, by relations, by sound, by feel, by appearance, but never alphabetically. He compares these experiences to the posts and holes in a pinball machine. We bounce and roll our problem off these posts and around these holes until we find the hole that fits. Our memory and subconscious do this, matching the problem to the experiences, fitting them together in various patterns until some suggestion of a plan arises.

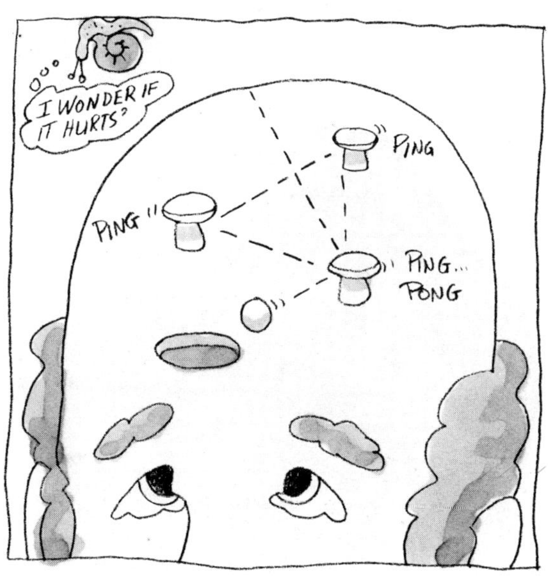

4 CARRY OUT YOUR PLAN Your plan may be a plausible hunch, and carrying it out may be combined with making the plan clear. Most people carry out a plan in steps. Was each step the one you meant to take? Are there any blunders? Was the step true or correct? Can you prove it? What is the relation between each step and the next? Some people can give you the product of two 6-digit numbers without having the faintest idea how they got the product. Even if psychic action or extrasensory perception is involved, it is good if you know what you are doing and why, and can explain the steps to someone else.

The effort invested in planning an attack is closely related to the effort spent in carrying out the plan. Usually the more effort spent planning, the less spent in carrying it out. Often the total effort is least if most of the effort is spent in planning.

5 OVERVIEW Check your result. Is there an answer given? Is your result approximately correct? Does your result satisfy the conditions given? Did you answer the question asked, or a related one some devil put in your mind? Does the argument flow smoothly from step to step? Can you now see another way to work the problem that is shorter? One that is longer? Can you see the problem at a glance? Should you do the problem over? Can you recall a similar problem? A related problem? Can you make up a problem like this one? Where else will this method work? Can you make up a different problem with the same numbers?

Of what value was this problem? What did you learn? What skill did you sharpen? What previous ideas were used and extended? Where does the problem lead?

Checking is working a problem another way. The easiest way is to look in the answer section. Other ways are to work the problem again with simpler numbers, simpler conditions so that the problem can be done mentally. The answers should be about the same. For example, the product 786×432 should be about $800 \times 400 = 320{,}000$. The area of [irregular shape] should be about the area of the square [square].

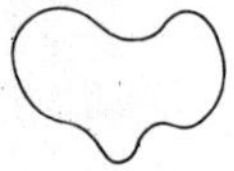

Checking is a vital part of problem solving that is frequently overlooked because people think they can learn a new idea or demonstrate a skill without checking. Often this is true. Without checking, however, one often comes to believe a false idea. Many students depend on the teacher to check their work and point out their mistakes. This is a sad pitfall, because no teacher can catch all mistakes and teacher corrections are not as useful as student corrections. Perhaps the greatest gain in power and knowledge for the effort spent comes during the overview. Here the student sees the problem as a whole, reviews the steps and sees how they fit together, imagines how these new techniques can be used again.

The after-the-answer-is-found activity may be likened to what a skilled pilot, doctor, or space scientist does. They go over their first solutions thoroughly, imagining what might go wrong as well as what is right, considering alternatives. Then they are able to make the best possible response when demands come.

6 WORKING CONDITIONS FOR PROBLEM SOLVING Industries faced with the discipline of making a product for a cost no greater than the selling price know that appropriate working conditions make valuable differences. Cluttered, dirty, cramped, stressful conditions and dull, lost tools are severe handicaps. Here are specific hints: (1) Use sharp pencils or thin lead mechanical pencils with soft leads that make a clear mark. (2) Have a soft fresh eraser that leaves no smudge (in smoggy climates erasers may get hard with age). (3) Avoid ink that cannot be erased. Rewrite work showing many changes. (4) Use clean paper with your work lined up horizontally and vertically. (5) Use no scratchwork calculations at odd angles in odd places on paper thrown away. Do calculations at the side of the main paper or else neatly on a separate sheet where it can easily be connected to the main sheet if desired. Doing calculations is not shameful, to be hidden, like stealing children's lunch money. Calculations should be readily available for checking.

7 ATTITUDES FOR PROBLEM SOLVING Are you curious? Do you want to know why things work? Are you adventurous? Are you lazy, always looking for an easier way of doing something? Do you like puzzles? If so, you are lucky, with a high probability of being a succesful problem solver.

Be proud of the problems you solve. Keep them in a notebook and index the notebook. It will impress you and your friends and be a good reference. Be glad that you can identify problems you can't solve and that you can get help on them. Talk about what you are doing with others. Explain problems to your friends. Tutor or help someone who knows less than you do.

8 HOW TO GET HIGH GRADES ON TESTS Problem solving requires deeper knowledge than merely taking the kind of tests often given. Many students feel they have been quite successful on tests by memorizing facts that they forget afterward. The best of these students resent tests and suggestions on taking tests, because they feel tests are useless and the suggestions irrelevant and a waste of time. Such students are a challenge to teachers who wish to give credit for knowledge other than recall.

While problem solving is more than working exercises on an examination, some hints on taking tests may be appropriate. The traditional advice of "Know the material, work fast, concentrate, and make no blunders" is cold comfort. The real question is how to do these things.

What will be asked on the test? You should find out the type of questions, the material to be covered, and the grading method used. What does the examiner want the student to know? It helps to prepare a list of questions you think might be asked. As you answer them, consider what you would do if the questions were changed slightly. Some people rely on pure memory, others combine this with learning situations and structures that reinforce memory and allow for flexibility in answering unforseen questions.

Will partial credit be given? Will informed guessing be good? The grading method will tell. On a famous examination designed to give credit for checking, the scoring was the number right minus the number wrong. It turned out that zero was an above average score. The students were seen to check, but they were so used to partial credit grading, they could not bear to omit a problem even if it did not check.

Review your past performance. Many people find their errors are of two types–ignorance and carelessness. Ignorance losses may be reduced by effective review. Often the best candidates for review topics are earlier examinations, illustrations in the text, and homework exercises. It is more valuable if you review these in the light of the ideas, theories, and relations discussed in the course, although some students prefer to try to memorize these exercises as separate facts.

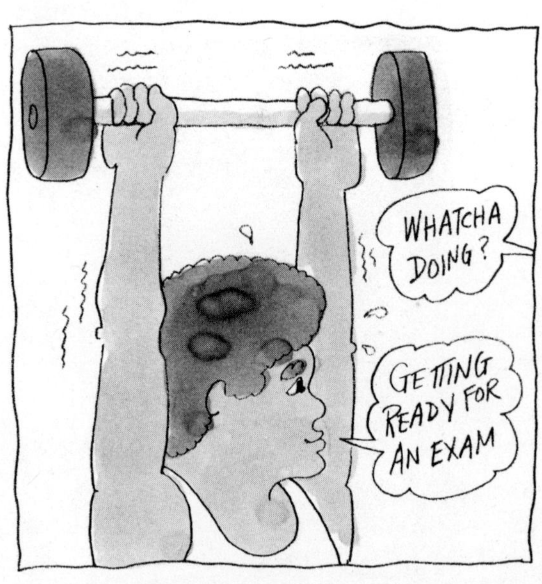

Careless errors are reduced very little by extra study of the material. Carelessness is related to distraction and to tension. You can minimize lost points in three ways:

1. Check thoroughly and then take appropriate action. In reviewing trial exercises, plan how you will check as well as work them.

2. Reduce distraction by doing work neatly, including computations with the problem. Have your working area uncluttered if possible.

3. Reduce tensions by being at your physical best, rested, and reasonably well fed. If tied up at a critical point, relax, let your mind go blank temporarily, let your muscles fall limp.

Plan a miniature time-and-motion study. Avoid spending precious exam time sharpening pencils, tying shoes, fixing hair, and so on. The best strategy during the test is to earn the most points per minute of exam time. This is more important than most points per problem. Hence, if you are stuck, it may be better to leave that problem and turn to another where your points gained per minute will be higher.

9 SUMMARY No perfect theory of solving problems and exercises has ever been designed. While successful techniques depend both on the nature of the problem and the personality of the solver, some general ideas can be given that are occasionally useful. We suggest that if stuck when working on a problem, you review these ideas to see if you get any help from them:

1. Understand the problem. Rephrase it. State explicitly what you know and what you want to know.

2. Plan an attack.

3. Carry out your plan.

4. Look over your results, methods, and answers.

Probably steps 1 and 4 are the most important.

In preparing to solve problems, one is helped by his physical fitness and attitudes as well as his background studies and experiences.

EXERCISES

Answer these questions for each of the problems 1 to 6. If the question is not relevant, so state. Working the problem is not important.

(a) Rewrite the problem in other words.
(b) What is wanted?
(c) List the given quantities.
(d) List the given relations.
(e) Draw a diagram of the problem.
(f) Can the problem be solved?

Example: Without finding the sum, is it obvious that $17 + 9 = 4 + 22$?

(a) Can I see at a glance without calculating that $17 + 9 = 4 + 22$?
(b) A personal judgment is wanted. Hinted but not posed is the related problem "Write $17 + 9 = 4 + 22$ in some way that the equality is obvious."
(c) The given quantities are 17, 9, 4, 22.
(d) The given relation is equals.
(e) A diagram is

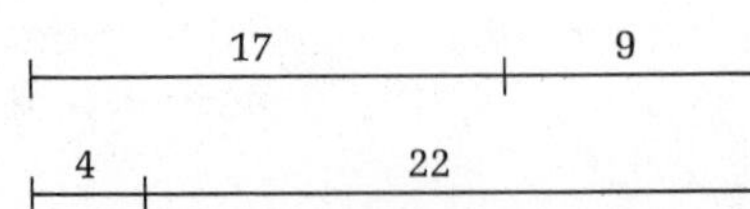

(f) Yes. I can always have a personal judgment. The related problem may not be solvable, since what is obvious to one person may not be obvious to another; but it should be possible to write the statement in a more obvious way.

1. Of the 100 basic subtraction facts in base ten, give the fact for which the sum of the three numbers is largest.

2. Multiply (XXVI) by (XXXXIII) in Roman numerals. Point out what corresponds to $6 \times 4 = 24$ in the layout.

3. Using the standard algorithm, find x if $204x = 7548$.

Answers to selected questions:

1. (*a*) Of the 100 missing-addend problems made from the addition facts, which one gives the largest sum of the three numbers? Other possibilities also are good; (*b*) of $a - b = c$, find the one for which $a + b + c$ is largest; (*c*) the 100 facts $0 - 0 = 0$, $1 - 0 = 1$, . . . $5 - 3 = 2$, . . . $15 - 7 = 8$, . . . $17 - 9 = 8$, . . . ; (*d*) unequals; (*e*) not relevant; (*f*) yes—compare a few numbers; **5.** (*a*) count the ways 3 dice can show a total of 10; (*b*) a number, a count; (*c*) sets of three numbers 1 – 6 where $r + q + w = 10$.

4. Suppose a law were passed making all numbers greater than 1000 illegal. Describe two ways you would be inconvenienced, and how you would meet your difficulty.

5. In how many ways can a total of 10 be rolled with a red, a green, and a white die?

6. Interpret $^{+}4 \times {}^{-}6$ in terms of receiving or giving stacks of chips with values.

Frequently several students make the same error. Understanding a problem is aided by seeing where mistakes are likely. In each of exercises 7–12

(*a*) Work the problem correctly.
(*b*) Describe an error you think might be made often.
(*c*) Give the resulting answer when this error is not caught.
(*d*) If possible, point out how the answer is obviously wrong and its wrongness caught in the overview check.

7. Problem stated in exercise 1.

8. Problem stated in exercise 2.

9. Problem stated in exercise 3.

10. Problem stated in exercise 4.

11. Problem stated in exercise 5.

12. Problem stated in exercise 6.

13. Interview a friend who has studied mathematics, and who agrees that at times his studying is more successful than at others. Describe, compare, and contrast the conditions when his studying was best and when it was only average.

14. Classify the errors on one of your graded examination papers, listing the points lost through carelessness and those lost through ignorance.

15. Describe, compare, and contrast strategies to reduce the points lost through carelessness and those lost through ignorance.

16. Classify the errors on someone else's examination or homework paper.

(*d*)

1 + 3 + 6 = 10	3 + 1 + 6 = 10	5 + 1 + 4 = 10
1 + 4 + 5 = 10	. . .	. . .
1 + 5 + 4 = 10	. . .	. . .
1 + 6 + 3 = 10	3 + 6 + 1 = 10	5 + 4 + 1 = 10
2 + 2 + 6 = 10	4 + 1 + 5 = 10	6 + 1 + 3 = 10
2 + 3 + 5 = 10	. . .	. . .
. . .	. . .	. . .
2 + 6 + 2 = 10	4 + 5 + 1 = 10	6 + 3 + 1 = 10

(*e*)

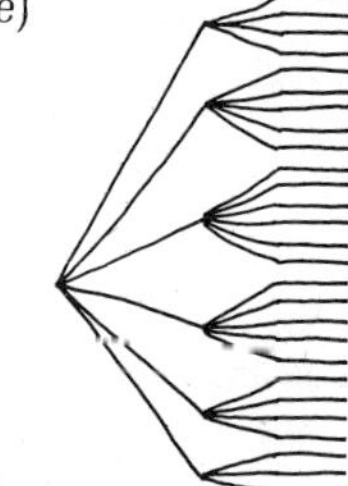

(*f*) yes—count the tips in the tree; **7.** (*a*) 18 − 9 = 9; (*b*) might think only single-digit numbers allowed; (*c*) 9 − 9 = 0, or 9 − 8 = 1, or 9 − 0 = 9; (*d*) the possibility of more than one answer, and the fact that there are not 100 subtractions with single-digit numbers are flags to suggest error; **11.** (*a*) 4 + 5 + 6 + 5 + 4 + 3 = 27; (*b*) Since there are 6 numbers on each die, the number of ways is 3 × 6; (*c*) 18; (*d*) a quick mental listing would show more ways.

Answers to selected questions, chapter 32:

1.

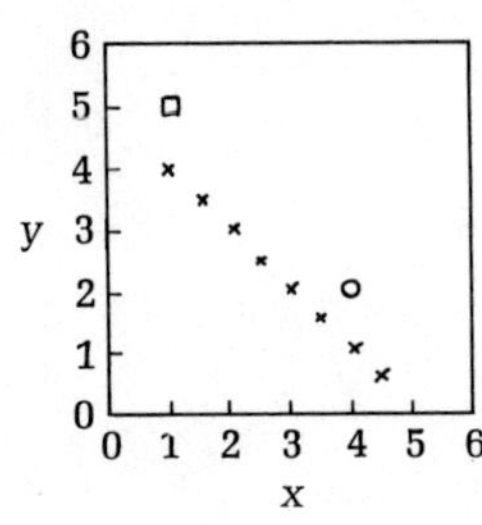

2. (*a*) C; (*d*) 450 meters; (*f*) about 22,000 square meters, 2.2 hectares; **3.** (*b*) E, F, T, Y.

4. (*a*)

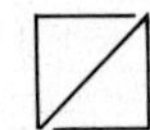

(*b*) not possible, 4 odd vertices.

(*c*)

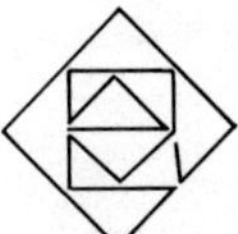

(*d*)

(*e*)

(*f*)

5. (*b*) body, nose; **8.** many ways; **11.** 55;
14. (*a*) 51.43°; (*b*) 51° 25′ 43″.

16.

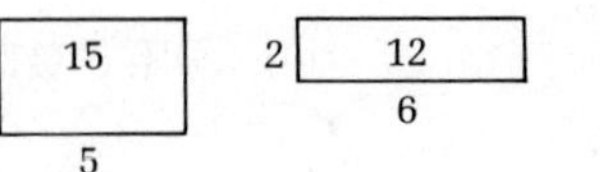

18. 3 axes; turns through 120°, 240°, 0°;
20. intersect at left.

22.

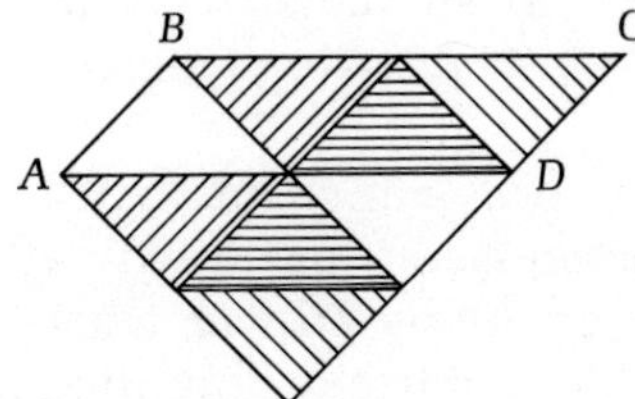

25. parallelogram

Index

This book was designed by Janet Bollow,
illustrated by Richard Cork,
with technical art by the House of Graphics of Palo Alto, California,
set in Melior by
Applied Typographic Systems of Mountain View, California,
printed and bound by R. R. Donnelley and Sons in Crawfordsville, Indiana,
sponsored by Karl Schmidt and
edited by George Oudyn.

56789/54321